Steel Design for Structural Engineers

CIVIL ENGINEERING AND
ENGINEERING MECHANICS SERIES

W. J. Hall, editor

BOGDAN O. KUZMANOVIĆ
Professor of Civil Engineering
University of Kansas

NICHOLAS WILLEMS
Professor of Civil Engineering
University of Kansas

Steel Design for Structural Engineers

SECOND EDITION

PRENTICE-HALL, INC., *Englewood Cliffs, New Jersey 07632*

Library of Congress Cataloging in Publication Data

Kuzmanović, Bogdan O., 1914–
   Steel design for structural engineers.

   Includes bibliographical references and index.
   1. Building, Iron and steel. 2. Steel, Structural.
I. Willems, Nicholas. II. Title.
TA684.K797 1983   624.1′821   82-18337
ISBN 0-13-846287-9

©1983, 1977 by Prentice-Hall, Inc.
Englewood Cliffs, New Jersey 07632

All rights reserved. No part of this book may be reproduced,
in any form or by any means, without permission in writing from
the publisher.

Printed in the United States of America

10 9 8 7 6 5 4 3 2 1

Editorial/production supervision and interior design: Paul Spencer
Cover design: Photo Plus Art/Celine Brandes
Manufacturing buyer: Anthony Caruso

ISBN 0-13-846287-9

Prentice-Hall International, Inc., *London*
Prentice-Hall of Australia Pty. Limited, *Sydney*
Editora Prentice-Hall do Brasil, Ltda., *Rio de Janeiro*
Prentice-Hall Canada Inc., *Toronto*
Prentice-Hall of India Private Limited, *New Delhi*
Prentice-Hall of Japan, Inc., *Tokyo*
Prentice-Hall of Southeast Asia Pte. Ltd., *Singapore*
Whitehall Books Limited, *Wellington, New Zealand*

To Our Families

Contents

Preface to the Second Edition   xv

1   Principles of Steel Design   1

1.1 THE STEEL STRUCTURE   1
     1.1.1 Types of Structures   1
         1.1.1.1 *Introduction*   1
         1.1.1.2 *Steel Buildings and Their Parts*   2
         1.1.1.3 *Steel Bridges and Their Parts*   10
         1.1.1.4 *Special Structures*   16
     1.1.2 Loads on Structures   19
         1.1.2.1 *Dead Loads*   21
         1.1.2.2 *Occupancy Loads on Buildings*   21
         1.1.2.3 *Snow Loads on Buildings*   21
         1.1.2.4 *Wind Loads*   22
         1.1.2.5 *Earthquake Loads*   23
         1.1.2.6 *Live Loads on Bridges*   26
         1.1.2.7 *Impact in Buildings and Bridges*   27
1.2 DESIGN CRITERIA   28
     1.2.1 Introduction   28
     1.2.2 Yield Criterion   30
     1.2.3 Plastic Criterion   30
     1.2.4 Deflection Criterion   33

		1.2.5 Dynamic Response Criterion	*34*
		1.2.6 Instability Criterion	*35*
		1.2.7 Fatigue Criterion	*35*
		1.2.8 Brittle Fracture Criterion	*36*
		1.2.9 Conclusions	*36*
	1.3	OPTIMUM DESIGN	*37*
	1.4	STRUCTURAL SAFETY	*37*
		1.4.1 Deterministic Approach	*38*
		1.4.1.1 *Factors of Safety in Working Stress Design*	*38*
		1.4.1.2 *Load Factors in Plastic Design*	*38*
		1.4.2 Probabilistic Approach	*41*
		1.4.2.1 *Introduction*	*41*
		1.4.2.2 *Load and Resistance Factor Design*	*42*
		1.4.2.3 *Current European Practice in Structural Safety*	*42*
	1.5	DESIGN TOOLS AND AIDS	*43*
		1.5.1 Design Tools	*43*
		1.5.2 Design Aids	*44*
	1.6	DESIGN PROCEDURES	*44*
		1.6.1 Planning and Site Exploration	*45*
		1.6.1.1 *Functional Considerations*	*45*
		1.6.1.2 *Economy, Strength, and Code Requirements*	*46*
		1.6.2 Preliminary Design	*47*
		1.6.3 Structural Analysis and the Accuracy of Computations	*47*
		1.6.4 Selection of Member Cross Sections	*48*
		1.6.5 Secondary Design Considerations	*48*
	NOTATIONS		*50*
	REFERENCES		*50*
2	***Structural Steels and Their Properties***		**53**
	2.1	STRUCTURAL STEELS	*53*
		2.1.1 Introduction	*53*
		2.1.2 Carbon Steels	*56*
		2.1.3 High-Strength, Low-Alloy Steels	*56*
	2.2	STRUCTURAL STEEL PRODUCTS	*57*
		2.2.1 Structural Shapes	*57*
		2.2.2 Plates and Strips	*58*
	2.3	MECHANICAL PROPERTIES	*58*
	2.4	FATIGUE	*63*
		2.4.1 Introduction and Historical Review	*63*
		2.4.2 Basic Aspects of Fatigue	*64*
		2.4.3 Design Concepts and Considerations	*70*
	2.5	BRITTLE FRACTURE	*71*
		2.5.1 Historical Review	*71*
		2.5.2 Basic Characteristics	*73*
		2.5.3 Design Considerations	*79*
	2.6	LAMELLAR TEARING	*86*
		2.6.1 Introduction	*86*
		2.6.2 Basic Characteristics	*87*
		2.6.3 Design Considerations	*88*

	2.7 CORROSION	90
	2.7.1 Introduction	90
	2.7.2 The Nature of Corrosion	91
	2.7.3 Corrosion-Resistant, Low-Alloy Steels	91
	2.7.4 Protective Coating	92
	2.8 THE CHOICE OF THE PROPER STEEL	93
	NOTATIONS	94
	REFERENCES	94
3	***Behavior of Structural Steel Members***	**98**
	3.1 STRUCTURAL BEHAVIOR OF MEMBERS	98
	3.1.1 Tension Members	99
	3.1.2 Compression Members	102
	3.1.3 Flexural Members	105
	3.1.4 Members in Combined Bending and Axial Compression	115
	3.1.5 Light-Gage Steel Members	116
	NOTATIONS	119
	REFERENCES	119
4	***Design of Tension Members***	**122**
	4.1 CROSS SECTION DESIGN	122
	4.1.1 Strength-Based Design	122
	4.1.1.1 *Net Section of Bolted Members*	123
	4.1.1.2 *Net Section of Welded Members*	127
	4.1.1.3 *Composition of the Cross Section:*	
	Simple Built-Up Sections	128
	4.1.2 Slenderness-Based Design	132
	4.2 SPECIFICATIONS FOR TENSION MEMBERS	133
	4.2.1 U.S. Specifications	133
	4.2.1.1 *AISC for Buildings*	133
	4.2.1.2 *AASHTO for Highway Bridges*	134
	4.2.1.3 *AREA for Railroad Bridges*	135
	4.2.2 Foreign Specifications	135
	NOTATIONS	136
	PROBLEMS	137
	REFERENCES	142
5	***Design of Compression Members***	**143**
	5.1 BUCKLING OF STRAIGHT PRISMATIC MEMBERS	143
	5.1.1 Axial Force	145
	5.1.2 Eccentric Force	152
	5.1.3 Effective Buckling Length	153
	5.1.4 Design Procedures	158
	5.2 BUCKLING OF INITIALLY CURVED MEMBERS	180
	5.3 BUCKLING OF THIN PLATES	184

		5.4 POSTBUCKLING STRENGTH OF PLATES	*187*
		NOTATIONS	*188*
		PROBLEMS	*189*
		REFERENCES	*191*

6 *Torsion of Beams* — *193*

6.1 GENERAL — *193*
    6.1.1 Shear Center — *193*
    6.1.2 Pure and Restrained Torsion — *194*
6.2 PURE TORSION OF PRISMATIC MEMBERS — *195*
    6.2.1 Introduction — *195*
    6.2.2 Circular Closed Sections — *195*
    6.2.3 Tubular Closed Sections — *196*
    6.2.4 Solid Rectangular Sections — *197*
    6.2.5 Open Sections — *198*
6.3 RESTRAINED TORSION — *200*
    6.3.1 Introduction — *200*
    6.3.2 I-sections — *201*
    6.3.3 Channels — *204*
    6.3.4 Box Sections — *205*
6.4 COMBINED TORSION AND BENDING — *205*
6.5 DESIGN FOR TORSION — *208*
NOTATIONS — *208*
PROBLEMS — *208*
REFERENCES — *211*

7 *Bending of Beams* — *212*

7.1 STRAIGHT PRISMATIC BEAMS — *212*
    7.1.1 Simple Bending — *212*
    7.1.2 Biaxial Bending — *214*
    7.1.3 Unsymmetrical Bending — *221*
    7.1.4 Lateral-Torsional Buckling — *224*
    7.1.5 Design Considerations and Procedures for the Working Stress Method — *228*
    7.1.6 Plastic Design of Beams — *247*
7.2 CURVED BEAMS — *253*
    7.2.1 Beams of Large Curvature — *253*
7.3 TAPERED BEAMS — *257*
NOTATIONS — *258*
PROBLEMS — *259*
REFERENCES — *262*

8 *Beam-Columns* — *264*

8.1 INTRODUCTION — *264*
    8.1.1 Ultimate Strength of Beam-Columns — *264*
    8.1.2 Differential Equation for Beam-Columns — *266*
    8.1.3 Moment Magnification Factor — *268*

8.2	INTERACTION EQUATIONS FOR BEAM-COLUMNS	268
	8.2.1 Zero-Length Members	268
	8.2.2 Instability in the Plane of Bending	269
	8.2.3 Instability by Lateral-Torsional Buckling	270
8.3	WORKING STRESS DESIGN METHOD FOR BEAM-COLUMNS	271
	8.3.1 AISC Working Stress Method	271
	8.3.2 AASHTO Working Stress Method	271
	8.3.3 Examples	272
8.4	PLASTIC (ULTIMATE) DESIGN METHOD FOR BEAM-COLUMNS	276
	8.4.1 AISC Plastic Design Method	276
	8.4.2 AASHTO Load Factor Design Method	277
	8.4.3 Examples	277
8.5	FOREIGN DESIGN PRACTICES FOR BEAM-COLUMNS	280
8.6	COMBINED BENDING AND AXIAL TENSION	280
	NOTATIONS	281
	PROBLEMS	281
	REFERENCES	285

9 Connections 287

9.1	INTRODUCTION	287
	9.1.1 Types of Connections	288
	9.1.2 Types of Fasteners	289
	9.1.2.1 *Unfinished (Rough) Bolts (ASTM A307)*	292
	9.1.2.2 *High-Strength Bolts*	293
	9.1.3 Electric-Arc Welding	296
	9.1.3.1 *Welding Procedures for Lowest Costs*	306
	9.1.3.2 *Weldability of Steels*	309
	9.1.3.3 *Weld Defects*	310
	9.1.3.4 *Distortions and Residual Stresses Due to the Welding*	311
9.2	CONNECTION PERFORMANCE	315
	9.2.1 Bolted Connections	315
	9.2.1.1 *Single Bolt in Concentric Shear-Bearing–Type Behavior*	316
	9.2.1.2 *Bolt Groups in Concentric Shear*	321
	9.2.1.3 *Bolt Groups in Torsion and Eccentric Shear*	321
	9.2.1.4 *Single Bolt in Tension*	325
	9.2.1.5 *Concentric Tension on Bolt Groups*	328
	9.2.1.6 *Bending and Eccentric Tensile Forces on Bolt Groups*	331
	9.2.1.7 *Combined Shear and Tension on Bolts*	335
	9.2.1.8 *Fatigue Strength of Bolted Connections*	336
	9.2.2 Welded Connections	339
	9.2.2.1 *Concentric Forces on Welds*	345
	9.2.2.2 *Torsion and Eccentric Shear Forces on Welds*	348
	9.2.2.3 *Bending of Welds*	350
	9.2.2.4 *Fatigue Strength of Welded Joints*	353
9.3	CONNECTION DESIGN	354
	9.3.1 Flexible (Shear) Beam Framing Connections	354
	9.3.1.1 *Seated Beam Connections*	368
	9.3.1.2 *Stiffened Seated Beam Connections*	373
	9.3.1.3 *End-Plate Shear Connections*	377

9.3.2 Rigid Moment Beam Connections	*379*
9.3.2.1 *Welded Connections*	*379*
9.3.2.2 *Bolted Connections*	*389*
9.3.3 Column Bracket Connection	*392*
9.3.4 Beam and Girder Splices	*397*
9.3.5 Truss Connections and Splices	*413*
NOTATIONS	*421*
PROBLEMS	*424*
REFERENCES	*429*

10 *Built-Up Beams*   **433**

10.1 PLATE GIRDERS	*433*
10.1.1 Load-Carrying Capacity of Plate Girders	*434*
10.1.2 Buckling Strength of Plate Girders	*446*
10.1.3 Optimization of Plate Girder Dimensions	*450*
10.2 OPEN-WEB JOISTS	*468*
10.3 COMPOSITE BEAMS	*469*
10.3.1 Introduction	*469*
10.3.2 Composite Steel-Concrete Construction	*471*
10.3.3 Shear Connectors	*474*
10.3.4 Design of Composite Beams	*477*
10.3.5 Concluding Remarks	*487*
10.4 LIGHT-GAGE STEEL MEMBERS	*489*
10.4.1 Design of Flexural Members	*489*
10.4.2 Steel Roof Decks	*491*
10.4.3 Wall Studs	*492*
NOTATIONS	*494*
PROBLEMS	*496*
REFERENCES	*499*

11 *Beam and Column Supports*   **501**

11.1 BEAM SUPPORTS	*501*
11.1.1 Tangential (Simple) Supports	*503*
11.1.2 Hinged Supports	*505*
11.1.3 Roller Supports	*505*
11.1.4 Beam-to-Beam Supports	*513*
11.2 COLUMN SUPPORTS	*513*
11.2.1 Pinned Column Base	*513*
11.2.2 Flat Column Base Plates	*514*
NOTATIONS	*520*
PROBLEMS	*520*
REFERENCES	*523*

12 *Summary: Design of Simple Structures*   **524**

12.1 INTRODUCTION	*524*
12.2 SINGLE-STORY BUILDINGS	*525*
12.2.1 Structural Concepts of Building Systems	*525*
12.2.2 Design Example 12.1	*525*

12.3 BRIDGES		*542*
12.3.1 Girder Bridges		*542*
12.3.2 Design Example 12.2		*542*
REFERENCES		*558*

Appendices *559*

APPENDIX I-1	Fortran IV Program for the Design of Simply Supported Beams, AISC Specification	*561*
APPENDIX I-2	Minicomputer Program for the Design of Simply Supported Beams, for HP9820A	
	I-2.1 Moments, Shears, and Modifiers, C_b	*563*
	I-2.2 Selection of WF Section	*565*
	I-2.3 Storage of Data	*566*
APPENDIX I-3	Program for Design of Simply Supported Beams for HP9830A	
	I-3.1 Structural Analysis	*567*
	I-3.2 Beam Design	*570*
APPENDIX II	Fortran IV Program for Optimization of Plate Girder Without Stiffeners (Example 10.3)	*573*
APPENDIX III	Optimization of Plate Girder with Stiffeners (Example 10.4)	*576*
APPENDIX IV	Minicomputer Optimization Program of Girders Without Stiffeners, Sufficiently Laterally Braced (Example 10.3)	
	IV-1 For HP9820A	*580*
	IV-2 For HP9830A	*583*
APPENDIX V	Minicomputer Optimization Program of Girders with Stiffeners (Example 10.4), for HP9830A	*586*
APPENDIX VI	Torsion: Tables 1 and 2	*588*
APPENDIX VII	Solutions for Beam-Column Cases	*591*

Index *593*

Preface to the Second Edition

The basic purpose of this book is unchanged: to demonstrate to the beginner how to make first assumptions, how to select initial sections, and what procedure to follow after making a first choice so as to arrive at a final design.

This new edition reflects recent changes in the various specifications governing the design of steel structures. Specifically the American Institute of Steel Construction (AISC) substantially revised its specifications, and the new AISC *Manual* was published in 1980. Major changes in allowable connection stresses took place, which are incorporated in this second edition. It also reflects several changes that came out from the AASHTO (American Association of State Highway Officials) and the AREA (American Railway Engineering Association) since the publication of the first edition. Included, too, are the most recent data published by the AWS (American Welding Society) and the ASTM (American Society for the Testing of Materials).

A new section on the various types of loadings acting on bridges and buildings now appears in the first chapter. Moreover, a new chapter (Chapter 3) has been added, which serves as an introduction to the behavior of steel members. The chapter on connections has been completely rewritten so as to discuss connection performance first, followed by the detailed design of the various types of connections. This chapter was also moved to a later point in the book, so as to follow the coverage of the basic design of tension members, columns, and beams. Computer programs have been updated, and several new ones have been added.

To give the instructor a wider selection of homework problems, their number has been doubled; and the Solutions Manual has been completely revised.

The authors are indebted to students, colleagues, and other users who have suggested improvements of wording, identified errors, and recommended items for inclusion or deletion. They specifically want to thank Dr. William J. Hall of the University of Illinois, who carefully read the whole manuscript and offered many valuable suggestions as to the content and rearrangement of this second edition.

In addition, the authors express their appreciation to the School of Engineering at the University of Kansas for allowing time to prepare this second edition, to Mrs. Betty Lane for typing the manuscript, and to Mrs. Carolyn Davis for her assistance.

Bogdan O. Kuzmanović
Nicholas Willems

Steel Design for Structural Engineers

1

Principles of Steel Design

1.1 THE STEEL STRUCTURE

1.1.1 Types of Structures

1.1.1.1 *Introduction*

For centuries the two main building materials were timber and stone. The design and construction of the "Iron Bridge" by Thomas Franolls Pritchard in 1779 across the Severn River in England began the era of iron and steel and marked the beginning of the Industrial Revolution. The technical developments in the production of structural iron since that time have been immense. In Europe, the improvement of the rolling mill in 1783 when Henry Court patented his grooved rollers; the production of puddling iron in 1784; the invention of the Bessemer converter in 1856; the introduction of the open-hearth process by Martin and Siemens in 1865; the manufacture of high tensile steels in 1920; and the use of welding are the six main events that stand out in the history of the development of structural steel. Developments and applications in the United States were equally phenomenal. In 1819 the first structural shapes, angle irons, were rolled. In 1884 William LeBaron Jenney designed the Home Insurance building in Chicago, IL, which was the first true skyscraper that made use of the concept of a fully developed skeletal construction. In 1909 the American Optical

Company building in Worcester, MA was the first building in which use was made of wide-flange beams. In 1927 wide-flange beams up to 36-in. depth were first produced. In 1974 the 1450-ft tall Sears building in Chicago, IL was completed (Fig. 1.1).

The main type of civil engineering steel structures are buildings and bridges but in addition there are many special structures such as towers, cranes, space structures, and so on which are of interest to the civil engineer. Although the main emphasis in this text is on the behavior and design of structural components, an introduction to the design of a few simple systems will be discussed in the last chapter.

1.1.1.2 Steel Buildings and Their Parts

Steel buildings, in spite of their name, usually are not built of steel only. They usually have a steel skelton which is covered or filled in by various architectural materials to achieve the required enclosed space. All the foreseeable loads acting on a steel building must be transferred to this steel skeleton or framing by the use of filling material. The framing system then transmits the loads to the foundations and finally to the ground. This "load path" represents the actual structural system and the designer must have a clear understanding of it to be able to design an effective system capable of carrying the loads in a most efficient way. As the applied loads (external forces) are vertical and/or horizontal, most buildings are composed of vertical and horizontal structural systems or framings. Nowadays only a few buildings, unusual in shape, are designed as space frames.

To give a beginner a better understanding of how structural systems behave, a very simple structure is shown schematically in Fig. 1.2. Only its steel frame is shown. This building is a one-story factory building, commonly called a mill-building. The names of the typical members of the structural systems are shown in the same figure.

First we will consider the path of the vertical loads consisting of the weight of the structure itself and snowloads. We see that the roofing deck transmits its loads to the purlins supporting the roof. Purlins transfer their own weight and the weight of the roofing as well as its vertical load to the roof trusses. They in turn transmit the purlin loads and their own weight to the main columns, which pass them on to the foundations. The girts transmit the load from the siding panels or walls and transfer them to secondary and main columns and through them to the ground.

Horizontal loads acting in the east-west direction (wind on gable walls, crane inertia forces, and seismic forces) have to be transferred to the lateral bracings in the end panels of the sidewalls. Top reactions of the columns in the gable wall under wind action are carried by the roof truss bracings between the first two and the last two bents. The chords of these bracing trusses coincide with the top chords of the main roof trusses, while their posts are the purlins. The only added elements are the diagonals. The reactions of roof bracings are transmitted by eave struts to the lateral bracing in both the sidewalls. These bracings finally transfer the wind load from the eave strut down to the main column foundations 1 and 9.

For horizontal load action in the north-south direction the bents composed of the columns, which are considered to be hinged at both ends, cannot safely transmit the horizontal loads acting in their own plane. To ensure lateral stiffness knee bracing

Figure 1.1 Chicago, Sears Building, 1972–74. 109 Story, 1450 Ft Above Street Level, with a Steel Frame of Only 29.7 psf (1422 Pa) of Steel (From 1.43).

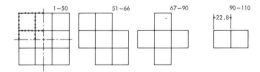

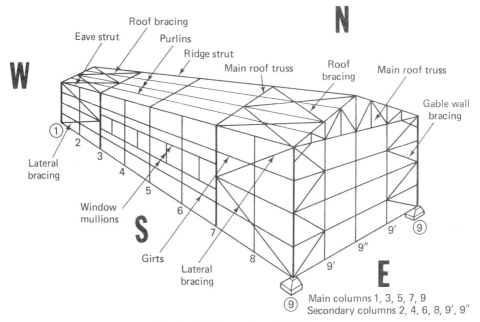

(a) Steel skeleton of a mill building

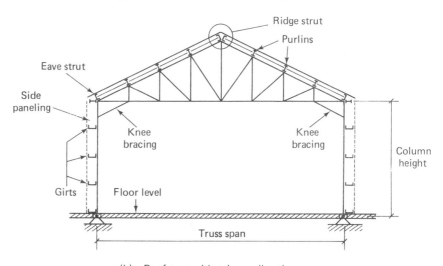

(b) Roof truss with columns (bent)

Columns considered as hinged top and bottom

Figure 1.2

is added. However, such knee bracing (Fig. 1.2b), which is customary in mill-buildings, cannot produce complete frame action, especially for resisting seismic forces. Therefore, in both gable walls two sets of bracings, called gable bracing, between the main and the adjacent secondary columns, transmit the horizontal forces acting on the longitudinal walls. For this reason bents 3, 5, and 7 are considered to be "supported bents" while 1 and 9 are called "braced bents." The side panels must be strong enough to carry loads acting normal to their planes and transfer them to the horizontal girts. Girts in their turn will transfer them finally to the two braced end bents, as they are more rigid than the intermediate, supported bents.

After this discussion of a simple framing system for a typical mill building we will next consider in more detail different types of steel buildings and their framing systems.

A building either will be used mainly by people or will serve for storage of some material or for industrial processing. Use by people can be residential or for business or institutions (schools, hospitals, assemblies, recreation facilities, etc.). Warehouses, garages, hangars, and sheds represent different types of storage building. Industrial buildings consist mainly of low-rise (single or two-to-three story) factories, mills, shops, power plants, or laboratories of various kinds.

Considering how the loads applied in a structure are transferred from the point of their application to the ground, i.e., considering its structural system, four different types of building construction systems are presently in use: wall-bearing, beam-column framing, shell, and suspension systems.

The wall-bearing type of building normally is used for low-rise, light-load buildings such as schools, stores, and offices. They are one- to three-story buildings where the columns are substituted for by bearing walls (Fig. 1.3a and 1.3b).

With new developments in reinforced brick masonry and prefabricated building elements, the wall-bearing type of construction has been revived and its use extended

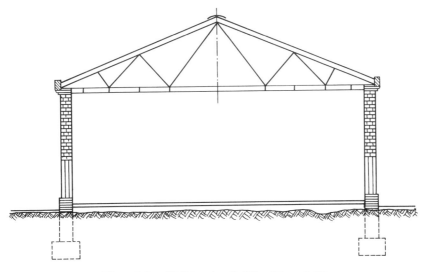

Figure 1.3a Wall-bearing Building (From 1.43).

Figure 1.3b IBM Research Center, Yorktown Heights, NY (From 1.43).

to about 20-story buildings, which are known as masonry tall buildings. Beam-column framing (Fig. 1.4a–e) is characterized by horizontal and vertical structural elements.

These elements and their connections are the main topic of this text. Vertical loads are carried by the floor deck and transmitted to the beams, which pass these loads on to the columns. Horizontal loads (wind and seismic) are carried either by the framing action of the beams and columns, by vertical bracings or shear walls (Fig. 1.5 a, b, c, d, e), or, in the case of high-rise buildings, by an interior tubular core. Floor decks act as very rigid diaphragms for any loads acting in their plane. They distribute the horizontal loads to the vertical bracing systems, which in turn transmit them to the ground (Fig. 1.6a, b).

In a shell-type system the skin, plating, or sheeting forms the main load-carrying element as well as serving as a functional covering element. This dual purpose results in weight savings which are partially offset by the accompanying increase in construction costs. In Fig. 1.7 a roof in the form of a hyperbolic-paraboloid (hypar) shell is shown. It is constructed of straight wide-flange ribs covered by light-gage corrugated steel sheets. It covers 36 ft^2, three corners have an elevation of 10.00 ft, and the fourth one 28.00 ft. Only two corners at both ends of a diagonal are supported on concrete abutments. During erection, the front and rear corners of the roof were temporarily supported until all the structural steel was in place, at which time the roof was completely stable.

For large spans and to avoid long span trusses (as shown in Fig. 1.4e) and/or interior columns, suspension cables and secondary framing systems are used in an attractive and economical solution (Fig. 1.8). Framing systems are required to take care of the reaction components of the individual cables. In Fig. 1.8 the inner or ten-

Figure 1.4a Beam and Column Framing (From 1.43).

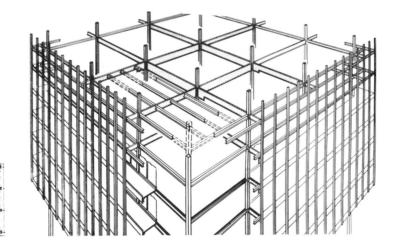

Figure 1.4b In this 12-story office building, square on plan, the columns are widely spaced at 36 Ft (11 m.) centers. Building is stiffened in both directions by two three-bay multi-story rigid frames. (From 1.43.)

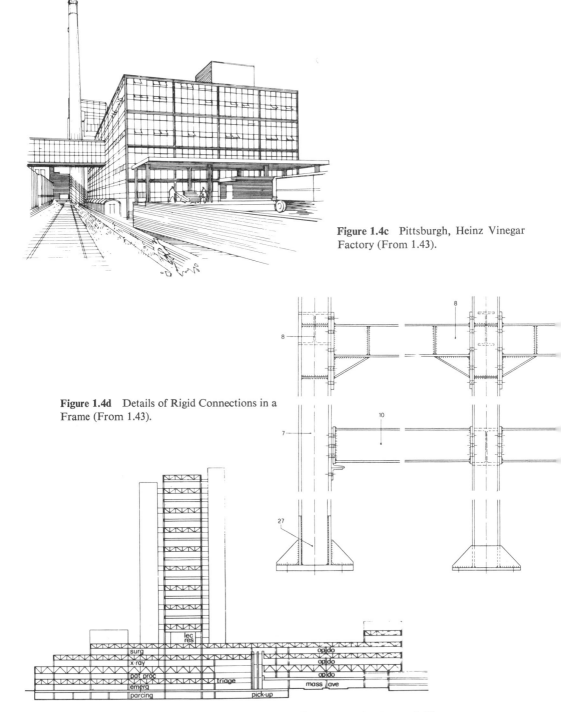

Figure 1.4c Pittsburgh, Heinz Vinegar Factory (From 1.43).

Figure 1.4d Details of Rigid Connections in a Frame (From 1.43).

Figure 1.4e Boston City Hospital, Cross Section (From 1.43).

Arrangement of vertical bracing systems

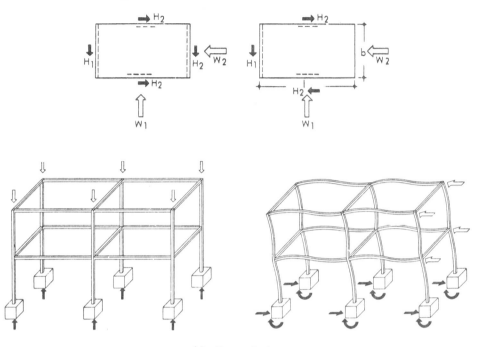

(a) Frame Action

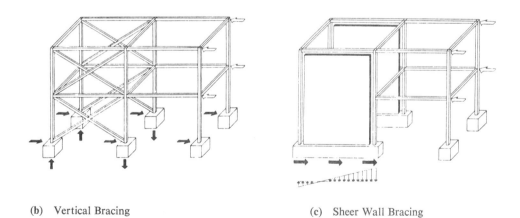

(b) Vertical Bracing

(c) Sheer Wall Bracing

Figure 1.5a, b, c Types of Vertical Bracing System (From 1.43).

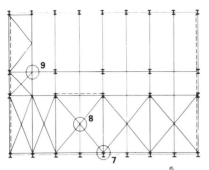

Figure 1.5d In this example, the vertical bracing is inserted in the two end walls and also between a pair of internal columns. The horizontal bracing in the longitudinal direction is installed between one internal row of main beams and the external columns, which are interconnected by beams forming the outer chord of the lattice girder. The transverse horizontal bracing is inserted between two floor beams. (From 1.43.)

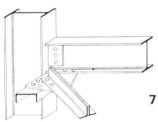

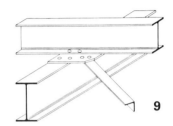

7 If the floor beams are all connected directly to columns, without main beams, a separate member (in this instance, a channel section) must be provided to form the chord of the lattice bracing.

8 Intersection of the wind-bracing members with a floor beam.

9 If the floor beams are supported on main beams, structurally the best solution is to attach the bracing members at the level of the bottom flanges of the floor beams.

Figure 1.5e Details of Points 7, 8, and 9 in Figure 1.5d (From 1.43).

sion ring, the exterior or compression ring, and a system of columns and floor beams comprise this framing system. Light and translucent acrylic plastic sheets are used usually as roofing.

1.1.1.3 *Steel Bridges and Their Parts*

The main function of a bridge is to carry traffic over any interruption such as a crossing with a river, a canyon, a ravine, or another line of traffic. The floor deck of a bridge has to receive the live load and transfer it to the main girders which span between the abutments and the piers. They form the main structural system as they transfer the total load on the bridge (dead and live load) to the abutments and piers and then to the foundation. The main girders, which are usually located in two or more parallel planes, must be interconnected by vertical and horizontal bracings so as to enable them to carry loads acting normal to their planes (centrifugal force, wind, and seismic loads). To illustrate this, in Fig. 1.9a the skeleton of a typical through-truss highway bridge is shown. The top chord in that figure is in compression and the top lateral bracing will provide the supports against out-of-plane buckling of this chord and acts as a horizontal truss carrying the main part of the wind load to the

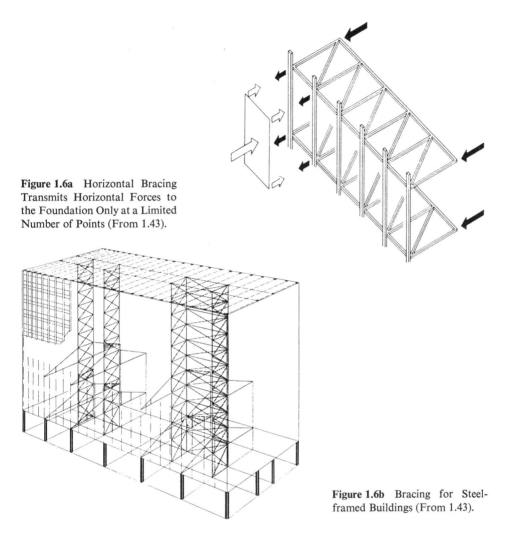

Figure 1.6a Horizontal Bracing Transmits Horizontal Forces to the Foundation Only at a Limited Number of Points (From 1.43).

Figure 1.6b Bracing for Steel-framed Buildings (From 1.43).

bridge portals. The top lateral bracing is composed of lateral struts and diagonals. The end portal frames located in the plane of the main end diagonals transfer the wind load reactions to the bridge supports. In the vertical planes across the bridge, sway bracing is provided to give lateral rigidity to the structure and equalize the loads acting on the two main trusses caused by asymmetrical vertical loads (Fig. 1.9b). This sway bracing also ensures torsional rigidity of the bridge.

The top and bottom lateral bracing as well as the sway bracing and both portal frames transform a bridge structure into a space or three-dimensional truss. It is very seldom analyzed as such unless use is made of already existing computer programs for the analysis of space trusses. Normally, loads are found for each truss in different planes and each truss is separately analyzed. The forces in members common to more than one truss are then added algebraically. An example of this is the top chord

The Steel Structure

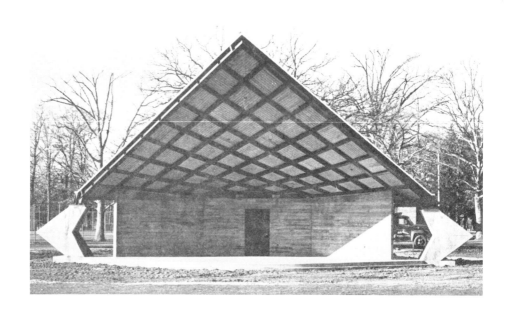

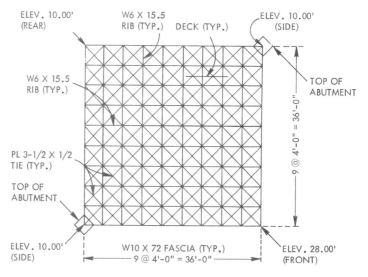

Figure 1.7 Hypar Band Shell (Courtesy AISC, *Modern Steel Construction*, 1978, Vol. XVIII, p. 10).

Figure 1.8 Madison Square Garden Sports and Entertainment Center During Construction (Courtesy Bethlehem Steel Corporation).

members which form part of the main truss and the top lateral bracing. In Fig. 1.9c several types of trusses used for bridges and roof trusses are shown.

That part of the bridge shown in Fig. 1.9 which is above the abutments and piers is called the superstructure. Several types of steel bridge superstructure exist, depending upon the classification criteria that are used. Bridges can be classified as follows:

1. According to their function
    a. Highway (Fig. 1.10)
    b. Railroad (Fig. 1.11)
    c. Pedestrian
    d. Material handling (conveyors, pipelines, coal or ore-unloaders)
2. According to the cross section of the main bridge element
    a. Girder type (simple section, Fig. 1.12, or box section, Fig. 1.13)
    b. Truss type (Fig. 1.14)

The Steel Structure **13**

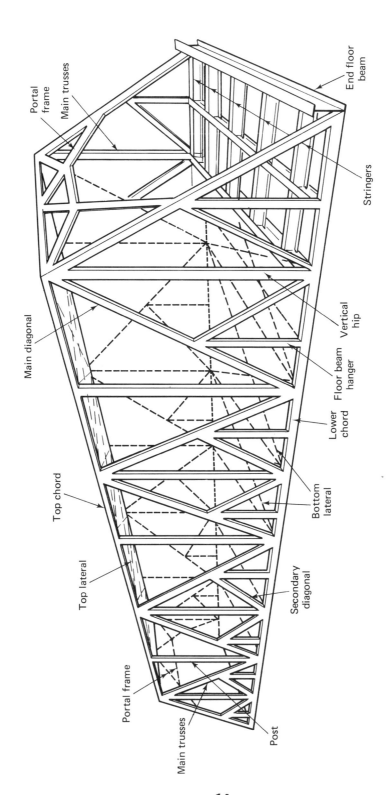

Figure 1.9

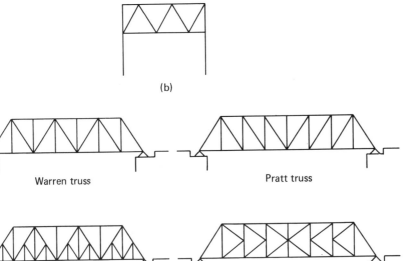

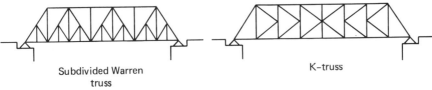

(c) Some typical bridge and roof trusses

Figure 1.9 (continued)

3. According to the form of the bridge structural axis in the vertical plane
   a. Beams (simple, continuous, or cantilever beams, straight axis)
   b. Frames (axis is polygonal)
   c. Arches (beam with curved axis)
   d. Suspension bridges

   and in the horizontal plane
   e. Straight and
   f. Curved
4. With respect to the position of the floor system in a bridge section
   a. Deck bridges (floor on top of bridge section, Fig. 1.14)
   b. Half-through bridges
   c. Through bridges (floor at the bottom of the bridge cross section)
5. With respect to the angle between the longitudinal bridge axis and the abutments and/or piers
   a. Normal
   b. Skew bridges

In Figure 1.15 a two-lane highway bridge and its cross section near the midpier are shown. The 7/16-in. thick steel deck plate is stiffened by closed 5/16-in. trough-shaped stiffeners, and carries the dead and traffic loads, transmitting them to the floor-beams. They are 15 ft-7½ in. apart and in turn pass this load on to two 54-in. deep main girders. These girders are acting as continuous beams across two 96-ft spans. The middle pier is a column frame pier with a box cross section.

The Steel Structure **15**

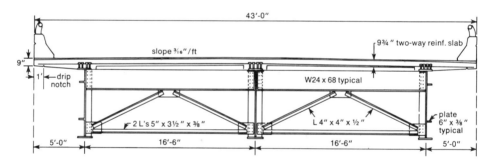

Figure 1.10 The structure is designed in accordance with AASHTO Standard Specifications for Highway Bridges. Design live load is HS-20 and the dead load includes a future wearing surface of 30 psf. (Courtesy Bethlehem Steel Corporation.)

1.1.1.4 *Special Structures*

Transmission poles and towers, either for the transmission of electrical power or for television, radio antennas, radio telescopes, towers in mines for vertical transportation, and many other structures built for some specific function are classified as special structures. As mentioned earlier, they will not be discussed in this text.

The all-welded box sections were frabricated in units ranging from 42 ft to 116 ft in length.

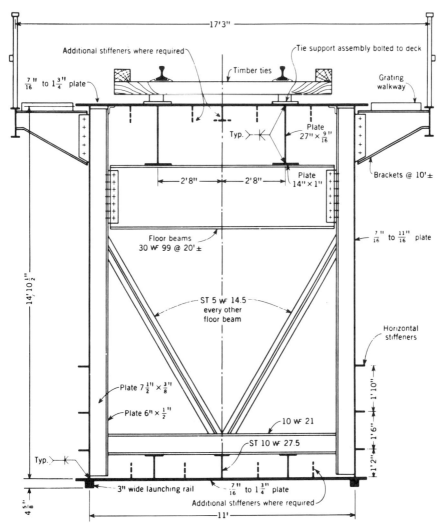

Figure 1.11 Typical Box Girder Cross Section and Fabricated Segments (Courtesy Bethlehem Steel Corporation).

17

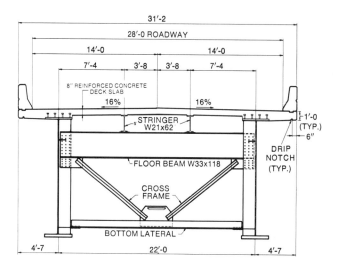

Typical Cross Section — Main Span Unit

Figure 1.12 Patuxent River Bridge (Courtesy Bethlehem Steel Corporation).

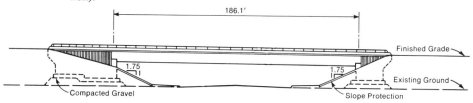

New Britain Avenue

Figure 1.13a Web depths for the New Britain Avenue structures are 80 in. (Courtesy Bethlehem Steel Corporation).

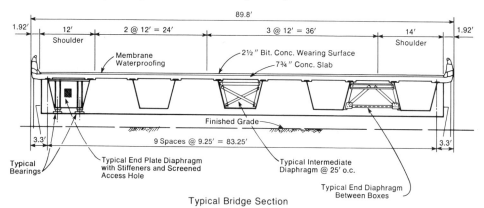

Typical Bridge Section

Figure 1.13b Both crossings consist of twin structures with five box girders in each structure. The two-coated field applied finish, applied to the exterior of the ASTM A588 material, is expected to provide a low maintenance structure. (Courtesy Bethlehem Steel Corporation).

Figure 1.14 (Courtesy Bethlehem Steel Corporation).

1.1.2 Loads on Structures

The analysis and design of any structure in general has to start with a determination of the acting loads to which the structure will be exposed. Under acting loads we include all the forces which are imposed by gravity, the environment, and the use of the structure. These are the external forces. The internal forces or stresses in the structure itself are determined by analysis. The external forces consist of the permanent (dead) load, or the weight of the structure itself and all the attachments to it; snow, wind, seismic, or other inertia forces; live or moving loads (occupancy loads for buildings, transportation loads for bridges); temperature effects; and dynamic effects

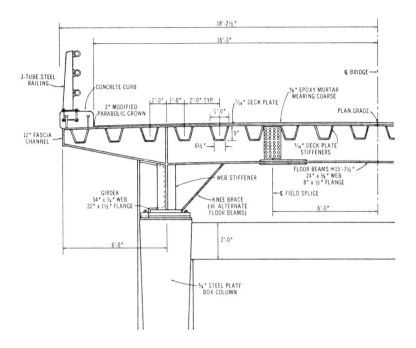

Figure 1.15 Orretz Road Overpass at Lansing, MI (Courtesy Bethlehem Steel Corporation).

of live loads. A static load is strictly defined by its magnitude, but the effect of a dynamic load is also determined by its mass and velocity or rate of application. Dynamic response in the form of vibration of the structure may result from dynamic loads, and such action needs to be studied carefully. A good designer knows that dynamic action is not simply a repetition of a set of static loads.

1.1.2.1 *Dead Loads*

Dead loads can be defined as vertical loads which are fixed in position and are produced by the weight of the elements of the structure or the whole structure with all its permanent components. Although these loads are known quite accurately once the design of the structure is completed, at the beginning of an analysis a dead load has to be assumed as close as possible. Previous design experience, available data, and weight tables, as well as some empirical formulas, are helpful in this stage of the design process.

Weights of building materials and of types of built-up roofs and bridge and building floors can be found in any good handbook (1.1)* or manual (1.2), and therefore will not be reproduced in this text.

1.1.2.2 *Occupancy Loads on Buildings*

All loads other than dead loads are live loads. The live loading of steel buildings is highly variable, depending upon the use of the building. Minimum magnitudes are usually specified by local or national building codes. Some types of live loads may be practically permanent in nature, although subject to removal or relocation. Movable partitions (of about 20 psf (960 Pa)), hung ceilings, and building equipment fall in this category.

To produce a safe design, occupancy loads are taken conservatively, derived more from experience and current practice than from accurately computed values from statistical data based on the probability of their occurrence. The American National Standard Building Code of the American National Standard Institute (ANSI) (1.3) gives the minimum values for such loads. To make these loads more realistic, most codes allow for some percentage reduction from the full loading. ANSI-1972 (1.3) in this respect allows, for live loads of 100 psf or less, a reduction of the design live load on any member supporting 150 ft² (14 m²) or more at the rate of 0.08% per ft² (0.9% per m²) of the tributary area of that member, except for public assembly areas, garages, and flat roofs. The maximum reduction is limited to

$$R_{max} = 23(1 + D/L) \leq 60 \qquad (1.1)$$

where R = reduction in percent, D = dead load in psf, and L = live load in the same units as dead load.

1.1.2.3 *Snow Loads on Buildings*

During the winter of 1979 more than 200 roofs collapsed in the northern counties of Illinois and Wisconsin, following more than 1,300 collapses in the northern U.S. in the winter of 1978. Subsequent investigations focused on two problems most

---

*Numbers in parentheses, as here (without any other indication), refer throughout the text to entries in the "References" sections at the end of each chapter.

common in heavy snow areas: unpredictable amounts of snow and the nonuniform distribution of it. Roof failure caused by snow usually does not occur as a result of a uniform load but from a localized drift or ponding load. Wind drifting of snow has been the root cause of many failures. At this point it is good to remind the reader that specification values are not always the desired values to be used in design. Also, the minimum loads prescribed by codes do not relieve the designer of responsibility for the safety of a particular design. The designer's judgment must lead to the proper load magnitudes. A major structural designer in California used for years 2 to 4 times the snow loads called for in the Sierra Nevadas for designing chalets and ski resorts. Another successful designer in Alaska used snow loadings as much as 50% higher than those specified, especially for unequal loadings on roofs of industrial buildings. In conclusion it can be stated that a good designer does not rely strictly on specified loadings. For that reason ANSI (1.3) is in the process of revising the snow load codes. The Snow Load Subcommittee of ANSI has drafted a new standard, which should be adopted shortly. The snow maps, generated from data at 9,000-plus locations, will provide more reliable information. Roof slope factors will depend on the thermal condition of a roof and the ability of its surface to shed snow by sliding. In the existing ANSI A58.1-72 snow load coefficients in some way are taking into consideration possible drift accumulations.

The current ANSI code gives a basic snow load, q, which is then multiplied by the appropriate coefficient, C_s. The basic snow load corresponds to the ground load in psf for 50 years mean recurrence interval. These loads are used for all permanent structures except those that are judged to represent an unusually high degree of hazard to life and property in case of failure. For those a 100-year mean recurrence interval must be used. If the risk to human life is negligible, a 25-year mean recurrence interval may be used. ANSI A58.1-72 gives in an appendix a map of the United States showing isotones of ground snow.

The snow load coefficients, C_s, depend on the wind speed and direction, the geometry of the structure, and the temperature gradient between the inside of the structure and the outside. The basic coefficient is $C_s = 0.8$, which may be decreased to reflect slide-off of snow on roofs with slopes exceeding 30° and must be increased to reflect nonuniform accumulation on pitched or curved roofs as well as in valleys formed by multiple series roofs.

The probabilistic approach to finding design snow loads in Load and Resistance Factor Design (LRFD) is discussed in Section 1.4.2.2.

1.1.2.4 *Wind Loads*

Improvements in the building and construction industry, as well as new structural materials and structural systems, have resulted in structures that are more responsive to significant loading influence by the wind. To design successfully and safely a designer must have detailed knowledge about static loading induced by steady wind action as well as the dynamics of flexible structures and their response to intermittent wind-induced forces. A good design ensures that the performance of structures exposed to the wind action will be adequate during their service life from the standpoint of both structural safety and serviceability.

The wind is a movement of free air caused, on a large scale, by thermal currents in the first 10 miles above the earth's surface (1.4). The variation with height follows the logarithmic law up to 300 ft (100 m) above the ground. Above that height the wind speed variation with height is negligible (1.5). In a storm, the cold air flow spreads horizontally over the ground. The first gust (or gust front), i.e., the wind occurring in a thunderstorm that changes considerably and relatively rapidly both speed and direction, is of importance. The gust size is the wind speed increase. It may vary approximately from 10 to 100 ft/sec (3 to 30 m/sec). The design wind speed is the maximum gust, which is taken as equal to 1.3 times the fastest-mile wind at 30 ft (10 m) above the ground, recorded at a station in the period 1924–1952. The Uniform Building Code (1.6) accepted that definition. ANSI A58.1 has as the design wind speed the 50-year fastest mile for most permanent structures. As a result of the wind speed a dynamic wind pressure q is formed

$$q = 0.00256 \, V^2 \quad (1.2)$$

where q is in psf and V is the wind speed velocity in miles per hour (Eq. 2 in ANSI (1.3)).

For usual types of buildings this dynamic pressure q is converted into an equivalent static pressure p

$$p = q \cdot C_e \, C_g \, C_p \quad (1.3)$$

where C_e is an exposure factor variable with the height, C_g is a gust factor, and C_p is a shape factor of the building as a whole (1.7).

In summary, wind loads (design pressures and suctions) on buildings and parts of them are obtained in four steps:

1. Selection of an appropriate mean recurrence interval
2. Determination of basic windspeed for specific geographic location
3. Determination of an effective velocity pressure for a specific structure and terrain
4. Determination of design pressures for the structure and its components

For load factors for wind in LRFD building design the reader is referred to Section 1.4.2.2.

In highway bridges, the AASHTO Specifications (1.8), section 1.2.14, specifies wind load forces per square foot of exposed area of all structures for a wind velocity of 100 mph (160 km/hr). In addition to this, for the design of the superstructure, a moving uniformly distributed wind load is applied horizontally at right angles to the longitudinal axis of the superstructure.

1.1.2.5 Earthquake Loads

The vibrations of the crust of the earth caused by sliding of earth on one side of a fault with respect to the earth on the opposite side constitute an earthquake. The dynamic movement of the earth during an earthquake is in all directions, and though its horizontal components are of major importance for earthquake resistant design, a rigorous dynamic design today involves consideration of the vertical excitation as well. When the ground under a structure suddenly moves, the mass inertial forces tend to resist this movement and a horizontal shear force V is developed between the ground and the mass of the structure. This lateral force, which is usually empirically prescribed by many codes, represents a static load supposedly equivalent to the actual dynamic load, which never can be true. For simple low-rise structures this approach may be adequate, but for buildings which are more complicated in plan and of larger height a full dynamic analysis of the earthquake-building interaction is required.

The Modified Mercalli intensity scale is widely used in the United States and in some other countries as a measure of the damage potential of an earthquake. Ratings on this intensity scale are based on people's reaction to earth movement, the observed structural damage, and some other physical effects.

The ground acceleration and duration of shaking are most important. A continuous record of the earthquake acceleration of the ground or of a building is produced by the strong motion accelerograph. Such a record, called an accelerogram, is shown in Fig. 1.16.

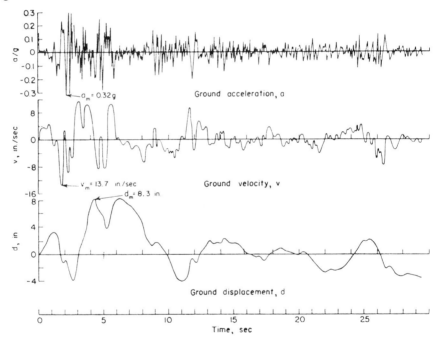

Figure 1.16 Accelerogram of El Centro, California, Earthquake of May 18, 1940, North-South Component (from 1.1).

For a rigid structure, rigidly coupled to its foundation, the force V equals the product of its mass M and the ground motion acceleration at any instant, as shown in Fig. 1.17. If the structure deforms lightly, then for short periods of time this force may be somewhat less, because deformation of the structure absorbs some of the energy, storing it for some later time. However, if a very flexible structure is subjected to a ground motion whose fundamental period is near that of the structure, a much greater force may result, especially if several cycles of ground motion occur.

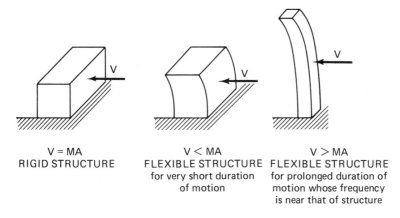

V = MA
RIGID STRUCTURE

V < MA
FLEXIBLE STRUCTURE
for very short duration
of motion

V > MA
FLEXIBLE STRUCTURE
for prolonged duration of
motion whose frequency
is near that of structure

Mass of structure is "M." Ground moves with acceleration "A."

Figure 1.17

The use of a *spectrum* for evaluating the effects of actual ground motions was proposed by M.A.Biot in 1933. This approach, which was greatly expanded by G.W. Housner (1.9), is much better than using an equivalent static load. For any given area or site (1.10) estimates need to be made of the maximum ground acceleration, the maximum ground velocity, and the maximum ground displacement. The lines representing these values can be drawn to different scales and therefore there are dozens of different types of response spectra. One of them, drawn on the tripartite logarithmic chart, is shown in Fig. 1.18.

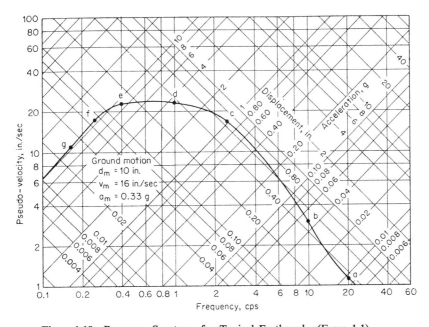

Figure 1.18 Response Spectrum for Typical Earthquake (From 1.1).

The Steel Structure

It is outside the scope of this text to go into more details of earthquake designs based on response spectra. Seismic design for buildings is based on the Uniform Building Code (UBC) Chapter 23, (1.6), which is essentially the same as the *Recommended Lateral Force Requirements*, 4th Edition, 1974, by the Structural Engineers Association of California (1974 SEAOC Code). The minimum total lateral seismic forces (shear) assumed to act nonconcurrently at the base in the direction of each of the main axes of the structure are given by the following formula

$$V = ZIKCSW \tag{1.4}$$

where Z = seismic-zone coefficient, varying from 3/16 to 1
I = occupancy importance factor, between 1 and 1.5
K = horizontal force factor, between 1 and 2
$C = \dfrac{1}{15\sqrt{T}} \leq 0.12$ where T is the fundamental period of vibration, in seconds, in the direction considered
S = site-structure resonance coefficient. The product CS need not exceed 0.14.
W = total dead load (for storage and warehouse occupancies total dead load plus 25 percent of floor live load)

The total lateral force V is distributed over the height of the structure in the following manner:

(1) A force F_t at the uppermost level n given by

$$F_t = 0.07\, TV \tag{1.5}$$

This force need not exceed $0.25V$ and may be taken as zero if $T \leq 0.7$ sec.

(2) A force F_x at each level, including the uppermost level n given by

$$F_x = (V - F_t)\, \dfrac{w_x h_x}{\sum\limits_{1-i}^{n} w_i h_i} \tag{1.6}$$

As several bridges were severely damaged in recent earthquakes, especially in California, the AASHTO Specifications (1.8) in section 1.2.20 has introduced earthquake design based on either equivalent static forces or the use of a response spectrum. Many obserbed damaged bridges, of the beam structure type, just slipped off their supports and fell down. To prevent this type of failure, the design of restraining features to limit the displacement of the superstructure, e.g., hinge ties, shear blocks, etc., are introduced.

1.1.2.6 *Live Loads on Bridges*

Live loadings on the roadways of highway bridges consist of standard truck or lane loads which are equivalent to truck trains. Both loadings are assumed to occupy a width of 10 ft. These loads are placed in 12-ft (3.66-m) wide design traffic lanes, spaced across the entire bridge roadway width, in numbers and positions required to produce the maximum stress in the member under consideration. The load distribution across the width of a bridge is taken according to semiempirical rules given in a table. They depend upon the kind of floor (timber, concrete, steel box girders, steel grid, etc.), the number of design lanes, and the distance between stringers.

As with building loads, there have been occasions where the number of vehicle crossings has been seriously underestimated using published specifications. The engineer cannot rely strictly on specified loadings and must make an estimate based on a projection of traffic loads and their frequency or risk running into difficulty with respect to fatigue.

For railroad bridges similar semiempirical loadings, which are called the Cooper E 80 train, are specified. They consist of a series of concentrated forces at fixed distances followed by a uniform loading. This loading is prescribed by the American Railway Engineering Association (AREA) (1.11).

1.1.2.7 Impact in Buildings and Bridges

All loads acting on a structure are either static or dynamic in nature. As mentioned earlier (see Section 1.1.2), the effect of dynamic loading is determined not only by its magnidute but by the mass and velocity or rate of application of the loads. Therefore, the effect of a dynamic force is larger than its magnitude. To compensate for this dynamic effect, many specifications increase some live loads by an empirical factor (larger than one), called the impact factor.

For building design, the AISC specification (1.3) for structures carrying live loads which induce impact specifies impact factors for support girders and their connections exposed to traveling cranes, motor-driven light machinery, reciprocating machinery or power-driven units, and floor and balcony hangers.

In the same way, the AISC specification gives horizontal lateral and longitudinal forces acting on a crane runway (Fig. 1.19), to provide for the effect of moving crane trolleys. They are expressed as percentages of the weights of the lifted load and the crane trolley and of the maximum wheel loads of the crane, respectively.

In the design of the superstructure of highway bridges (including steel or concrete supporting columns, steel towers, and legs of rigid frames) live load stresses produced by truck or lane loading have to be increased to compensate for dynamic, vibratory, and impact effects. This increase is obtained from an impact formula and expressed as a fraction of the live load stresses. The current formula is

$$I = \frac{50}{L + 125} \leq 0.30 \qquad (1.6)$$

where I is the impact fraction (maximum 0.30) and L is the length in feet of the loaded portion of the span that produces maximum stress in the member. In the case of roadway floors and floor beams the span length is to be used for L. In computing truckload moments, except for cantilevers, again the span length is used. The AREA specifications use similar formulas.

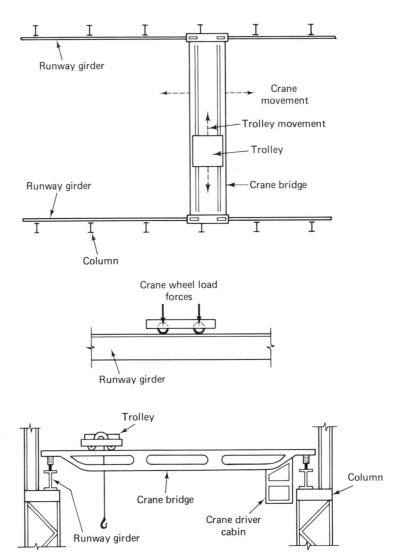

Figure 1.19

1.2 DESIGN CRITERIA

1.2.1 Introduction

For many years civil engineering design was mainly an art based on intuition and experience. The development of applied mechanics toward the end of the nineteenth century gradually introduced the use of mathematical operations and more and more sophisticated mathematical models into design. Most engineering design in the past was deterministic in character. Loads, structural strength (resistance), and safety, the three basic elements of structural design, usually are considered to be deterministic quantities without any uncertainty with regard to their value. With the development of probability theories, and the realization that design parameters are statistical in nature, a new type of design has developed called probabilistic design, As this approach is still in its developmental stage and lacks sufficient data for a rigorous mathematical treatment, most of the design followed in this text is still of the classical deterministic type. More details about the use of probability concepts in design will be given later in Section 1.4, which deals with structural safety.

The design of any civil engineering structure consists of creating a final design that represents the optimum of all possible solutions. A solution can be considered optimum if it represents the most economical one among all those which are aesthetically and technically acceptable. The aesthetic features of a structure are very important, and a designer must realize that beauty of a structure is not automatic and that technical perfection alone is not enough. Good designers are not slaves of their formulas; rather they are artists who use their calculations as tools to create working shapes as inevitable and harmonious in their appearance as the natural laws behind them. The art of a structural design lies in the development of the beauty latent in those structural forms that most effectively use the strength and special properties of the material they are composed of.

A structure is technically acceptable if it meets all safety and performance requirements. Safety requires that the design loads be safely supported for the duration of the projected life of the structure. Design loads may be prescribed by codes and specifications or dictated by actual anticipated service loadings. The final design, as well as the procedure used to achieve such a design, are greatly influenced by the loading. For example, a design is either static or dynamic depending on the loads. The useful life of a specific structure is terminated if the service conditions have changed so as to render the structure uneconomical or unsafe or if a failure occurs. Failures can be classified in terms of either serviceability or strength. Serviceability failure is related to criteria governing normal use with respect to unacceptable deformations, displacements, vibrations, stresses, or other undesirable damage. Strength failure is associated with collapse or inelastic deformations of an unacceptable magnitude. Different types of strength failure are possible depending on the loading conditions and the type of structure. In many cases design criteria and also possibly design procedures depend on the type of failure that is expected to occur. Usually several different types of failure are possible. Consequently several design critera have to be considered simultaneously. Distinction can also be made between different types of failure, such as local failure, sectional failure, and failure of the structure as a whole, or structural failure for short. Local failure in a particular member or connection may or may not result in sectional failure or failure of the structure as a whole depending on whether such failure results in one or more sectional failures and finally in instability or collapse of the structure as a whole.

A discussion of the various design criteria that need to be considered by the engineer follows next.

1.2.2 Yield Criterion

One of the oldest methods of steel design is based on the criterion that a structure must be considered failing if nominal yielding stresses are reached at any location in a structure. In this method the reaching of the yield-point stress at such a location is considered critical for the structure. Based on ideal linear elastic behavior, according to Hooke's law, the design is thus based on the assumption that stresses anywhere in the structure may not exceed a given allowable working stress. The determination of allowable working stresses is based on past experience, making due allowance for variations in material and sectional properties, possible working load variation, secondary stresses, residual stresses, and so forth. Most codes and specifications prescribe maximum allowable stresses by dividing the yield-point stress by an appropriate "factor of safety." This method of design usually is referred to as the "elastic design method" but more appropriately could be called the allowable stress or working stress design method.

1.2.3 Plastic Criterion

The plastic design method makes use of the fact that when yielding stress is reached at a certain location in a structure, this does not necessarily imply that the structural member or connection concerned is failing as a whole. Local yielding at a point does not automatically result in an unrestrained plastic flow and sectional failure at that point. In addition to this phenomenon, constraints in the structure as a whole and support conditions may be such that even if sectional failure occurs no structural failure or collapse will result. Thus three different types of so-called coaction phenomena (1.12) may occur either separately, consecutively, or simultaneously before structural collapse, excessive deformations, or instability result in structural failure. These three coaction phenomena are discussed next.

1. The first phenomenon that occurs when steel is used as a structural material is that of the adjustment or adaptation of stresses within a cross section or a member as a whole. This equalizing of stresses before sectional failure takes place has long been recognized by engineers and accepted as a working basis for elastic design. The assumption of uniform stress distribution in axially loaded members, equal bolt loads in bolted joints, and the neglecting of stress concentrations and residual stresses are all simplifying assumptions normally used in allowable stress design, which are based on stress equalization as a result of localized plastic yielding. In the case of an axially loaded bar with a round hole, the assumption of a uniform stress distribution (Fig. 1.20) across the net section is known to be incorrect. However, an allowable stress design based on a uniform working stress distribution is justified in the sense that the section will only fail when the yield stress is reached across the whole section. This latter stress distribution is uniform, and therefore a factor of safety is applied to the ultimate load. At working load, yielding may have taken place in spite of the assumption made in allowable stress design that first yielding represents failure. (The example of an axially loaded bar also demonstrates that, provided the factor

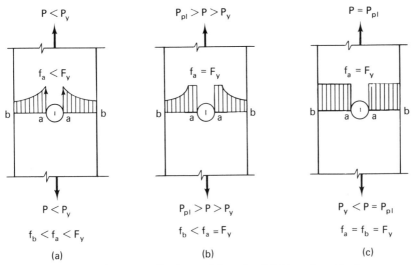

Figure 1.20 Stress Equalization or Adaptation Within a Cross Section.

of safety and load factor are taken as equal, no economic gain results by considering the more correct ultimate load concept.) The equalization of stresses in sections subjected to pure bending resulting in the formation of plastic hinges represents a logical extension of this phenomenon of adaptation of stresses (Fig. 1.21). The example of a bent section, in contrast to the one discussed above for an axially loaded bar, demonstrates that by considering the plastic moment capacity (M_{pl}) rather than the moment at first yield (M_y) more economic sections can be used. The additional bending moment capacity obtained by taking into

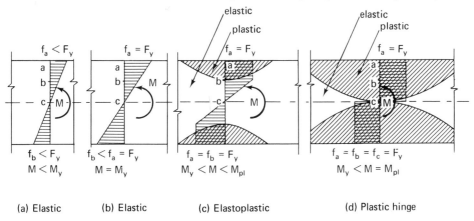

Figure 1.21 Stress Equalization in a Cross Section of a Beam in Bending.

account full plastification of a bent section is often referred to as the "first reserve capacity," and is expressed in terms of the "shape factor," $f = M_{pl}/M_y$.

2. A second phenomenon which might occur in a steel structure as a result of sectional failure is the redistribution of internal forces. In continuous beams

and frames this phenomenon is known as the redistribution of moments. If a structure is redundant, sectional failure will cause other sections to start carrying any additional loadings that are imposed, resulting in additional sectional failures (plastic hinges) (Fig. 1.22). Only when the structure has lost all its redun-

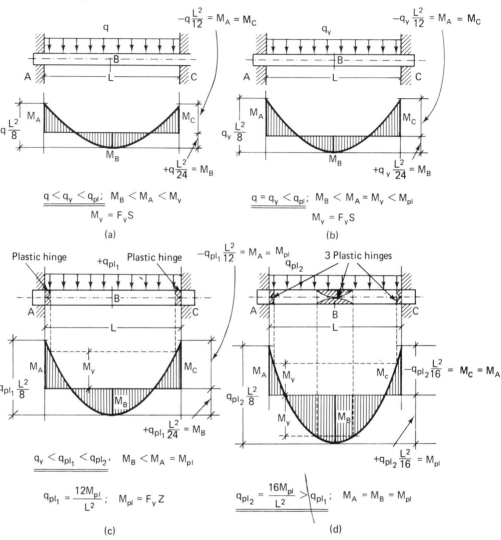

Figure 1.22 Built-in Beam of Uniform Section under Various Loadings.

dancy as a result of the formation of one or more plastic hinges will one additional sectional failure cause collapse or instability. The additional load capacity that may exist in a redundant structure over and above the capacity reached at the time of first sectional failure is known as its "second reserve capacity."

3. The third phenomenon, which is less well known, is that which results from the development of additional axial secondary stresses due to support conditions. Such secondary stresses may or may not result in a "third reserve capacity." For example, if a continuous beam has two adjacent hinge supports, restraining of horizontal elongation along the bottom flange as a result of a predominant positive moment will result in the development of horizontal thrust. Such thrust will change the stress distribution not only at the location of plastic hinges but throughout the beam (Fig. 1.23).

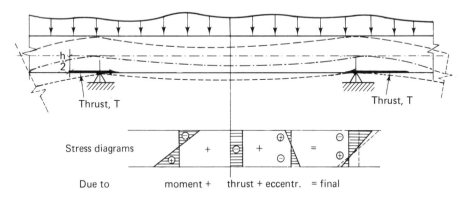

Figure 1.23 Third Reserve Capacity.

1.2.4 Deflection Criterion

The recent development of high-strength steels has resulted in stronger members which posses less rigidity than would be the case if lower-strength members were used. In many cases nowadays it is found that this results in deflections and deformations which are objectionable and as such constitute a controlling design criterion. It must be realized, however, that some of the deflection criteria used for lower-strength materials and old-fashioned structural systems are not necessarily any longer relevant. In many cases large deflections are still acceptable in spite of existing prescribed limits. Limitations on deflections in the past were often based on limiting values that applied to adjoining structural elements or materials. A good example of this is the use of plastered ceilings in conjunction with floor beams. The replacement of such "older" materials by other structural materials which can undergo larger deformations often invalidates the deflection restrictions used in the past. For highway bridges, for instance, it is good engineering practice to counteract aesthetically objectionable large downward deflections by building negative cambers into the main members. Nevertheless, the effect of large load deflections has to be carefully considered.

In addition to elastic deflections, which are recoverable, some structures undergo creep deflections as a result of long-term loadings. If these deformations are significant, such creep effects need to be considered as part of the design.

1.2.5 Dynamic Response Criterion

When structures are loaded, the loading is either static or dynamic. Although the majority of loadings are dynamic, experience has shown that in most cases the dynamic effect is minor and can be neglected or can be represented by introducing an increased equivalent static load, a reduced factor of safety, a dynamic load factor, or an impact factor. In other words, an analysis based on static loads can be used with confidence in most cases with or without compensation for dynamic loading.

The main purpose of this section is to point out that there are loadings for which a static analysis is insufficient not only in view of structural strength but often because of other effects such as response deflection and what is known as disturbance or annoyance to occupants. Dynamic effects that may require modification or redesign of a structure evaluated on the assumption of static loading may be grouped as follows.

1. Vibrations produced by human or vehicular traffic. Many floor systems, for instance, exhibit annoying vibrations when persons jump on or walk over them.
2. Motions produced by blasts, wind gusts, or waves.
3. Motions due to support movements caused by earthquakes.
4. Vibrations and possible resonance caused by operating machinery supported by the structure.
5. Vibrations induced in slender structures due to static wind loading in a direction lateral to the wind direction.

Among the various problems listed above, the first group has become significant only during recent years. This is mainly due to the use of lighter structures because of the availability of stronger materials, the design of more efficient sections and systems, and the design of larger span lengths, in some cases combined with heavier loads.

Problems in the second and third groups have always existed, but the design of larger and higher structures has made it necessary to study their dynamic response as part of the design.

The fourth group represents classical vibration problems; however, in the past these were often overlooked by structural engineers and were consequently treated on a remedial basis. As will be obvious from the discussions on fatigue in Section 2.4, failures due to alternating or varying loading are a distinct possibility. Also, questions of resonance, damping, and annoying effects may have to be considered.

The fifth group of vibrations occurs when slender structures start vibrating due to a static wind loading in a lateral direction. Such vibration, which is caused by periodic vortex shedding, is now recognized as being the main cause of the Tacoma, WA, bridge disaster in 1940. Particularly for structures such as suspension bridges and slender towers or poles, the possibility of wind-induced vibrations needs to be considered as part of their structural design.

1.2.6 Instability Criterion

Stability of individual members such as columns or the stability of a structure as a whole in many cases controls a design. If the instability criterion is controlling, modification of allowable stresses or ultimate load criteria is necessary. Distinction can be made between local instability, member instability, and structural instability.

1.2.7 Fatigue Criterion

Fatigue design deals with repetitive dynamic stresses. Under the action of repeated loadings, fatigue cracking may occur. This phenomenon depends not only on the type of loading but also to a large extent on the properties of the steel used and its physical detailing. Sharp stress concentrations due to structural or metallurgical stress raisers (discontinuities) at any location are normally starting points for cracks. If brittle fracture is excluded, such cracks may slowly propagate under further load repetitions. The increased use of welding during the last decades often has resulted in unexpected or premature fatigue cracking due to weld defects which eventually act as sharp stress raisers. A thorough understanding of fatigue cracking and subsequent crack propagation is essential for the modern engineer.

There are two major approaches to fatigue design (1.13). One is based on the concept of "safe-life," while the other is based on that of "fail-safe." The first approach requires that no fatigue cracks develop during the whole prescribed life of the structure. This implies that the life of a structure can be predicted and that before the end of such a period the structure can be repaired, replaced, or retired. To succeed in "safe-life" design, several elements of the design matrix must be properly considered, such as

1. The modes and frequencies of applications of working loads during the whole operational life of the structure.
2. The configuration and sizing of all the components and their connections, which must be planned so as to provide sufficient reserve fatigue strength as compared with the real loading.
3. Systematic fatigue tests of structural groups, which must be performed to prove "safe-life" experimentally.
4. Additional measures against the environmental effects on fatigue strength.

The second, "fail-safe" or "damage-tolerant," method of design realistically assumes that total protection against all cracks is impossible. Therefore, the design must guarantee that dynamic propagation of cracks will not be the cause of complete structural failure. There are several ways to achieve this, such as

1. Performing a timely inspection to detect a crack of a predetermined minimum size so that its propagation will be slow.

2. Arresting a crack before it can propagate completely through a life-important structural part.
3. Making additional stress paths available and thus guarding against damage due to failure of one structural member by providing structural redundancy.

For a more detailed treatment of fatigue the reader is referred to Section 2.4.

1.2.8 Brittle Fracture Criterion

Cracking of steel members or connections under moderate nominal stresses can also be caused by brittle fracture. A brittle fracture may be defined as one that absorbs a relatively small amount of energy in propagation. Failure by cleavage of a large fraction of individual grains is the most common form in which brittle fracture occurs in structural steels. One way to prevent brittle fracture and particularly its propagation, when using the allowable stress design method, is to reduce allowable stresses. This, however, may result in exceptionally low values for allowable stresses and expensive or impractical designs. The tendency to prevent brittle fracture by using special design details and types of steel that are sufficiently tough is usually a more fruitful approach to this problem. Again, in order to prevent brittle fracture from occurring, it is important to understand its causes and subsequent behavior. The concepts of fracture mechanics have been used to analyze the fracture behavior of structural members used in steel structures. These will be discussed in more detail in Section 2.5.

1.2.9 Conclusions

For a given design, very seldom does only one failure criterion control. Yielding, deflections, dynamic response, instability, fatigue, and brittle fracture criteria could well apply to the design of a single structure. In most cases a combination of these criteria has to studied to achieve a safe design.

In spite of the fact that many different criteria could apply to any one design, the two major methods of structural steel design are the "allowable stress" and the "plastic" design methods. Both methods emphasize the strength aspect. Initial sizing and proportioning based on these methods need to be checked and modified to meet any additional criteria that are relevant.

The allowable stress method uses nominal local yielding as its strength criterion and makes use of allowable stresses by dividing the yield stress by a factor of safety. Although most civil engineering structures behave sufficiently linearly elastic under working loads, it should be recognized that this loading when multiplied with the same factor of safety quantitatively and qualitatively is different from the loading that would cause structural collapse. For reasons of economy and a better understanding of the real behavior of a steel structure, the plastic design method is superior to the allowable stress method. Its development during the past three decades has made great strides, although additional research is required. Both methods of design, where appropriate, are treated in this text.

1.3 OPTIMUM DESIGN

As mentioned previously, it must be the engineer's goal to produce an optimum design that is aesthetically acceptable. The search for an optimum and complete solution is known as the systems design approach. Systems design has developed tremendously since the 1960s, mainly due to the fact that most of the various search techniques can only be executed on electronic computers. Unfortunately, there are many optimization aspects which at this time cannot be handled satisfactorily with a systems design approach. For instance, a systems design can be applied to a high-rise building to determine floor layouts, column spacings, and the design of structural elements (1.14), but in bridge design it is yet impossible to determine by this method the best type of structure and the optimum design of structural elements. Available programs are still restricted to structures for which the configuration is given or assumed, and they are mainly concerned with the sizing of the constituent elements so as to minimize the costs of a particular type of structure (1.15), normally by using minimum weight design.

Because this volume deals with the design of structural members, design optimization of an overall structure will not be discussed. It must be realized, however, that in any systems design the optimum design of the structural members forms part of the overall optimization program (1.16).

The plastic design method is more suitable for design optimization than the allowable stress method. In the latter method, when dealing with statically indeterminate structures, the solution is directly dependent on the sectional size initially chosen for the main supporting members. To arrive at an optimum allowable stress design it is necessary to try and compare many possible combinations, which is more a matter of analysis than of design. For the plastic design method, sectional sizes can be modified as the analysis proceeds, thereby leading directly to an optimum design. The design procedure in this case consists of a typical interplay of analysis and design. If, for example, the weight of a structure is considered to be its economic criterion, it is usually possible to express such a criterion as a linear function of the shape parameters. The Simplex or Danzig method (1.7) as applied to optimization, based on satisfying all restricting conditions and using linear programming, leads directly to an optimum solution. In many cases the plastic moment distribution method (1.18) can be employed and programmed more easily to obtain an optimum.

Systems design is very much in the stage of development (1.19–24), and it is expected that the structural engineer will make more and more use of this approach in the future.

1.4 STRUCTURAL SAFETY

When discussing design criteria it was pointed out that a successful design must satisfy aesthetic, safety, performance, and economic requirements in the best possible way. In other words, structures must be designed so as to have sufficient safety against any load they may be subjected to during their operational life and at the same time adequately and economically serve their function (1.25). To construct a structure as strong as possible without losing economical validity is one of the fundamental conditions in the design of structures. It is easy to understand how these two conditions, economy and strength, are in most cases contradictory. Therefore, the three basic elements of a structural design—the applied load, the strength of the structure, and its safety—must be carefully balanced. To be able to do this a thorough knowledge of the characteristics of load and strength distribution functions is necessary. Unfortunately, such data are still insufficiently understood at the present time. Therefore, a general treatment of structural safety is still in the developing stages. At present the classical procedure of using safety margins is still the one most used; however, the new probabilistic approach using these margins as probable values is rapidly developing (1.26). Load and resistance factor design method belongs to this category.

Because at present practically all specifications employ fixed safety factors and load factors in some form or another, the classical method will be discussed first.

1.4.1 Deterministic Approach

1.4.1.1 *Factors of Safety in Working Stress Design*

Working loads are usually specified by codes or specifications, or they are specific service loads. Such loads usually represent an idealization of actual loading conditions.

The requirement that a structure be designed such that its calculated minimum strength (or capacity) is larger than the estimated maximum design load is common among all classical design methods. Therefore, a factor of safety larger than unity is introduced. This factor is selected without a precise knowledge of the true values of load and safety and often represent a factor of ignorance.

In the allowable stress or elastic design method the necessary margin of safety against local or complete yielding is formed by allowing stresses which are equal to the yield-point stress divided by a factor of safety. Thus

$$F_{\text{all}} = \frac{F_y}{\text{F.S.}} \qquad (1.7)$$

1.4.1.2 *Load Factors in Plastic Design*

In the plastic design method safety against collapse is formed by multiplying the working loads by a load factor. Thus

$$L_u = (\text{L.F}) \times L_w \qquad (1.8)$$

The introduction of a safety margin either as a factor of safety or as a load factor needs some additional examination. The idea that such safety margins serve only to protect a structure against overloading represents an oversimplification of the combined action of several factors which may imperil the safety of a structure. These factors will be discussed in more detail next.

Methods of analysis

Methods of analysis and design for practical reasons are in most cases rather approximate and limited. This fact introduces the possibility of deviations from the calculated design stresses or the allowable loads.

Material properties

The assumed properties of any material always represent simplifying idealizations. Steel is assumed to be a continuous, homogeneous, isotropic material that behaves ideally linearly elastic or perfectly elastic-plastic. A knowledge of the basic structure of any steel material makes it evident that these assumptions are all approximations. Steel is a polycrystalline material consisting of cubic space lattices of particles oriented in different directions within each crystal grain (see Section 2.1, "Structural Steels"). It can only be considered a continuous, homogeneous, and isotropic material in a statistical sense. In addition, for any specific type of steel the physical properties such as yield-point stress, durability, and modulus of elasticity vary considerably. Consequently standards required by specifications usually represent extreme values (maximum or minimum) rather than average values. Material idealization and variations in physical properties thus require some additional margin of safety to insure that a design is safe.

Sectional dimensions

Due to variations inherent in any manufacturing process, the cross-sectional dimensions of any rolled steel member may be less than those specified (underrun). Maximum tolerances are allowed by most specifications [see AISC *Manual* (1.2), p. 1-121]. Even if these are not exceeded due to careful manufacturing or acceptance controls, stresses may exceed these calculated based on a "true" section.

Workmanship

Particularly in built-up sections, poor workmanship may result in undersized sections, high-stress raisers, and locked-in stresses.

Residual stresses

Any rolled steel section, due to the manufacturing process, possesses residual stresses. Unequal cooling rates cause compressive residual stresses in those parts of a cross section which cool first. In an I-shaped section these areas usually are located at the flange tips and in the midweb region. The intensity of residual stresses can be very high; for normal structural steels compressive residual stresses of about 13 ksi (83 MPa) ($= \text{kips/in.}^2 = 1,000 \text{ lb/in.}^2$) are quite normal (1.27).

For welded sections, apart from residual stresses already present in the composing elements, very high residual stress patterns can develop. Depending on the welding technique and sequence employed, such patterns may be quite different from those usually observed in rolled beams.

Although the internal forces resulting from residual stresses in a section are in equilibrium, the stresses caused by external loading act together with the residual stresses already present in a section. The resulting stress pattern may differ so much from that which would occur without residual stresses that, as will be shown later, special design considerations must be introduced.

Length variations

Especially in structural systems which exhibit internal redundancy, stresses may result from lack of fit due to fabrication variations or poor workmanship.

Service conditions

The intended use of a structure, whether it is temporary or permanent, for private or public use, and the duration and frequency of the expected loads will influence the magnitude of the required safety margin. Environmental conditions such as extreme temperatures, moisture, or the presence of corrosive agents may also require an increase in the safety margin due to the possibilities of reduction in cross-sectional areas, creation of brittle fracture, and excessive temperature stresses.

Type of failure

Even in allowable stress design it has been recognized that the seriousness of a possible failure should influence the magnitude of the factor of safety. The factor of safety against instability, for instance, is larger than that for most other types of failure. The factor of safety against uniaxial tensile failure, which can be considered the basic factor of safety, gives an idea of how conservative is any specific design standard. These factors of safety as required by different current specifications in various countries are shown in Table 1.1.

Table 1.1 FACTORS OF SAFETY AGAINST TENSION FAILURE

Country	U.S.			U.K.		France		Germany		USSR	
Spec.	AISC	AASHTO	AREA	BS 449 Build.	BS 153 Bridg.	CM- Build.	56 Bridg.	DIN 1050 Build.	DV 804 Bridg.	TUPM Build.	Bridg.
Ref.*	(1.2)	(1.8)		(1.33)	(1.34)	(1.35)	(1.35)	(1.36)	(1.37)	(1.38)	(1.38)
F.S.	1.67	1.80	1.80	1.68	1.68	1.50	1.50	1.50	1.50	1.36	1.36

*Consult the appropriate entry in this chapter's "References" section.

The magnitude of the F. S. given in the American Institute of Steel Construction (AISC) specifications can be simply derived by writing an inequality stating that to prevent failure the minimum resistance $(R - \Delta R)$ must be larger than the maximum loads $(L + \Delta L)$; thus

$$R - \Delta R \geq L + \Delta L \tag{1.9}$$

If the possible maximum variations of both resistance and loads are each taken as 25%, then the minimum F.S. is obtained from Eq. (1.9) by equating the two sides after substituting variations of $0.25R$ and $0.25L$ respectively and solving for the F.S. as the ratio R/L. This yields

$$\text{F.S.} = \frac{R}{L} = \frac{1.25}{0.75} = 1.67 \tag{1.10}$$

The magnitude of the variations is generally estimated on the basis of subjective judgment rather than objective fact. The discrepancy between the highly refined procedure of modern design and the rather arbitrary manner of choosing the safety factor and its possible influence on the final success of creating a balanced design is obvious. The classical F.S. as used is based rather more on ignorance than on guarding against chance, while it should represent the degree of uncertainty or probability. To improve the present situation a rational method of evaluating the magnitude of the F.S. has to be introduced based on the statistical character of the main design parameters. Failure cannot be prevented with certainty, only to a certain degree of probability (1.28).

As early as 1946 Normy i Tekhnicheskie Usloviia 1 (NiTU1-46), the Soviet specifications have used a factor of safety derived on the basis of probability theory (1.29).

Load factors (L.F.) for plastic design in various countries are not as well defined as safety factors, with the exception of those in the United States and the Soviet Union. The AISC specification for beams and frames requires a value of 1.70 for gravity loading and 1.30 for wind loading and/or earthquake forces (1.2). The American Association of State Highway and Transportation Officials (AASHTO) specifications (1.8) gives for "Group I" (dead and live loads) an overload factor of $\frac{5}{3}$ to be applied to live loads and impact, and then another factor of 1.3 to be applied to the total loads

$$\text{Group I: L.F.} = 1.3[\text{D.L.} + \tfrac{5}{3}(\text{L.L.} + \text{I.L.})] \tag{1.11}$$

1.4.2 Probabilistic Approach

1.4.2.1 *Introduction*

The factor of safety or load factor concept is normally adequate for static design of very simple structures in which the limit load will occur at most a few times during their operational life. With dynamic loading this concept is less adequate mainly due to various kinds of scatter. Fatigue design, for example, is more a probabilistic concept; therefore, a "probable factor of safety" must be used—that is, the ratio of the probable resistance to the probable critical load. Such factors are a function of the desired reliability and the variations (scatter) in the loads and resistance.

As already mentioned, the use of probability theory has resulted in a new design approach called probabilistic design as opposed to classical deterministic design (1.26, 1.30, 1.31, 1.32). The main difficulty in the latter is the lack of distribution functions of load and resistance. The parameters of these distributions can be estimated from the sample of observed values, which are results of repeated experiments.

Due to the fact that most loads, $L(t)$, and resistances, $R(t)$ (where t is time), are random in nature, the safe (operational) life of a structure (structural reliability) can be represented by a random variable, T. The basic problem of structural reliability (1.31), then, is to find the probability of survival (reliability), $S_T(t)$, which is decreasing with time. We have

$$S_T(t) = P(T > t) = P\{0 \leq T \leq t; R(T) > L(T)\} \tag{1.12}$$

where t and T are different time intervals.

The complement of the probability of survival is the probability of failure, $P_f(t)$, which usually can be expressed in terms of a design factor of safety and thus is used to provide for a probabilistic interpretation of the factor of safety.

1.4.2.2 Load and Resistance Factor Design

The load factors used for plastic design in Part 2 of the AISC Specification, as well as those used in the AASHTO Specifications, represent known constants consistent and in good balance with the factors of safety used in allowable working stress design. There is no doubt or uncertainty about their magnitude. A further improvement in the plastic design method is a newer method called the *load and resistance factor design method* (LRFD) (1.39–1.41) or *limit states design*. This is a procedure that aims to make full use of available test information, design experience, and engineering judgment, applied by the use of probabilistic analysis. It proportions structural members and connections so that strength and serviceability limit states exceed the factored load combinations. In general the LRFD principle can be expressed as:

$$\phi R_n \geq \gamma_0 \Sigma \gamma_i L_i \quad i = (D.L.), (L.L.), W, S \tag{1.13}$$

where R_n = nominal resistance of the member (bending strength, stiffness, etc.), ϕ = resistance factor < 1, γ_0 and γ_i are load factors < 1 (normally), L_i = mean load effect, and the right side of Eq. (1.13) sums the products of mean load effects and corresponding load factors $\gamma_{D.L.}, \gamma_{L.L.}, \gamma_W, \gamma_S$, all multiplied by γ_0, the analysis factor. The resistance factor ϕ, a number less than 1.0, accounts for resistance uncertainties for a particular limit state. It depends upon variability in material properties like yield stress or ultimate tensile strength; geometric deviations in depth, thickness, and straightness as the result of milling fabrication, and erection practice; and statistical measures of agreement between design models predicting resistance for a particular limit state versus experimental data for that same limit state. The load factors γ_i, numbers larger than 1 (except when a load type increases resistance for limit states like uplift or stability against overturning) provide for load variations with time, as well as for uncertainty about their location on the structure. The dead load factor, $\gamma_{D.L.}$, is smaller than the live load factor, $\gamma_{L.L.}$. The analysis factor, γ_0, also larger than 1.0, accounts for uncertainties in structural analysis. The proposed criteria (1.44) prescribe separate numerical values for all resistance and load factors. LRFD is expected to be introduced in building design in the near future.

1.4.2.3 Current European Practice in Structural Safety

Because all the data necessary for a rigorous probability approach to the treatment of safety are not available, and also because excessively complicated design calculations must be avoided, the following procedure is recommended. The CECM

(Convention Européene des Associations de la Construction Métallique) has endorsed this procedure, and it is expected to gain universal acceptance. In this method "characteristic values" of the strength are utilized to define the mechanical properties of the materials. The same is done for loads based upon a fixed probability that the actual values will be either less or greater than the characteristic values selected. To cover the remaining uncertainty factors these "characteristic values" are transformed into "design values" by the introduction of certain coefficients, the values of which depend on the limit state being considered, the behavior of the construction material and the structure itself, and the probability of combinations of load occurring (1.26).

Thus the treatment of the safety aspect in structural design lies in the definition of three so-called partial factors of safety γ_m, γ_l, and γ_c, which are introduced into the design calculations in the treatment of the various limit states. By the assignment of appropriate values to these partial factors of safety for each limit state, it is possible to provide reasonable and adequate protection against the structure becoming unfit for use during its design life. It is recognized that this approach is not consistent with a probabilistic treatment of safety, since the individual factors cannot be treated separately. However, for practical purposes this is the most convenient approach at the present time, and obviously it can be modified as knowledge improves.

1.5 DESIGN TOOLS AND AIDS

The design of any major structure is a creative process. The designer must have not only a good knowledge of design techniques but in addition creativity, originality, and an appreciation of the aesthetic aspects.

The designing process itself can be rationalized and broken down into a sequence of logical steps. To execute these various steps, a knowledge of all the available design tools and aids is essential.

1.5.1 Design Tools

A knowledge of available materials and their properties, design methods, and relevant codes and specifications are indispensable for any design.

One of the basic decisions that has to be made right at the beginning of any design is the choice of the main structural material. In order to be able to make such a choice, the designer must be familiar with the different physical properties of the available structural materials and their behavior. A general knowledge on the engineering or macrolevel is not sufficient. In order to evaluate the expected response under various loading conditions of any material, its atomic structure, or rather its microstructure, must be known. A detailed discussion of the properties of structural steels is given in the next chapter.

The engineer must also be aware of the most recent developments in the design field so that he can produce an economical and competitive design.

Specifications and codes are necessary design tools. They serve as a guide to load intensities and safety margins. It must be realized, however, that a design based only on the information given in codes and specifications is unimaginative and not necessarily correct. Specifications and codes must be viewed rather as a summary of and/or compromise among a wide variety of design practices and theories to protect the designer and the users alike.

The main difference between codes and specifications lies in the fact that codes are legal regulations which must be satisfied for certain structures or areas. This text uses the various codes and specifications of the American Institute of Steel Construction (AISC), the American Association of State Highway and Transportation Officials (AASHTO), the American Railway Engineering Association (AREA), the American Welding Society (AWS), and the American Society for the Testing of Materials (ASTM). In addition, where relevant, reference will be made to various foreign specifications.

A designer must always use the latest edition of any code or specification, as they are periodically revised and updated.

1.5.2 Design Aids

Various design aids are available for different types of civil engineering structures and their members. These consist of manuals, nomograms, tables, charts, computer programs, and so forth. Their main purpose is to save time and allow the designer to spend more time on design decisions.

When using the various design aids and throughout the design process, it must be realized that the level of precision must be uniform and not excessively high. Apart from preventing cumulative calculation errors, there is no sense in using precisions up to three or four decimals in view of the variation in mechanical properties, actual loads, and so on. In addition, the analysis or design assumptions are often idealizations made to simplify the work involved and therefore can be considered as rough approximations of the actual conditions. A knowledge of resulting possible errors and their order of magnitude, however, is important to the designer in evaluating his work and its required precision. For example, stresses in ksi instead of psi (pounds per square inch) are more than sufficiently accurate. Graphical methods in many cases are satisfactory for many design calculations. Naturally the engineer should use methods which most closely represent the actual behavior from an analysis point of view. However, this does not exclude the use of practical engineering methods to execute such an analysis.

1.6 DESIGN PROCEDURES

Although each design problem of a particular structure is different and has its own features, the solution process passes through several sequential steps or phases which are similar in nature from problem to problem. To give the student or young engineer a better idea of what is involved in a design process, these steps are next briefly discussed. They should not be taken as rigid rules or as always applicable.

From the first concept of a structure a feasibility study including its financing should be performed. After that, as the second stage, is the actual design, which consists of the following steps.

1.6.1 Planning and Site Exploration

The design of any conventional structure must start with the planning of that structure and learning about the local conditions at the actual site. This needs to be done irrespective of whether the structure is a bridge or a building.

Planning for a bridge structure often represents a somewhat simpler problem than planning for a building. For a highway bridge, the importance of the highway itself will determine the width of the bridge (how many lanes) and its loading class (intensity and load frequency). The alignment of the bridge is usually defined by the highway's alignment. Selecting a favorable crossing should be considered as part of the preliminary route location so as to minimize construction, maintenance, and replacement costs. Hydraulic or traffic studies of the bridge site will determine the total bridge length, size of eventual interim spans, and the necessary clearances below the superstructure. For bridges crossing over rivers, site exploration has to provide information data on existing bridges, their performance during past floods, available highwater marks with dates of occurrence, information on ice, and so on. The geological profile of the water bed along the bridge axis or not far from it will affect the position of the abutments and river piers and their type of foundation. Soil mechanics analysis will decide or find dimensions of the foundations. If there are any expected settlements, they will determine which structural systems should be avoided.

For a building, especially of large dimensions, preliminary planning is a much more involved process.

1.6.1.1 *Functional Considerations*

An optimum structural design results in a structure that satisfies best the four requirements of function, safety, economy, and aesthetics. These requirements are all equally important, but the functional considerations, i.e., the kind and the magnitude of the services to be performed, are the basis of any building design. The character of building use, integration of building functions, location, and environment will dictate the plan layout, the shape of the structure, and its main exterior dimensions. If the building is of the industrial type the provision of the best plan for mechanical layout and operation and for the rational use of space is the most important factor. A designer also should know that no building should be planned with such restricted space that the equipment and layout cannot be modified at all in the future.

On the basis of these considerations the general layout with the main broad dimensions of the building under consideration will emerge. A further detailed study of the integrated services or processes foreseen for the building use will produce the internal geometrical parameters such as location of columns, minimum clear spans, clearances for crane operations (if required) or other vehicles, maximum depths for beams and girders, as well as size of columns.

The development of building technology, the rationalization of construction, the increasing use of prefabricated components, and the advancing industrialization of building construction in general point to the necessity of the standardization of components and of universally valid rules for the coordination of structural dimensions with one another (1.43). An agreement of this kind, possessing international validity, exists in what is known in Europe as modular coordination (1.44). Coordinating dimensions, i.e., structural dimensions which determine the location of components relative to one another, should be modular dimensions, i.e., whole multiples of the respective modulus. The basic module is 100 mm (3.94 in. $\approx$ 4 in.). One multimodule of 600 mm $\approx$ 2 ft is a very useful dimension for a planning grid. The Building Industry agreement on American Standards Association Standard A62.1, sponsored by the American Institute of Architects, Associated General Contractors of America, National Association of Homebuilders, and the Producers' Council Inc., established 4-in. as the basic module for United States building materials and products.

Actually for steel structures, compliance with a modular coordination system is not very important, as the fabrication of steel components is not dependent on fixed increments. However, for the space enclosing components of steel structures, usually materials are employed which are prefabricated, and which should, for this reason, conform to modular dimensions. Consequently, the steel work to which such components are fixed should similarly comply.

In addition to these considerations, fire hazards, maintenance cost, future use of the building, structural materials, and systems also have to be studied. Therefore, the final cost of the building is greatly affected by this functional study. Even if we do not consider the financing of the project, the legal process for acquiring the site, and the human relations between owner, engineer, and builder after functional considerations there is still a long succession of events involved in planning a building. As already pointed out, on the basis of functional considerations a decision must be made as to the purpose to be accomplished by the structure. Its form and enclosed space must be chosen to fulfill this function, and costs need to be studied in comparison with the probable length of the useful life of the structure. This study will lead to a final choice of the building outline and a final choice of the structural materials.

1.6.1.2 *Economy, Strength, and Code Requirements*

As already mentioned in Section 1.4 on structural safety, the economy and strength of a structure must be carefully balanced to satisfy the current codes and specifications.

Good specifications are indispensable to practical design. Where they can be employed to relate sound theoretical and experimental findings to the proportioning of actual structures, they furnish in part the medium for proceeding from theories to practical conclusions (1.45). Because of the required procedures to amend any part of a specification (careful review and approval by a committee of the proposed change and the comments on it by the industry) there is always a lag of several years between research, the state-of-the-art, and current specifications. A good designer must be a permanent student of the developments in his or her particular field of activity to be always up-to-date and able to produce good competitive designs.

1.6.2 Preliminary Design

The basic form of a structure, its main geometrical dimensions, and clearance requirements have been defined in functional considerations. Next, several first rough iterative designs have to be made to select the best suited structural analytical system and model. Sometimes, the local site conditions will narrow such choice to very few possible systems, but in other cases several comparative rough designs have to be executed to arrive at one or two favored systems. Here, as well as in many other design stages, the cut-and-try method is applied. The creativity, knowledge, and, above all, experience of the designer are essential to reach an adequate, fruitful design through preliminary designs.

Preliminary sizes, or ratios of sizes, have to be assumed in order to enable analysis. Size ratios depend on the type of structure, its redundancy, and the major design procedure that is adopted. Sizes refer to cross-sectional areas, or to moments of inertia for use in the chosen method of design: allowable stress or plastic steel design: static or dynamic. Again, experience and comparison with similar designs and use of available empirical rules combined with some rough calculations are the main tools used to arrive at a preliminary design. Whenever applicable, such empirical rules will be discussed in this text.

1.6.3 Structural Analysis and the Accuracy of Computations

On the basis of values obtained in the preliminary design and the chosen mathematical model which is sufficiently close to the physical structure and its behavior under loading, the next step in the design procedure, the structural analysis, can be executed. Depending upon the complexity of the structural system and the number of loading cases, use will be made of a longhand calculation method with some available design aids or of a digital computer. In any case, some computational machine—hand-held calculator, programmable desk calculator, minicomputer, or even full-size electronic computer—will be used. Then the question arises how many significant digits should be kept in the interim steps of the calculation and in the final results. The accuracy of the results certainly depend upon the accuracy of the input data (no output is better than its input) of the structural model and the interim calculations. Therefore, the interim calculations should avoid truncations which would accumulate errors and make the results below the required accuracy. Normally, to carry out the interim steps to three decimal places is sufficient. The final results should be adjusted to the proper overall accuracy of the whole analysis. Loads are an idealization of actual loadings, and material constants and dimensions also vary. The chosen structural analytical model represents another approximation. Finally, the analysis itself of the chosen model will affect the results. If this combined effect is within 3 to 5% of the correct value, such an analysis can be considered as extremely good.

Normally the discrepancy is much larger, from 10 to 15%. In view of this, results should always be in kips (tons) for forces, and in ksi (N/m²) for stresses. The number of decimal places should correspond to the above expected percentage errors.

For computerized structural analysis, there are at present many good structural analysis programs for different electronic machines.

1.6.4 Selection of Member Cross Sections

In most cases strength (first yield or structural collapse) is the initial design criterion, while all other criteria (see Section 1.2, Design Criteria) are used to carry out checks in the next design step. On the basis of the analysis thus performed, a revised selection of sizes and shapes and their ratios takes place. These are compared with those initially assumed. To arrive at an economical design it is usually necessary now to repeat the analysis with the revised sizes and shapes. This trial-and-error method of assuming dimensions and checking the resulting stress by means of an analysis is typical of any design. With computer usage, at least the time factor in calculations is eliminated or drastically reduced so that optimization of all dimensions can be attempted and achieved rapidly.

Member design, the subject of this text, is mainly concerned with the last two design phases, the selection of sizes and the checking for secondary criteria. Iterative procedures can be applied in this process and much time saved by making use of short computer programs. Several such programs have been developed by the authors for programmable desk calculators as well as full-size digital computers and the reader is referred to the appropriate chapters in the text.

1.6.5 Secondary Design Considerations

Once all necessary sizes of the structural members are obtained, some additional design checks usually are required. Overall structural stability, deformations, dynamic behavior, the occurrence of secondary stresses due to temperature changes, support movements, fatigue limitations, and the design and detailing of connections could all be classified as the applying of secondary design criteria. These are secondary only in the sense that they follow an initial design based on strength criteria. It may well be that, upon investigating of these so-called secondary effects, they prove to be major effects requiring modifications or drastic revisions of the initial strength design. On the basis of experience the engineer may be able to recognize any such governing criteria beforehand and develop an initial design from these rather than on the basis of strength, which now instead is treated as a secondary criterion. A good example for this is the design of a plate girder with stiffeners exposed to moving loads, where fatigue considerations (allowable stress ranges) are often very restrictive.

It is obvious that matters concerning the overall structure, such as its stability and total deformations, do not play a role in the initial design of individual members. It must be recognized, however, that member design does affect the overall structural behavior and that any restrictions placed on the latter therefore will indirectly affect the design of the individual members. Figure 1.24 shows the various steps usually involved in any steel design, in the form of a macroflowchart.

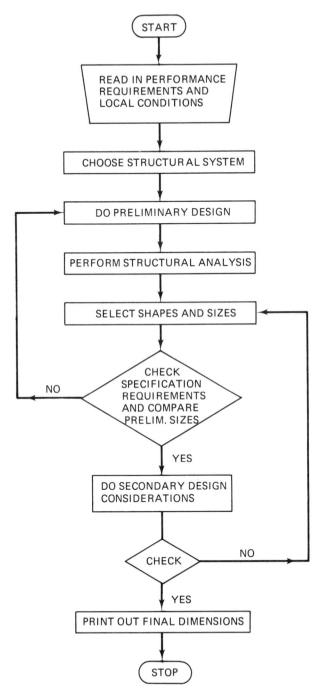

Figure 1.24 Design Flowchart.

NOTATIONS

f = actual stress or shape factor
k = coefficient depending on probability
t = time
γ = partial factor of safety, strain
γ_c = structural coefficient
γ_m = strength reduction coefficient
γ_l = load increase coefficient
D.L. = dead load
F_{all} = allowable working stress
F.S. = factor of safety
F_y = yield point of stress
I.L. = impact load
L = load

L.F. = load factor or factored load
L.L. = live load
L_u = ultimate load
L_w = working load
M_{pl} = full plastic moment
M_y = moment at first yield
P = probability
$P_f(t)$ = probability of failure
R = resistance
$S_T(t)$ = probability of survival
S = elastic section modulus
T = time at specific moment
Z = plastic section modulus

REFERENCES

1.1. Gaylord, E. H., and C. N. Gaylord, (eds.), *Structural Engineering Handbook*, 2nd ed., McGraw-Hill, New York, 1979.

1.2. American Institute of Steel Construction, *Manual of Steel Construction*, 8th ed., Chicago, 1980.

1.3. *American National Standard Building Code Requirements for Minimum Design Loads in Buildings and Other Structures*, American National Standards Institute, ANSI A58.1–1972, New York.

1.4. Sachs, P., *Wind Forces in Engineering*, 2nd ed., Pergamon, Oxford, 1978.

1.5. Simiu, E. and R. H. Scanlan, *Wind Effects on Structures: An Introduction to Wind Engineering*, J. Wiley, New York 1978.

1.6. International Conference of Building Officials, *Uniform Building Code Standards*, 1978 ed., Whittier, CA, 1978.

1.7. Associate Committee on the National Building Code, National Research Council of Canada, *National Building Code of Canada*, Ottawa, 1977.

1.8. American Association of State Highway and Transportation Officials, *Standard Specifications for Highway Bridges*, 12th ed., Washington, DC, 1977 (with 1978, 1979, 1980, and 1981 Interim Bridge Specifications).

1.9. Housner, G. W., "Behavior of Structures During Earthquakes," *Journal of the Engineering Mechanics Division*, ASCE, Vol. 85, No. EM4, Oct. 1959, pp. 109–29.

1.10. Newmark, N. M. and W. J. Hall, "A Rational Approach to Seismic Design Standards for Structures," *Proc. Fifth World Conference on Earthquake Engineering*, Rome, 1973, Vol. 2, pp. 2266–75.

1.11. American Railway Engineering Association, *Manual for Railway Engineering*, Chapter 15 (Specifications for Steel Railway Bridges), Chicago, 1980.

1.12. Colonetti, G., "Les Phénomènes de Coactions Élasto-Plastique et L'Adaptation à la

Résistance des Materiaux," *Annales de l'Institut Technique du Bâtiment et des Travaux Publics*, No. 99, Nov. 1949.

1.13. Osgood, C. C., *Fatigue Design*, Wiley-Interscience, New York, 1970, p. 17.

1.14. Khan, F., "Systems Approach to the Design of High-Rise Concrete Buildings," *Proc. 14th Struc. Conf.*, University of Kansas, April 9, 1969, pp. 1–4.

1.15. Massonnet, C., "The Use of Digital Computers in Civil Engineering," in *Applications of Digital Computers*, W. F. Freiberger and W. Prager, eds., Ginn, Boston, 1963, pp. 138–57.

1.16. Wasiutynski, Z., and A. Brandt, "The Present State of Knowledge in the Field of Optimum Design of Structures," *AMR*, Vol. 16, No. 5, May 1968, pp. 341–50.

1.17. Dantzig, G. D., *Linear Programming and Extension*, Princeton University Press, Princeton, NJ, 1963.

1.18. Kuzmanović, B. O., and N. Willems, "Optimum Plastic Design of Steel Frames," *J. Struc. Div., Proc. ASCE*, Vol. 98, No. ST8, Aug. 1972, pp. 1692–723.

1.19. Dorn, W. S., et al., "Automatic Design of Optimal Structures," *Journal de Mechanique*, Vol. 3, No. 1, 1966, pp. 223–45.

1.20. Hill, Jr., L. A., "Automated Optimum Cost Building Design," *J. Struc. Div. Proc. ASCE*, Vol. 92, No. ST6, Dec. 1966, pp. 247–63.

1.21. Megarefs, G. J., and H. S. Sudhu, "Simplification in Minimal Design of Frames," *Proc. ASCE*, Paper No. 6308, Vol. 94, No. ST12, Dec. 1968, pp. 2985–98.

1.22. Moses, F., "Optimum Structural Design Using Linear Programming," *J. Struc. Div., Proc. ASCE*, Paper no. 4163, Dec. 1964, pp. 89–104.

1.23. Thürlimann, B., *Optimum Design of Structures: Plastic Design of Multistory Frames*, Fritz Engr. Lab. Report nos. 273.20, 273.24, Lehigh University, Bethlehem, 1965.

1.24. Bigelow, R. H., and E. H. Gaylord, "Design of Steel Frames for Minimum Weight," *J. Struc. Div., Proc. ASCE*, Vol. 93, No. ST6, June 1967, pp. 109–31.

1.25. Tetsuo Ikeda, "On Safety of Structures," *Proc. Symposium on Safety of Structures*, Tokyo, Dec. 1956, pp. 1–9.

1.26. Rowe, R. E., "Current European Views on Structural Safety," *J. Struc. Div., Proc. ASCE*, Vol. 96, No. ST3, March 1970, pp. 461–67.

1.27. Beedle, L. S., and L. Tall, "Basic Column Strength," *J. Struc. Div., Proc. ASCE*, Vol. 86, No. ST7, July 1960, p. 139.

1.28. Freudenthal, A. M., "The Safety of Structures," *Proc. ASCE*, Vol. 71, No. 8, Oct. 1945, pp. 1157–91.

1.29. Strelietskii, N. S., *Osnovy Staticheskogo Uchota Koefitsienta Zapasa Prochnosti Sooruzhenii*, Stroĭizdat, Moscow, 1947.

1.30. Benjamin, J. R., "Probabilistic Structural Analysis and Design," *J. Struc. Div., Proc. ASCE*, Vol. 94, No. ST7, July 1968, pp. 1665–79.

1.31. Ang, A. H. S., and M. Amin, "Safety Factors and Probability in Structural Design," *J. Struc. Div., Proc. ASCE*, Vol. 95, No. ST8, July 1969, pp. 1389–405.

1.32. Task Committee on Structural Safety of the Administrative Committee on Analysis and Design of the Structural Division, "Structural Safety—A Literature Review," *J. Struc. Div., Proc. ASCE*, Vol. 98, No. ST4, April 1972, pp. 845–84.

1.33. British Standards Institution, *Specifications for the Use of Structural Steel in Building*, British Standard 449: 1959 (incorporating British Standard Code of Practice CP 113), London, U.K., reset and reprinted 1965.

1.34. British Standards Institution, *Specifications for Steel Girder Bridges*, British Standard 153: parts 3B, "Stresses," and 4, "Design and construction," London, U.K., reset and reprinted 1966.

1.35. La Chambre Syndicale des Entrepreneurs de Constructions Métalliques de France, *Règles pour le calcul et l'execution des constructions métalliques*, Règles C. M. 1956, Documentation Technique du Bâtiment et des Travaux Publics, Paris, 1956.

1.36. Deutsche Industrie Normen, *Stahl im Hochbau, Berechnung, und bauliche Durchbildung*, DIN 1050, Berlin, June 1968.

1.37. Deutsche Industrie Normen, *Die vorläufigen Vorschriften für geschweisste, vollwandige Eisenbahnbrücken*, DV 848, 1955 *Geschweisste Stählerne Strassenbrücken*, DIN 4101, Feb. 1970; *Stählerne Strassenbrücken, Berechnungsgrundlagen*, DIN 1073, 1956.

1.38. *Stroitelnye Normy i Pravila*, part 2, section 5, chapter 3: "Stalnye Konstrukcii, Normy Proektirovania," Gosstroiizdt, 1962.

1.39. Ravindra, M. K., and T. V. Galambos, "Load and Resistance Factor Design for Steel," *ASCE Journal of the Structural Division*, Vol. 104, No. ST9, Sept. 1978, pp. 1337–54.

1.40. Bjorhovde, R., T. V. Galambos, and M. K. Ravindra, "LRFD Criteria for Steel Beam-Columns," *ASCE Journal of the Structural Division*, Vol. 104, No. ST9, Sept. 1978, pp. 1389–408.

1.41. Yura, J. A., T. V. Galambos, and M. K. Ravindra, "The Bending Resistance of Steel Beams," *ASCE Journal of the Structural Division*, Vol. 104, No. ST9, Sept. 1978, pp. 1355–71.

1.42. American Iron and Steel Institute, "Proposed Criteria for Load and Resistance Factor Design of Steel Building Structures," *Bulletin* No. 27, Washington, DC, Jan. 1978.

1.43. Hart, F., W. Henn, and H. Sontag, *Multi-Story Buildings in Steel*, Halsted Press, J. Wiley, New York, 1978.

1.44. The German Standard, DIN 18000, *Modulordnung im Bauwesen*, Parts 1 to 3, Beuth-Vertrieb, November 1973 and March, 1976, Berlin and Cologne.

1.45. McGuire, W., *Steel Structures*, Prentice-Hall, Englewood Cliffs, NJ, 1968, p. 17.

2

Structural Steels and Their Properties

2.1 STRUCTURAL STEELS

2.1.1 Introduction

Steel is mainly an alloy of metallic iron and nonmetallic carbon. Several other alloying elements are always present.

Different varieties of structural steel (or mild steel) are used for civil engineering structures. The strength of a particular steel depends to a large extent on its carbon content. The ASTM standards specify the exact percentage of carbon and other alloying elements such as manganese, silicon, copper, and so forth. Percentages are specified for molten steel prior to rolling (in ladle) or after manufacturing (check analysis). Carbon increases the hardness and strength of a steel; however, it also reduces its ductility. The presence of phosphorous and sulfur also adversely affects the ductility. In order to ensure a minimum amount of ductility, maximum percentages of carbon, phosphorous, and sulfur are specified.

Although the chemical composition of a steel mainly determines its physical and mechanical properties, these are also influenced by the rolling process and in general by any stress history or heat treatment the material is subjected to. Moreover, manufacturing processes affect the atomic structure of a steel and consequently also those properties which are directly related to its crystalline structure. Some of these properties which are important to a steel designer, such as flow and fracture, are very sensitive to imperfections in the atomic arrangement within crystals. In a perfect crystal a certain number of atoms are arranged in some kind of a pattern which repeats itself in three directions in space. Careful X-ray investigations of assumedly perfect single crystals have shown that a stress-relieved (annealed) crystal is composed of blocks of perfect atomic arrangement with linear dimensions from 1,000 to 10,000Å [1 angstrom (Å) = 10^{-10} m] which are tilted with respect to one another by about 10 to 15 minutes. Practically all crystals do exhibit a certain degree of disorder in the arrangement of their blocks. The atoms in iron crystals form cubic lattices. The type of lattice depends on the temperature, and up to 1,670°F (910°C) under normal atmospheric pressure the atoms form a body-centered cubic (b.c.c.) lattice called the α-iron lattice (Fig. 2.1). When the temperature reaches 1,670°F α-iron transforms to

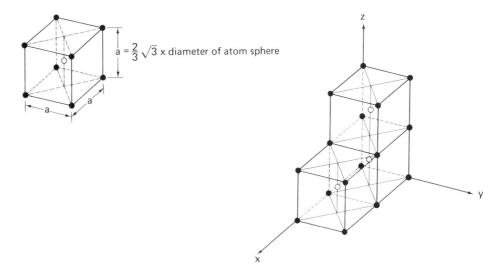

Figure 2.1 Body-centered Cubic Structure.

γ-iron, which has a face-centered cubic (f.c.c.) lattice (Fig. 2.2). At 2,530°F (1,390°C) the lattice transforms back to a body-centered cubic lattice.

In a crystalline solid atoms are fairly closely spaced. Atoms consist of a nucleus surrounded by a certain number of electrons. Assuming that all electrons are located within a sphere of a certain size with the nucleus as its center, we recognize that each type of atom has a given size. Iron atoms of the b.c.c. or f.c.c. type are touching one another. The magnitude of the side a of a cube of α-iron is 2.9Å. The number of atoms per unit cube, also called a cell, is called the atom density, which for α-iron equals 2. The eight atoms at the cube corners are each shared by eight cells, while the single atom at the cube center is shared by no other cells. When the temperature increases to transform the lattice to the f.c.c. type, the cube side a becomes 3.6Å, while the atom density changes to 4. Each atom on a face center is shared by two cells. Comparing the f.c.c. and b.c.c. types, we see that the ratio of the atom densities is 2 to 1, while the volume ratio is $(3.6)^3$ to $(2.9)^3$ or 1.91 to 1.0. This means that a compaction at the atom

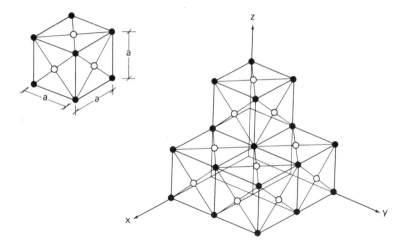

Figure 2.2 Face-centered Cubic Structure.

level is introduced during this increase of temperature. Conversely, changes from the γ- to the α-iron are therefore associated with an abrupt volume expansion accompanied by the creation of internal tensile forces which may affect the mechanical properties considerably (2.8).

Apart from imperfections within a crystal, at the boundaries between crystals, because of their different orientation, some cells are destroyed. Atoms are usually located at points of mutual equilibrium due to the attraction of surrounding atoms. The bond between such atoms becomes weaker when the atom distances become larger and the potential energy increases. Such imperfections (anomalies) act as stress raisers, and internal strains occur first at the grain boundaries.

From the above discussion it is clear that steel is not a homogeneous, isotropic continuum. If equivalent average behavior is assumed (homogeneity condition), and also mechanically equivalent directions (isotropy condition) for the lattice points only, then steel could be considered as a quasi-homogeneous, statistically isotropic continuum.

According to the ASTM A6 standard (2.9), steels used in the United States for buildings, bridges, and other civil engineering structures are classified as

1. Carbon steels (A36, A529).
2. High-strength, low-alloy steels (A242, A441, A572, A588, A606, A607, A618)
3. Alloy steels (A514).

For fasteners (connectors), filler metal for welding, forgings, and cast steel beams, different steel alloys are used (see Chapters 9 and 11). In addition to the high-strength steels specified by ASTM, several steel producers have their own designations for their high-strength steels, such as USS-T-1, J & L Cor-Ten, and Jalten.

2.1.2 Carbon Steels

Structural steel, ASTM A36 (AASHTO designation M-183), is widely used for general structural purposes in bolted and welded steel buildings and bridges. It has replaced (since 1960) both A-373 and A-7 grades of steel with yield point stresses of 32 and 33 ksi, respectively. The guaranteed minimum yield stress for A-36 steel is 36 ksi (250 MPa). The actual values lie between 43 and 48 ksi (300 to 330 MPa). Plate thicknesses used are to 8 in. (200 mm) inclusive. For welded and seamless steel pipe ASTM A-53, Grade B steel is used. Its minimum yield point is 30 ksi. The main mechanical properties of these two steels are given in Table 2.2.

Rimmed steel is steel containing sufficient oxygen to give a continuous evolution of carbon monoxide while the ingot is solidifying, resulting in a case or rim of metal virtually free of voids. *Semikilled steel* is incompletely deoxidized steel containing sufficient oxygen to form enough carbon monoxide during solidification to offset solidification shrinkage. *Capped steel* is rimmed steel in which the rimming action is limited by an early capping operation. Capping may be carried out mechanically by using a heavy metal cap on a bottle-type mold or it may be carried out chemically by an addition of aluminum or ferrosilicon to the top of the molten steel in an open-top mold. *Killed steel* is steel which is deoxidized either by the addition of strong deoxidizing agents or by vacuum treatment, to reduce the oxygen content to such a level that no reaction occurs between carbon and oxygen during solidification.

2.1.3 High Strength, Low-Alloy Steels

To this group belong, among others, A441 and A572 steels, as well as corrosion resistant A242 and A588 steels. These steels cover several yield strengths with different grades, such as 40, 42, 46, 50, 60, and 65 ksi for the corresponding guaranteed minimum yield stress in ksi. Generally, their yield stress depends upon plate or flange thickness. Plate thickness is shown in Table 1 (p. 1-5) of the AISC *Manual of Steel Construction* (2.10) under the heading Plates and Bars. The steel producers also have classified wide flange shapes into five groups, depending on their flange thickness, and these are compatible with the steel grade. For instance, in A572-Grade 50 all five groups and plates up to 2 in. (50 mm) are available. A242 has enhanced atmospheric corrosion resistance of at least two times that of carbon structural steels with copper. Use is limited to 4 in. (100 mm), inclusive, in thickness.

A588 is a high-strength low-alloy structural steel with 50 ksi (345 MPa) minimum yield point to 4-in. (100-mm) thick. It is also called "weathering steel," and is used for bridges and buildings (or parts of) exposed to the atmosphere (see more in Section 2.7.3, Corrosion-Resistant, Low-Alloy Steels).

ANSI/ASTM A709-77 standard specification for structural steel for bridges covers carbon and high-strength, low-alloy steel for structural shapes, plates, and bars; and quenched and tempered alloy steel for structured plates intended for use in bridges. Five grades are available in three strength levels (36, 50, and 100 ksi). Grades 50W and 100W have enhanced atmospheric corrosion resistance.

AASHTO gives high-strength, low-alloy steels their designations A572 Grade

50 (OM-223) and A588 (M-222). For high-yield-strength quenched and tempered alloy steel AASHTO uses A517 (M-244).

2.2 STRUCTURAL STEEL PRODUCTS

For the fabrication of structural elements mainly rolled steel products such as plates, structural shapes, and bars are used. Various manuals (for example, 2.10) give dimensions, sectional properties, and other values pertinent to a design. Dimensions usually are either actual (for design) or nominal (for detailing). The ASTM A6 standard (2.9) covering the requirements for delivery of structural steel gives permissible variations in dimensions and shapes from specified values (tables 1–16 are for plates, tables 17–26 for shapes, and tables 27–32 for bars). For instance, the cross-sectional area or weight of any structural shape is not allowed to vary more than 2.5% from the theoretical or specified values. The AISC *Manual* also gives these permissible tolerances (2.10).

2.2.1 Structural Shapes

Structural shapes mostly used in civil engineering structures are (Fig. 2.3) wide-flange beams (W); bearing pile shapes (HP); American standard beams (S); shapes that cannot be classified as W, HP, or S, called miscellaneous shapes (M); T-sections (WT, MT, and ST); channels (C); angles (L); pipes; and tubes. Designations for some of these shapes are given below.

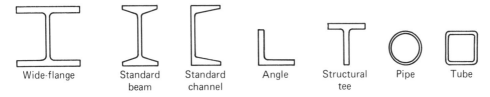

Wide-flange   Standard beam   Standard channel   Angle   Structural tee   Pipe   Tube

Figure 2.3 Structural Shapes.

1. *Wide-flange Beams:* W36 × 160 has a 36-in. nominal depth and weighs 160 lb/ft.
2. *Bearing Pile Shapes:* HP12 × 74 has a 12-in. nominal depth and weighs 74 lb/ft.
3. *S-shapes:* S10 × 35 has a 10-in. depth and weighs 35 lb/ft.
4. *M-shapes:* M6 × 20.
5. *Structural Tees:* (from W-shapes) WT5 × 12.5.
                (from S-shapes) ST9 × 35.
                (from M-shapes) MT3 × 2.2.
6. *American Standard Channels:* C6 × 10.5.
7. *Angles, Equal Legs:* L4 × 4 × $\frac{1}{2}$—1 ft-$\frac{1}{2}$ in. has legs 4 in. wide of $\frac{1}{2}$ in. thickness and a length of 12$\frac{1}{2}$ in.
8. *Angles, Unequal Legs:* L6 × 4 × $\frac{3}{8}$—10 ft-0 in. has legs 6 in. and 4 in. wide both of $\frac{3}{8}$ in. thickness and a length of 120 in.

Structural steel pipes and tubes (square or rectangular) are made from ASTM A500 steel by cold-forming or from A53 and A501 steel as hot-formed welded or seamless carbon steel structural tubing.

2.2.2 Plates and Strips

Plates, depending on their shape, are classified as strips or sheets. Bars are produced as rounds, squares, and hexagons of all sizes. To distinguish between different bars and plates consult Table 2.1, reproduced from the AISC *Manual* (p.

Table 2.1

Thickness (Inches)	Width (Inches)					
	To $3^1/_2$ incl.	Over $3^1/_2$ to 6	Over 6 to 8	Over 8 to 12	Over 12 to 48	Over 48
0.2300 & thicker	Bar	Bar	Bar	Plate	Plate	Plate
0.2299 to 0.2031	Bar	Bar	Strip	Strip	Sheet	Plate
0.2030 to 0.1800	Strip	Strip	Strip	Strip	Sheet	Plate
0.1799 to 0.0449	Strip	Strip	Strip	Strip	Sheet	Sheet
0.0448 to 0.0344	Strip	Strip				
0.0343 to 0.0255	Strip		Hot rolled sheet and strip not generally produced in these widths and thicknesses			
0.0254 & thinner						

NOTE: U.S. Standard Gage is officially a weight gage, in oz per ft² as tabulated. The Approx. Thickness shown is the "Manufacturers' Standard" of the American Iron and Steel Institute, based on steel as weighing 501.81 lb per cu ft (489.6 true weight plus 2.5 percent for average over-run in area and thickness). The AISI standard nomenclature for flat rolled carbon steel is as above.

6-3). Widths and thicknesses are preferably specified in increments of $\frac{1}{4}$ in. and $\frac{1}{8}$ in., respectively.

2.3 MECHANICAL PROPERTIES

To explain the mechanical properties of structural steels, the most suitable test is a tension test of a specimen under static loading. Compression tests are not used for three reasons. Firstly, buckling instability affects the size of the specimen; secondly, material collapse as such cannot be observed, even if buckling does not occur, an ultimate compressive strength is not obtained; thirdly, it is desirable to calculate stresses with reference to the original cross-sectional area of the specimen. In the tension test when the maximum tensile stress is reached, local yielding and so-called necking take place, thereby making the true stresses higher than maximum nominal stresses calculated on the basis of the original cross section. In the compression test the cross section keeps on increasing, and as a result the load rate or strain rate increase has to be much higher than for a tension test.

In Figs. 2.4 and 2.5 the results of a typical technical statical tension test are shown for A36 steel, carried out at room temperature and a normal strain rate. The stresses (F)* are "technical" or "nominal" stresses, as explained above, and are plotted vertically. The strains (ϵ) represent the relative elongations $(\Delta l/l)$ and are plotted horizontally. Figure 2.4 shows the results of a complete test up to physical collapse of the specimen, while Fig. 2.5 shows in more detail the behavior up to a strain of 2%. The characteristic points on the stress-strain $(F\text{-}\epsilon)$ curve or diagram are: the

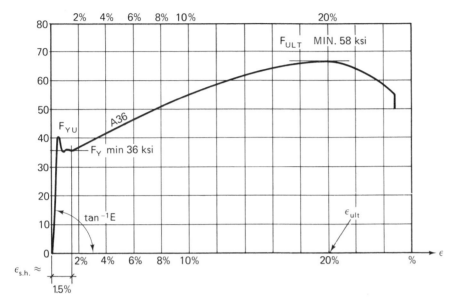

Figure 2.4 $(F - \epsilon)$ Curve of A36 Steel at Room Temperature and Normal Speed of Strain Increase.

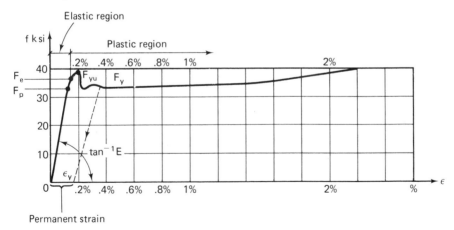

Figure 2.5 Part of $(F - \epsilon)$ Curve in Figure 2.4 Shown on a Larger Scale.

*In accordance with the two major specifications referred to in this text, the familiar symbol for stress, σ, is replaced by F or f.

proportionality limit, F_p; the elastic limit, F_e; the upper and lower yield stresses, F_{yu} and F_y; the ultimate stress, F_{ult}, also called the tensile strength; the work- or strain-hardening strain, ϵ_{sh}; and finally the strain at ultimate stress, ϵ_{ult}. These characteristic points define the following phases of behavior

1. Linear proportionality (Hooke's law) between 0 and F_p.
2. Elastic behavior between 0 and F_e.
3. Unrestricted plastic yielding between strains of approximately .2% and 1.2% to 1.5% (ϵ_{sh}).
4. Work- or strain-hardening between ϵ_{sh} and ϵ_{ult} (about 20%).

The characteristic shape of the F-ϵ curve actually illustrates certain mechanical properties. From a design point of view, the three most important ones are

1. The elastic or Young's modulus, E, which represents the slope of the initial straight portion of the curve passing through the origin. Thus

$$\epsilon = \frac{F}{E} \tag{2.1}$$

2. The occurrence of yield (F_{yu} and F_y).
3. The unrestricted flow between ϵ_y and ϵ_{sh} at an almost constant stress level, F_y. This phase is often referred to as the plastic yielding zone and this mechanical property as the plastic capacity.

The capacity of structural steel to flow plastically is probably its most important characteristic. Both the allowable stress and the plastic design methods make use of this property. As discussed in Chapter 1, in the allowable stress design method the yield stress (F_y) is used as the critical stress in lieu of the ultimate stress (F_{ult}). The assumption in this method of uniformity of stresses in simple tension which neglects residual stresses, stress concentrations, and other stress raisers, is justified only because local yielding will tend to equalize stresses and to reduce theoretical stress peaks. As a result of yielding, stress distributions just prior to strain hardening will in almost all cases be uniform, a fact which forms the basis for the allowable stress design method. The coaction phenomena resulting from plastic flow, as discussed in Section 1.2.3 form the basis and justification for the plastic design method.

The elastic modulus is the same for most structural steels: $E = 29,000$ ksi (200,000 MPa). Other properties common to all structural steels are: the coefficient of thermal dilatation, which is the linear expansion per °F per unit length, $\alpha_t = 6.5 \times 10^{-6}$ (12×10^{-6} per °C per unit length); the unit weight, which is 490 lb/ft³ (157,100 kilo newtons/m³); and Poisson's ratio, μ, which is 0.30. A convenient way of calculating the weight of a section in pounds per linear foot (or newtons per linear meter) is to multiply the cross-sectional area in square inches (or square centimeters) by a conversion factor of 3.4 (or 7.7).

The horizontal portion of the F-ϵ curve and its length is a measure of the ductility of the steel. Some authors define the ductility as the ratio of ϵ_{sh} to ϵ_y—that is

$$\psi = \frac{\epsilon_{sh}}{\epsilon_y}. \qquad (2.2)$$

The ductility of various steels differs. High-strength steels have a lower ductility than, for instance, A36 (Fig. 2.6). Some of the higher-strength steels have a ductility close to unity, indicating almost no horizontal portion in the F-ϵ curve. Also, no sharply defined yield stress (F_y) and corresponding yield strain (ϵ_y) exist for these steels. Upon stress reduction when unloading, the stress-strain curve approximates a straight line with a slope equal to the initial slope of the F-ϵ curve at point 0. For high-strength steels the stress resulting in a permanent strain of 0.2% is defined as the yield stress, or rather the yield-point stress, F_y, of the material. Lack of ductility makes a structural steel much more sensitive to the presence of residual stresses and also increases the danger of brittle fracture. Special care has to be taken during fabrication when using a less ductile steel, particularly when welding, in order to avoid cracks at points of restrained flow under the action of small nominal stresses (see Section 2.5 on brittle fracture).

The stress-strain diagram shown in Figs. 2.4 and 2.5 applies only if the loading rate is kept constant and slow throughout the test. It is also assumed that if unloading and reloading take place during testing, the yield stress is not reached. If the yield stress is reached prior to unloading, the F-ϵ curve might change, particularly if a rest period is allowed before reloading (about five days at room temperature). Upon reloading, a higher yield stress, F_y, a reduced ductility, and a slightly higher tensile strength, F_{ult}, might occur (Fig. 2.6). This phenomenon is known as the "aging" of steel. Aging can be used to increase a material's yield stress, provided the loss of ductility is not detrimental to the structural behavior. In reality yield stresses are reached more often than predicted by most standard methods of analysis. It is important for the engineer to recognize the possibility of aging and the material's subsequent lack of ductility at points with high stress raisers. If reloading is carried out immediately after unloading, no change in the F-ϵ curve occurs (Fig. 2.7).

Another typical characteristic of structural steel which is not obvious from the normal F-ϵ diagram is that of "elastic hysteresis." This occurs when a steel is subjected to alternating tensile and compressive stresses. Under an increasing number of loadings a hysteresis loop is formed (Fig. 2.8), which is stabilized at a certain number of

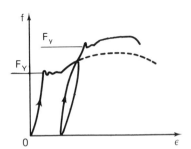

Figure 2.6 Repeated Loading after a Long Rest Period.

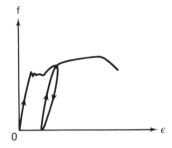

Figure 2.7 Repeated Loading after a Short Rest Period.

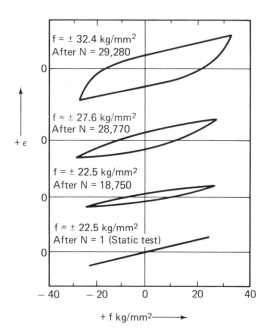

Figure 2.8 Hysteresis Loops (2.10).

loadings. The area of the loop is proportional to the energy dissipated and transformed into heat.

The shape of the F-ϵ diagram is also affected by variations in temperature. An increase in temperature results in a reduced modulus of elasticity. Also, the sudden increase in strain at F_y becomes less pronounced as the temperature increases, and it disappears practically completely at 570°F (300°C). In addition, the yield point of carbon steels starts decreasing when they are heated over 390°F (200°C), and at about 750°F (400°C) it is half of its value at room temperature. The tensile strength changes very little up to 570°F (300°C), but beyond that temperature it starts dropping rapidly. The ultimate stress of carbon structural steel at 930°F (500°C) is below its allowable working stress at room temperature, and for all practical purposes the steel will have lost its load-carrying capacity at this temperature.

To explain the mechanical behavior of steel several hypotheses have been developed. The dislocation theory (2.11) and the phenomenon of stress-raising anomalies in boundary layers as proposed by W. Engelhardt (2.12) gives a satisfactory explanation of the mechanism of plastic slippage and also enables its prediction. This modified "cloud" hypothesis not only explains the existence of the yield point (2.13), but also the aging phenomenon as well as the effects of temperature. This hypothesis is briefly discussed next.

Any steel contains chemical impurities. These are often referred to as "foreign atoms" or "strangers." These strangers are not part of the basic space lattice, and in the case of a nonhomogeneous stress field they are concentrated in regions of high stress, as if attracted by these. The foreign atoms thus form a cloud around the basic space lattice particles and are able to move through this lattice at the speed of diffusion, which is dependent upon the temperature. If a mechanical disturbance occurs in the form of a deformation, it travels at sonic speed. The difference between these two speeds forms the basis for explaining aging, the yield point, and also the effects of temperature. When loading is increased, stresses are intensified and the corresponding stress field in general will be nonhomogeneous. Clouds of foreign atoms are created around the "lattice atoms" located in regions of high stress. As soon as the critical stress is reached, these highly stressed lattice atoms will try to move. If this occurs at room temperature, the diffusion speed of the foreign atoms is too low to follow any deformational slippage, and the moving lattice atoms will break away from the surrounding cloud and its braking influence. This results in a large deformation which tries to make up for the braking effect initially exerted by the cloud. This large deformation manifests itself as plastic yielding beyond the yield point. If the temperature is now raised, the speed of diffusion of the foreign atoms increases and may equal the speed of the lattice atoms when moving at critical stress. If this is the case, no breaking away from the cloud occurs, and the latter keeps on exerting its braking influence; consequently a marked yield point followed by plastic yielding is not exhibited. In case of plastic yielding at room temperature, upon unloading, the foreign atoms will try to catch up with the lattice atoms. If there is insufficient time, however—that is, if there is fast reloading—no second yield point is possible. If on the other hand, sufficient time is available between unloading and reloading (a rest period), the foreign atoms catch up with the lattice atoms, thereby reestablishing their braking influence. If the material is now reloaded, the next breaking away will occur at a higher yield stress ($F'_y > F_y$).

Apart from specifying the chemical composition of structural steels, ASTM as well as several steel producers have specified their main mechanical properties. The main mechanical properties for some steels are shown in Table 2.2.

2.4 FATIGUE

2.4.1 Introduction and Historical Review

For more than a century it has been realized that steel structures and their constituent elements when subjected to variable or repeated loads may fail at stresses far below the static stresses necessary to start cracking.

In this section only high-cycle fatigue will be discussed. Other types of fatigue, such as low-cycle, corrosion, fretting, and static fatigue are not considered because they occur less frequently in most engineering structures.

J. V. Poncelet in 1839 in the second edition of his book *Industrial Mechanics* discussed the strength of materials under repeated cyclic loading and first introduced the concept of "fatigue." W. J. Macquorn Rankine was probably the first person to publish, in 1843, a research paper in English about the fatigue failures of "puddle or weld-iron" used in railroad axles (2.14). Contrary to the generally accepted assumption at that time which considered a change from a fibrous structure to a crystalline structure to be the reason of crack nucleation, Rankine showed that the crack starts with a very small, smooth, and regular crack at the surface. Such a crack then slowly

Table 2.2 MAIN MECHANICAL PROPERTIES OF SOME STRUCTURAL STEELS

Steel Type	Designation	F_y Minimum Yield Point or Yield Stress (ksi)	F_{ult} Tensile Strength (ksi)	Minimum Elongation in 8 in. (%)	Minimum Reduction of Area (%)
Struc. Steel	A36	36	58–80	20	—
Steel Pipe	A53	30	48	—	—
High-strength Struc. Steel	A440	42–50	63–70	18	—
High-strength, Low-alloy Struc. Steel	A441	42–50	63–70	18	—
	A500 (A & B)	39–46	45–58	23–25 (in 2″)	—
Struc. Steel	A501	36	58	20	—
High-yield-strength	A514	90–100	105–135	17–18 (in 2″)	—
High-strength Alloy Steel	A517	100	115–135	16 (in 2″)	—
Struc. Steel	A529	42	60–85	19	—
Struc. Steel	A570 (All Grades)	25–42	45–58	14–20	—
High-strength, Low-alloy	A572 (All Grades)	42–65	60–80	15–20	—
High-strength, Low-alloy	A588	50	63–70	18	—
High-strength, Low-alloy	A606	45	65	22 (in 2″)	—
Low-alloy Columbium and/or Vanadium	A607	45–70	60–85	14–25 (in 2″, Hot-rolled) 14–25 (in 2″, Cold-rolled)	
Low-carbon-alloy Steel	T-1	90–100	105–135	16–18 (in 2″)	45–50
	Cor-Ten A & B	50	70	19	—
	Jalten #1	50	70	18	—
	Jalten #3	50	70	18	—

propagates into the interior of the axle, resulting in weakening and subsequent failure. He recommended that axle edges be rounded off to avoid the sudden cutting of fibers resulting in local inelastic behavior. Others at the time also studied fatigue failures in railway axles (e.g., 2.15).

A detailed study of the fatigue of railroad axles in Germany was carried out from 1858 to 1870 by A. Wöhler (2.16). The first fatigue machine for the testing of iron specimens was developed in 1848 by James and Galton. The work by Wöhler can be considered the first scientific research in this field; his testing machines also represented a great improvement at that time.

2.4.2 Basic Aspects of Fatigue

Although fatigue is usually associated with parts of moving machinery or aircraft, civil engineering steel structures can also develop fatigue failure. Loading is frequently dynamic or cyclic in nature. Bridges, crane girders and booms, buildings loaded by moving machinery, and unbalanced rotating machine parts are examples of structures which may be in danger of fatigue failures. In addition, nowadays welding is used predominantly in the fabrication of steel structures, as a result of which large residual stresses may occur, while defects in weldments may introduce sharp stress raisers and initial microscopic cracks. The modern steel designer should be fully aware of the danger of fatigue failure, which may render an otherwise good design unsuccessful (2.17, 2.18).

The main characteristic of a fatigue crack is its practically complete lack of plastic deformation prior to subsequent failure. The absence of apparent large deformations prior to failure makes it very difficult to discover such cracks in time. It has been observed that cracks usually develop in regions of high stress concentrations at the surface (about 90% of all cracks) followed by subsequent propagation into the section. Sometimes cracks are initiated at weld defects, which do not necessarily occur at the surface. In all cases the region surrounding the initial crack exhibits a smooth silky appearance which extends to the limits of the fatigue fracture region. Careful examination of the fracture surface frequently reveals the existence of concentric rings (also called beach markings) around the fracture nucleus and radial lines emanating from it (Fig. 2.9). These markings can be very helpful in locating the fracture origin.

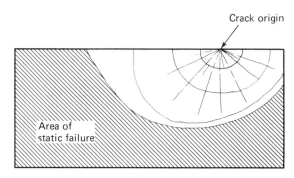

Figure 2.9 Fatigue Failure.

Fatigue-testing notation

Fatigue life depends not only on the maximum stress but also on the minimum stress occurring under cyclic loading (2.19).

Several parameters are used to describe all possible types of cyclic loading:

1. Maximum stress, f_{max}
2. Minimum stress, f_{min}
3. Average stress, $f_{av} = (f_{max} + f_{min})/2$
4. Stress range, $f_{sr} = f_{max} - f_{min}$
5. Stress ratio, $R = f_{min}/f_{max}$

Compressive stresses are taken as negative and tensile stresses as positive.

Some types of cycling stress can be recognized, as shown in Fig. 2.10.

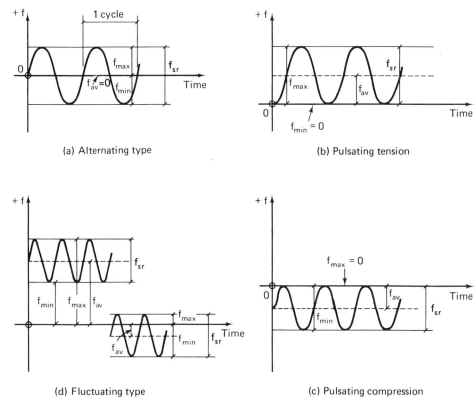

Figure 2.10 Types of Cyclic Stress.

Both the AISC and the AASHTO specifications use notations that are slightly different from those given above. The main difference is their use of allowable stress ranges (F_{sr}) instead of actual maximum or minimum stresses (f_{max} and f_{min}). As a result of the recent changes in the AASHTO specifications (Interim 8, Fatigue Stresses, Washington, 1974) both specifications now use the same procedures. These specifications and their use will be discussed in greater detail later in the text wherever applicable.

Presentation of test results

Wöhler first introduced the concept of limiting stress (2.20). This stress is known in the modern literature as the "fatigue limit" of a material. It is the maximum stress which can be repeated indefinitely without causing fatigue failure when applying alternating bending stresses. Wöhler's curve represents the relationship between the number of cycles (N) required for failure and the maximum stress (F_{max}) (Fig. 2.11). Most fatigue testing of steel is stopped at $N = 2 \times 10^6$, and the fatigue limit is approximated as the fatigue strength at that number of cycles. For most steels no sizable reduction in fatigue strength occurs for $N > 2 \times 10^6$ cycles.

The main disadvantage of Wöhler's curve is that it is limited to one type of

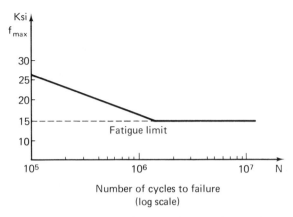

Figure 2.11 Wöhler's Curve (Log Scale)—Rotating Bending Test.

cycling. In order fully to define the fatigue strength for a particular type of cycling and specimen, several F-N curves for different stress conditions would have to be drawn. The results of such an investigation can be presented diagrammatically in several ways; two of these are the Smith diagram and the modified Goodman diagram.

By plotting both the extreme stresses as ordinates versus the mean stresses, the Smith diagram is obtained (Fig. 2.12). The stress range is represented by the vertical distance between the two curves. The line AA' represents the case of alternating cycling, and $OA = OA'$. Pulsating tension is represented by line BB', and $OB' = \frac{1}{2}BB'$. A half–tensile stress cycle is represented by the line CC', and $F_{min} = \frac{1}{2}F_{max}$.

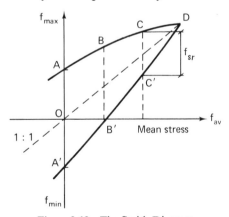

Figure 2.12 The Smith Diagram.

Point D represents a static tensile stress, where $F_{max} = F_{ult}$. The main parameters in the Smith diagram are thus the stress range and the mean stress. For design purposes it is more convenient to have as main parameters the maximum and the minimum stress. For this reason the modified Goodman diagram shown in Fig. 2.13 is mostly used. The maximum stresses are plotted as vertical and the minimum stresses as horizontal ordinates. The line OA represents alternating cycling ($R = -1$), line OB

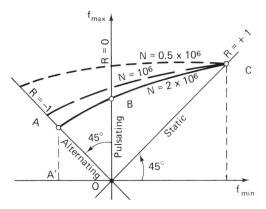

Figure 2.13 The Modified Goodman Diagram.

pulsating cycling ($R = 0$), and point C statical tensile loading ($R = +1$). Different curves for different values of N can be drawn through point C representing the fatigue life for various types of cycling. The vertical distance (Fig. 2.14) between a point on the N-curve and a line at 45° through the origin (provided the horizontal and vertical scales are the same) represents the stress range. The mean stress can also be determined graphically, as shown in the same figure.

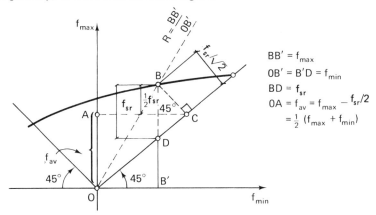

$BB' = f_{max}$
$OB' = B'D = f_{min}$
$BD = f_{sr}$
$OA = f_{av} = f_{max} - f_{sr}/2$
$\quad = \frac{1}{2}(f_{max} + f_{min})$

Figure 2.14 Maximum, Minimum, and Mean Stress Determination.

Correlation between test results and the fatigue behavior of actual structures

An abundance of test results are available for the more simple types of loading, such as tension, compression, bending, and torsion. Comparison between different published test results often proves difficult because of differences in test execution, the specimen size and surface finish, and even the type and construction of the testing apparatus. In addition, test results often cannot be compared because different types of steels have been used. It has been stated that, between 0 and 5,000 cpm (cycles per minute), fatigue test results are independent of the actual testing frequency. The authors believe that at higher frequencies heat exchange effects play a role, while corrosion effects are also known to be a function of the testing frequency. For structural steel elements actual frequencies lie between 10 and 600 cpm (2.21).

The relationship between the fatigue strengths of different stress states is still insufficiently known. Actual structural steel members are known to have stress peaks at stress raisers or in general nonhomogeneous or restrained (incomplete) stress states. Relationships between the fatigue stress and the ultimate tensile strength or yield stress can be determined for test specimens only. To extrapolate results obtained from small, smoothly finished specimens to actual structures in many cases is questionable.

In most cases the prediction of actual fatigue behavior from test specimens is difficult to achieve. One way of gaining an almost complete insight into actual fatigue behavior is to test full-sized members or components or even complete structural systems under simulated service loadings. This method, which for instance is being used extensively in the aircraft industry, can also be applied economically for the testing of machinery and structural components which are mass-produced. To test an actual civil engineering structure such as a crane or a bridge is not feasible.

In spite of several existing theories (2.22), at present no acceptable complete theory or hypothesis exists that can explain satisfactorily the nucleation of a fatigue crack or the fatigue failure mechanism itself.

The danger of fatigue failure of course has to be evaluated relative to other possible contributing types of failure or design criteria. Increases in dynamic loadings during the life of a structure (railroad bridges, for instance) and the loss of sectional area due to corrosion are typical examples of factors causing premature fatigue failure. A careful examination of the fatigue behavior of existing structures, and specifically of certain standard details, combined with a sound evaluation of new construction details will in most cases yield a design which will be safeguarded against fatigue failure.

Additional factors affecting fatigue strength

The character of a dynamic loading acting on a structure is basically different from a statical type of loading. A dynamic loading cannot simply be considered a repetition of a certain number of static loadings. When stresses vary with time, it must be realized that each loading has to be superimposed on the previous stress and strain states. Each time a new loading cycle is acting, it is superimposed on a structure with a stress-strain record or history that changes with time. It can be said that a structure exhibits a "memory" of its past loading history which affects its behavior. The assumption of a series of static load repetitions is for this reason inadequate (2.23).

The fatigue strength of a steel is affected by the variation of strain with time. The resulting stresses, which usually are three-dimensional, will also be time-dependent. In addition, the surface finish of a specimen and its physical and chemical changes with time affect aging during testing, as well as its fatigue strength. To obtain the same degree of aging as occurring in a statical test at room temperature, the temperature in a fatigue test has to be higher (2.24).

It is obvious that the fatigue strength depends on many factors which usually differ from test to test and that it is impossible to predict fatigue strength exactly or to express it in one simple equation. Both manufacturing and fatigue loading introduce these variables.

2.4.3 Design Concepts and Considerations

Wherever dynamic or cyclic loading takes place, making provision for safety only against yielding under static loading is no longer valid. During the rapid development of steel as a structural material and its application, factors of safety in the past were mainly based on intuition and service experience. By expressing working stress as the static yield-point stress divided by a factor of safety, a wide range of values were used (between 2 and 10). In spite of these high values, in some cases failures occurred under fairly light loadings, indicating that the consideration of static yield stress in cases of dynamic or cyclic loading is an improper criterion. The recognition of fatigue strength as opposed to static yield strength led to the development of fatigue testing. Initially the factor of safety for fatigue was simply based on the fatigue limit. This was later revised to take into account the life of the structure which, depending on the number of cycles expected during its life, may lead to a higher fatigue strength.

For example (2.25), the fatigue strengths of two different steels are shown in Fig. 2.15, assuming that $F_{max} = 50$ ksi for a life of $N = 2 \times 10^5$ is wanted. The factor of safety for materials (a) and (b) is different depending on whether it is calculated on

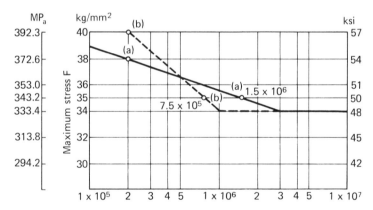

Figure 2.15 Definition of the Safety Factor in the Range of Fatigue.

the basis of stress corresponding to a given number of N, or on the basis of the number of cycles of the life of structure to cause failure for F_{max}.

Factor of Safety Based on:

1. *Fatigue strength*   2. *Number of cycles N (life)*

$$\text{material } (a) = \frac{54}{50} = 1.08 \qquad \frac{1.5 \times 10^6}{2 \times 10^5} = 7.5$$

$$\text{material } (b) = \frac{57}{50} = 1.14 \qquad \frac{7.5 \times 10^5}{2 \times 10^5} = 3.8$$

Both the allowable maximum stress, F_{max}, and the number of cycles, N, that can be expected in a structure are very difficult to determine. The magnitude of the maximum stress is only partly affected by the dead weight and mainly by the magnitude of the moving loads, impact, wind and snow loads, and inertia forces. The number of load repetitions and the relative magnitudes of the loads that cause these are also of importance. For this reason both types of factors of safety are usually considered. Which of the two types is more relevant or important depends on actual conditions, and no general rule can be given.

Another difficulty arises when making use of any type of programmed loadings. The validity of the assumption that in such cases damage is linearly cumulative has proved questionable in many cases. To use a percentage of total damage as a measure of a structure's safety is therefore not always correct. The contribution of different fatigue effects also varies from case to case, thus making generalizations or extrapolations based on individual effects often unrealistic.

The present state of the art and the vast amount of available test results, although not sufficient to enable the use of a general valid prediction theory, have nevertheless been useful in giving the designer guidance in designing connection details and certain typical members or structural details. Such presently accepted guidelines for designing to avoid fatigue failures as adopted in the AISC and AASHTO specifications will be discussed wherever applicable in this text. The emergence of fracture mechanics and its fast development appears to offer the best potential for a better evaluation of fatigue parameters and more rational design techniques. For more information on these developments and proposed new design techniques the reader is referred to two of the most recent publications in this field (2.26 and 2.27).

2.5 BRITTLE FRACTURE

As mentioned previously, the occurrence of brittle fracture can be a controlling design criterion (Section 1.2.8). As early as 1879 (2.28) the problem of brittle fracture was recognized, and a discussion of the phenomenon took place at a meeting of the Iron and Steel Institute in the United States in 1886. The importance of brittle fracture in the design and subsequent service behavior of large structures such as ships and bridges, however, has been fully recognized only since the 1940s.

2.5.1 Historical Review

The first recorded brittle fracture failure took place in October 1886. At Gravesend, Long Island, New York, during a hydrostatic acceptance test of a large riveted steel water standpipe, 1-in.-thick plates in the lower section of the pipe cracked suddenly along a vertical crack of 20-ft. length. The cracked plates were made of a very brittle steel. In 1898 a large gas holder in New York City failed in brittle fracture. In January 1919 a molasses tank in Boston failed. Initial failures mainly took place in storage tanks or pipes, all of which are structures exposed to uniform high stresses. Until 1920 bridges were mostly riveted and comprised of several rather thin plates primarily subjected to uniaxial stress states. Although some of the riveted steels used in these bridges were fairly brittle (2.31), the presence of predominantly uniaxial stress states seems to be the reason that very few brittle fractures took place. After the First World War, in Europe and especially in Germany the construction of welded bridges developed rapidly. Flange plates in these bridges were thicker than for riveted sections. Figure 2.16 shows typical cross sections of welded girders built in 1932 (Fig. 2.16a), 1936 (Fig. 2.16b), and 1951 (Fig. 2.16c). The earlier structures were merely imitations of their riveted counterparts, consisting of several thin plates. As

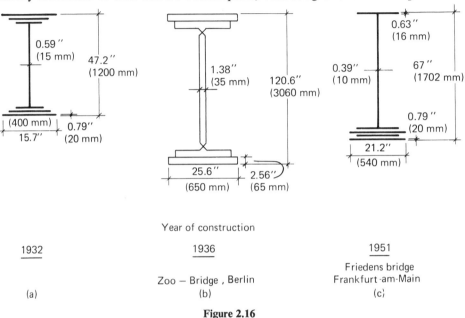

Figure 2.16

plates became thicker brittle fractures occurred more frequently. For this very reason after the Second World War a return to flanges built up from several thinner plates took place (Fig. 2.16c). The same trend is today expressed by the increased use of box girder type of bridge cross sections for large bridge spans in order to avoid the use of thicker plates.

In March 1938 the Hasselt Bridge in Belgium collapsed completely. The Duplessis Bridge in Canada collapsed in January 1951, three years after its construction (2.32), in 30-degree-below-zero weather. This structure had a history of prior cracking during its fabrication. The steel was rimmed and of poor quality, supplied and accepted in error. After the Zoo Brücke in Berlin collapsed in December 1939 it was concluded that high residual stresses caused by the welding of the extremely thick flange plates were the major cause. During the Second World War brittle failures occurred in welded tankers and Liberty ships. In 1943, the S.S. *Schenectady*, an all-welded tanker, broke in two while being moored. The upper deck plate at that time was calculated to have a stress of only 10.8 ksi (74.5 MPa). During the next ten-year period over 200 of these ships built during World War II collapsed. The majority of these ships were welded, and most of the fractures started at points of high stress concentrations such as hatch corners.

2.5.2 Basic Characteristics

The brittle fracture failures that took place in the above-mentioned bridges and ships cannot be explained from overloading or discrepancies between actual and calculated stresses. From these low level stresses it appears that reducing the allowable stress is not the proper approach. Also, the use of more sophisticated and more detailed methods of analysis does not seem to provide a safer design against brittle fracture. In view of the competition that now exists between steel and other construction materials such as prestressed concrete, as well as the trend toward welding versus bolting or riveting, it is essential that the underlying causes be recognized and prevented.

In most fractures a lack of ductility of the fractured material after failure has been found to exist as compared with good initial ductility of the same material at the same temperature. This is one of the most significant characteristics of brittle fracture. However, many of the fractured materials did not possess an exceptionally high tensile strength or hardness, both of which are indicative of brittleness. Although in some instances fatigue created initial cracking, the kind of failure is different. Of all ships that have failed due to brittle fracture, only two authentic cases are known to be partially caused by fatigue.

When brittle fracture occurs in welded structures, the following common features have been noticed (2.33)

1. There existed residual stresses which in some portions of the welds caused triaxial tensions.
2. There were cracks in plate girders which occurred in areas where no plastic deformation had taken place, starting at the longitudinal welds (fillet or double-butt) between the web and the tensile flange.
3. The steels used were susceptible to aging and in some cases had large quantities of nonmetallic inclusions.
4. The plates were at least $\frac{3}{4}$-in. thick.
5. Brittle fracture occurred not immediately after welding was completed, but later under the action of a static external load approximately equal to that caused by the dead load of the structure, and not at the maximum design load.
6. Brittle fracture in all cases occurred very suddenly.
7. Quite a number of failures occurred at low temperatures.
8. In most cases no special heat treatment such as preheating was applied before welding.
9. The steels used had a limited capacity to absorb energy without crack initiation (notch-sensitive).

Cleavage and shear fracture

L. Prandtl (2.34) in 1907 was the first to state the fact that there are only two types of failure of solid bodies: cleavage (brittle) and shear (ductile). Brittle fracture is a type of catastrophic failure in structural materials that usually occurs without prior plastic deformation and at extremely high speeds [as high as 7,000 ft/s (200 m/s) in steels]. The fracture is usually characterized by a flat fracture surface (cleavage) and average stress level below those of general yielding. Steel has randomly oriented but uniformly distributed microscopic flaws. Since the flaws are randomly oriented and uniformly distributed, one is always oriented so that the fracture is initiated at a point which has the largest normal stress, in the direction of that stress. Brittle fracture occurs with little or no deformation or reduction in area and with very little energy absorption.

A material will behave in a brittle manner when the resistance against slipping or sliding exceeds that against separation. On the other hand, a material will behave in a ductile manner when the resistance against separation is larger than the resistance against sliding. Ductile failure is characterized by a material's undergoing large deformations and reduction in area with a large amount of energy absorption. Yielding results from slippage along critically oriented planes. The two types of resistance for a given material are not constants but vary with temperature, deformation velocity, stress state or triaxiality, and the material's prior stress and strain history. In Fig. 2.17 three of these four factors are plotted against fracture energy. An increase in temperature reduces the brittle fracture probability and vice versa. An increase in strain rate and triaxiality increases the possibility of brittle fracture. As several of these factors usually act simultaneously, a sharp transition point between cleavage and shear does not exist, but rather a transition zone.

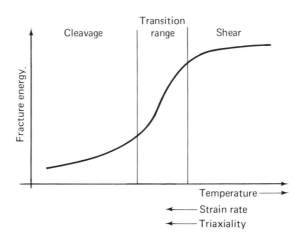

Figure 2.17 Generalized Transition Curve.

The influence of temperature is dependent on the orientation of the crystal faces. With decreasing temperatures the energy required for cleavage fracture becomes less than the amount required for ductile shear fracture.

Although there may be no direct evidence that high strain rates initiate cleavage fracture, they certainly are involved in their propagation (2.35). It is considered that a relationship exists between temperature and strain rate (2.36).

From experience it is known that the stress state has an important effect on the mode of fracture. Even in a uniaxial tension test of a ductile mild steel round bar, when large local plastic deformation (necking) takes place uniform axial stress is replaced (Fig. 2.18) by a three-dimensional rotational-symmetric stress state (2.37).

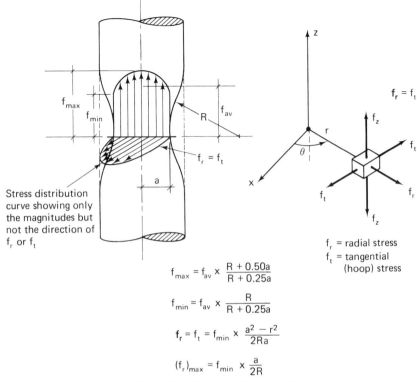

Figure 2.18 Stress Distribution in Neck of a Tensile Bar.

MacGregor (2.38) proved by taking X-rays that the crack in a tension test starts at the center of the neck, where the tensile stress is at its highest, and then propagates radially. In the presence of a notch, three-dimensional stress states similar to those for a tensile bar occur, but far worse. Depending on the type and geometry of the notch and the so-called notch sensitivity of the material, crack initiation will take place. Notch sensitivity is determined by the limited ability of a material to absorb energy before the initiation of a crack. The stress gradient is determined by the sharpness of a notch and the stress intensity by its depth. Neuber (2.39) has given many stress concentration factors taking into account the possibility of plastic yielding. Hill (2.40), assuming ideal plastic behavior and no work hardening, has shown that in deeply notched material bars the maximum attainable stress may be up to 2.57 times the yield stress in uniaxial tension due to restraining of local yielding. If now, in addition, a high strain rate is present resulting in raising the yield point and the whole stress-strain curve at the expense of ductility, brittle fracture is likely to occur. The influence of the previous history of stress and strain upon cleavage is mainly caused by the residual stresses from past loadings. Such loading includes cold-working, cold-bending, partial heating or welding, and manufacturing processes. The initiation of cracks is not clearly understood. It is likely that some degree of plastic deformation causes it. Such plastic deformation may lead to normal ductile failure or brittle cleavage. Plastic deformation alone, however, does not lead to cleavage in ductile states, even at low temperatures.

Energy criteria for brittle fracture
and its propagation

The total energy absorbed during fracture consists of three main parts

E_{ed} = elastic recoverable energy
E_{pd} = plastic dissipated energy
E_{fs} = energy required to separate free surface (crack propagation)

About 10% of the energy is stored as elastic energy, E_{ed} (2.41). The magnitude of E_{pd} is an indication of the degree of brittleness of a fracture. For brittle fractures E_{pd} is small. The values of E_{fs} are higher for ductile materials than for brittle materials. The E_{fs} is small because this energy is actually supplied by the stored elastic energy (2.42).

Whether an initiated fracture will propagate depends on (2.43)

1. The relationship between the minimum size of the original notch or crack and the stored elastic energy required for propagation, and
2. The conditions required to arrest a propagating fracture.

The oldest theoretical treatment of crack propagation was done by A. A. Griffith (2.44). According to his formula, cracks of the order of 10^{-4} cm (4×10^{-5} in.) are required to account for the observed brittle fracture strength of steel. This critical size of crack of course is unrealistic, being too small. Therefore, his expression has to be modified for ductile materials, which require inclusion of the plastic work as well. Irwin (2.45) and Orowan (2.46) have corrected Griffith's formula to include ductile materials by introducing a term p, the effective surface energy, which includes plastic work. This effective surface energy has been estimated to be about 10^6 ergs/cm²·sec (10^3 watts/m²·sec). Thus, the principal difficulty in the equation

$$F = \sqrt{\frac{2Ep}{c}} \qquad (2.3)$$

where F is tensile stress and $2c$ the crack length, is in accounting for the temperature dependence, and it must be assumed that the term p varies strongly with temperature.

If dW is the work increment required during crack propagation, dA the increase in fracture area surface, and dE the energy released per unit crack area, then as long as dW/dA is greater than dE/dA the fracture will not propagate. If dE/dA equals or exceeds dW/dA, a fracture will start and will not be arrested as long as this relationship holds. The larger the crack, the lower the stress required to propagate it.

The question now is how large may cracks grow in structures without leading to an unstable crack and brittle failure? The answer can be given with the aid of fracture mechanics, a new branch of applied mechanics. Fracture mechanics is a method of characterizing fracture behavior in structural parameters that can be used by the engineer—stress and flaw (or crack) size.

In fracture mechanics the fundamental assumption is made that cracks can be present in all welded structures even after all inspection and weld repairs are finished (2.47).

The basis of fracture mechanics is that the stress distribution ahead of a sharp crack can be characterized in terms of a single parameter, K_I, the stress intensity factor, with units of ksi·$\sqrt{\text{in}}$. For various crack geometries the theoretical expressions for K_I in terms of applied stress and flaw size are given in Fig. 2.19. In all cases K_I is a function of the nominal stress and the square root of flaw size. By knowing the critical value of K_I at failure, K_c, for a given steel at a specific temperature and loading rate, the tolerable flaw and crack sizes can be determined for each predetermined stress level. The opposite is also true; that is, for an assumed flaw size, which is dependent upon the inspection precision, or better to say upon the economic considerations of the inspection cost as compared to the overall cost of the structure, the corresponding stress level can be calculated. A designer would be very realistic to assume that he will not be able to eliminate flaws smaller than $\frac{1}{4}$ in. (6 mm).

For example, assume that laboratory tests have determined that for a particular steel with yielding stress of 80 ksi, at service temperature, loading rate, and the plate thickness used the critical stress intensity factor is $K_c = 60$ ksi·$\sqrt{\text{in}}$. If the working (allowable) stress is 45 ksi, for a through-thickness crack in a wide plate (Fig. 2.19a, where $K_c = F\sqrt{\pi a}$) the maximum tolerable flaw or crack size would be

$$a = \frac{1}{\pi}\left(\frac{K_c}{F}\right)^2 = \frac{1}{\pi}\left(\frac{60}{45}\right)^2 \approx 0.6 \text{ in. (1.5 cm)}$$

This would be quite satisfactory; but if residual stresses are present and a stress concentration factor of, say, 2.0 magnitude is expected, then the actual stress, f, may reach yield stress, in this case 80 ksi. The tolerable flaw size in that case is reduced to only about 0.2 in. (0.5 cm), which is less than the practical limiting size of $\frac{1}{4}$ in. (6 mm). The cost of structural steels generally increases with their ability to perform satisfactorily under actual operating conditions. The designer, on the basis of an economic analysis, has to choose the proper type of steel with the adequate toughness, which in this general context can be defined as its ability to deform plastically. The inherent level of toughness for the normal steels used today (A36, A441, A572,

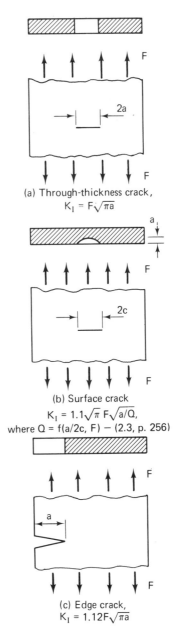

(a) Through-thickness crack,
$K_I = F\sqrt{\pi a}$

(b) Surface crack
$K_I = 1.1\sqrt{\pi}\, F\sqrt{a/Q}$,
where $Q = f(a/2c, F) - (2.3, \text{p. } 256)$

(c) Edge crack,
$K_I = 1.12 F\sqrt{\pi a}$

Figure 2.19 K_I-values for Various Crack Geometries.

A514, etc.) usually is adequate for most structural applications. As the toughness depends upon manufacturing variables, it might happen that even though the material meets an existing material specification with regard to the simple tension test, the toughness level is not adequate. Therefore, it seems that specifications should require a minimum steel toughness. Next to this, a fracture control plan, basically a systems approach to prevent fractures in structures, should be outlined.

As higher-strength steels are used, the yield point gets closer and closer to its breaking point, resulting in a smaller margin of ductility (2.51). The lower-strength steels generally have more ductility and usually are less notch-sensitive than higher-strength steels.

For plates used in civil engineering structures up to about 2 in. in thickness at minimum service temperatures, typical toughness criteria would be

1. Plates of all thicknesses and strength levels will develop through-thickness yielding ahead of a sharp crack, even at yield stress loadings. That means ductile behavior rather than brittle pop.
2. At design stress loads, large cracks equal to the thickness of the plate or greater (depending upon whether the crack is an edge crack or a surface crack) can be tolerated by the steel.

Although toughness criteria have been considered for various structural applications, it was not until 1973 that AASHTO adopted a notch toughness requirement for bridge steels (2.48). Two references (2.49 and 2.27) describe these requirements and the background leading to their development; others (2.50–2.56) deal with brittle fracture.

Tests for brittleness

The various principles involved in brittle fracture have been applied to develop a standard test to measure resistance to crack propagation.

A truly brittle material will fracture during elastic deformation. Most low-carbon steels, however, are capable of deforming plastically before fracture. All specifications for structural steels include required minimum values for ductility as measured in the standard tensile test. The standard tensile test as well as the standard bending test have proved satisfactory for the purpose of eliminating this kind of brittleness.

Testing for cleavage or notch sensitivity is directed toward a different problem—namely, that of resistance to cleavage fracture.

Due to the current disagreement on the required margin of safety and how such a margin should be assessed, many tests have been proposed. The test that is probably used most is the Charpy V-notch (CVN) impact test, with a swinging-pendulum type of loading (Fig. 2.20). Although Charpy tests do leave quite a bit to be desired, they will at least identify the brittle steels. Some other tests which employ impact loadings and notched specimens are the Izod, Schnadt, Mesnager, Charpy keyhole, and Pellini drop-weight tests.

Another group of tests are those using slow bending. Figure 2.21 shows the

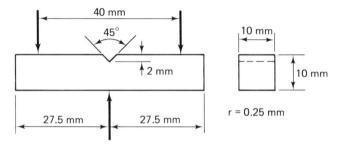

Figure 2.20 Charpy V-notch Specimen.

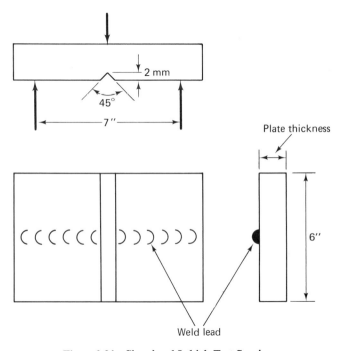

Figure 2.21 Slow-bend Lehigh Test Specimen.

specimen for the slow-bend Lehigh test. This test was developed from the Kinzel test incorporating a weld bead and is almost identical to tests recommended by Kommerel used in Europe and particularly in Germany. The fact that relatively few brittle fracture failures have occurred in Germany with steels tested with Schnadt's and especially Kommerel's tests seems to indicate that this type of test is very valuable.

2.5.3 Design Considerations

This section will be concerned with welded steel structures because of their increased importance and their different type of construction.

Method of design

There is no need to change the analysis or design procedure when using welded structure and connections. The real design problem consists of choosing the appropriate type of steel. To reduce the notch brittleness of steel at operating temperatures, approaches to be considered are

1. The use of alloy steels.
2. Using killed steel (deoxidized steels to prevent gas bubbles and reduce nitrogen content).
3. Lowering the carbon content and raising the manganese content.
4. Increasing the rate of cooling during rolling, or rolling at a lower temperature, or even applying a heat treatment involving quenching.

The chemical composition and the grain size are the most important factors when determining notch sensitivity of steels. The grain size is influenced by the rolling process and consequently also by the thickness of the product. The thicker the plate the more careful one has to be.

Residual stresses caused by welds

Although brittle fractures have occurred in the absence of welds, it is recognized that welds often are potential sources of crack initiation and also may reduce the possibility of crack arrest.

It was believed in the past that a welded structure that has not developed cracks by the time it cools down after the welding process will last forever. The failure of some welded bridges already in service exposed to dead loading and low temperatures has proved the opposite. One should distinguish between two basically different cases in weld cracking due to stress states caused by welding; a study of these states is necessary.

1. *Stress States Without Real Residual Stresses:* Such stress states are caused by welding in the presence of unyielding supports (Fig. 2.22) resulting in cross-sectional forces.

2. *Stress States With Real Residual Stresses:* These occur when there are no boundary restraints and no resulting cross-sectional forces. Figure 2.23 shows the residual stresses in a welded girder along its axis. Of necessity, the stress distribution is nonhomogeneous, and stresses must change their signs at least twice. Stresses of this type are difficult to avoid and usually can only be changed slightly by use of a particular welding order. They are referred to as "forced" stresses. If now a crack starts at the location of the highest residual stress, the fact that all other stresses are smaller may prevent its propagation. Nevertheless, the danger of collapse at a later date is more or less hidden in the structure. It must also be realized that in this particular case a high longitudinal forced tensile stress exists all along the joint between the web and the flange. Any imperfections in this weld or in adjacent material indicates a high crack probability. Whether such an initiated crack will propagate depends upon several conditions,

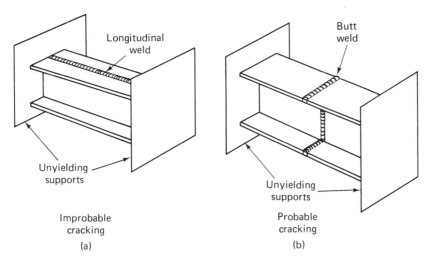

Figure 2.22 Different Behavior of Longitudinal and Butt Welds—No True Residual Stresses, Only Sectional Forces (2.33).

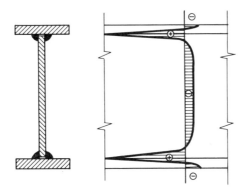

Figure 2.23 Residual or Forced Stresses in the Direction of Girder Axis Due to the Fillet Welds.

such as type of steel, temperature, aging, moving loads, structural system, and amount of stress relieved as a result of cracking. When cooled, longitudinal welds between web and flange, as shown in Fig. 2.24, may result in three-dimensional tensile stresses in the mutually perpendicular planes F, C, and L shown in Fig. 2.25.

A parallelepiped cut out from the weld when red hot is in compression, while after cooling it is under tension in all three planes. The largest of these three tensile stresses is that one whose direction coincided with the direction of greatest opposition to expansion during welding. Usually this is in the direction of the longitudinal axis (x-axis). In plane C tensile stress increases with an increase in the flange thickness but at a faster rate. Transverse stiffeners welded to the web even further increase these stresses. If the weld between web and

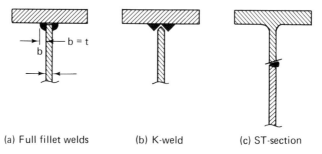

(a) Full fillet welds   (b) K-weld   (c) ST-section

Figure 2.24 Different Forms of Weld between Flange and Web.

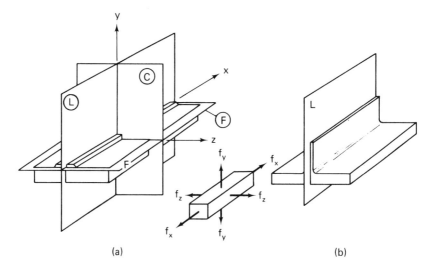

Figure 2.25 Forced Stresses in Weld between Flange and Web.

flange is moved into the web (Fig. 2.25b), no stresses occur in planes C and F. The type of longitudinal weld also influences the brittleness sensitivity. For example, the fillet welds shown in Fig. 2.24a combined with an air space between flange and web have proved to result very seldom in a brittle fracture. This type of weld also has a low production sensitivity (discussed below) as compared with the K-butt weld shown in Fig. 2.24b.

The effects of residual stresses when service loads are acting, statical, or cyclic need to be considered next. In general it can be stated that they have no measurable effect on the fatigue limit of test specimens with bead welds because residual stresses are reduced due to cyclic loading (Fig. 2.26).

A second effect is the reduction of residual stresses due to yielding at the location of the second largest stress peak. When static service loading causes yielding here, such yielding will cause new residual stresses. This will always result in a reduction of the original maximum peak stress (Fig. 2.27).

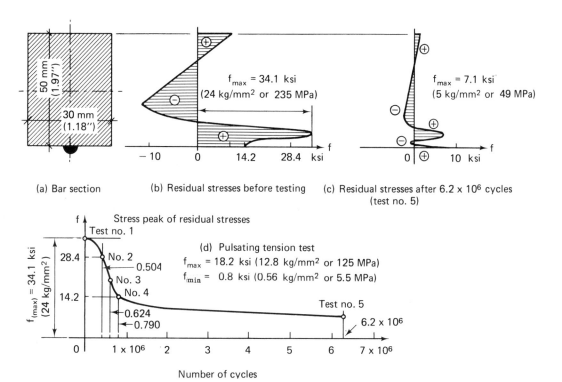

Figure 2.26 Stress Peak Reduction of Forced Stresses Due to Cyclic Loading.

Flange thickness

The effect of flange thickness on brittle fracture has already been discussed. Apart from a technical stress evaluation point of view, other factors may also contribute to the danger of brittle fracture. Some of these factors are metallurgical, crystal-structural, and methods of production such as rolling and cooling. All these contributing factors may act negatively, thereby causing a serious danger of brittle fracture. A clear demonstration of the effect of flange thickness is shown in Fig. 2.28. Only the average stresses are shown. The maximum stress is several times larger than the average stress.

The yield stress usually is reduced when the thickness of the material increases. In addition, it is obvious that cooling speeds become more critical. As the thickness increases, the danger of hardness and core brittleness also increases. By selecting a fine-grained steel killed with added aluminum, or by normalizing the material, these dangers can usually be eliminated. For example, in the United States it is required that, if the plate thickness of a steel tank wall reaches 1.18 in. (30 mm), residual stresses be reduced by heating the plates.

High temperatures reduce brittleness of structural steel, while low temperatures have a negative effect. For riveted structures, a stress concentration factor, α_k, of

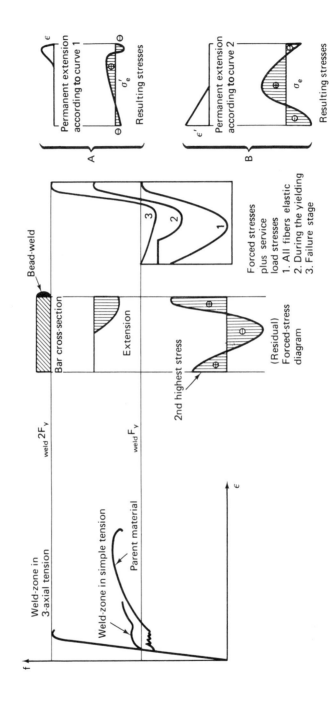

Figure 2.27 Combined Effect of Working and Forced Stresses.

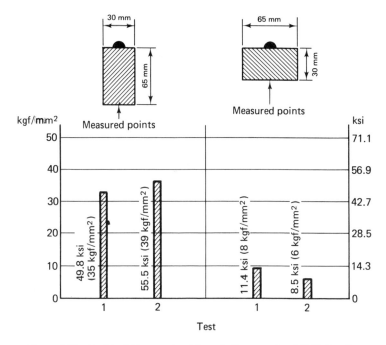

Figure 2.28 Residual Stresses in Prismatic Bars made of St52 for Two Different Bar Thicknesses (Average Stresses).

3.5 combined with comparatively thin plates (¾ in. or 20 mm) and operating temperatures of around freezing point in general do not lead to brittle fracture. The fact that in welded bridges brittle fractures have occurred at higher operating temperatures indicate the possibility of stress concentration factors in excess of those assumed for riveted structures. The so-called cold-weather effect is emphasized when thick plates are used. The highest increase in brittleness due to low temperature occurs with rimmed steels and the smallest with fine-grained killed steels.

In addition to thickness, the character of the loading has to be considered. Dead loading, because of its permanent character, has to be considered in combination with any possible short cold spell that may occur. The same criterion applies to live loads applied at a high frequency.

Degree of danger to the whole structure

The danger resulting from cracking depends on the location of the crack, the structural details at such a location, the possibility of arresting a crack, and the degree of static indeterminacy of the structure. For example, if the tension flange of a plate girder bridge is built up of several plates instead of using one single plate (Fig. 2.29), then the chances of arresting a crack are much better, resulting in a less dangerous situation when compared with a single flange plate that starts cracking. As already mentioned, the tendency to use thinner plates in plate and box girders therefore reduces the degree of danger resulting from cracking.

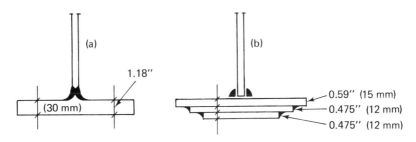

Figure 2.29 Flange Cross-Sectional Development and Its Welding to Web.

Fabrication sensitivity with respect to weld defects

The sensitivity of longitudinal welds is particularly variable according to the type of weld and welding process. The two fillet welds shown in Fig. 2.24a with an air space in between are far less sensitive than the K-weld shown in Fig. 2.24b. In addition to lessening sensitivity to weld defects the air space makes it easier to arrest a crack and also proves to be easier for X-ray quality control.

Circumferential welds in pipes and tanks must be considered equivalent to a longitudinal weld in a flanged section.

Cold-working of steel products

Aging is aggravated by cold-working of the material. Frequently welding takes place in areas where the maximum amount of cold-working has taken place. The stresses introduced in the material by a three-roller mill are of the same order of magnitude as those used for the artificial aging of steel. Again, the thickness of plates has an unfavorable effect.

2.6 LAMELLAR TEARING

2.6.1 Introduction

The increased use of welding and the accompanying use of heavier members has in many instances led to conditions of high joint restraint (2.57, 2.58). The occurrence of lamellar tearing in some highly restrained joints has been observed. Lamellar tearing appears as separation in the base or parent material as a result of through-thickness strains induced by weld metal shrinkage. A typical example of such a lamellar tear is shown in Figure 2.30.

In this section only the causes underlying lamellar tearing will be examined. Chapter 9

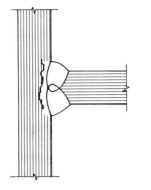

Figure 2.30 A T-Cross Section Showing Typical Lamellar Tear Resulting from Weld Shrinkage and Thick Materials.

includes a discussion of proper weld connection design so as to avoid lamellar tearing.

2.6.2 Basic Characteristics

To recognize conditions that may induce lamellar tearing it must be understood that local strains due to weld metal shrinkage are often many times higher than yield-point strains and far in excess of those caused by applied loading. No cases are known of lamellar tears being initiated or propagated by design loads. As stated above, the combination of high localized strains and internal restraint may cause lamellar tearing. "Internal restraint" is not intended to mean properly designed joint restraint, such as in a full bending moment connection, but internal restraint that inhibits large localized weld metal shrinkage strains.

The property of being able to deform inelastically is ductility. Among the commonly used structural materials only steel loaded parallel or transverse to the rolling direction exhibits this ductility (Fig. 2.31). In the through-thickness direction steel has great strength, but it may have limited capacity to be strained in excess of the elastic limit strains.

The cross section of a lamellar tear is steplike, with horizontal terraces that are markedly longer than the vertical portions (Fig. 2.32). The fracture has a fibrous

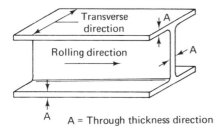

Figure 2.31 Terminology Related to Rolling Direction.

Figure 2.32 Typical Lamellar Tear due to Large Through-thickness Strain—Note Horizontal Terraces with Relatively Short Vertical Shear Planes.

appearance, and this characteristic, along with the terraced profile and location within the base material away from the weld fusion line, is the best way of distinguishing the lamellar tear from cracks in the heat-affected zone caused by hydrogen. The tearing occurs principally in T- and corner-type joints, where a sufficient degree of restraint exists such that weld shrinkage strains imposed on the parent metal in the through-thickness direction cannot be accommodated because of reduced ductility. During the welding procedure, after a sufficient number of passes have been deposited, the weld shrinkage strains increase in magnitude as the weld cools, to a degree where decohesion occurs at the interface between nonmetallic inclusions and the steel. As more weld metal is deposited, additional microscopic tears form. Since the non-metallic inclusions and strains are dispersed throughout the steel in an irregular manner, the tear takes the most susceptible path (Fig. 2.33). Subsequent completion of the

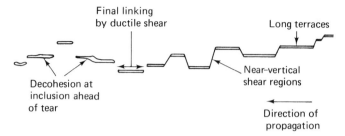

Figure 2.33 Diagram of a Partially Developed Lamellar Tear During Welding and Subsequent Cooling Stages.

weld followed by cooling to ambient temperature increases strains, so that terraces resulting from decohesion link together by shearing failure to form the completed lamellar tear (Fig. 2.34).

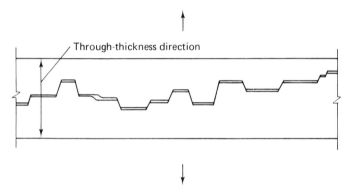

Figure 2.34 Cross Section of Parent Material Showing Complete Development of the Incomplete Tear from Figure 2.33.

2.6.3 Design Considerations

Connected material properties

Joints which stress the steel in the through-thickness direction do not necessarily cause difficulty. However, in a highly restrained design, if the weld shrinkage strains tend to pull the steel apart in the through-thickness direction, the joint will exhibit a greater tearing tendency than if the shrinkage forces were oriented in the plane of the member.

The reduction in through-thickness ductility can be largely attributed to microscopic nonmetallic inclusions. These inclusions consist primarily of residues from additions which must be made to liquid steel to improve the product by reducing the oxygen content and refining the grain structure. The inclusions, as cast, consist primarily of sulfides, oxides, and silicates which are progressively elongated longitudinally and spread laterally parallel to the rolled surface as the steel is rolled into a plate or shape.

Efforts are being made to improve techniques for the production of steel and to reduce the incidence of nonmetallic inclusions. However, there are both economical and technical difficulties involved which will require time to remove. Although continual progress is being made, it is unlikely that there will be a breakthrough that would change the anisotropic character of economical structural steels.

Weld metal properties

The requirements for electrodes, wires, and fluxes for use with "matching" base metals are well defined in the AWS *Structural Welding Code* (D 1.1-80) Miami, FL, 1980, and in AISC specifications. In general, the "matching" of electrodes to base metals is made on the basis of ultimate tensile strengths. When the designer selects an electrode that will match closely the ultimate tensile strengths of structural-grade steels, the weld yield points are generally significantly higher. Thus the total strain is forced to take place in the connected material. Overmatching of electrodes to connected material compounds the problem. Lower-yield electrodes aid in redistribution of strains.

Restraint

Lamellar tearing may occur when welds are made to the face of a plate or flange where high restraint exists. Such restraint, which may be caused by thickness of material, rigidity of a particular connection, volume of weld metal, or concentration of strains in localized areas and connections, will be discussed again in Section 9.3 on welded connections.

Connection details

A reduction in the amount of weld metal will help to prevent lamellar tearing. However, the most effective prevention results from good detailing of welded connections. Transverse weld shrinkage in detail (a) of Fig. 2.35 acts on a single line of inclusions which may form a tear. In Fig. 2.35b this action is applied to several layers of inclusions by slanting the weld, which will definitely decrease the danger of lamellar tearing. Figure 2.36a shows a usual detail of a beam-to-column connection with col-

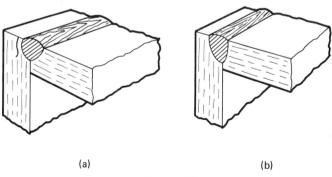

(a)   (b)

Figure 2.35

umn web stiffeners. Transverse shrinkage of the top beam flange groove weld and the groove weld of the top stiffener could produce lamellar tearing in the column flange lying between these welds. If the top beam flange (Fig. 2.36b) is slit for the column web thickness and stretched through the slot in the column flange, it will substitute for the top stiffeners. The shrinkage of groove welds now located in the column flange will act through the beam top flange, which is usually of a lesser thickness than the column flange. Therefore, this will reduce the danger of tears, which, if they still occur, will now be parallel to the tension force in the top beam flange. Lamellar tears in the detail (a) eventually could cause separation of the beam from the column. Therefore, detail (b), although more difficult to fabricate, is much safer.

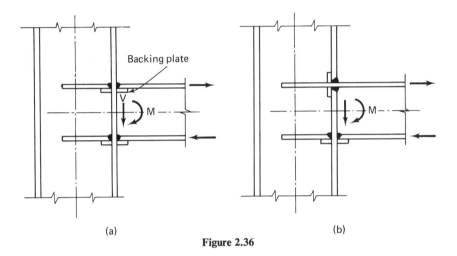

Figure 2.36

2.7 CORROSION

2.7.1 Introduction

Corrosion occurs when metals react with nonmetallic elements of their environment. The chemical compounds formed, called corrosion products, are either oxides or salts, the nature of which often has considerable influence on the course of the reaction, thus determining whether or not supplementary protection is required. Iron derives little or no protection from the corrosion products formed under ordinary circumstances. Therefore, it nearly always requires some means of corrosion prevention (2.59)—either by using corrosion-resistant alloys, or by applying protective coatings, or by controlling the environment. For civil engineering structures environmental control will be rarely used, thus necessitating the other methods of protection.

Metallic corrosion represents a tremendous annual loss. An estimate of this loss in the United States alone is given as over $5 billion. This figure includes the cost of preventive measures, such as the material and labor costs of painting and the additional cost of corrosion-resistant materials, as well as the cost of replacement of corroded equipment.

2.7.2 The Nature of Corrosion

Most metals, if exposed to natural environments without protection, will react with constituents in the environment to form corrosion products typical of the ones from which they originally were extracted. The tendency for this to occur is measured thermodynamically by the amount of free energy released during the reaction. The study of corrosion thus must consider which reactions between a metal and its environment are thermodynamically possible and those factors that may affect the reaction or corrosion rate.

There are two general classes of corrosion reaction: those in which there is a direct combination of metals with nonmetallic elements; and those in which the metal first dissolves, usually in an aqueous environment, later to combine with nonmetallic constituents in the environment to form corrosion products. The former is referred to as "dry corrosion" and is exemplified by oxidation, halogenation, or sulfidation reactions; the second is often called "wet corrosion." When metals corrode in the atmosphere, either or both of these processes may be involved. Both types of corrosion are electrolytic in character and depend upon the operation of electrochemical cells at the metal surface.

The dissolution of a metal in a liquid environment occurs at discrete locations, anodes, where the ions are formed and the metal is left with more electrons. To maintain electrical neutrality in the metal, there must be a simultaneous cathodic reaction in which electrons are consumed and either hydrogen or, in near-neutral solution, hydroxyl ions are formed in the electrolyte.

A corrosion cell consists of an anode and a cathode in contact with one another and with a common electrolyte present. The driving force for the corrosion reaction is determined by the difference in electric potential between the anode and cathode.

The nature of the oxide film thus formed and hence the rate of oxidation is strongly dependent on the metal's composition—both its minor constituents and major alloying elements. The addition of small amounts of alloying elements to steel can greatly increase the protective nature of the oxide formed at ordinary temperatures. One sample of 1-in. "high-strength, low-alloy" steel lost an average 3×10^{-4} in. in thickness during 16-year exposure in an industrial atmosphere, compared with 39×10^{-4} in. for a carbon steel (2.60).

2.7.3 Corrosion-Resistant, Low-Alloy Steels

Natural protective coatings are, in reality, initial corrosion products that are first, highly insoluble, and secondly, formed in intimate contact with the metal to provide continuous coverage. The passivity exhibited under suitable conditions by certain metals—for example, chromium, nickel, and stainless steel—is due to such a surface film, often of monomolecular thickness. Passivity often may also be induced by adding an appropriate agent to the metal or to the environment. Copper and copper alloys have long been used in applications where the corrosive environment consists of water or salt air. Other metals, such as stainless steel, monel metal, and lead, are used in special environments.

The high-strength, low-alloy steels have added copper and nickel and achieve an atmospheric corrosion resistance recognized to be 4 to 6 times that of carbon steel. The high initial cost of such metals can be more than compensated for by the increased service life and resulting lower annual replacement costs.

More recently, highly corrosion-resistant steels without any protective coating have been manufactured. These are called "weathering steels" or "self-painting steels" —for instance, A588 steel (from 1969). The cost of weathering steels is approximately 40% higher than that of carbon steels, but their strength is also 50 to 80% higher. The cost of fabricating structures of weathering steels has been found to range from about the same as that using carbon steels to about 10% higher. Figure 2.37 indicates the calculated average reduction in thickness (in mils) over time of carbon and weathering steel in an industrial atmosphere.

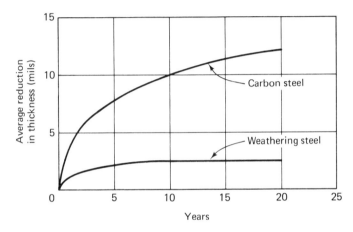

Figure 2.37 Comparison between the Anticorrosion Behavior of Carbon and Weathering Steel.

Studies have been carried out to consider the use of corrosion-resistant low-alloy steels for short- and medium-length highway bridges (2.61). Comparative designs were made using A7, A373, A242, A441, and T-1 steel for highway bridges with single spans from 61 to 120 ft. (19.5 to 36.6 m) in length, both composite and noncomposite, with HS20-44 AASHTO loading. The results indicate that corrosion-resistant, high-strength, low-alloy steels have definite economic possibilities for such use. It is recommended that a nickel-copper type of high-strength, low-alloy steel be used due to its higher resistance to atmospheric corrosion.

2.7.4 Protective Coating

It is frequently impractical to use the most corrosion-resistant materials because of high cost, lack of strength, or some other limitation. An alternative is the use of protective coatings. Coatings can be classified as those offering purely mechanical protection by separating the electrode from the electrolyte or atmosphere; those offering galvanic protection by serving as an anode to the base metal; and "passivators," which in effect shift the base metal toward the cathodic end of the electromotive series (2.62).

The first category includes organic coatings such as paints and asphalt compounds. The most important considerations in their use are the passivity of the coating and its flexibility and adhering qualities. New plastic coatings have been developed which can be applied easily in the field.

The most familiar example of the second category is the zinc coating yielding galvanized iron. At least part of the overall protection results from the fact the coating is anodic to the base metal. At holes and cut edges the coating becomes the anode and the base metal the cathode, and in electrolytic action the coating metal is deposited on the exposed base metal, adding to its protection. The coating also provides a mechanical protection, which is also often a major part of the total protection.

Finally, various oxide coatings whose protection is largely mechanical are classified as "passivators."

This discussion of the various causes of corrosion, its mechanism, and the different methods of protecting against it should be sufficient to make the designer aware of the problem as it may affect his design. It is outside the scope of this text to discuss the details of coating production and application. If the possible corrosion of a structure proves a major problem, the designer should study all available information in detail and preferably obtain advice from experts in this field.

2.8 CHOICE OF THE PROPER STEEL

As discussed in Section 2.1, besides A36, the basic structural steel, there is a variety of steels of different mechanical and physical properties. Their price per unit weight differs, and the type chosen for a particular job will be an important factor in determining the final cost of the whole structure. The designer's responsibility is to decide which one to use on the grounds of fabrication, performance, end cost, and maintenance requirements.

A major decision is to decide whether to use basic structural steel or one of the high-strength steels. High-strength steel will reduce the weight of the structure. In the case of a multistory building, it may also reduce the total height of it for the same number of stories by reducing the depth of each floor between stories. In the case of a bridge of long span, where the dead load is of primary concern, the use of high-strength steel may provide the only possibility to build a bridge for given spans and available construction height. An economical comparison in each case has to be performed before a rational decision can be made. This comparison is not limited solely to the weight of the structure, but must include the cost of fabrication and maintenance. For this reason, in a highly corrosive environment, special steels with enhanced resistance against corrosion, although of higher unit price, might be the proper choice. For welded structures the weldability of steels should be the main concern, together with the choice of proper electrodes (see Section 9.1.3.2).

The danger of brittle fracture will influence the choice between rimmed, capped, semikilled, and killed steel of a chosen grade. The possible dynamic character of the loading, stress levels, specific plate thicknesses, particular details of the structural design, redundancy of the structure, expected temperatures during its service life, as well as the projected inspection plan for crack detection (smallest crack size to be tolerated) will indicate which type has to be used.

A beginner should realize that in most cases rimmed A36 steel is the steel to be used. The problem of the proper steel grade choice arises only for cases having unusual design aspects such as heavy loads on low-rise or mill buildings, high-rise buildings, special aesthetics considerations, longspan bridges, special environmental conditions (very corrosive atmosphere, violent storms, high earthquake risks, very low operating temperatures), or special stringent operation and stiffness conditions. Only in such cases does the choice of a grade of steel other than a rimmed A36 steel arise. Such choice represents a complex problem which is best solved in steps. First, the decision is made that a special steel grade has to be used. Next, taking into consideration all the design parameters, the proper grade is chosen, based on experience, steel market prices, performance conditions, and several comparative designs.

NOTATIONS

a = side of cubic space lattice or flaw size
c = half-crack length
f_{av} = average stress
f_{max} = stress with maximum algebraic value
f_{min} = stress with minimum algebraic value
f_{sr} = stress range
p = effective surface energy
t = thickness of elements
α_k = stress concentration factor
α_t = thermal coefficient
ϵ_{sh} = strain at the beginning of strain hardening
ϵ_{ult} = strain at fracture
ψ = ductility ratio

μ = Poisson's ratio
Å = angstrom
E_{ed} = elastic recoverable energy
E_{fs} = energy required for crack propagation
E_{pd} = plastic dissipated energy
E = Young's modulus
F.S. = factor of safety
F_e = elastic limit
F_p = proportionality limit
F_y, F_y' = yield-point stress
F_{yu} = upper yield stress
F_{ult} = ultimate tensile stress
K_c = critical stress intensity factor
K_I = stress intensity factor
N = number of load cycles
R = stress ratio

REFERENCES

2.1. VAN VLECK, L. H., *Elements of Materials Science*, Addison-Wesley, Reading, MA, 1964.

2.2. MARIN, J., *Mechanical Behavior of Engineering Materials*, Prentice-Hall, Englewood Cliffs, NJ, 1962.

2.3. POLAKOWSKI, N. H., and E. J. RIPLING, *Strength and Structure of Engineering Materials*, Prentice-Hall, Englewood Cliffs, NJ, 1966.

2.4. ROSENTHAL, D., and R. M. ASIMOW, *Introduction to Properties of Materials*, Van Nostrand Reinhold, New York, 1971.

2.5. GOLDMAN, J. E., *The Science of Engineering Materials*, J. Wiley, New York, 1957.

2.6. BARRETT, C. R., W. D. NIX, and A. S. TETELMAN, *The Principles of Engineering Materials*, Prentice-Hall, Englewood Cliffs, NJ, 1973.

2.7. PASK, J. A., *An Atomistic Approach to the Nature and Properties of Materials*, J. Wiley, New York, 1967, p. 74.

2.8. FREUDENTHAL, A. M., *The Inelastic Behavior of Engineering Materials and Structures*, Wiley, New York, 1950, p. 93.

2.9. *ASTM Standards*, part 4: "Structural Steel [etc.]," Philadelphia, 1979, p. 19.

2.10. AISC, *Manual of Steel Construction*, 8th ed., Chicago, 1980.

2.11. COTTRELL, A. H., *Dislocations and Plastic Flow in Crystals*, Clarendon Press, Oxford, 1953.

2.12. ENGELHARDT, W., "Die Ursache der natürlichen Streckgrenze," *Die Technik*, vol. 7, no. 9, 1952, p. 515; vol. 7, no. 11, 1952, p. 659.

2.13. MASING, G., "Streckgrenze und Alterung bei weichem Stahl," *Archiv für Eisenhüttenwesen*, vol. 21, no. 9/10, 1950, pp. 315–23.

2.14. RANKINE, W. J. M., *Proc. Inst. Civ. Engrs.* (London), vol. 2, 1843, p. 105.

2.15. MCCONNEL, J. E., "On Railway Axles," *Proc. Inst. Mech. Engrs.*, 1847–1849, Oct. 24, 1849.

2.16. HEROLD, W., *Wechselfestigkeit metallischer Werkstoffe*, Vienna, 1934, p. 91.

2.17. GURNEY, R. T., *Fatigue of Welded Structures*, Cambridge University Press, Cambridge, 1968, p. 1.

2.18. MUNSE, W. H., and L. H. GROVER, *Fatigue of Welded Steel Structures*, Welding Research Council, New York, 1964.

2.19. HALL, D. H., and I. M. VIEST, "Design of Steel Structures for Fatigue," *Proc. ASCE Nat. Meeting on Water Resources Eng.*, New York, Oct. 16–20, 1967, p. 2.

2.20. TIMOSHENKO, S. P., *Istoriya Nauki O Soprotivlenii Materialov*, Gosizdat, Moscow, 1957, p. 205.

2.21. SCHLEICHER, F., *Taschenbuch für Bauingenieure*, 2nd ed., Springer, Berlin, 1959, vol. 1, p. 578.

2.22. LIN, T. H., and Y. M. ITO, "Mechanics of a Fatigue Crack Nucleation Mechanism," *J. Mech. Phys. Solids*, vol. 17, 1969, pp. 511–23.

2.23. ROŠ, M., and A. EICHINGER, "Die Bruchgefahr fester Körper bei wiederholter Beanspruchung," *EMPA Bericht* (Zurich), no. 173, 1950, pp. 3–5.

2.24. ROŠ, M., and A. EICHINGER, "Festigkeitseigenschaften der Stähle bei hohen Temperaturen," *Diskussion Berichte EMPA* (Zurich), no. 87, April 1934; no. 138, Nov. 1941.

2.25. Ibid. (2.24), pp. 10, 108.

2.26. BARSOM, J. M., "Fatigue Behavior of Pressure-vessel Steels," *Welding Res. Counc. Bull.*, no. 194, May 1974.

2.27. ROLFE, S. T., and J. M. BARSOM, *Fracture and Fatigue Control in Structures: Applications of Fracture Mechanics*, Prentice-Hall, Englewood Cliffs, N.J, 1976.

2.28. TIPPER, C. T., *The Brittle Fracture Story*, Cambridge University Press, Cambridge, 1962, pp. 12, 21.

2.29. SHANK, M. E., *A Critical Survey of Brittle Failure in Carbon Plate Steel Structures, Other Than Ships*, prepared for the Ship Structure Committee, Dept. of the Navy, Bureau of Ships, Washington, DC, Dec. 1, 1953.

2.30. Boyd, A. M., *Conf. on Brittle Fracture*, Report no. P.3, Cambridge, H.M.S. Office, London, 1959.

2.31. Klöppel, K., "Gemeinschaftsversuche zur Bestimmung der Schwellzugfestigkeit voller, gelochter, und genieteter Stäbe aus St 37 and St 52," *Stahlbau*, vol. 9, 1936, p. 97.

2.32. Merritt, F. S., *Eng. News Rec.*, vol. 146, no. 6, Feb. 8, 1951, pp. 23–24; no. 7, Feb. 15, 1951, p. 21.

2.33. Klöppel, K., "Sicherheit and Güteanforderungen bei geschweissten Konstruktionen," *Stahlbau*, 2nd ed., Stahlbau, Cologne 1971, vol. 1, pp. 61–93.

2.34. Prandtl, L., *Verhandlungen Deutscher Naturforscher und Ärzte*, Dresden, 1907.

2.35. Felbeck, D. K., and E. Orowan, *Am. Weld. J.*, vol. 34, 1955, p. 570.

2.36. Rolfe and Barsom, *Fracture and Control* (2.27).

2.37. Davidenkov, N. N., and N. T. Spiridonova, "Analiz napryazhennogo sostayaniya v sheike rastyanutogo obraztsa," *Zhurn. Zavodskaya laboratoriya*, vol. 11, Moscow, 1945, p. 6.

2.38. MacGregor, C. W., *Relations Between Stress and Reduction in Area for Tensile Tests of Metals*, AIME Tech. Pub. no. 805, New York, 1937.

2.39. Neuber, H., *Kerbsspannungslehre*, 2nd ed., Springer, Berlin, 1958.

2.40. Hill, R., *Matematicheskaia Teoria Plastichnosti*, Gosizdat, Moscow, 1956, pp. 281–90.

2.41. Hort, H., G. I. Taylor, W. S. Farren, and H. Quinney, *Proc. Roy. Soc.* (London), ser. A, vol. 107, 1925, p. 422; vol. 143, 1934, p. 307; vol. 163, 1937, p. 157.

2.42. "Conference on Brittle Fracture in Steel," *J. West of Scotland Iron and Steel Inst.*, vol. 60, 1953, p. 143.

2.43. Szczepanski, M., *The Brittleness of Steel*, J. Wiley, New York, p. 22.

2.44. Griffith, A. A., "The Phenomena of Rupture and Flow in Solids," *Phil. Trans. Roy. Soc.* (London), vol. 221, ser. A, 1921, p. 163.

2.45. Irwin, G., "Fracture of Metals," *Symp. Am. Soc. Metals*, Chicago, 1947, p. 147.

2.46. Orowan, E., ed., *Proceedings of the Symposium on Fatigue and Fracture of Metals*, M.I.T. Press and J. Wiley, New York, 1950, p. 139.

2.47. Rolfe, S. T., "Fracture Mechanics in Bridge Design," *Civ. Eng. ASCE*, vol. 42, no. 8, August 1972, pp. 37–41.

2.48. *AASHTO Material Specifications*, rev. ed., Washington, DC, 1973.

2.49. *Proc. ASCE Speciality Conf. on Safety and Reliability of Metal Structures*, Pittsburgh, Nov. 2, 3, 1972.

2.50. Liebowitz, H., *Fracture: An Advanced Treatise*, Academic Press, New York, 1969, vol. 1: *Microscopic and Macroscopic Fundamentals;* vol. 2: *Mathematical Fundamentals;* vol. 3: *Engineering Fundamentals and Environmental Effects;* vol. 4: *Engineering Fracture Design;* vol. 5: *Fracture Design of Structures*, vol. 6: *Fracture of Metals;* vol. 7: *Fracture of Nonmetals and Composites*.

2.51. Parker, E. R., *Brittle Behavior of Engineering Structures*, J. Wiley, New York, 1957.

2.52. Boyd, G. M., ed., *Brittle Fracture in Steel Structures*, Butterworths, London, 1970.

2.53. Hall, W. J., H. Kihara, W. Soete, and A. A. Wells, *Brittle Fracture of Welded Plate*, Prentice-Hall, Englewood Cliffs, NJ, 1967.

2.54. Kanazawa, T., and A. S. Kobayashi, eds., *Significance of Defects in Welded Structures*, Proceedings of the Japan-U.S, Seminar, 1973, Tokyo, University of Tokyo Press, Tokyo, 1974.

2.55. Pellini, W. S., and P. P. Puzak, *Fracture Analysis Diagram Procedures for the Fracturesafe Engineering Design of Steel Structures*, Welding Res. Counc. Bulletin no. 88, New York, May 1963, p. 28.

2.56. Rolfe, S. T., "The New AASHTO Material Toughness Requirements," *Proc. ASCE Speciality Conf. on Metal Bridges*, St. Louis, Nov. 12, 13, 1974, pp. 156–63.

2.57. Adams, F. S., "Design of Highly Restrained Joints," *Proc. 18th Ann. Struc. Eng. Conf.*, University of Kansas, March 1973, pp. 1–21.

2.58. "Commentary on Highly Restrained Welded Connections," *AISC Eng. J.*, vol. 10, no. 3, 1973, pp. 61–73.

2.59. Burns, R. M., and W. W. Bradley, *Protective Coating for Metals*, 3rd ed., Van Nostrand Reinhold, New York, 1967.

2.60. Larrabee, C. P., *Corrosion*, vol. 9, 1953, p. 259.

2.61. Hayes, J. M., and S. P. Maggard, *Economic Possibilities of Corrosion-Resistant Low-alloy Steel in Highway Bridges*, ASCE Struct. Eng. Conf. (Miami, 1966) reprint no. 266, New York, 1966.

2.62. Richards, C. W., *Engineering Materials Science*, Wadsworth, San Francisco, 1961, p. 499.

3

Behavior of Structural Steel Members

3.1 STRUCTURAL BEHAVIOR OF MEMBERS

Depending upon the type of loading acting on them, structural steel members are divided into

1. Tension members or ties
2. Compression members or columns
3. Flexural members or beams, and
4. Combined compression and bending members or beam-columns

In the first two categories forces are acting along the member axis, while in the case of beams they act in a transverse direction, and in the case of beam-columns they act in both directions.

For each category of structural members in this chapter, the main characteristics of stress state, the design criteria, and the types of cross section of the individual members will be discussed. For all members, cross-sectional dimensions are small compared with their length.

3.1.1 Tension Members

A tension member is a basic structural element with a nominal uniform cross-sectional stress field. As was shown in Section 1.2.3, local stress concentrations are eliminated as a result of the local yielding of steel and when full plastification of each cross-section is reached, the stress field actually becomes uniform. Structurally the first requirement is that the cross-sectional area of tension members be sufficiently large to carry the design load with an appropriate safety margin. According to AISC specification (3.1) the stress actually should not exceed $0.60F_y$ on the gross area or $0.50F_u$ on the effective net area in working load design. In plastic design maximum stress is yield-point stress, F_y. For the design of tension members, see Chapter 4. The second requirement is that they be sufficiently stiff to prevent excessive sagging or harmful vibration.

It is assumed that the force action which has to be transmitted from one end of a tension member to the other coincides with its centroidal axis. This is only a working hypothesis which simplifies the design, because some eccentricity in the force action will always exist. Such eccentricity can be introduced either by the application of the force itself or by the initial curvature of the member. The first eccentricity is caused by the difference between the centroidal axis of the end connections and that of the member. If the section of a member is composed of one or two angles, this eccentricity can be disregarded (Fig. 3.1). Again, as with other stress concentrations, the plastic

Figure 3.1

behavior of steel, which is acting before any actual strength collapse of tension members takes place, eliminates the initial eccentricity over almost the whole length of the member (Fig. 3.2) and the moment of eccentricity has a limited local effect. Therefore, the AISC specification (3.1), in section 1.15.3, Placement of Rivets, Bolts, and Welds (p. 5-46), allows the designer to neglect the eccentricity between the centroidal axis of such members and the gage lines of their end connections for static loads. Caution has to be exercised if dynamic loading with repeated variation in stress is expected.

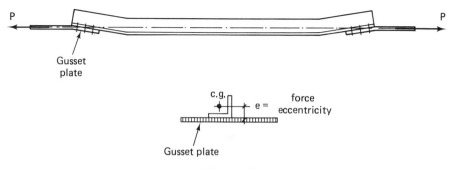

Figure 3.2

Generally speaking, there are four types of tension members

1. Wire ropes or cables
2. Eye bars or pin-connected plates
3. Rods and bars
4. Structural shapes

Only the last type will be discussed in this book, because cables are used only for special structures not discussed here; types 2 and 3 are outdated and are no longer used.

Simple cross sections

Tension members are mainly used as truss members. As long as the forces to be transmitted and the available space allows, these members are composed of simple rolled sections. When the forces become too large, built-up sections are more economical because their end-connections at truss joints are easier.

Figure 3.3 shows some simple cross sections. Single (a) or double angles, in a pair (b) or in a star (c) configuration, are usually sections of truss members with smaller forces such as in roof trusses. Structural tees (d) are often used for diagonals in lateral bracings of bridge trusses. Wide-flange sections (e) and multi-angle I-sections (f) are used for either diagonals or verticals (hangers) of bridge trusses. The dotted line in Fig. 3.3 represents the axis of the gusset plates of the end connection. Sections (a), (b), (c), and (d) in this figure are called single-plane sections and those in (e) and (f) double-plane sections. The magnitude of the force in the members will determine if a single- or double-plane section should be chosen. When designing a truss, the feasi-

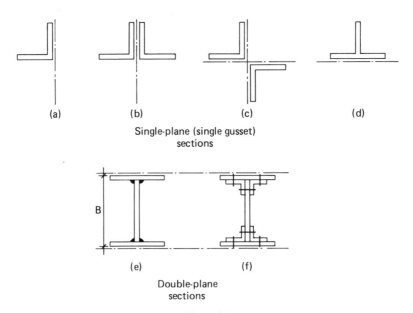

Figure 3.3

bility of accommodating the various end connections in the joints of a truss is very important. A structure is no better than its connections. For this reason, the same consideration for connection design must be given as for structural members. The performance of a structure under earthquake or severe wind loading depends mainly upon the end connections of the frame and bracing members, provided that the members are properly designed. The stress flow from the member to the joint and through the connection should be disturbed as little as possible. In general, single-plane sections are easier to handle, have smaller surfaces, and statically are easier to understand. Although for statical reasons there is no need to design tension members with stiff cross sections, it is not recommended that flat plates be used as tension members.

Built-up cross sections

If a cross section has at least one centroidal axis not passing through the material, it is said that this is a built-up section. In Figure 3.4a, b, and c the vertical axis is a nonmaterial axis, while the next part (Fig. 3.4d) shows a section with two such axes.

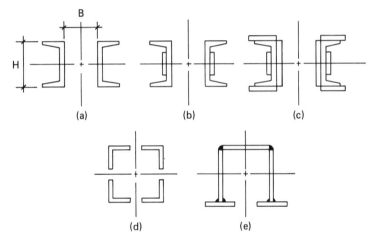

Figure 3.4

The sequence of the three sections given in Fig. 3.4a, b, and c shows how such sections can be strengthened in a simple way by the addition of vertical and horizontal plates. For trusses with straight chords it is of importance that the positions of the center of gravity of all the chord members coincide as closely as possible. For the case illustrated in Fig. 3.4a, b, and c, a zero change was achieved.

Double-plane sections are usual for built-up members. For truss bridges they are used for spans larger than 100 ft (about 30 m). Single-plane sections are used up to a cross-sectional area of about 100 in.^{2} (around 650 cm^2). If a larger cross section is needed for the truss member, double-panel built-up members are used. It is considered that an area of about 1,550 in.^{2} (10,000 cm^2) is a practical maximum that still can be achieved reasonably.

The main dimensions of a built-up section are the height (depth) of the chord H (Fig. 3.4) and the width of the "box," which is the distance between the walls, B.

A good depth H for the chord cross section for single- or double-panel sections is given by the following practical formula, in which l is the span length in feet or meters.

$$H_{in} = 0.12 l_{ft} - \frac{l_{ft}^2}{11{,}000} \quad \text{or} \quad H_{cm} = l_m - \frac{l_m^2}{400} \tag{3.1}$$

Similarly, for the width B of trusses of medium span

$$B_{in} = H_{in} - 0.012 l_{ft} \quad \text{or} \quad B_{cm} = H_{cm} - 0.1 l_m \tag{3.2}$$

and for trusses of larger spans, where l is again the span length and H is the depth obtained from Eq. (3.1).

$$B_{in} = H_{in} - 0.0025 l_{ft} \quad \text{or} \quad B_{cm} = H_{cm} - 0.2 l_m \tag{3.3}$$

3.1.2 Compression Members

When an axial compressive force is acting on a structural member, a compressive uniform field stress is created in any of its perpendicular cross sections. A short compressive member is called a strut: a longer one that can develop buckling problems is called a column.

For smaller forces and shorter buckling lengths, simple or single cross sections (Fig. 3.5) can be chosen. Built-up or double sections (Fig. 3.6) must be selected when either forces and/or buckling lengths become substantial or when local conditions require.

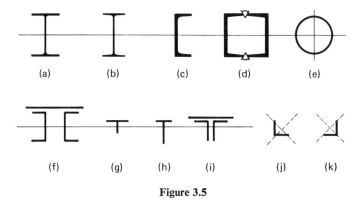

Figure 3.5

Usually, columns are prismatic members, i.e., the cross section perpendicular to the column axis is constant. As stated before, cross-sectional dimensions are small compared to the column length.

For any column cross-section there are two mutually perpendicular principal axes, the maximum and the minimum one. With adequate bracing of the column in these two principal planes whenever possible, the column properties can be equalized so as to prevent a penalty being paid for buckling about a "weak" axis. When differential bracing is impossible, in order to have an economical design there should be as little difference as possible in the column buckling strength in the two orthogonal principal directions.

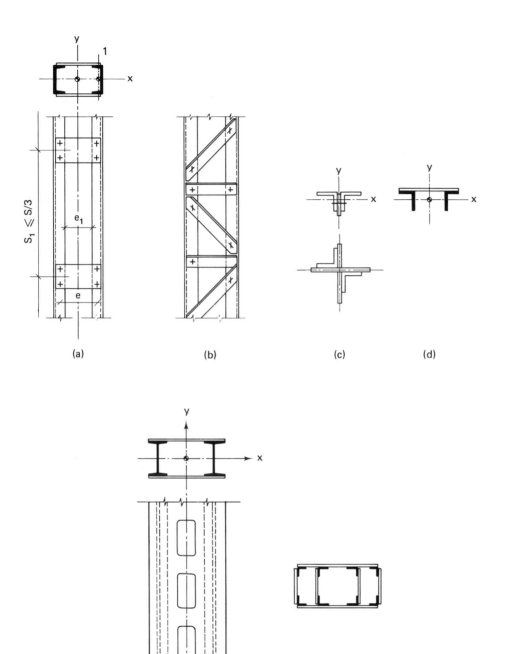

Figure 3.6

Simple cross sections

Compressive members must be designed mainly from stability considerations, because economy usually excludes columns for which buckling is no longer a design criterion. On the basis of stability the design of the cross section will tend to produce a column for which the radii of gyration with respect to both principal axes are balanced. It is clear that among the cross sections shown in Fig. 3.5 only (d) and (e), i.e., tube-like sections, have this characteristic. Wide flange sections, either rolled or welded, as shown in (a), approach this goal if sizes 12 and 14 are chosen. Other sections can be useful if for some reasons the buckling lengths for the two principal directions are different. Structural tees and angle sections (g, h, i, j, and k) are used mainly for bracing members in buildings or bridges. Tubular sections (d, e) together with wide-flange sections (a) are often used for building columns. Tubular sections, when hollow (without concrete filling), can be used as ducts (for electricity, heating, air conditioning, water, and sewage systems), which can result in sizable space savings. In some buildings column pipes are filled with water to reduce the hazard of fire. If sections such as shown in Fig. 3.5d and e are filled with concrete, provided certain requirements are satisfied (3.2), the composite action of steel and concrete can be taken into account.

Compression members of the cross-sectional type shown in Fig. 3.6c are connected by separators (see Section 5.1) to avoid damage during transportation and also to ensure that such members act as a unit and can sustain wind and vibration effects.

Built-up cross sections

In built-up sections it is fairly easy not only to produce equal radii of gyration about both the principal axes, but also to have a radius of gyration about the nonmaterial (open) axis 10–20% larger than about the other axis. For instance, the British Specifications BS-449 (3.3) for laced columns require that a radius of gyration about the axis perpendicular to the plane of the lacing (the nonmaterial axis) be not less than the radius of gyration at right angles to that axis. For battened columns they suggest that the ratio of the slendernesses about the open axis (or axis perpendicular to the battens) should be at most 80% of the slenderness about the axis at right angles to it (the material or closed axis) so as to achieve a maximum spacing of battens.

Two or more sections (Fig. 3.6e and f) of a built-up section must be "tied" together. This can be done using lacing (b), stay plates or battens (a), or perforated cover plates (e).

In the past, lacing was predominantly used on the open sides of compression built-up members. The AISC Specification (3.1, Section 1.18, Built-up Members) gives guidance for the design of lacing members (flat bars, angles, channels, or other shapes).

Battened columns are used very extensively worldwide, owing to their economy and ease in fabrication. At present, American building and bridge specifications have no specific regulations for their design. Battened columns are already used extensively in this country and it is expected that their use will increase. Perforated cover plates are usually heavy and their use is only economical for very high compressive forces and long buckling lengths, as is the case in pylons of bridges.

In deciding the distance e (Fig. 3.6a) it is a good design practice to have $I_y \geq 1.10 I_x$ and a distance e_1 sufficient for accessibility inside the column for fabrication, repair, and maintenance.

3.1.3 Flexural Members

A structural member which is exposed to loads acting mainly perpendicular to its axis, with cross-sectional dimensions small in comparison with its member length, is called a beam.

Beams can be fabricated using rolled sections with or without cover plates (Fig. 3.7a, b). In buildings, open-web joists are very economical as roof or floor beams (Fig. 3.8). When the dimensions of a required beam section are beyond those available in rolled sections, plate girders (Fig. 3.9) are used.

If reinforced concrete slabs overlaying steel beams are acting with such steel beams or girders (Fig. 3.10), by employment of special mechanical connectors (in this case studs), both materials are forced to perform in a composite action under certain

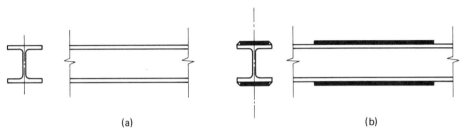

Figure 3.7

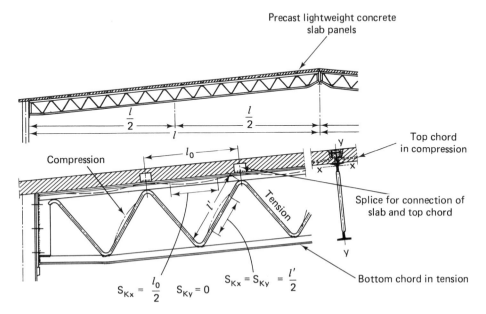

Figure 3.8

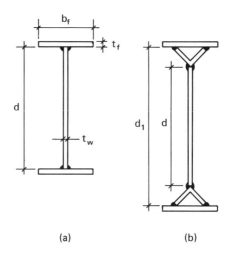

Figure 3.9

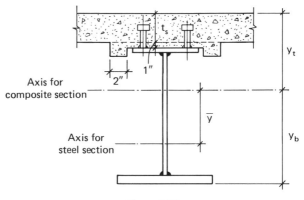

Figure 3.10

loading conditions. This usually proves to be economical, because owing to fairly large slab dimensions, the section modulus of a composite section is quite large and the moment capacity of the girder is greatly improved.

Beams in general have to satisfy strength, stability, and stiffness requirements. For rolled sections the dimensions are such that local stability as well as shear strength conditions are normally satisfied. Details of beam design are discussed in Chapter 7 under the heading of Bending of Beams.

Rolled beams

The most economical beams are rolled sections. Fabrication work consists mainly in cutting to size and installing bearing stiffeners and bearing plates at the supports (Fig. 3.11). If a beam is part of a floor system, connecting angles are provided, usually at the ends of the beam. Figure 3.12 shows several types of connections between rolled beams in floor systems. Which one is chosen depends upon relative height posi-

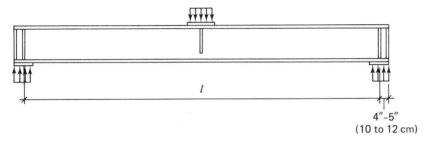

Figure 3.11

tions and the need for continuity. Figure 3.12a shows "shear" connections between two beams (stringer and cross or floor beams). "Rigid" connections are shown in Fig. 3.12b. A tie plate over the top of the beams provides continuity between the connected beams. In Fig 3.12c and d, two beam-to-column web connections are presented. In the first figure (c) the column web is not cut through for a continuity plate because the precision needed in its fabrication and erection will upset any material gain due to the reduced field moment. Instead, continuity is established using two short pieces cut out of a wide flange beam of a slightly smaller section than the column and connected to the top flanges of the beam at each beam end and to the column flanges from the inside. In the second figure (d), the column is interrupted at the beam flanges and a tie plate is used. The top column, which has a different section, is connected with a bearing plate on top of the tie plate.

Open-web joists

Open-web steel joists (Fig. 3.8 and 3.13) are used for floor and roof structures. These are open web parallel chord load-carrying members utilizing hot-rolled (Fig. 3.8) or cold-formed steel (Fig. 3.13). According to the steel quality they are classified as J-series when built of material with a yield point of 36 ksi (around 250 MPa) and H-series when fabricated with a yield point of 50 ksi (345 MPa). The design of the J- and H-series is regulated by the American Institute of Steel Construction (3.5). Longspan steel joists (3.6) and deep longspan steel joists (3.7) are specified by the AISC together with the Steel Joist Institute.

Joists are usually designed as simply supported uniformly loaded trusses with a roof or floor deck so constructed as to brace the top chord of the joists against lateral buckling. Normally the clear span of joists does not exceed 24 times their nominal depth (the depth at midspan). The clear spans covered by open web joists range from 8 to 128 ft. For roof trusses a standard top chord pitch of 1/8 in. per foot (about 1%) is required.

The diagonals of open-web joints are made of one continuous round bar acting alternatively in tension and compression. The German practice in calculating the buckling length for R-joists (round bar joists) is shown in Fig. 3.8, where S_{kx} and S_{ky} are the effective buckling lengths, which are half the panel length for the top chord member and half the diagonal length for the diagonals. The American specification gives various formulas to calculate the allowable compressive stress for corresponding slenderness ratios. In these slenderness calculations either the unbraced length is taken

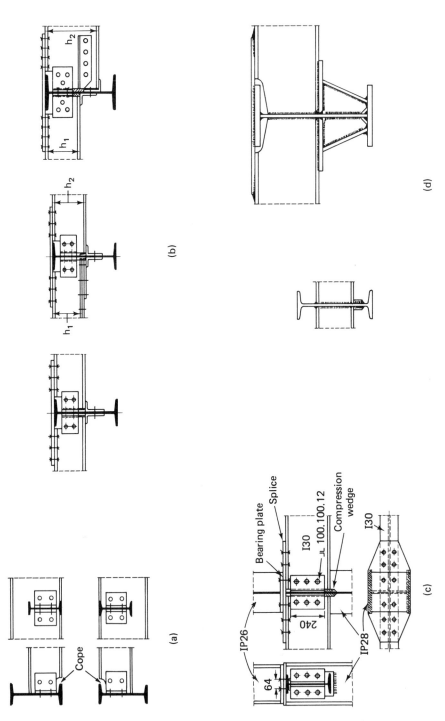

Figure 3.12

Dimensions and Properties

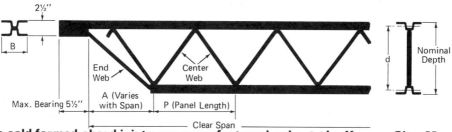

These cold formed chord joists are manufactured only at the Kansas City, Mo. plant.

Joist Designation ***	Nominal Depth	Effective Depth "d"	Section Number		A Varies With Span		Bar Web Diameter		B Bearing Width	P Panel Length	Approx. Weight per ft. *	Joist Moment of Inertia **
			Top Chord	Bottom Chord	Min.	Max.	End	Center				
J-Series	in.	in.			in.	in.	in.	in.	in.	in.	lbs.	in.⁴
8 J 2	8	7.06	2T	2B	20	31	13/16	5/8	3 5/8	24	4.2	13
10 J 2	10	9.06	2T	2B	20	31	13/16	5/8	3 5/8	24	4.2	21
10 J 3	10	9.10	3T	3B	20	31	13/16	5/8	3 5/8	24	5.0	22
10 J 4	10	9.14	4T	4B	20	31	13/16	5/8	4	24	6.1	26
12 J 2	12	11.06	2T	2B	19	30	13/16	5/8	3 5/8	24	4.5	31
12 J 3	12	11.10	3T	3B	19	30	13/16	5/8	3 5/8	24	5.2	34
12 J 4	12	11.14	4T	4B	19	30	13/16	11/16	4	24	6.2	39
12 J 5	12	11.15	5X	5B	19	30	13/16	11/16	4 7/8	24	7.1	47
12 J 6	12	11.13	6X	6D	19	30	13/16	11/16	5 3/8	24	8.2	56
14 J 3	14	13.10	3T	3B	18	29	13/16	5/8	3 5/8	24	5.5	46
14 J 4	14	13.14	4T	4B	18	29	13/16	11/16	4	24	6.5	54
14 J 5	14	13.15	5X	5B	18	29	13/16	11/16	4 7/8	24	7.4	65
14 J 6	14	13.13	6X	6B	18	29	13/16	25/32	5 3/8	24	8.6	77
14 J 7	14	13.24	7T	7B	18	29	13/16	3/4	5 3/8	24	10.0	94
16 J 4	16	15.14	4T	4B	19	30	13/16	11/16	4	24	6.6	72
16 J 5	16	15.14	5T	5B	19	30	13/16	25/32	4 7/8	24	7.8	85
16 J 6	16	15.12	6T	6B	19	30	13/16	25/32	5 3/8	24	8.6	101
16 J 7	16	15.24	7T	7B	19	30	13/16	3/4	5 3/8	24	10.4	125
16 J 8	16	15.24	8T	8B	19	30	13/16	25/32	5 3/8	24	11.6	141
18 J 5	18	17.14	5T	5B	21	32	13/16	25/32	4 7/8	24	8.0	109
18 J 6	18	17.12	6T	6B	21	32	13/16	3/4	5 3/8	24	9.2	130
18 J 7	18	17.24	7T	7B	21	32	13/16	25/32	5 3/8	24	10.4	160
18 J 8	18	17.24	8T	8B	21	32	13/16	25/32	5 3/8	24	11.6	181
20 J 5	20	19.14	5T	5B	23	34	13/16	3/4	4 7/8	24	8.4	136
20 J 6	20	19.12	6T	6B	23	34	13/16	25/32	5 3/8	24	9.6	162
20 J 7	20	19.24	7T	7B	23	34	13/16	25/32	5 3/8	24	10.7	199
20 J 8	20	19.24	8T	8B	23	34	13/16	13/16	5 3/8	24	12.2	225
22 J 6	22	21.12	6T	6B	26	37	13/16	25/32	5 3/8	24	9.7	197
22 J 7	22	21.24	7T	7B	26	37	13/16	13/16	5 3/8	24	10.7	243
22 J 8	22	21.24	8T	8B	26	37	13/16	13/16	5 3/8	24	12.0	274
24 J 6	24	23.12	6T	6B	28	39	13/16	13/16	5 3/8	24	10.3	237
24 J 7	24	23.24	7T	7B	28	39	13/32	5 3/8		24	11.5	290
24 J 8	24	23.24	8T	8B	28	39	13/16	27/32	5 3/8	24	12.7	328

*The weights per foot as shown in these tables are approximate only. Such weights are shown only for the convenience of the designer. They cannot be used in figuring prices or determining shipping weights.
**See page 7 for computation of deflections due to uniform loading.
***Steel for J-Series joist chords and webs is tested in accordance with "Section 3. Materials" of the SJI Specifications for Open Web Steel Joists, J- and H-Series (36,000 psi, minimum yield strength).

Chord properties

Sec. No.	Area sq. in.	t in.	D in.	y in.	w in.	X-X Axis		Y-Y Axis		Sec. No.
						I	r	I	r	
2T	.500	.125	1.125	.47	2 13/16	.08	.40	.28	.75	2T
3T	.562	.125	1.125	.43	3 3/16	.09	.40	.43	.87	3T
4T	.712	.156	1.156	.43	3 7/16	.11	.40	.59	.91	4T
5T	.843	.188	1.188	.44	3 7/16	.14	.40	.71	.92	5T
5X	.878	.188	1.188	.43	3 5/8	.14	.40	.82	.97	5X
6T	.997	.219	1.219	.45	3 5/8	.16	.41	.93	.96	6T
6X	1.025	.219	1.219	.44	3 3/4	.17	.40	1.02	1.00	6X
7T	1.243	.219	1.219	.38	4 3/4	.19	.39	2.01	1.27	7T
8T	1.406	.250	1.250	.39	4 13/16	.21	.39	2.36	1.29	8T
2B	.500	.125	1.125	.47	2 13/16	.08	.40	.28	.75	2B
3B	.500	.125	1.125	.47	2 13/16	.08	.40	.28	.75	3B
4B	.554	.125	1.125	.43	3 1/4	.09	.40	.41	.86	4B
5B	.659	.141	1.141	.42	3 5/16	.10	.40	.58	.94	5B
6B	.795	.172	1.172	.43	3 5/8	.13	.40	.70	.94	6B
7B	.945	.172	1.172	.38	4 7/16	.14	.39	1.31	1.18	7B
8B	1.066	.188	1.188	.37	4 3/8	.16	.39	1.63	1.24	8B

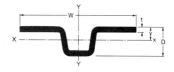

Figure 3.13 (From *Armco Steel Joists*, p. 12. Courtesy Armco Steel Corporation. No longer commercially available.)

as the clear distance between attachments for chord and web members divided by the least radius of gyration or the buckling length is taken as the distance between panel points for chord members and the unbraced length, clear of attachments, for web members, together with the corresponding least radius of gyration.

Plate girders

As pointed out earlier, when the required bending capacity cannot be supplied by existing wide flange beams, the designer has to design a unique three-plate beam by welding plates together (Fig. 3.9).

In designing a welded plate girder the following rules must be obeyed (3.4, p. 58).

1. Discontinuous welds in dynamically stressed girders should be avoided because end points of welds, apart from the bad quality of welds (craters), are normally starting points of fatigue cracks.
2. Avoid cross-cutting of welds, because such points are sharp stress-raisers.
3. In a tension flange, welds perpendicular to it should be avoided if possible. If such cross-welds are impossible to avoid, the smallest permitted size of fillet welds should be used.
4. There should be always some transition section when going from a stronger to a weaker section or vice versa.
5. The number of welds, as well as the number of web or flange joints, should be kept at a minimum.
6. Maximum plate lengths should be used when fabricating, in units which still can be transported without costly special measures, even if this involves paying additional cost to the steel mills for abnormal lengths.
7. To reduce shrinkage stresses to a minimum, the order of welding should be such that first the webs are jointed together, if they are made of several pieces, then the web stiffeners are welded, and last the flanges are welded to the web plates.
8. As the safety against cracking of welded flanges depends greatly upon the plate thickness, this thickness has to be kept to a minimum. Up to 3/4-in. (20-mm) thickness poses no problems. For greater thicknesses (especially in tensioned parts of flanges and webs) killed steel should be used or instead of one thick plate several thinner plates fillet-welded together should be used. The actual cost of better-quality steel and one thick flange as compared to using normal steel quality with several thinner plates will decide which direction to go.

If the height of a girder is not limited by specific constraints, the optimum height (see Section 10.1.3) will be chosen. Because deflection limitations are not as stringent as in the past, and because the dead load deflections can be completely compensated for by camber, a welded plate girder with the web height of $l/15$ (where l = span length) for simple beams and of $l/25$ for continuous beams can be used quite economically. These measurements can still further be reduced if a box girder is used (Fig. 3.14).

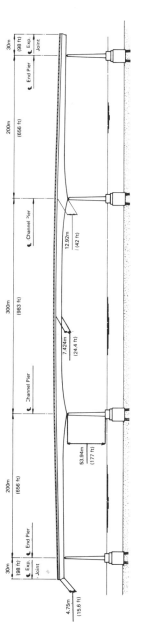

ELEVATION OF MAIN SPANS

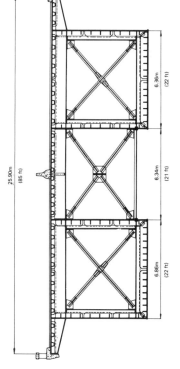

CROSS SECTION

Figure 3.14 Rio-Niteroi Bridge, World Record Orthotropic Deck Box Girder (From American Society of Civil Engineers, *Journal of the Structural Division*, Sept. 1972).

The web thickness must be sufficient to withstand the shear force and have also a sufficient moment of inertia to match that of the flanges and allow a satisfactory connection by fillet welds to the web.

Several typical cross sections are shown in Fig. 3.15, with and without special flange sections.

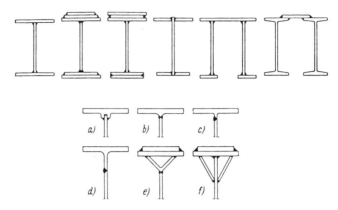

Figure 3.15 (From *Stahlbau*, Vol. 2, p. 111. Courtesy Stahlbau-Verlags GmbH.)

Composite beams

Reinforced concrete slabs resting on steel beams or girders have long been used as floor decks in buildings or bridges. Initially, the design concept was that each element, the slab or the beam, is acting separately under the loads (Fig. 3.16) with each element bent about its own neutral axis. In reality, owing to the friction and some bond between the slab and top flange under a positive moment, even if no mechanical connectors are provided, both elements, the slab and the beam, behave like one unit. The reinforced concrete slab is actually working as a flange cover plate of the steel member (Fig. 3.10). Composite construction started in 1914 in Switzerland (3.8). Also, a highway bridge floor system was designed and built as a composite system in the same country as early as 1920. The first intentional use of composite construction in buildings was in 1935 on a highway bridge near Rüdersdorf in Germany (3.1), where a reinforced concrete slab was poured on steel beams without any connectors. The slab

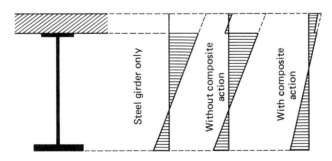

Figure 3.16 Composite Girder-System.

112   BEHAVIOR OF STRUCTURAL STEEL MEMBERS

was also subdivided in regular intervals by joints and only at midspan was it connected to steel beams with one single connector to enable the transfer of longitudinal horizontal movable forces (braking and inertia forces). Deflection studies revealed that composite action still took place. The next logical step was to enhance this composite action by using specially designed mechanical shear connectors (Fig. 3.10).

During the 1932 Congress of the International Association for Bridge and Structural Engineering (IABSE) in Paris, a separate discussion session was held on composite design. The first large bridge built in composite construction was the bridge across the River Sava in Zagreb in Yugoslavia, built in 1938-39 (3.10). Figure 3.17 shows the results of test measurements taken on this four-span, continuous girder bridge. For comparison, the values calculated without composite action are also shown. The first extensive use of composite girders took place mainly in Germany after World War Two (3.11, 3.12).

Composite action is based on the assumption that the horizontal shear forces, acting in the contact plane between the steel beam and concrete slab, can be trans-

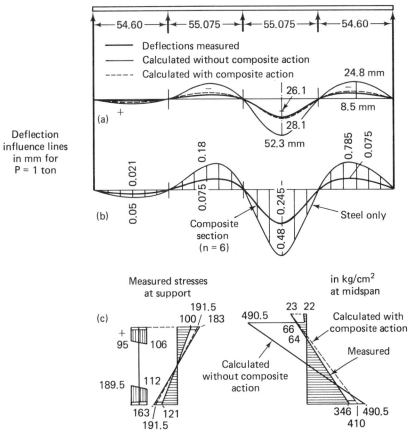

Figure 3.17 Bridge Across Sava River in Zagreb, Yugoslavia. The First Larger Composite Highway Bridge. (From F. Schleicher, *Taschenbuch für Bauingenieure,* 2nd ed., Springer, Berlin, 1959, Vol. 1, p. 716.)

Structural Behavior of Members

mitted. Normally, the concrete slab is placed directly on the top flange of the girder. If special shear connectors are used (Fig. 3.18), no relative slip is assumed to occur. This results in "rigid" composite action. The elastic deformation of the studs and the local deformation of both the concrete slab and the steel girder are not taken into account.

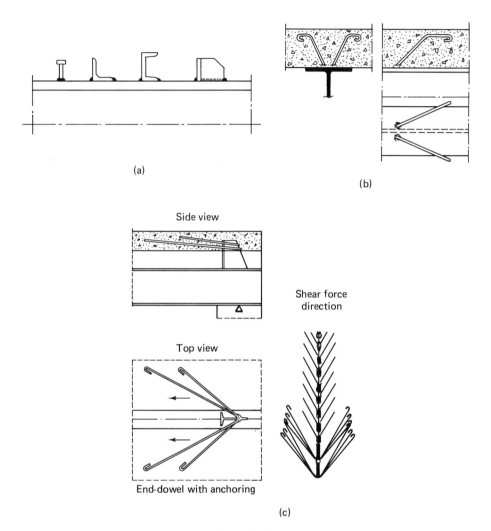

Figure 3.18

When rigid connectors are assumed, the horizontal shear forces transmitted by individual connectors are proportional to the vertical shear force and to the distance between the connectors. The values calculated on that assumption are never realized, owing to the fact that the connectors give way. If relative movements between the steel girder and the concrete slab occur, significant force redistribution takes place, which will result in large reductions of the extreme shear force values.

In trying to calculate stresses it should be remembered that for long-duration loading concrete shrinkage and creep will change the calculated values substantially. The best approach for composite design is limit design (see Section 10.3). Tedious calculations of stresses as time-dependent variables can then be reduced to checking of the stresses in the steel girders due to the long-duration loadings such as dead load weight at the working stress (service) level.

3.1.4 Members in Combined Bending and Axial Compression

Beam-columns are structural members which are subject to forces producing both bending and compression. Figure 3.19 shows beam-columns with and without transverse loads bent about one or two axes.

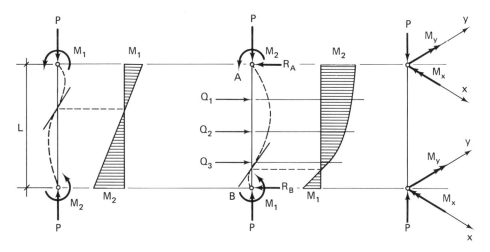

Figure 3.19

In recent years, beam-columns have been investigated extensively by many researchers. Most of this research has taken place in this country at the Fritz Engineering Laboratory of the Department of Civil Engineering at Lehigh University. Some recent references are given at the end of the chapter (3.13, 3.14, 3.15, 3.16, and 3.17).

Beam-columns represent a complicated nonlinear problem for analysis as the principle of superposition cannot be applied. So-called "second-order" effects of the deflected shape as well as elasto-plastic behavior of some parts of the column are complicating features of this problem.

Three major approaches in designing beam-columns have been developed. The oldest method applies superposition of stress without taking into account the effect of deflections due to bending, and is based on maximum stresses never reaching yielding. The second approach makes use of empirically determined and experimentally tested "interaction equations." This approach is by far the most popular, and is used in

American specifications [AREA (3.18), AISC (3.1), and AASHTO (3.20)], as well as in British [(BS 449 (3.3; p. 24)], Russian (3.21; p. 15), and German (DIN 4114) (3.19) specifications.

The third and most recent approach uses theoretically developed ultimate strength interaction equations. The AISC beam-column design rules for plastic design use this method. As a matter of fact, all three approaches are based on ultimate bearing capacity with the application of appropriate load factors.

The cross sections of beam-columns are similar to those used for columns. To avoid repetition, the reader is referred to Section 3.1.2 for the shape of cross sections and to Chapter 8 for their design.

3.1.5 Light-Gage Steel Members

The design of light-gage steel members composed of sections of cold-formed steel sheets or strips is no longer a temporary solution used only during periods when steel was scarce. It now represents a direction of development that has as a goal to design lighter structures which have the same safety as normal structures made of hot-rolled steel shapes and plates. Such development is based on the application of the latest knowledge in the fields of material sciences, strength of materials, and stability theory.

In the past, light-gage sections were used extensively in industry (for cars, planes, boats, and various types of equipment, etc.) but not for building construction (Fig. 3.20). This was the case in spite of the fact that the use of light-gage steel members in building construction began in about the 1850s in both the United States and England (3.22). In recent years, the use and the development of light-gage, cold-formed steel construction in this country has increased, as reflected in the publication of several specifications and design manuals (3.23, 3.24, 3.25, and 3.26).

Light-gage, cold-formed steel structural members provide several advantages in building construction. For instance:

1. More economical beams for relatively light loads and short span length are achieved.
2. Large strength-weight ratio can be obtained.
3. Load-carrying panels and decks provide economical floors and roofs, and they also act as shear diaphragms.

There are two major types (3.22) of light-gage, cold-formed steel members:

1. Individual structural members
2. Panels and decks (Fig. 3.21)

The type of buildings in which they are used are

1. Standardized single-story buildings either made entirely of cold-formed sections (fairly small buildings) or (for relatively larger buildings) with cold-

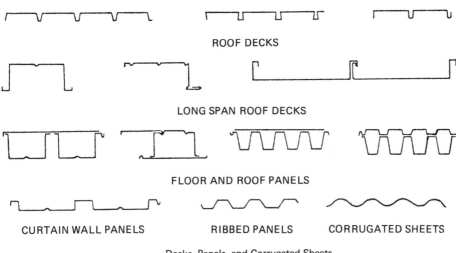

Decks, Panels, and Corrugated Sheets

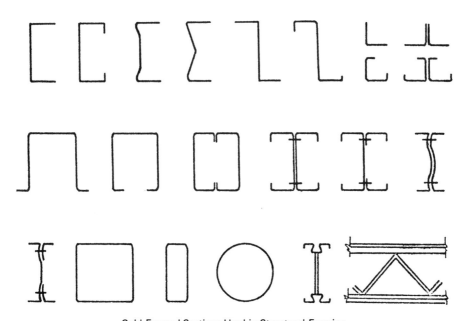

Cold-Formed Sections Used in Structural Framing

Figure 3.20 (From 3.22, pp. 3 and 5.)

Figure 3.21 Exterior Curtain Wall Panels Employing Corrugated Steel Sheets (From 3.22).

formed sections used for girts, purlins, roofs, and walls together with welded steel plate girder frames (Fig. 3.22)

Figure 3.22 Standardized Building Made Entirely of Cold-Formed Sections (From 3.22. Courtesy National Steel Products Company).

2. Steel roof decks used in folded plate or shell construction

The AISI specification (3.23) specifies six different structural qualities of steel sheet and strip to be used for light-gage, cold-formed steel members flush:

1. Steel Sheet, Zinc-coated (Galvanized) by the Hot-Dip Process, Structural (Physical) Quality, ASTM A446-72 (Grades A, B, C, D, and F)
2. Hot-Rolled Carbon Steel Sheet and Strip, Structural Quality, ASTM A570-79
3. Steel Sheet and Strip, Hot-Rolled and Cold-Rolled, High Strength, Low Alloy, with Improved Corrosion, Resistance, ASTM A606-75
4. Steel Sheet and Strip, Hot-Rolled and Cold-Rolled, High Strength, Low Alloy Columbian and/or Vanadium, ASTM A607-75
5. Steel, Cold-Rolled Sheet, Carbon, Structural, ASTM, A611-72 (1979) (Grades A, B, C, and D)
6. Sheet Steel and Strip, Hot-Rolled, High Strength, Low-Alloy, With Improved Formability, ASTM A717-75 (Grades 50 and 60)

The design of these members is mainly concerned with their stiffness and stability. Recent developments in research have made it possible to have reliable design methods that are in accordance with conventional methods of structural design. In this country, research (from 1939) was performed mainly at Cornell University (3.27). In Europe most of the research was done in England (3.28), Germany (3.29), and Switzerland (3.30).

NOTATIONS

b_f, t_f	= flange width and thickness	$B_{in, cm}$	= box width of double-panel truss member in inches or centimeters
d, t_w	= web depth and thickness		
e	= distance between centroidal axes of a built-up member	F_{all}	= allowable stress
		$H_{in, cm}$	= depth of chord member in inches or centimeters
e_1	= net inside width of an open section	I_x, I_y	= moment of inertia
h	= beam height or depth	S_{kx}, S_{ky}	= effective buckling length with respect to x- and y-axis, respectively
l_{ft}	= span length in feet		
l_m	= span length in meters		
n	= E_s/E_c = modular ratio		

REFERENCES

3.1. American Institute of Steel Construction, *Specification for the Design, Fabrication, and Erection of Structural Steel for Buildings*, Nov. 1978, New York, NY.

3.2. American Concrete Institute, *Building Code Requirements for Reinforced Concrete* (ACI 318–77), Section 10.14—Composite Compression Members, p. 37–8.

3.3. British Standard 449, *Specification for the Use of Structural Steel in Building*, British Standards Institution, 1965.

3.4. SCHAPER, G., *Feste Stählerne Brücken*, Wilhelm Ernst & Sohn, Berlin, 1934, p. 73.

3.5. *Standard Specifications and Load Tables for Open-Web Steel Joists*, American Institute of Steel Construction and Steel Joist Institute, March 1, 1965.

3.6. *Standard Specifications, Load Tables and Weight Tables*, Steel Joist Institute, 1981.

3.7. *Deep Longspan Steel Joists*, American Institute of Steel Construction and Steel Joist Institute, Feb. 1, 1970.

3.8. Roš, M., "Träger im Verbundbauweise," *EMPA—Bericht Nr.* 149, Zurich, 1944.

3.9. Tischer, "Autobahnbrücke bei Rüdersdorf," *Zeitschrift des Vereins Deutscher Ingenieure*, 1936, p. 205.

3.10. Erega, J., *Bauingenieur*, Vol. 22, 1941, p. 1.

3.11. Stahlverbund-Bauweise, Arbeitstagung, Hanover 1949 and 1950, *Der Bauingenieur*, Vol. 25, 1950, No. 3 and No. 8.

3.12. DIN 4239, Verbundträger-Hochbau, Sept. 1956 and DIN 1078, Verbundträger-Strassenbrücken, Sept. 1956; Subcommittee of the Task Committee on Composite Construction, *Composite Steel-Concrete Construction*, Journal of the Structural Division, ASCE, Vol. 100, No. ST-5, Proc. Paper 10561, May 1974.

3.13. Galambos, T. V., and J. Prasad, "Ultimate Strength Tables for Beam Columns," *Fritz Engineering Laboratory Report No. 287.3*, Lehigh University, Jan. 1962.

3.14. Lu, L. W. and H. Kamalvand, "Ultimate Strength of Laterally Loaded Columns," Journal of the Structural Division, ASCE, Vol. 94, No. ST-6, *Proc.*, Paper 6009, Jun. 1968, pp. 1505–23.

3.15. Chen, W. F., "General Solution of Inelastic Beam-Column Problems," *Journal of the Engineering Mechanics Division*, ASCE, Vol. 96, No. EM-4, Proc. Paper 7482, Aug. 1970, pp. 421–2.

3.16. Chen, W. F., "Simple Interaction Equations for Beam-Columns," *Journal of the Structural Division*, ASCE, Vo., 98, No. ST-7, Proc. Paper 9020, July 1972, pp. 1413–26.

3.17. Chen, W. F. and T. Atsuta, "Ultimate Strength of Biaxially Loaded Steel H-Columns," *Journal of the Structural Division*, ASCE, Vol. 99, No. ST-3, Proc. Paper 9613, Mar. 1973, pp. 469–89.

3.18. American Railway Engineering Association, *Manual for Railway Engineering*, Vol. 11, Chap. 15, Steel Structures, AREA, Chicago, 1980.

3.19. DIN 4114, "Berechnungsgrundlagen für Stabilitätsfälle in Stahlbau," (Knickung, Kippung, Beulung), July 1952, Section 10.

3.20. American Association of State Highway and Transportation Officials, *Standard Specifications for Highway Bridges*, 12th ed., Washington, DC, 1977, p. 189.

3.21. Gosudarstvennyĭ Komitet Sovyeta Ministrov USSR, "Stroitelynye normy i pravila," Part II. Section V, Chapter 3, Steel Structures, Moscow, 1963.

3.22. Yu, W. W., "Design of Light-Gage Cold-Formed Steel Structures," Engineering Experiment Station, West Virginia University, 1965, p. 1.

3.23. American Iron and Steel Institute, *Specification for the Design of Cold-Formed Steel Structural Members*, Washington, DC, 1980.

3.24. Winter, G., "Commentary on the 1968 Edition of the Specification for the Design of Cold-Formed Steel Structural Members," American Iron and Steel Institute, New York, 1970, pp. 151–71.

3.25. American Iron and Steel Institute, *Cold-Formed Steel Design Manual*, Washington, DC, 1977.

3.26. American Iron and Steel Institute, "Illustrative Examples, Based on the 1968 Edition of Specification for the Design of Cold-Formed Steel Structural Members," New York, 1973.

3.27. Winter, G., "Cold-Formed Light-Gage Steel Construction," *Journal of Structural Division*, ASCE, Proc. Vol. 85, No. ST-9, Nov. 1959, pp. 151–73.

3.28. Institution of Structural Engineers, "The Contribution of Research in Developing the Use of Cold-Rolled Sections and Buildings," *Symposium on Applications of Sheet and Strip Metals in Building*, London, 1959.

3.29. Jungbluth, O., "Stahlleichtbau," *Stahlbau, ein Handbuch für Studium und Praxis*, Vol. 2, Stahlbau-Verlag, Koln 1957, pp. 424–37.

3.30. Kollbrunner, C. F. and N. Hajdin, "*Wölbkrafttorsion Dünnwandiger Stäbe mit Offenem Profil*, Teil 1, Schweizer Stahlbau-Vereinigung, No. 29, 1964.

4

Design of Tension Members

4.1 CROSS SECTION DESIGN

The design of tension members is based on strength and stiffness criteria.

4.1.1 Strength-Based Design

In Section 3.1.1, Tension Members, it was pointed out that only tension members composed of structural shapes are treated in this text. Some simple cross sections as well as some built-up sections serving as tension members were discussed in connection with Fig. 3.3 and 3.4. It was also said that failure of such members under static loading might occur either by yielding of the gross section or by fracture in the net section. The stress distribution across a gross or net section at yielding or at tension failure usually is assumed uniform. However, it is important to consider the effect of discontinuities such as those that would occur at connections such as weldments, bolt holes, or sudden changes in section. Such discontinuities can lead to local failure and subsequent member failure due to localized high stresses. If local yielding occurs, a redistribution of stresses may take place, depending on the ductility of the material and particularly its notch toughness in the vicinity of discontinuities. In view of the above it is desirable for the designer to consider the load distribution or stress flow through the member, including its connections, so as to accommodate a smooth transition of load.

In the working (service) load design method of the AISC for static loading, for both types of failure, the same basic factor of safety of $\frac{3}{5}$ is applied to the yield stress, F_y, and the tensile strength, F_u, to obtain the working or allowable design stress. In the case of ultimate tensile stress, F_u, as is customary, an additional strength reduction factor ϕ is applied. Therefore, the design stress on the gross area is

$$F_t = F_y/(5/3) = (3/5)F_y = 0.60F_y \qquad (4.1)$$

and on the net area

$$F_t = \phi\, F_u(5/3) = 0.85 F_u/(5/3) = 0.50 F_u \qquad (4.2)$$

For A36 material the ratio of allowable stresses on gross and net sections is

$$\frac{0.60 \times 36}{0.50 \times 58} = 0.75$$

Therefore, for A36 steel when the loss of area due to holes is more than 25% the design will be controlled by the net area. For steel with yield stress of 50 ksi and $F_u = 70$ ksi the loss must be larger than 14 percent, as the ratio is

$$\frac{0.60 \times 50}{0.50 \times 70} = 0.86$$

When failure occurs in the net area of tension members by fracture, except in the case of flat plates or bars, the failure load divided by the net area is generally less than the specimen tensile strength of the steel, unless all of the segments comprising the profile are connected so as to provide a uniformly distributed transfer of stress (*Manual*, Commentary on the AISC Specification, p. 5-142). Therefore, the effective net area is introduced. Where the load is transmitted by bolts through some but not all of the cross-sectional elements of the tension member, the effective net area, A_e, is still smaller than the net area, A_n, i.e.

$$A_e = C_t A_n \qquad (4.3)$$

in which A_n = net area of the member, and C_t = a reduction coefficient, from 0.75 to 0.90 (see AISC *Manual*, Section 1.14, p. 5-43). Also, for the purpose of design calculations the net area of a bolted splice and gusset plate when subject to tensile force cannot be taken as greater than 85 percent of the gross area. This means that for a splice or gusset plate

$$(A_n)_{\max} = 0.85 A \qquad (4.4)$$

The designer should have these two possible limitations in mind when designing tension members. Next, the net-area calculations will be discussed.

4.1.1.1 Net Section of Bolted Members

If a tension member has any holes at any cross section over its entire active length, the loss of area must be calculated and subtracted from the gross area to obtain its net section, A_n. Therefore, the net section of a riveted or bolted tension built-up member is the sum of the net sections of its component parts. The net section of a part is the product of the thickness, t, of the part and its least net width, b_n. For each component part of the cross section the least area has to be found. In computing the net area, the diameter of a rivet or bolt hole shall be taken as $\frac{1}{16}$ in. (1.6 mm) greater than the nominal dimension of the hole normal to the direction of the applied stress, if designing for buildings [AISC (4.1), section 1.14.4]. For bridges, either highway [AASHTO (4.2), section 1.7.44(M)] or railway [AREA (4.3), section 1.5.8], the increase is the same, $\frac{1}{8}$ in. (3.2 mm) greater than the nominal diameter of the high-strength bolt. AISC stipulates that in the case of splices and gusset plates the net section taken through a hole must never be considered as more than 85 percent of the corresponding gross section; that is, the loss of area is at least

$$(\Delta A)_{\min} = 0.15A \tag{4.5}$$

The net width for any chain of holes extending progressively across a part must be obtained by deducting from the gross width the sum of the diameters of all the holes in the chain and adding, for each gage space in the chain, the quantity

$$\frac{s^2}{4g} \tag{4.6}$$

where s = pitch of any two successive holes in the chain and g = gage (transverse spacing), both in inches. The AISC *Manual* (4.4) gives graphically the values for Eq. (4.6).

The following example will illustrate the procedure in finding the critical chain of holes and the corresponding net area of a plate.

Example 4.1 Find the net area of a plate $\frac{1}{2}$ × 15 in. (12 × 380 mm) having holes for $\frac{3}{4}$-in. (20-mm) diameter high-strength bolts in the pattern shown in Fig. 4.1. Use the AISC specification (4.1).

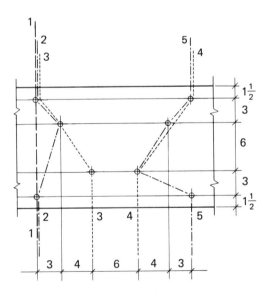

Figure 4.1 Least Chain Section (Example 4.1).

Solution First, inspect how many different reasonable chains can be drawn. Section 1-1 is a straight section with 2 holes. Section 2-2 passes through 3 holes, but it will have an addition to the net width, b_n, for 2 zigzag lines of

$$\frac{(3)^2}{4 \times 3} + \frac{(3)^2}{4(6+3)} = 1.00 \text{ in.} \tag{4.7}$$

Section 3-3 passes also through 3 holes, but it will have an addition to the net plate width, b_n, of

$$\frac{(3)^2}{4 \times 3} + \frac{(4)^2}{4 \times 6} = 1.42 > 1.00 \text{ in.} \tag{4.8}$$

Therefore, section 3-3 need not be investigated. Section 4-4 also passes through 2 holes, as does section 1-1, and the addition will be

$$\frac{(7)^2}{4(3+6)} = 1.36 \text{ in.} > 1.00 \text{ in.}$$

Therefore, it should be dropped.

Section 5-5 passes through 4 holes. No more reasonable chains can be drawn. The net width for different chain sections is, in general,

$$b_n = b - m(d + \tfrac{1}{8}) + \sum_{i}^{k} \frac{s_i^2}{4g_i} \tag{4.9}$$

where m = number of holes in the chain, d = bolt diameter, k = number of gage spaces in the chain, and s = pitch.

Section 1-1: $b_n = 15.0 - 2(\tfrac{3}{4} + \tfrac{1}{8}) = 13.25$ in. (33.6 cm) (4.10)

Section 2-2: $b_n = 15.0 - 3(\tfrac{3}{4} + \tfrac{1}{8}) + 1.00 = 13.38$ in. (34.0 cm) (4.11)

Section 5-5: $b_n = 15.0 - 4(\tfrac{3}{4} + \tfrac{1}{8}) + \frac{(4)^2}{4 \times 6} + \frac{(7)^2}{4 \times 3} + \frac{(3)^2}{4 \times 3}$

$$= 17.00 \text{ in. } (43.18 \text{ cm}) \tag{4.12}$$

The least net width is $b_n = 13.25$ in. of section 1-1 as shown in Fig. (4.1). The net cross-sectional area is

$$A_n = b_n \cdot t = 13.25 \times \tfrac{1}{2} = 6.6 \text{ in.}^2 \ (42.7 \text{ cm}^2)$$

Therefore, the true minimum net area to be used in the design is $A_n = 6.6$ in.² (42.7 cm²).

If an angle is part of a cross section, a slightly different procedure should be followed. First, the gross width, b, of an angle is equal to the sum of both leg sizes minus one leg thickness. This can also be obtained graphically as shown in Fig. 4.2 using point A as the center of rotation. Thus

$$b = a_1 + a_2 - t \tag{4.13}$$

Section 1-1 is a straight section passing through 2 holes. The net width, b_n, is (Eq. 4.9)

Section 1-1: $b_n = b - 2(d + \tfrac{1}{8})$ (4.14)

Section 2-2 is passing through 3 holes and has 2 gage spaces. Using the notation from Fig. 4.2, the net width of this section is

Section 2-2: $b_n = b - 3(d + \tfrac{1}{8}) + \frac{s^2}{4g_2} + \frac{s^2}{4(g_1 + g_3 - t)}$ (4.15)

The smaller of these two is chosen, and the net section of an angle is then

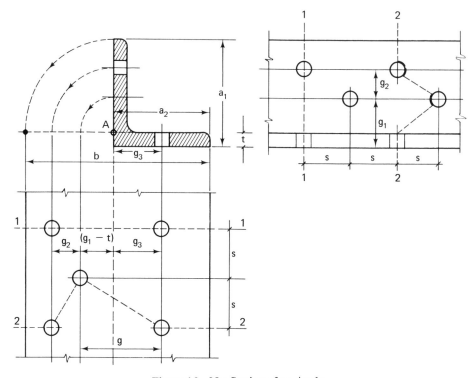

Figure 4.2 Net Section of an Angle.

$$A_n = b_n t$$

In case of an angle splice, this again has to be checked for $(\Delta A)_{\min} = 0.15A$ — that is:

$$(A_n)_{\max} = 0.85bt \quad (4.4)$$

and the smaller A_n of the two should be used.

Example 4.2, below, will illustrate the outlined procedure.

Example 4.2 Find the net area of an angle $\angle 8 \times 4 \times \frac{1}{2}$ ($\angle 203 \times 102 \times 13$ mm) if $\frac{3}{4}$-in. (20-mm) bolts are used with a hole pattern as indicated in Fig. 4.3. Use the AISC specification (4.1).

Solution Gross width, b, is

$$b = 8 + 4 - \tfrac{1}{2} = 11.5 \text{ in. (29.2 cm)}$$

Gross area is

$$A = 11.5 \times 0.5 = 5.75 \text{ in.}^2 \text{ (37.1 cm)}^2$$

Net width of section 1-1 (Eq. 4.14)

$$b_n = 11.5 - 2(\tfrac{3}{4} + \tfrac{1}{8}) = 9.75 \text{ in. (24.8 cm)}$$

Net width of section 2-2 (Eq. 4.15)

$$b_n = 11.5 - 3(\tfrac{3}{4} + \tfrac{1}{8}) + \frac{(2)^2}{4 \times 3} + \frac{(2)^2}{4 \times 5} = 9.41 \text{ in.} < 9.75 \text{ in.}$$

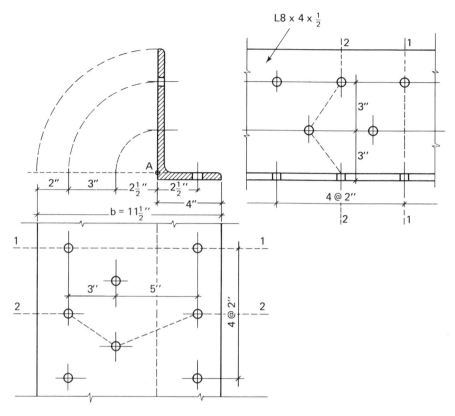

Figure 4.3 Net Section of an Angle (Example 4.2).

Therefore,
$$A_n = 9.41 \times 0.5 = 4.7 \text{ in.}^2 \, (30.4 \text{ cm}^2)$$

Check $(\Delta A)_{\min}$, in case that this angle is used as a splice
$$A_n = 0.85 \times 5.75 = 4.89 \text{ in.} > 4.7 \text{ in.}$$

Use $A_n = 4.7 \text{ in.}^2 \, (30.4 \text{ cm}^2)$.

Note: Use appropriate reduction coefficient, C_t, to get effective net area, A_e, if required (in case of an end connection).

4.1.1.2 Net Section of Welded Members

The net area of tension members, fabricated and end-connected by the use of welding, is equal to their gross area if there are no holes for erection bolts at the critical section. A careful design can locate such holes in the region of end connections where the welds of the joint have already transferred some of the force to the gusset plates. In that way the force in the member is reduced and the loss of area due to holes normally can be completely compensated for using such a reduction. If for some reason this is not possible, the material thicknesses for some length around (in front of and behind) such holes can be increased to compensate for the lost area. This compensation results in an economy of welded tensile members of 15–20% as compared to members that are bolted or riveted.

4.1.1.3 Composition of the Cross Section: Simple Built-Up Sections

The most economical tension members for the same required sectional area are fabricated from rolled shapes—single or double angles, T-sections, I-sections, single or double channels, or sometimes single plates. The fabrication process is fairly simple: checking for straightness (camber and sweep), cutting to size, grinding the cut surface, and making holes for end connections. Simple built-up sections (Fig. 3.3(e) and (f) and Fig. 3.4) require more work, because of the need to assemble the member, connecting either by bolts or welds its composing elements into a unit. However, built-up sections need to be used when the required sectional area is larger than available rolled (structural) shapes of comparable outline. Example 4.3 will illustrate the design process for static loading.

Example 4.3 Design a tension chord member of a simple roof truss of 170-ft (51.8-m) span if the force in the member is 1020 kips (4.537 MN). Use 1-in. (25-mm) diameter bolts, A36 material, and AISC specification (4.1).

Solution

GROSS AREA. As stated earlier, the net area of an A36 steel tension member will control the design if the loss by bolt holes is larger than 25 percent. As the T-section will be connected directly to the gusset plate with its web and angles, but the flanges will be connected only indirectly, a reduction coefficient, C_t, of 0.85 has to be applied to get the effective net area. Therefore, a loss of only $0.85 \times 0.25 = 0.21$ will control the design. As the hole diameter is fairly large, 1 in., it might easily happen that the effective net area will control the design. Therefore, we shall base our design on the gross area increased by $1/C_t = 1.18$ and only check the stress in the net area to make sure that it is equal to or less than $0.5F_u$. The reduction coefficient, C_t, must be taken into account.

The allowable stress according to Eq. (4.1) is $0.6 F_y = 22$ ksi (151.7 MPa) (see also AISC *Manual*, Table 1, p. 5–72). Therefore, the required gross area, A_{rqd} is

$$A_{rqd} = \frac{P}{C_t F_t} = \frac{1.18 \times 1020}{22} = 54.71 \text{ in.}^2 \ (353.0 \text{ cm}^2) \quad (4.16)$$

This is a fairly large area. For instance, the two heaviest American standard channels (*Manual*, p. 1–36) have an area of only 29.4 in.² (189.7 cm²). If a wide flange section could be used [which depends upon whether one- or two-plane (double-webbed) sections are used], then the least section would be a W14 × 167. Therefore, in this case a built-up section has to be designed by combining several plates and angles. As the required area is smaller than 100 in.² and the roof span smaller than 200 ft, assume that a one-wall section is adequate. In that case, a T-shaped section will be the probable answer. To get the height of the web plate, we shall use the following empirical equations (Section 3.1.1).

$$H_{\text{in.}} = 0.12 l_{\text{ft}} - \frac{l_{\text{ft}}^2}{11,000} \quad \left(\text{or} \quad H_{\text{cm}} = l_{\text{m}} - \frac{l_{\text{m}}^2}{400}\right) \quad (4.17)$$

or

$$H = 0.12 \times 170 - \frac{(170)^2}{11,000} = 17.8 \text{ in. (45.1 cm)} \quad (4.18)$$

This indicates that a web of 18 in. (Fig. 4.4) could be used. If a material thickness of $\frac{3}{4}$ in. (20 mm) and, say, angles $\angle 6 \times 4 \times \frac{3}{4}$ ($\angle 152 \times 102 \times 20$ mm) are chosen, the total area provided by web and angles is 27.38 in.^{2} (176.6 cm^2). This means that flange plates have to supply an area of 27.33 in.^{2} (176.3 cm^2). A width of 13 in. is adequate, because 2 angle legs plus web thickness require a minimum width of $12\frac{3}{4}$ in. One flange plate PL-$\frac{3}{4} \times 13$ has an area of 9.75 in.^{2} (62.9 cm^2). Therefore, the required number of plates is $27.33/9.75 = 2.8$—that is, 3 plates (Fig. 4.4). The total gross area is now $A = 56.63$ in.^{2} (365.3 cm^2) > 54.71 in.^{2} <u>OK</u>

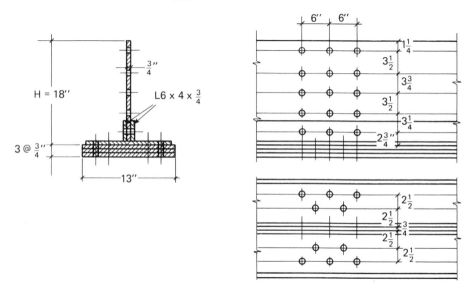

Figure 4.4 Cross Section of Tensile Member (Example 4.3).

The net section will be obtained by adding together net sections of each part of the section. The holes in the members are taken as developed for splicing.

WEB PLATE. A straight section passes through 5 holes in the web and there is no other reasonable section to be tried out. Therefore, the net width is

$$b_n = 18 - 5(1 + \tfrac{1}{8}) = 12.38 \text{ in. (31.4 cm)}$$

The net area of the web is thus $12.38 \times \frac{3}{4} = 9.29$ in.^{2} (59.9 cm^2). Check: $0.85A = 0.85 \times 12.38 = 10.52$ in. > 9.29 in.^{2}. Use 9.29 in.^{2}.

ANGLES. The gross width of one angle is

$$b = 6 + 4 - \tfrac{3}{4} = 9.25 \text{ in. (23.5 cm)}$$

A straight section passes through 2 holes; therefore, the net width is

$$b_n = 9.25 - 2(1 + \tfrac{1}{8}) = 7.00 \text{ in. (17.8 cm)}$$

Cross Section Design

Another section passing through 3 holes gives the net width

$$b_n = 9.25 - 3(1 + \tfrac{1}{8}) + \frac{(3)^2}{4 \times 2.5} + \frac{(3)^2}{4(2.5 + 2.75 - 0.75)}$$

$$= 7.28 \text{ in.} > 7.00 \text{ in.}$$

Therefore, the net area of one angle is

$$A_n = 7.00 \times 0.75 = 5.25 \text{ in.}^2 \, (33.9 \text{ cm}^2)$$

Check: The gross area of one angle is 6.94 in.² $0.85A = 0.85 \times 6.94 = 5.90$ in.² > 5.25 in. Use $A_n = 5.25$ in.².

FLANGE PLATES. A straight section passes through 2 holes in the flange plates. The net width is now

$$b_n = 13.0 - 2(1 + \tfrac{1}{8}) = 10.75 \text{ in.} \, (27.3 \text{ cm})$$

Another section, which passes through 4 holes, gives the following net width:

$$b_n = 13.00 - 4(1 + \tfrac{1}{8}) + 2\left(\frac{3^2}{4 \times 2.5}\right) = 10.30 \text{ in.} < 10.75 \text{ in.}$$

Therefore, the net area of one flange plate is

$$A_n = 10.30 \times 0.75 = 7.73 \text{ in.}^2 \, (49.8 \text{ cm}^2)$$

Use $A_n = 7.73$ in.². The total net area of the whole section is therefore

$$A_n = 9.29 + 2(5.25) + 3(7.73) = 43.0 \text{ in.}^2 \, (277.4 \text{ cm}^2)$$

Required net area is $P/(0.5\, C_t F_u) = 41.4$ in.². Because 43.0 (supplied) > 41.4 (required), the section is OK. Percentage loss due to holes is $[(56.63 - 43.00)/56.63](100) = 24.1\%$.

DYNAMIC LOADING. If a tension member is subjected to repeated variations of live load stresses, the design should incorporate fatigue considerations.

In buildings and bridges, the number of expected stress cycles, their range, and the type and location of the material part under consideration will affect the fatigue design.

In buildings, the loading condition has first to be established using AISC *Manual* table B1 (4.4, p. 5-86). Then, from *Manual* table B3 and corresponding category ["Illustrative Examples," fig. B1 and table B2 (4.4), pp. 5-89], the appropriate allowable range of stress, F_{sr} (ksi), has to be found. For given maximum and minimum forces and designed section, the stress range has to be calculated and compared with the allowable.

Example 4.4 Check the section designed in Example 4.3 if the maximum tensile force is 1020 kips (4.537 MN) and the minimum tensile force is 430 kips (1.92 MN). The number of loading cycles is about 10^6.

Solution From AISC *Manual* table B1 (4.4, p. 5-86) the loading condition is 3. From table B2 and fig. B1 in the AISC *Manual* the stress category is B for high-strength friction-type bolts (example 8). Then from table B3 the allowable stress range is

$$F_{sr} = 18 \text{ ksi} \, (165.5 \text{ MPa})$$

As the effective net section is $A_e = 43.00$ in.² (277.4 cm²), the calculated stress range is

$$f_{sr} = \frac{1020 - 430}{43.00} = 13.7 \text{ ksi (94.2 MPa)} < 18 \text{ ksi}$$

The design is OK for the given dynamic loading.

In highway bridge design, the allowable fatigue stress range for the actual category and according to the type and location of the material and number of cycles has to be obtained from Fig. 9.115 and from AASHTO tables 1.7.241 and 1.7.242 (4.2, pp 144–149).

Example 4.5 Redesign the section from Example 4.3 for the dynamic conditions as given in Example 4.4, but now applying AASHTO (4.2 fatigue design) (truck loading).

Solution For the number of cycles 10^6 and H loading (only truck load) from AASHTO table 1.7.2B (4.2, p. 151) it is found that this loading type corresponds to case I, which allows up to 2×10^6 cycles. From Fig. 9.115 the illustrative example is 18 and from AASHTO table 1.7.2A2 (4.2, p. 149) category B is chosen (base metal adjacent to friction-type fasteners). In AASHTO table 1.7.2A1 (4.2, p. 144) the allowable range of stress is 18 ksi > 13.97 ksi: OK. No changes in the design due to the fatigue consideration are thus required. (The calculated stress range, $f_{sr} = 13.97$ ksi, is found at the end of this example.)

The net area of the section given in Fig. 4.4 must now be calculated according to AASHTO specifications.

First, the diameter of the bolt has to be increased by $\frac{1}{8}$ in. as for buildings. In addition, section 1.7.8 of the AASHTO specifications (p.154) must be considered. In our case (single plane member with angles on opposite sides of a gusset plate) the full net area of the shapes will be considered effective, i.e., $C_t = 1.00$.

The net areas are as follows.

WEB PLATE. Because AASHTO does not require a minimum 15% reduction, the net width is

$$b_n = 18 - 5(1 + \tfrac{1}{8}) = 12.38 \text{ in. (31.4 cm)}$$

The net area is

$$A_n = 12.38 \times 0.75 = 9.28 \text{ in.}^2 \text{ (59.9 cm}^2\text{)}$$

ANGLES. Using a straight section, the net width is

$$b_n = 9.25 - 2(1 + \tfrac{1}{8}) = 7.00 \text{ in. (17.8 cm)}$$

The net area of one angle is

$$A_n = 7.00 \times 0.75 = 5.25 \text{ in.}^2 \text{ (33.9 cm}^2\text{)}$$

FLANGE PLATES. For a section passing through 4 holes, the net width is

$$b_n = 13.00 - 4(1 + \tfrac{1}{8}) + 2\left(\frac{3^2}{4 \times 2.5}\right) = 10.3 \text{ in. (26.2 cm)}$$

The net section of one plate is $A_n = 10.3 \times 0.75 = 7.72$ in.² (49.8 cm²) Therefore, the total net section is

$$A = 9.28 + 2(5.25) + 3(7.72) = 43.0 \text{ in.}^2 \text{ (277.4 cm}^2\text{)}$$

Maximum stress is

$$f_{t\text{-max}} = \frac{1020}{43.0} = 23.7 > 20 \text{ ksi} \quad \underline{\text{NG}} \ (18.6\% \text{ overstress})$$

The minimum stress is $f_{t\text{-min}} = 430/43.0 = 10.0$ ksi. The stress range is

$$f_{sr} = 13.7 \text{ ksi} < 18 \quad \underline{\text{OK}}$$

The section is not adequate for static loading and has to be strengthened. For instance, if flange, angle, and web thickness is increased to $\frac{7}{8}$ in. (22 mm), the stress is 20.33 ksi or 1.7% overstress—which is satisfactory.

4.1.2 Slenderness-Based Design

A tensioned member under a static loading does not involve stability considerations. But if there is a change of live load stresses (dynamic loading), or if there is a transverse loading on the member due to weight or wind, "flutter" can develop. If the member is too slender, this can become a major problem and actually the design criterion itself. A restriction is made by limiting the slenderness ratio, which is the ratio of the laterally unbraced length of the member to the least radius of gyration of the cross section.

The AISC specification (4.1, section 1.8.4, p. 5-29) states that for tension members, other than rods, the slenderness ratio, l/r, should not exceed

For main members	240
For bracing and other secondary members	300

The AASHTO specifications (4.2, section 1.7.5, p. 152) states that for tension members other than rods, eyebars, cables, and plates, the ratio of unsupported length to radius of gyration must not exceed

For main members	200
For bracing members	240
For main members subject to reversal of stress	140

The AREA specifications (4.3, section 1.5.1, p 15-1-19) require that this ratio not exceed

For tension members	200

Flutter is very often a problem in slender hangers in bridge structures due to their usual I-shape. Apart from the noise produced by their vibrations, fatigue often can become a major problem resulting in the cracking of the end connections.

The normal procedure in designing a section of a tension member is to find the section using strength design and then to check it for necessary stiffness.

Example 4.6 If a tension chord member (main truss member) of Example 4.3 has a length (between the lateral bracings) of 52 ft, check to see whether the condition of the AISC specification (4.1) concerning the minimum slenderness ratio is satisfied.

Solution The least radius of gyration (about the vertical, y-axis) is calculated in Table 4.1. The result is

$$r = 3.10 \text{ in.}$$

Therefore, the slenderness ratio is

$$\frac{l}{r} = \frac{52 \times 12}{3.10} = 201.3 < 240 \quad \underline{\text{OK}}$$

Table 4.1 LEAST RADIUS OF GYRATION

Sketch	Part Name (in.)	Area (in.)	Own Moment of Inertia (in.²)	Position Moment of Inertia (in.⁴)	Total Moment of Inertia (in.⁴)
	Web PL-¾ × 18	13.5	0	0	0
	Angles 2 ∠6 × 4 × ¾	13.88	49.0	83.6	132.6
	Flange plates 3 PL-¾ × 13	29.2	411.9	0	411.9
	Sum	56.63	460.9	83.6	544.5

$$r = \sqrt{\frac{544.5}{56.63}} = 3.10''$$

4.2 SPECIFICATIONS FOR TENSION MEMBERS

4.2.1 U.S. Specifications

4.2.1.1 *AISC for Buildings*

The main characteristic of the AISC specification is the basic assumption that tension members may fail either by the yielding of the gross section or by breaking at the critical net section. For the working load design method this means that the allowable stresses for tension members are given with respect to the yield point stress and the tensile strength. Therefore, the nominal tensile stress on the gross area

$$f_t = \frac{P}{A} \leq 0.6 F_y \tag{4.19}$$

and on the effective net area

$$f_t = \frac{P}{A_e} \leq 0.6 \times 0.85 F_u = 0.50 F_u \tag{4.20}$$

needs to be checked.

To find the net area, the diameter of a bolt hole (already increased by $\frac{1}{16}$ in. beyond the nominal bolt diameter) is increased by another $\frac{1}{16}$ in., i.e., diameter of the lost area is

$$d_b + 1/8 \text{ in.} \tag{4.21}$$

in which d_b = nominal bolt diameter in inches.

The maximum net area of splices or gusset plates is restricted to

$$(A_n)_{\max} = 0.85A \tag{4.4}$$

in which A_n = net area, and A = gross area.

The effective net area is obtained by multiplying the net area by a reduction factor, C_t, which is 0.90 for W, M, or S shapes with flange width not less than $\frac{2}{3}$ of their depth, and structural tees cut from these shapes, provided the connection is to the flanges and has no fewer than three fasteners per line in the direction of stress. Built-up cross section and W, M, and S shapes not meeting the above criterion, i.e.,

$$b_f < \tfrac{2}{3}d \tag{4.22}$$

require a $C_t = 0.85$. Finally, all members connected with only two fasteners per line in the direction of stress must have a $C_t = 0.75$. The AISC specification (4.1) in Section 1.18.3 states that the longitudinal spacing of fasteners and intermittent fillet welds connecting a plate and a rolled shape in a built-up tension member, or two plate components in contact with one another, shall not exceed 24 times the thickness of the thinner plate or 12 in. The longitudinal spacing of fasteners and intermittent welds connecting two or more shapes in contact with one another and in tension shall not exceed 24 in. Tension members composed of two or more shapes or plates separated from one another by intermittent fillers (Fig. 3.7c) shall be connected to one another at these fillers at intervals such that the slenderness ratio of either component between the fasteners does not exceed 240.

In double-plane (double-webbed) sections (Fig. 3.4) on the open sides, either perforated cover plates (Fig. 3.7e) or tie plates (Fig. 3.7a) may be used. The length of tie plates shall be not less than two-thirds the distance between the lines of fasteners or welds connecting them to the components of the member. The tie plate thickness shall not be less than 1/50 of the distance between these lines. The longitudinal spacing of fasteners or intermittent welds at tie plates shall not exceed 6 in. The spacing of tie plates shall be such that the slenderness ratio of any component in the length between tie plates will not exceed 240.

4.2.1.2 AASHTO for Highway Bridges

The AASHTO specifications (4.2) base the design of tension members under static loading conditions on their net section with an allowable axial tension stress either

$$0.55F_y \quad \text{or} \quad 0.46F_u \tag{4.23}$$

whichever is smaller.

In Section 1.7.8, effective area of angles and T sections in tension, for a single angle tension member, a T-section, or each angle of a double angle tension member in which the angles are connected back to back on the same side of a gusset plate, AASHTO requires that the effective net area shall be the net area of the connected leg or flange plus one-half of the area of the outstanding leg. This means that the reduction factor, C_t, is actually taken as 0.75.

If a double angle or T-section tension member is connected with the angles or flanges back to back on opposite sides of a gusset plate, the full net area of the shapes shall be considered as effective.

When angles connect to separate gusset plates, as in the case of a double-webbed truss (double-plane truss), and the angles are connected by stay plates located as near the gusset as practicable, the full net area of angles is counted. If the angles are not so connected, $C_t = 0.80$.

Lug angles may be considered as effective in transmitting stress, provided they are connected with at least one-third more fasteners than required by the stress to be carried by the lug angle.

The calculation of the net section of riveted or high strength bolted tension members is the same as explained for the AISC specifications. The diameter of the hole shall be taken as 1/8 in. (3.2 mm) greater than the nominal diameter of the rivet or high-strength bolt.

4.2.1.3 AREA for Railroad Bridges

The AREA *Manual for Railway Engineering* (4.3), in Chapter 15, Steel Structures, Section 1.5.8 for designing tension members, follows in general the AASHTO specifications. The diameter of a hole is also increased by 1/8 in. (3.2 mm), and the net width of a section through a chain of holes is increased in the same way as in both the AISC and AASHTO specifications [see Eq. (4.6)]. For the effective net area in Section 1.6.5, AREA gives the following: If angles or tees in tension are so connected that bending cannot occur in any direction, the effective section shall be the net section of the member. If such members are connected on one side of a gusset plate, the effective section shall be the net section of the connected element plus one-half the section of the connected element. This is the same regulation as in AASHTO, Section 1.7.8.

4.2.2 Foreign Specifications

Of the foreign specifications, the British specifications for buildings, BS449 (4.6), and for bridges, BS153 (parts 3B and 4) (4.7), are closest to the American specifications. They are similar in respect to the increase of the diameter of the hole in excess of the nominal diameter of the rivet or bolt ($\frac{1}{16}$ in), in the reduced deduction for any zigzag line (for each gage space) in a chain of holes, and in limiting the ratio of unsupported length to the least radius of gyration (250 for railway and 300 for highway bridges).

The German specifications for buildings (4.8), section 4.2, ("Cross-Sectional Values and Hole Deduction") do not increase the hole diameter and do not reduce deduction for one zigzag line in a chain of holes. If a tension member has only a small tensile force but can be subjected to compression as a result of a slight change in loading position, then it must also be designed for a reasonable compressive force and cannot have a larger slenderness ratio than 250. Otherwise, there is no specific limitation for the slenderness ratio of tension members.

The French specifications (4.9), section 3.1 (Members Exposed to Simple Tension) require that the tensile normal stress of structural members be based on the net section. This section is obtained by subtraction from the gross section of all the holes in the most unfavorable section (a normal, oblique, or broken rupture line). No specific limitation of slenderness ratios for tension members is given.

The Soviet specifications (4.10), like the previous specifications, require stress calculation based on the net section, but they are more specific in limiting maximum slenderness ratios. This information is given in Table 4.2.

Table 4.2 LIMITING SLENDERNESS RATIOS, λ, OF TENSION MEMBERS (SOVIET SPECS)

	Maximum allowable slenderness ratios, λ	
Name of Structural Member	Dynamic Loading Acting Directly on the Structure	Static Loading Acting on the Structure
Chords and first diagonal of trapezoidal trusses	250	400
Other elements of trusses	350	400
Bottom flanges or bottom chords of crane girders and trusses	150	—
Structural elements of vertical bracings between columns (below crane girders)	300	300
Other bracing elements	400	400

NOTATIONS

a_1, a_2	= leg sizes of an angle	t	= thickness of a member
b	= gross width of a member	A	= gross sectional area
b_n	= net width of a member	ΔA	= loss of area due to the holes
d, d_b	= nominal bolt diameter	A_e	= effective net area
f_t	= calculated nominal tensile stress	A_n	= net area
		A_{rqd}	= required area
f_{sr}	= calculated stress range	C_t	= reduction coefficient
g, g_1, g_2, g_3	= gage lines	F_{all}	= allowable stress
k	= number of gage spaces in a chain of holes	F_{sr}	= allowable stress range
		F_t	= allowable stress in tension
l	= unsupported length of a tension member, span	F_y	= yield point stress
		F_u	= minimum tensile strength of material
m	= number of holes in one chain of holes	H	= height (depth) of a web plate
p	= pitch, spacing in the direction of force action	P	= tensile force in a member
		λ	= slenderness ration, l/r
r	= least radius of gyration	ϕ	= strength reduction factor
s	= pitch of bolts or holes		

PROBLEMS

4.1. Using the AISC specification (4.1) and A36 steel, design a diagonal of a roof truss. The force in the diagonal is 145 kips and its length 12 ft. Use a double-angle section, single gusset plate of $\frac{3}{8}$ in, thickness, one gage line in angle legs connected to gusset plate, and $\frac{3}{4}$-in. high-strength bolts of friction type, A325-F. Detail the connection.

4.2. A tension member 15-ft long is composed of 2 angles ∠ 5 in. × 5 in. × $\frac{1}{2}$ of A36 steel, with holes in both pairs of legs for $\frac{7}{8}$-in.-diameter bolts (see Fig. P-4.2). Gage lines (one in each leg) are $g = 3$ in., and pitch is 3 in. Find the maximum allowable force in that member. Use the AISC specification (4.1).

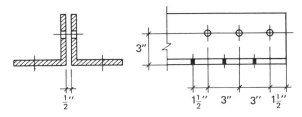

Figure P-4.2

4.3. Select an economical single angle of A36 steel for a tension member of 10-ft length and exposed to a tensile force of 8.0 kips. One leg is connected to a gusset plate with bolts of $\frac{5}{8}$-in. diameter. Perform your calculations twice, once using AISC specification (4.1) and once using AASHTO specification (4.2).

4.4. For the plate with holes for 1-in. bolts as shown in Fig. P.4.4 find the net area. Use ASIC specification (4.1).

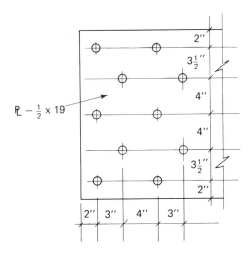

Figure P-4.4

4.5. In a highway bridge with a main truss girder made of ASTM A441 steel a tie member is made of W12 × 40. If bolts of 1 in. diameter were used to connect both flanges to gusset plates, what is the maximum force that the tie can take according to AASHTO specification (4.2), if fatigue is of no concern for that member?

4.6. A built-up tension member composed of two channels of A36 steel is connected by A325-X bolts of $\frac{7}{8}$-in. diameter to gusset plates, as shown in Fig. P-4.6. The force to be carried is 290 kips. The length of the member is 30 ft. Tie plates are used at member ends; therefore, flanges have also holes. Using A36 steel and AISC specification (4.1) design this member.

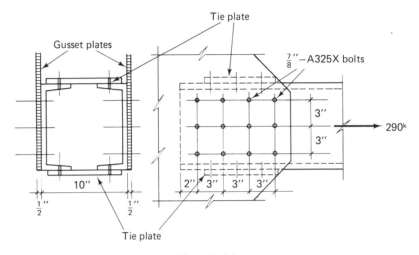

Figure P-4.6

4.7. In a small bridge truss a diagonal member composed of two equal-leg angles, fillet-welded to a gusset plate of $\frac{1}{2}$-in. thickness, carries a maximum tensile force of 90 kips and a minimum force of 15 kips. The member length is 14 ft. If the member is exposed to 10^5 stress cycles, using A36 steel and AASHTO specification (4.2) design this member and detail the connection.

4.8. A round steel bar is used as a hanger 20-ft long and is welded on both ends. Design the rod in A36 steel for a force of 100 kips using AISC specification (4.1). The maximum elongation of the rod has to be less than $\frac{3}{8}$ in.

4.9. In a roof of a factory building sag rods are used at one-third points of the purlins spanning 21 ft between roof trusses. If the purlins are spaced as shown in Fig. P-4.9 and roofing is made of precast concrete cellular units 8 in. deep and 55 lb/ft² weight, design the rods using A36 steel and AISC specification (4.1). Snow-load is 25 lb/ft² on horizontal projection.

4.10. The bottom chord of a roof truss is composed of two angles ∟4 × 4 × 3/8 and carries a tensile force of 105 kips. The truss panel is 10 ft long. Both angle legs in the truss plane and out of it have holes for 3/4-in. bolts staggered at $1\frac{1}{2}$ in. If the steel is A36 material, check whether the design is satisfactory according to AISC specification (4.1).

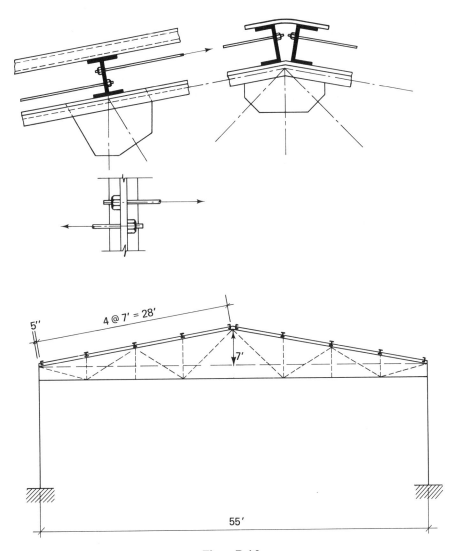

Figure P-4.9

4.11. A hanger rod with threaded ends, not upset, supports a load of 5.0 kips. It is required to design a round rod by AISC specification for steel A36.

4.12. An angle ∠ 8 × 6 × 3/4 in. with holes for 7/8-in. bolts as shown in Fig. P 4.12. Find the net section by the AISC specification.

4.13. A tension member of A36 steel consisting of two angles ∠ 6 × 4 × $\frac{1}{2}$-in. is connected at the ends by both legs, as shown in Fig. P 4.13, with 7/8-in. high-strength bolts A325-N. Find the maximum force the angles can take according to AISC specification and considering the angles only.

4.14. A tension member is composed of two C12 × 20.7 channels connected to a single gusset plate. If the material is A36 steel and holes for this end con-

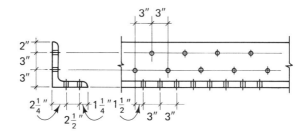

Figure P-4.12

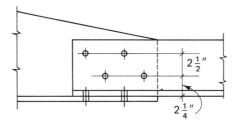 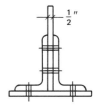

Figure P-4.13

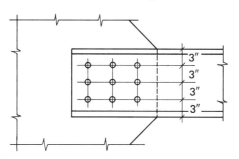 

Figure P-4.14

nection are for 3/4-in. bolts and of the pattern as shown in Fig. P 4.14, find the maximum force this member can take according to the AISC specification.

4.15. A tower is framed with main verticals, 8 × 8-in., and horizontal members, 3 × 3-in. single angles (Fig. P-4.15). Using AISC specification, A36 steel, design the diagonals for a tension load of 8 kips. Diagonals are single angles bolted by 3/4-in. bolts to the vertical members and they are connected to each other at point of intersection.

4.16. Design the diagonals from P-4.15 considering tie rods with clevis attachments (AISC *Manual*, p. 4-142).

4.17. If the load in horizontal members of the tower in P-4.15 has a tensile force of 4 kips, design them using A36 steel, AISC specification, and 3 × 3-in. single angles.

4.18. Each end of a ∠ 6 × 6 × ½-in. angle member is connected along one leg due to a 3/4-in. gusset plate, using one line of 7/8-in. bolts. Determine allowable tension load for the member considering A36 steel and AISC specification.

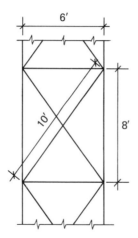

Figure P-4.15

4.19. A tension member in a bridge truss is a welded H shape 43 ft long and 16 in. deep overall. The maximum load on the member is 340 kips tension, and the minimum load is 120 kips tension. Select dimensions for the plates forming the member using A588 steel and AASHTO specifications. Disregard bending due to the weight of the member.

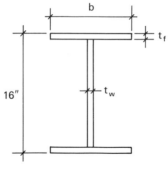

Figure P-4.19

4.20. A built-up tension member is spliced as shown in Fig. P-4.20 with 7/8-in. bolts. If the material is A36 and AASHTO specifications control, design this member for a tensile force of 350 kips.

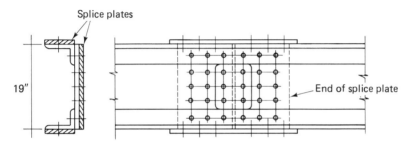

Figure P-4.20

REFERENCES

4.1. AISC, *Specification for the Design, Fabrication, and Erection of Structural Steel for Buildings*, New York, Nov. 1978.

4.2. AASHTO, *Standard Specifications for Highway Bridges*, 12th ed., Washington, DC. 1977 with 1978, 1979, 1980 and 1981 Interim Bridge Specifications.

4.3. AREA, *Manual for Railway Engineering*, Chapter 15, Steel Structures, Chicago, 1980.

4.4. AISC, *Manual of Steel Construction*, 8th ed., Chicago, 1980, p. 4–97.

4.5. AWS, *Structural Welding Code, Steel*, Miami, 1980.

4.6. British Standards Institution, *Specification for the Use of Structural Steel in Buildings*, BS449, London, 1965.

4.7. British Standards Institution, *Specification for Steel Girder Bridges*, BS153, parts 3B and 4, London, 1966.

4.8. Deutsche Industrie Normen, *Stahl im Hochbau, Berechnung und bauliche Durchbildung*, DIN1050, Berlin, 1968.

4.9. Le Centre Technique Industriel de la Construction Métallique de France, *Règles de Calcul des Constructions en Acier*, Règles CM66, ed. Eyrolles, Paris, 1974.

4.10. Gosudarstvenyĭ Komitet Soveta Ministrov SSSR po delam stroitelstva, *Stroitelnye Normy i Pravila*, part II, sec. V, Chapter 3, Stalnye Konstruktzii, SN i P II-V, 3–62, Moscow, 1963.

5

Design of Compression Members

5.1 BUCKLING OF STRAIGHT PRISMATIC MEMBERS

Although columns were used by our forefathers before the time of any written history, tests and practical investigations of the column-buckling problem first took place in 1729 (5.1). Petrus Van Musschenbroek, Professor of Physics in Utrecht, Holland, developed the first formula for the capacity of rectangular columns on the basis of empirical column curves (Fig. 5.1) in the form

$$P = k \frac{bd^2}{L^2} \tag{5.1}$$

where P = column capacity, k = empirical factor, b = width, d = depth of the rectangular column cross section, and L = length of the column. This formula is surprisingly close to those now in use.

In 1744 Euler (5.2) obtained analytically the expression for the buckling load of a column with one end free and the other end fixed (column AB in Fig. 5.2)*:

$$P = \frac{C\pi^2}{4L^2} \tag{5.2}$$

where C is defined as the "absolute elasticity" and P and L as in Eq. (5.1). Euler had actually studied different cases of bending (Fig. 5.2), classifying them according to

---

*Euler used variational calculus, which he had developed.

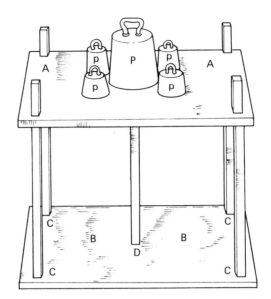

Figure 5.1 Van Musschenbroek's Tests on Buckling (From S. P. Timoshenko, *History of Strength*, (in Russian), Moscow, 1957).

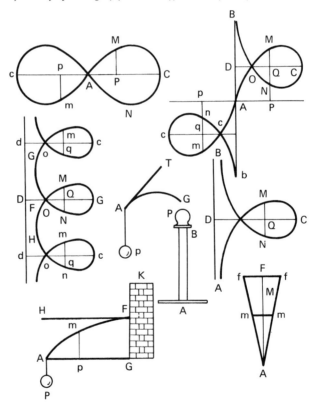

Figure 5.2 Elastic Curves Discussed by Euler (From S. P. Timoshenko, *History of Strength* (in Russian), Gosizdat, Moscow, 1957).

the magnitude of the angle contained between the force and the tangent on the curve at the point of force application. If this angle was very small, then this would represent the case of buckling of a column under an axial compressive force. In discussing the above formula, he concluded: "If the force P is not larger than [the quantity given by Eq. (5.2)], then maybe there will be no deflections; quite opposite, if the load is larger, the column cannot oppose the deflection" (5.2).

Both of the above formulas were developed for elastic buckling—that is, when the stresses due to the "critical" column force are still below the proportional limit stress of the material. One-and-one-half century later, Engesser (5.3) extended Euler's formula into the range of inelastic buckling by introducing the tangent-modulus theory in 1889. Under the influence of sharp criticism, Engesser in 1895 introduced instead the double-modulus or reduced-modulus theory. The reduced-modulus theory was considered correct until 1946, when F. R. Shanley (5.4) proved that the tangent-modulus theory is actually correct in predicting the load at which buckling of a perfect column might begin. The generalized Euler formula for elastic or inelastic buckling can therefore be written in the form

$$P_e = \frac{\pi^2 E_t I}{(KL)^2} \qquad (5.3)$$

where E_t is the tangent modulus, I is the moment of inertia of the column, K is a coefficient depending upon the effects of end restraints (see Section 5.1.3), and P and L are as defined in Eq. (5.1).

5.1.1 Axial Force

The equilibrium configurations of elastic structures are normally stable. If the equilibrium of such an elastic structure is disturbed by a force or impact for a short period of time, small vibrations around the original equilibrium configuration will develop. As a result of damping the system soon will stop vibrating and will resume its previous equilibrium configuration provided that this configuration was stable and the disturbance was small. If the elastic system does not return to its original shape, then it is unstable (Fig. 5.3). If, between certain limits, insensitivity exists against changes of the equilibrium configuration, then this is a neutral equilibrium. In Fig. 5.4 the three possible equilibrium positions of a body W are shown schematically. Configuration a corresponds to a minimum and c to a maximum of the potential energy, but in case b the potential energy close to the original position is constant.

Simple sections with one axis of symmetry

Elastic buckling

Consider an ideal or "perfect" straight column which is centrally loaded, with a uniform cross section, and made of a homogeneous material of unlimited elasticity. There are no residual stresses and no end moments (Fig. 5.3). When $P = P_e$, any imposed bent configuration will remain after a disturbance is removed; that is, the column is in neutral equilibrium. This means that for this critical load, P_e, besides a

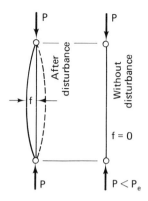

Figure 5.3 "Perfect" Column.

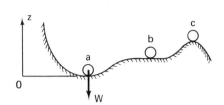

Figure 5.4 (a) Stable, (b) Neutral, and (c) Unstable Equilibrium Configurations.

straight equilibrium configuration with $f = 0$, (Fig. 5.3) also a slightly bent equilibrium position is possible with $f \neq 0$.

If the force is very slowly increased and applied without disturbances, the column can take loads with values $P' > P_e$ without bending, $f = 0$ (Fig. 5.5a). The straight form nevertheless is still unstable, and in response to the slightest disturbance it will snap into a new stable position with $f' \neq 0$. This means that when $P < P_e$, only one stable equilibrium, actually the straight form of the column, is possible. For $P' > P_e$ several equilibrium positions are possible, with the original straight form being always unstable. In other words, at the critical load, P_e, equilibrium bifurcation occurs (Fig. 5.5.a). When a change of equilibrium from $f = 0$ to $f = f'$ occurs, large bending stresses will result, and in practical cases this leads to failure of the column. Therefore, in most cases the diagram in Fig. 5.5(b) is valid.

On the basis of the usual linear theory, it is impossible to determine the magnitude of deflections or the degree to which they depend upon the intensity of loading.

Analytically, buckling loads are obtained as characteristic values of a differential

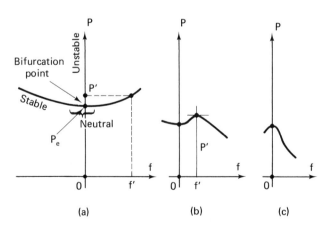

Figure 5.5 Bifurcation Diagram.

146 DESIGN OF COMPRESSION MEMBERS

equation, and the deflected buckled forms are the corresponding characteristic functions of the form

$$v(z) = f \sin \frac{n\pi z}{L} \tag{5.4}$$

where $v(z)$ are the deflections (Fig. 5.6), z is the distance from one column end, f is an integration constant, and L is the column height or its length.

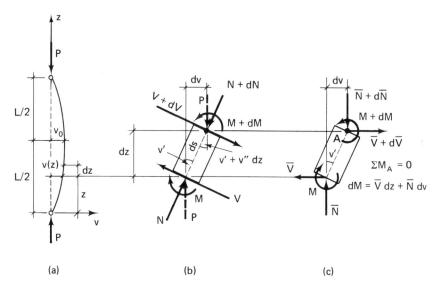

Figure 5.6 Column Element.

In practice, columns are not perfect. Practically, therefore, it is impossible for an elastic structure under a load beyond the critical limit ever to be in a stable state—that is, without buckling out. The theoretical buckling force therefore represents the upper limit of loading for which an unbuckled form ($f = 0$) is possible.

In practice various disturbances lead to visible deflections ($f \neq 0$) far below the critical load, P_e. Therefore, in the bifurcation diagram the P-f curve shows either an immediate drop (Fig. 5.5c) or a drop after a certain limit of the deflection, f', is reached (Fig. 5.5b).

Euler load

To derive the formula for the Euler buckling load, all previous assumptions about a "perfect" column are assumed to be valid.

We shall consider the column between two consecutive sections at a distance dz apart along the elastic line (Fig. 5.6a). Because the slope of the tangent, v', is small, the difference in length between ds and dz is of the second order of magnitude. The same is true for the difference between the slopes of the tangents at both ends (Fig. 5.6b). To derive the differential equation we shall use Fig. 5.6(c) instead of Fig. 5.6(b). If there is no transverse loading, $\bar{V} = 0$. The moment equilibrium equation, written

for point A using dz instead of ds, yields
$$dM = \bar{V}\,dz + \bar{N}\,dv$$

After dividing by dz and substituting the values for $\bar{N} = P$ and $\bar{V} = 0$ differentiating yields

$$M'' - Pv'' = 0 \tag{5.5}$$

Next, using Bernoulli's rule

$$M = -\frac{EI}{R} \tag{5.6}$$

and substituting in Eq. (5.6) for the curvature ($1/R$) the usual approximation $1/R \approx v''$, the following expression is obtained

$$M = -EIv'' \tag{5.7}$$

The final differential equation for a bent column from Eqs. (5.5) and (5.7) in the case of a constant flexural stiffness is now

$$EIv^{IV} + Pv'' = 0 \tag{5.8}$$

The general solution of Eq. (5.8) is

$$v(z) = A \sin \alpha z + B \cos \alpha z + Cz + D \tag{5.9}$$

where A, B, C, and D are four constants of integration and $\alpha = \sqrt{P/EI}$. The boundary conditions are (Fig. 5.6)

$$\begin{aligned} v = 0 \quad \text{and} \quad M = 0 \quad \text{for} \quad z = 0 \\ v = 0 \quad \text{and} \quad M = 0 \quad \text{for} \quad z = L \end{aligned} \tag{5.10}$$

These conditions will yield a system of four linear homogeneous equations with four constants of integration as unknowns. One solution of this system is the so-called trivial solution with $A = B = C = D = 0$, corresponding to $v(z) = 0$. The undeflected column form with $v(z) = 0$ is the usual equilibrium state for any force value smaller than P_e.

The equations have one solution which is not trivial when the determinant Δ

$$\Delta = \begin{vmatrix} 0 & 1 & 0 & 1 \\ \alpha & 0 & 1 & 0 \\ \sin \alpha L & \cos \alpha L & L & 1 \\ \alpha^2 \sin \alpha L & \alpha^2 \cos \alpha L & 0 & 0 \end{vmatrix} = 0 \tag{5.10a}$$

of the coefficients A, B, C, and D disappears. From this "buckling condition" we have

$$\alpha^4 \cdot L \cdot \sin \alpha L = 0 \tag{5.11}$$

and from this expression the critical values of the force, P_e, can be calculated. Equation (5.11) is only satisfied when $\sin \alpha L = 0$—that is, for

$$\alpha_{em} \cdot L = m\pi, \quad \text{with } m = 1, 2, 3, \ldots \tag{5.12}$$

where $\alpha_{em} = m\sqrt{P_e/EI}$. For each value of m a definite value for the buckling load is obtained. For design purposes we are only interested in the smallest critical value—that is, $m = 1$

$$P_{e1} = P_e = EI\left(\frac{\pi}{L}\right)^2 \tag{5.13}$$

Using the exact differential equation of the elastic curve for large deflections it is possible (5.8) to obtain equilibrium positions ("elastica") higher than P_e corresponding to the applied force, P. From Fig. 5.7 it is seen that for a column with a force applied at its free end and with the other end built in, the angle α between the force and the end tangent to the column should be 20° when $P = 1.015P_e$—that is, for an increase of only 1.5% over the critical force. Usually the material cannot resist such curvature, and bending failure occurs.

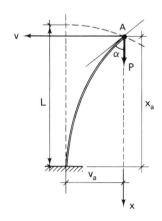

Load — Deflection Data

α	0°	20°	40°	60°	80°	100°	120°	140°	160°	176°
P/P_e	1	1.015	1.063	1.152	1.293	1.518	1.884	2.541	4.029	9.116
x_a/L	1	0.970	0.881	0.741	0.560	0.349	0.123	−0.107	−0.340	−0.577
v_a/L	0	0.220	0.422	0.593	0.719	0.792	0.803	0.750	0.625	0.421

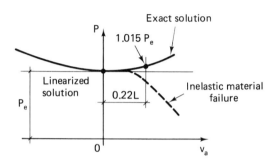

Figure 5.7 Supercritical Loads.

A more general case occurs when combined buckling and bending is taking place (Fig. 5.8). Bending results from both the force, P, and the transverse load, $w(z)$. If, as before, one portion, ds, of the column is isolated and the free body diagram is drawn, then, using the same reasoning as before, the following differential equation for a bent column with a constant EI is obtained

$$EI \cdot v^{IV} + Pv'' = w \tag{5.14}$$

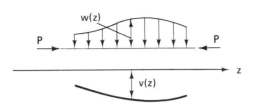

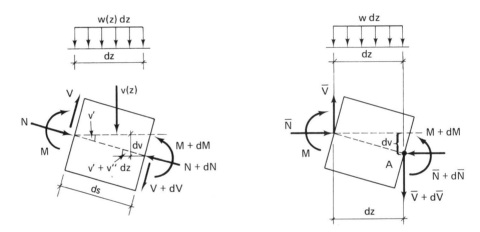

Figure 5.8 Transverse Loads.

Inelastic buckling

In practive there are very few, it any, columns so slender that the critical stress would be in the elastic range—that is, governed by Euler's formula.

If both sides of Eq. (5.13) are divided by the cross-sectional area of the column A, and the moment of inertia, I, is replaced by Ar^2, where r = radius of gyration, then the critical stress is

$$F_e = \frac{\pi^2 E}{(KL/r)^2} \tag{5.15}$$

where KL is called the effective buckling length (see Sect. 5.1.3) and the ratio $KL/r = \lambda$ the slenderness ratio. For slenderness ratios sufficiently small to make the critical stress, F_e, larger than the limit of proportionality, Eq. (5.15) becomes (Fig. 5.9)

$$F_e = \frac{\pi^2 E_t}{(KL/r)^2} \tag{5.16}$$

The lines in Fig. 5.9 between the yield-point stress and the Euler critical stress are called "transition" lines. Various curves, usually parabolas or even straight lines, semiempirical in nature, are used by different design specifications. They will be discussed, along with design specifications, in Section 5.1.4, Design Procedures.

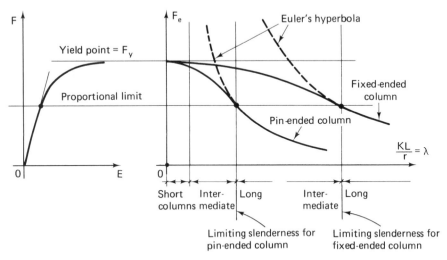

Figure 5.9 Inelastic Buckling.

Built-up sections

Built-up sections are used because of the limited sizes of available rolled sections and when an adequate column cannot be designed with single sections. As the effect of shearing force on the buckling strength of a built-up column cannot be disregarded, this effect will be discussed first (5.11).

It can be shown that the total curvature of the deflection curve due to the bending moment and shearing force is

$$\frac{d^2v}{dz^2} = -\frac{Pv}{E_tI} + \frac{\beta P}{G_tA} \cdot \frac{d^2v}{dz^2} \tag{5.17}$$

where β is a correction factor for using the average shear stresses instead of their actual value; E_t and G_t are the tangent moduli for Young's modulus and the shearing modulus, respectively, which are introduced to make this equation valid for the entire range of buckling (elastic and inelastic). Using the abbreviation

$$\alpha^2 = \frac{P}{E_tI} \cdot \frac{1}{1 - \beta P/G_tA}$$

the final form of the differential equation, taking into account both effects—that is, of bending and shear—is

$$v'' + \alpha^2 v = 0 \tag{5.18}$$

which is of the same form as in the case of simple sections except that now α^2 has a

Buckling of Straight Prismatic Members   **151**

different value. Applying the same reasoning as in the preceding discussions, the critical force, P'_e, is

$$P'_e = \frac{\pi^2 E_t I}{L^2} \cdot \frac{1}{1 + (\beta/G_t A) \cdot (\pi^2 E_t I/L^2)} \qquad (5.19)$$

This also can be written

$$P'_e = P_e \cdot \frac{1}{1 + (\beta/G_t A) \cdot P_e} \qquad (5.20)$$

As the denominator, $1 + (\beta/G_t A) \cdot P_e$, is always larger than 1 the force P'_e is always smaller than the force P_e.

5.1.2 Eccentric Force

In the preceding section an ideal or "perfect" column was used as the physical model. In practice such a column does not exist. Even if geometrically perfect conditions for a column were satisfied (straight and prismatic), the column load is usually acting eccentrically, although it may be parallel to the column axis.

To generalize, the case now discussed is shown in Fig. 5.10, with different eccentricities, $v(0)$ and $v(L)$, at the ends of an originally straight column. The other assumptions are the same as before (as in Section 5.1.1). The boundary conditions now are

$$\begin{aligned} \text{At } z &= 0, & v &= v(0): & M &= -EI \cdot v'' = P \cdot v(0) \\ \text{At } z &= L, & v &= v(L): & M &= -EI \cdot v'' = P \cdot v(L) \end{aligned} \qquad (5.21)$$

The displacements, $v(z)$, have finite values and are single-values as long as $P < P_e$ and $\Delta \neq 0$. In this region $(0 < P < P_e)$ the problem is reduced from one of stability to a case of equilibrium.

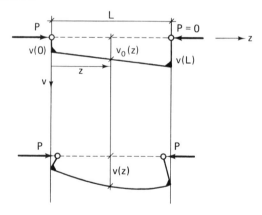

Figure 5.10 Eccentrically Loaded Straight Column.

The elastic line for $0 < P < P_e$ is

$$v(z) = v(0)\frac{\sin \alpha(L - z)}{\sin \alpha L} + v(L)\frac{\sin \alpha z}{\sin \alpha L} \qquad (5.22)$$

1. In the case of equal eccentricities, $v(0) = v(L) = e$. The elastic line is

$$v(z) = \frac{e}{\sin \alpha L}[\sin \alpha z + \sin \alpha(L-z)] \qquad (5.23)$$

If the safety factor against buckling is $FS_e = P_e/P$, then $\alpha L = \pi/\sqrt{FS_e}$.

The maximum stress at the column midheight is

$$f_{max} = -\frac{P}{A}\left[1 + \frac{ec}{r^2} \cdot \sec\left(\pi/2\sqrt{FS_e}\right)\right] \qquad (5.24)$$

where c is the distance of the outer fiber from the center line and r the radius of gyration. This equation is known as the secant formula.

2. In the case of equal and opposite eccentricities, $v(0) = -v(L) = e$ (Fig. 5.11), the deflection at the column midpoint remains zero as long as $P < P_e$ (Fig. 5.11c). When $P = P_e$, the value of $v(L/2)$ becomes $0/0$. A closer investigation shows that for $\sin \alpha L = 0$ the coefficient determinant of the system disappears and the equations are satisfied for any values of A.

In this case of antisymmetric eccentricities, we have a change of equilibrium. Figure 5.12 shows the bifurcation diagram. For comparison the displacements at the quarter-points, v_q, are also shown for the case when the eccentricities, $v(L) = \zeta v(0)$ (with $\zeta = -0.99$), are not quite equal.

5.1.3 Effective Buckling Length

Basis for the effective buckling length

It is customary to express various critical (Euler) loads for corresponding columns in terms of the Euler loads for the "classical" (pin-ended) column. This is achieved by adjusting the length, L, of the column (the actual buckling length) by a

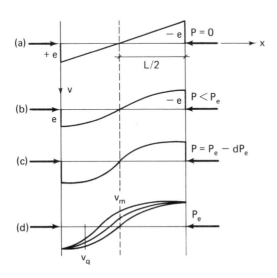

Figure 5.11 Equal and Opposite Eccentricities.

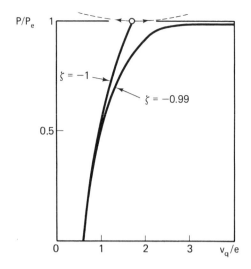

Figure 5.12 Bifurcation Diagram in Case of Eccentric Buckling.

factor K. The product, KL, is called the effective buckling length, which is then introduced in Eq. (5.3) instead of the actual length.

The effective-length factor, K, can be larger or lesser than 1.0 depending upon the column end restraints. As columns may be used either as isolated compression members or as parts of frames or trusses, K-factors will be derived in that order: first for individual columns and then for columns as integral parts of a structure.

Single columns

The theoretical K-values for individual columns can be obtained by comparison of the true solution of the actual case with the classical solution. The same can be achieved by finding the distance between the inflection points on the elastic curves or hinges, if there are any (Fig. 5.13). This distance is the effective buckling length.

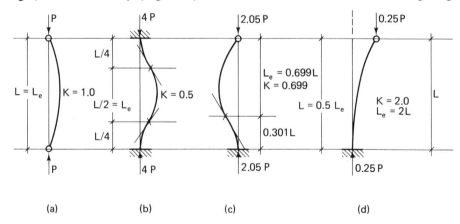

Figure 5.13 Equal Length Columns with Different Boundary Conditions (Same Actual Length, but Different Effective Length).

In the AISC commentary on specifications, table C 1.8.1, a total of six different cases are given with theoretical and recommended design K-values (5.12).

Columns in frames

In analyzing the buckling strength of frames it is customary to replace any distributed or concentrated loads acting on horizontal frame members by two equivalent joint loads, P, acting directly at the tops of columns or at the beam supports.

Half-frames

Consider first a half-frame (Fig. 5.14a) with a roller at support C. Even in a buckled position (Fig. 5.14b) there is no horizontal reaction at the column base at A. The vertical load, P_e, is distributed according to the lever-arm rule, where the left-

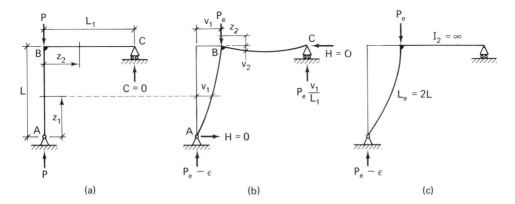

Figure 5.14 Buckling of a Half-Frame.

hand reaction can be approximated with P_e. The bending moment at B is $M_B = P_e \cdot v_1$. Applying the boundary conditions, the buckling criterion is obtained as

$$\alpha L \tan \alpha L = 3 \frac{L}{L_1} \cdot \frac{EI_2}{E_t I_1} \qquad (5.25)$$

If $E_t I_1 = E_2 I_2$ and $L = L_1$, the above expression simplifies to

$$\alpha L \cdot \tan \alpha L = 3 \qquad (5.26)$$

Figure 5.15 shows the solution of this transcendental equation (for details, see references 5.6, p. 994). The intersections between the two families of curves represent the solution. For the first mode of buckling the smallest value of αL is used, and the buckling load is

$$P_e = \frac{1.192^2 \cdot E_t I_1}{L^2} \qquad (5.27)$$

The effective buckling length is $L_e = (\pi/1.192)L = 2.63L$. If the right-hand beam support is not a roller, but a hinge (only rotation possible and no translation)—that is, if $v_1 = 0$ at $z_1 = L$—then the effective buckling length, L_e, is always less than L, depending upon the amount of the elastic restraint provided by the beam.

Buckling of Straight Prismatic Members

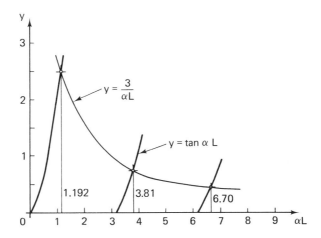

Figure 5.15 Solving Eq. (5.26): $\tan \alpha L = 3/\alpha L$.

Figure 5.14(c) shows the limiting case when the beam moment of inertia can be considered infinite. From $\alpha L \cdot \tan \alpha L = \infty$ it follows that $\alpha L = \pi/2$ and $P_e = \pi^2 E_t I/(2L)^2$; that is, this case is identical to that shown in Fig. 5.13(d).

Portal frames

Assuming that sufficient bracing is provided perpendicular to the plane of the frame or truss to prevent out-of-plane buckling, there are two modes of instability for a single frame:

1. The symmetrical mode (without sidesway) (Fig. 5.16a, b).
2. The antisymmetrical mode (with sidesway) (Fig. 5.17a, b).

From Fig. 5.16(c) it can be seen that the load necessary for sidesway collapse is always smaller than if sidesway is prevented.

In case sidesway is prevented, the effective-length factor, K, is always less than 1.0. Therefore, if no special analysis is performed to obtain a smaller value, K should be taken as unity (5.7, section 1.8.2, p. 5-29).

If sidesway is not prevented, the calculation of the stability limits and K-factors using the same analytical methods would be cumbersome. To avoid that, the "chart method" outlined in the AISC commentary (5.12) will be discussed (Fig. 5.18). To determine the effective-length factor, K, the relative stiffness, I/L, of the adjacent girders must be known or closely estimated. To enter the chart, G-values at he top and foot of the column must be computed from

$$G = \frac{\sum (I_c/L_c)}{\sum (I_g/L_g)} \tag{5.28}$$

in which the summation is applicable to all members connected to the joint lying in the plane in which buckling of the column is to occur. The use of this chart will be demonstrated in Example 5.3.

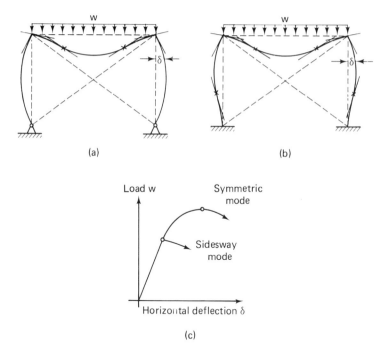

Figure 5.16 Symmetrical Failure Mode.

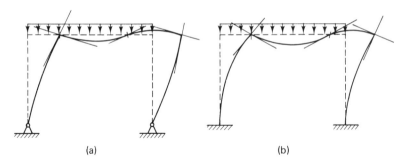

Figure 5.17 Sidesway Failure Mode.

Columns as truss members

When analyzing a truss the usual assumption is that any truss member is either a tie or an axially loaded column with pinned ends. Due to the rigidity of joints produced by the flexural stiffness of the gusset plates, as well as the stiffness of end connections themselves, each member will be subjected to an axial force and a bending moment. In addition, member ends will have a certain amount of rotational restraints, which in the case of compression members will reduce the actual buckling length. Normally the effect of bending on a truss member is disregarded but the rotational restraint is taken into account. To determine the effective length of a member it is necessary to solve the complete buckling of a truss as a whole, which is a time-consuming process. Therefore, in most practical design cases some recommended effective length factors are as follows (5.13)

Buckling of Straight Prismatic Members **157**

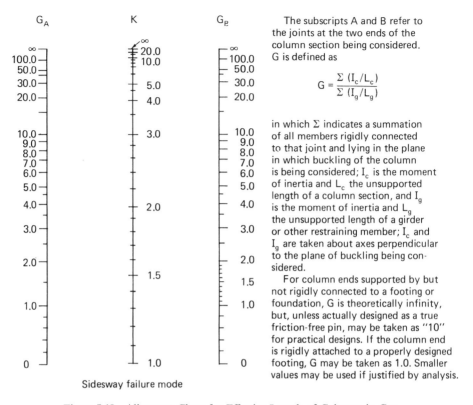

Figure 5.18 Alignment Chart for Effective Length of Columns in Continuous Frames (From AISC, *Manual of Steel Construction*, 8th ed., Chicago, 1980).

1. For the compression chord members of a truss which is designed economically, such as in bridges, K is taken as 1 for both out-of-plane or in-plane buckling. For a roof truss where often the complete compression chord is of constant cross section, K may be taken as 0.9.
2. In a continuous truss adjacent to the panel point where the chord stress changes sign, K may be taken as 0.85 for the compression chord continuous with the adjacent tension chord.
3. Web truss members (diagonals and verticals) designed for moving loads may be designed with $K = 0.85$.
4. For out-of-plane buckling of a main truss the K-factor is 1.0.

5.1.4 Design Procedures

In the following sections first the AISC and the AASHTO design recommendations will be discussed. Next some foreign design specifications will be reviewed. Finally longhand and computer-aided design methods will be discussed.

Design specifications

AISC SPECIFICATION (5.7). The AISC specification requires that in determining the slenderness ratio of an axially loaded compression member (except for bracing or secondary members) the length must be taken as its effective length, KL, and r as the minimum corresponding radius of gyration. The maximum slenderness ratio, KL/r, must not exceed 200.

To prevent local buckling, projecting elements under compression due either to axial compression or to compression due to bending must have a geometrical slenderness (ratio of width to thickness) not greater than the following values given in Part One of the AISC specification (1.9.1.2)

Single-angle or double-angle struts with separators $\quad\dfrac{76}{\sqrt{F_y}}$

Struts comprising double angles in contact;
Angles or plates projecting from girders, columns, $\quad\dfrac{95}{\sqrt{F_y}}$
  or compression flanges, of beams;
Stiffeners on plate girders

Stems of tees $\quad\dfrac{127}{\sqrt{F_y}}$

In Part Two of the AISC specifications dealing with plastic design, different width-to-thickness ratios are given to ensure section stability under full plastification. When the actual width-to-thickness ratio exceeds these values, the design is governed by appendix C of the specifications. According to this appendix—"Slender Compression Elements" (5.14, pp, 5.94–5.96)—the stress on unstiffened compression elements must be subject to a reduction factor, Q_s. The value of Q_s should be determined from given formulas according to the shape and the quality of the steel.

Built-up compression members composed of two or more rolled shapes separated from one another only by intermittent fillers (separators) must be connected to one another at these fillers at intervals such that the slenderness ratio, l/r, of either shape, between the fasteners, does not exceed the governing slenderness ratio of the built-up member (Fig. 5.19). The least radius of gyration, r, must be used in computing the slenderness ratio of each component part.

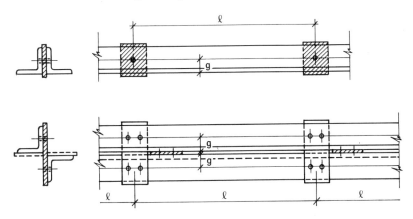

Figure 5.19 Built-up Members Composed of Two Angles.

Open sides of compression members built up from plates or shapes must be provided with lacing (Fig. 5.20) having batten plates as near the ends as practicable. In main members carrying calculated stresses, the end plates must have a length (along the column axis) of not less than the distance between the lines of bolts or welds connecting them to the components of the member. The thickness of plates must be not less than 2% of the distance (perpendicular to the column axis) between the lines of bolts or welds connecting them to the segments of the members. The pitch of bolts must be not more than 6 diameters, with a minimum of three fasteners per segment, and the total weld length should not be less than one-third its length.

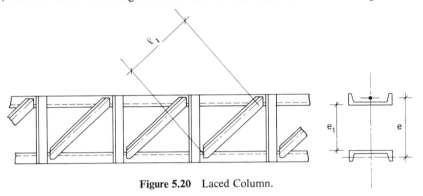

Figure 5.20 Laced Column.

Lacing must be spaced so that the ratio, l/r, of the flange included between their connections does not exceed KL/r, the governing ratio for the member as a whole. A nominal shear force (normal to the column axis) has to be taken into account equal to 2% of the total compressive stress in the column. The 2% is an old measure of the quality of workmanship expected in practice; that is, it is anticipated that, in a column, deviations from a straight line are about $\frac{1}{4}$ in. in 1 ft (2 cm in 1 m) (Fig. 5.21). The ratio, l/r for lacing bars arranged in a single system must not exceed 140.

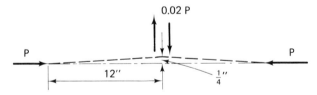

Figure 5.21 Shear Force Due to Fabrication Error.

Lacing bars in compression may be treated as secondary members, l_1 being taken as the unsupported length of the lacing bar between bolts or welds of its end connection (Fig. 5.22). The inclination of lacing bars to the axis of the member should preferably be not less than 60°. If the distance $a > 15$ in., the specifications require double lacing instead of single lacing. If single lacing is preferred, angles have to be used.

As an alternative to lacing, AISC specification allows the use of continuous cover plates perforated with a succession of access holes (Fig. 5.23.) It should be noted that perforated cover plates are only used for heavy columns, such as pylons, where large bending moments are acting together with large axial forces. Laced columns

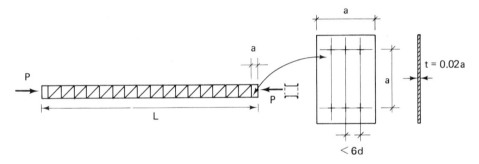

Figure 5.22 End Tie-plates.

Figure 5.23 Perforated Plate.

have been abandoned in Europe for some time as uneconomical. The amount of labor per unit weight for a finished laced column considerably exceeds that needed for a "batten" column (Fig. 5.24). As AISC specification does not consider such columns, the details of their design will be discussed below in the section dealing with foreign specifications.

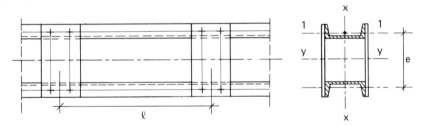

Figure 5.24 Batten Column.

The design based on AISC specification does not differentiate between a simple solid column or a built-up column, which contradicts Eq. (5.20).

For intermediate columns for a chosen section (Fig. 5.9), equation (1.5-1) in the AISC specification (5.14, p. 5-19) has to be used to obtain the allowable stress in compression:

$$F_a = \frac{(1 - \tfrac{1}{2}(\lambda/C_c)^2)F_y}{\text{F.S.}} \tag{5.29}$$

where λ = slenderness ratio, KL/r; F_y = yield-point stress; C_c = limiting slenderness ratio for which the critical (Euler) stress is equal to one-half the yield point stress—that is:

$$C_c = \lambda_c = \pi\sqrt{\frac{2E}{F_y}} \tag{5.30}$$

The factor of safety (F.S.) is

$$\text{F.S.} = \frac{5}{3} + \frac{3}{8}\left(\frac{\lambda}{C_c}\right) - \frac{1}{8}\left(\frac{\lambda}{C_c}\right)^3 \tag{5.31}$$

Buckling of Straight Prismatic Members

Equation (5.29) can be applied only if $0 < \lambda < C_c$ (characteristics of an intermediate column). The factor of safety F.S., is taken for "zero-length" columns equal to that required for members axially loaded in tension. For columns with $\lambda = C_c$ this factor of safety is increased by 15%—that is, from $\frac{5}{3} = 1.67$ to $\frac{23}{12} = 1.92$. In between, the F.S. is varied according to a quarter-wave of a sine function expressed in polynomial form as given in Eq. (5.31).

For "long" columns ($\lambda > C_c$) the F.S. is constant and taken as $\frac{23}{12} = 1.92$. As buckling in this range is elastic, Euler's formula does apply and

$$F_a = \frac{\pi^2 E}{\text{F.S.} \times \lambda^2} \tag{5.32}$$

where $\lambda = KL/r$ and F.S. $= \frac{23}{12}$. Equation (5.32) is the same as the AISC equation (1.5-2) (5.14, p. 5-19).

When the slenderness ratio, λ, exceeds 120 (for bracing and secondary members) the following formula [AISC equation 1.5-3 (5.7)] has to be used:

$$F_{as} = \frac{F_a}{1.6 - (L/200r)} \tag{5.33}$$

where F_a is determined from AISC equation (1.5-1) or (1.5-2) (5.7). In the AISC Manual (5.14), appendix A, tables 3-36–50 give the solution to Eqs. (5.29), (5.32), and (5.33). Illustrations are given in Examples 5.1 to 5.4.

AASHTO SPECIFICATIONS (5.15). In Section 7, *Structural Steel Design*, AASHTO (Table 1.7.1A) has introduced two formulas (Fig. 5.25), one for inelastic buckling ($KL/r = \lambda \leq C_c$):

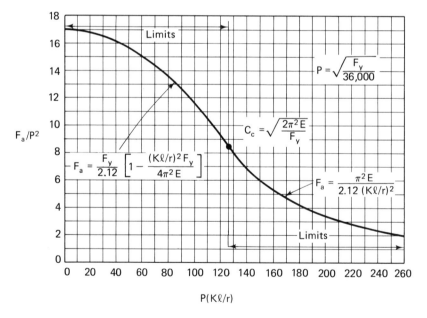

Figure 5.25 AASHTO Column Formulas (From AASHTO, *Standard Specifications for Highway Bridges*, 12th ed., Washington, D.C., 1977).

$$F_a = \frac{F_y}{\text{F.S.}}\left[1 - \frac{1}{2}\left(\frac{\lambda}{C_c}\right)^2\right] \qquad (5.34)$$

and another for elastic buckling $\lambda > C_c$

$$F_a = \frac{\pi^2 E}{\text{F.S.} \times \lambda^2} = \frac{135{,}000}{\lambda^2} \; (\text{lb/in.}^2) \qquad (5.35)$$

where F.S. = a constant factor of safety equal to 2.12 and

$$C_c = \left(\frac{2\pi^2 E}{F_y}\right)^{1/2} \qquad (5.36)$$

As the limiting slenderness ratio AASHTO gives 120 for main members and 140 for secondary members.

The effective-length factors, K, are the same as before. Under "Load Factor design" (5.15, section 1.7.52, pp. 215–39) the maximum strength of concentrically loaded columns is given as

$$P_u = 0.85 A F_{\text{crit}} \qquad (5.37)$$

where A = gross effective area of the column cross section and F_{crit} = critical stress determined by Eq. (5.34) or Eq. (5.35) without an F.S. This shows that "load factor design" is considered a limit design subjected to multiples of the design (service) loads without any factor of safety. These specifications also ignore batten columns and do not differentiate between columns with solid or built-up cross sections.

For the value of the shearing force, V, normal to the member in the planes of lacing or continuous perforated plates the AASHTO specifications (5.15, section 1.7.44(I), p. 183) give

$$V = \frac{P}{100}\left(\frac{100}{L/r + 10} + \frac{L/r}{3{,}300/F_y}\right) \qquad (5.38)$$

where P = allowable compressive axial load on members in kips, L = length of member in inches, r = radius of gyration of section about the axis perpendicular to plane of lacing or perforated plate in inches, and F_y = specified minimum yield point of type of steel being used in ksi. The force, V, has to be equally divided among all parallel planes. This formula obviously was developed for the old A7 steel with $F_y = 33$ ksi. For this type of steel and a "zero-length" column—that is, for $L/r = 0$—we see from Eq. (5.38) that V is 10% of the axial compressive load. A column should have a slenderness ratio L/r of 63 or 127 to have a fictitious shear force equal to $V = 0.02P$, the same as the constant-percentage shear force given by the AISC specifications. The AASHTO specifications are the only ones known to the authors to provide a variable-percentage shear force. In Fig. 5.26 the value of the percentage (the expression in parentheses) in Eq. (5.38) is plotted against the slenderness ratio, L/r. It is seen that this function has a minimum value of 1.90% for $L/r = 90$. There could be no rational explanation for such behavior, and this formula is certainly of dubious value. Anyway, this function from Fig. 5.26 practically approaches the value of 2% in the range of columns existing in practice, so that a much simpler assumption using $V = 0.02P$ would be in place.

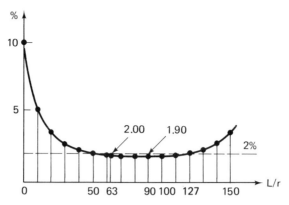

Figure 5.26 Ideal Shear Force (From AASHTO, *Standard Specifications for Highway Bridges*, 12th ed., Washington, D.C., 1977).

Foreign specifications

GERMAN SPECIFICATIONS—DIN 4114 (5.16). These specifications, like the American ones, use the expression for "classical" pin-ended columns and the effective buckling length—that is, by calculating for any particular column the "equivalent" length of a pin-ended column. According to this length the slenderness ratio, λ, is obtained and a coefficient, ω (buckling number), is calculated. The nominal compression stress, P/A, is then multiplied by ω and compared with the allowable stress, which depends upon the loading case and the type of steel used. Thus

$$\omega \cdot \frac{P}{A} \leq F_a \qquad (5.39)$$

For built-up columns the slenderness ratio, λ, which cannot be larger than 250, is calculated for buckling around a material axis as usual (Fig. 5.27) as

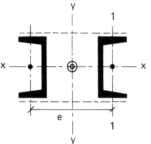

Figure 5.27 Built-up Section with $x-x$ = Material Axis, $y-y$ = Imaterial (Open) Axis.

$$\lambda_x = \frac{KL_x}{r_x} \qquad (5.40)$$

For the open axis, y-y, an "ideal" slenderness ratio, λ_{yi}, is obtained which is different for laced columns than for batten columns.

A value for λ_1 is first calculated. For a batten column (Fig. 5.24)

$$\lambda_1 = \frac{l}{r_1} \qquad (5.41)$$

For a laced column this value is

$$\lambda_1 = \pi \sqrt{\frac{A}{zA_d} \cdot \frac{l_1^3}{l \cdot e^2}} \qquad (5.42)$$

where A = gross cross-sectional area of the whole column
A_d = gross cross-sectional area of one diagonal
z = number of parallel bracing planes
l_1 = theoretical diagonal length

e = distance between the centroidal axes of the two sides,
l = largest panel of the column

To take into account the sensitivity of built-up columns to shear, DIN specifications introduce the ideal slenderness ratio for buckling in the plane parallel to lacing or batten planes. Thus for buckling normal to the principal axis, y-y, the ideal slenderness ratio, λ_{yi}, is calculated from

$$\lambda_{yi} = \sqrt{\lambda_y + \frac{m}{2}\lambda_1^2} \tag{5.43}$$

where m = number of sections connected by lacing or plates to make a built-up column. The column is stable if

$$\omega_x \frac{P}{A} \leq F_a \quad \text{and} \quad \omega_{yi} \frac{P}{A} \leq F_a \tag{5.44}$$

where ω_{yi} is the buckling number corresponding to the ideal slenderness, λ_{yi}, and F_a is the allowable compressive stress.

For buildings the slenderness of each section of a built-up column must also satisfy the condition

$$\frac{l}{r_1} \leq \frac{1}{2}\lambda_x \left(4 - 3\frac{\omega_{yi} \cdot P}{A \cdot F_a}\right) \tag{5.45}$$

When $\frac{1}{2}\lambda_x < 50$, instead of $\frac{1}{2}\lambda_x$ 50 can be taken.

In bridges and cranes the condition is

$$\frac{l}{r_1} \leq \frac{1}{2}\lambda_x \tag{5.46}$$

but not to exceed 50.

For the design of lacing or batten plates an ideal constant (independent of slenderness ratio) shear force, V_i, is used. In bridges and cranes this force is

$$V_i = \frac{A \cdot F_a}{80} = \frac{1.25}{100}(A \cdot F_a) \tag{5.47}$$

and in buildings

$$V_i = \frac{\omega_{yi} P}{80} = \frac{1.25}{100}(\omega_{yi} P) \tag{5.48}$$

In the case of batten columns, if the distance, e (Figs. 5.27 and 5.28), is larger than $20r_1$, this ideal shear force has to be increased by

$$5\left(\frac{e}{r} - 20\right)\% \tag{5.49}$$

For laced columns, which are preferred when the distance between the column branches is large, this increase in shear force is not required.

To design a batten column a simplification of the actual Vierendeel girder (5.17) is introduced. It is assumed that the inflection points are at the midpoints of the chords and the vertical members (Fig. 5.29). From the isolated girder part 1-2-3 (Fig. 5.29), summing the moments about point A and solving for T, the shear force in the batten plates is (Fig. 5.30)

$$T = V_i \frac{l}{e} \tag{5.50}$$

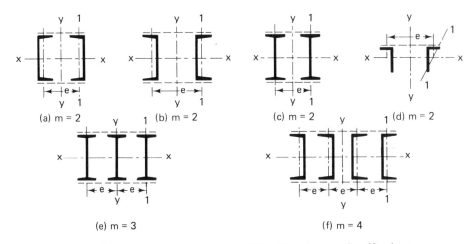

Figure 5.28 Batten Columns (From DIN 4114, German Specifications for Instability, Part 1, Section 8.212, fig. 5, Berlin, 1952).

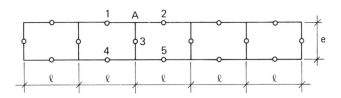

Figure 5.29 Assumed Inflection Points in a Vierendeel Girder.

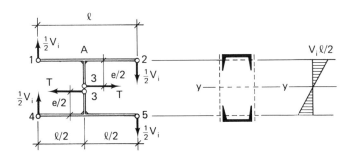

Figure 5.30 Shear in Battens.

This force, T, in the case of two-part built-up columns (Fig. 5.28a, b, c, and d) is carried by two battens. In case of three-part columns (Fig. 5.28e),

$$T = \frac{V_i l}{2e} \tag{5.51}$$

For design examples see Examples 5.1 to 5.4.

BRITISH SPECIFICATIONS—BS 449 (5.18) These specifications are quite similar to the AISC specifications (5.7) for columns of simple or built-up sections using a classical column and effective length. There is no provision for λ_{yi} as in other European specifications.

However, in addition to laced columns the British specifications make provision for batten columns. The battens must be placed opposite one another at each end of the member and at points where the member is stayed in its length and must, as far as practicable, be spaced and proportioned uniformly throughout. The number of battens must be such that the member is divided into not fewer than three bays within its actual length center-to-center of connections.

In batten columns in which the slenderness ratio, λ_y, about the open axis is not more than 80% of that about the *x*-axis (Fig. 5.27), the spacing of battens center-to-center of end fastenings must be such that the ratio of slenderness, $\lambda_1 = l/r$, is not greater than 50 or $0.7\lambda_x$, whichever is smaller. If $\lambda_y > 0.8\lambda_x$, then the spacing of battens must be such that $\lambda_1 \leq 40$ or $0.6\lambda_y$ (where the *y*-axis is the weaker axis), whichever is smaller.

The ideal shear force, V_i, is taken as 2.5% of the total axial force on the whole compression member, divided equally among parallel planes of battens.

FRENCH SPECIFICATIONS—CM66 (5.19). These specifications for solid (simple) and built-up section columns use the ω-procedure identical to the German (DIN) specifications.

U.S.S.R SPECIFICATIONS—SN AND PII-V.3-62 (5.20). The evaluation of column stability is similar to the procedure used in the German DIN 4114 specifications. Instead of finding a number which is larger than 1 ($\omega > 1$) for multiplication with the nominal compressive stress and comparing this value with the allowable compressive buckling stress, Soviet specifications require the determination of a coefficient of buckling, $\phi < 1$, as a function of the maximum slenderness ratio. Thus a column is stable if

$$\frac{P}{\phi A} \leq F_a \qquad (5.52)$$

where F_a = allowable compressive stress, which depends upon the type of steel, the structural element, and the loading condition (working load or ultimate). The maximum allowable slenderness is given as $\lambda = 220$. The reciprocal value of ϕ when compared with corresponding values of ω shows that (for mild structural steel) for $\lambda < 100$ they are very close, the Soviet values being slightly less conservative. For $\lambda > 100$ the difference becomes larger, and for $\lambda = 220$ it is about 25%, the Soviet values being smaller.

Iterative method for the design of columns using American specifications

It is possible to develop an algorithm for longhand computation which will lead the designer to the correct rolled section almost automatically. To get optimum results the same algorithm has to be applied to several different types of rolled sections; the optimum value is then chosen from the results of these various algorithms. Usually the optimum is considered to be the section producing the minimum total column weight.

The algorithm actually consists of a Do-Loop of computer instructions which in some cases becomes unstable, resulting in a closed do-loop. In such a case, the designer has to intervene by assuming a new "mean-value" choice. The reason for instability lies in the discontinuity in the spectrum of available sections. For economy, a simple (solid) section column should be tried first, provided of course that there are no specific design constraints requiring a built-up column. To start, a trial section must be initiated. If the buckling lengths for buckling around the x-axis (strong axis—that is, buckling occurring in the plane yoz) and around the y-axis (weak axis, plane of buckling zox) are the same (Fig. 5.31), then $KL_y/r_y > KL_x/r_x$ and the controlling slenderness is $\lambda_y = KL_y/r_y$. Wide-flange sections which have the smallest ratio, r_x/r_y, are most economical. A check from the AISC *Manual* tables designated "Dimensions and Properties" (5.14, pp. 1.14-29) will show that for columns only W8, W10, W12, and W14 come into consideration. For relatively small loads, $P < 200$ kips, and small buckling lengths (approximately $L < 12$ ft), W8 or W10 sections should be tried first. For moderate loads (say, 200 kips $< P < 600$ kips) and moderate length (12 ft $< L < 20$ ft), W12 and W14 should be tried. For larger loads ($P > 600$ kips) and lengths ($L > 20$ ft), W14 should be tried out. Only when no sections are available or only very heavy sections will satisfy should batten built-up columns be considered.

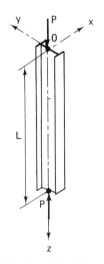

Figure 5.31 Equal Buckling Lengths.

The algorithm shown (Fig. 5.32) applies to one specific initially chosen W-designation. The steps are self-explanatory. As a starting value a slenderness ratio of $\lambda = 80$ is chosen as a useful average. For steps 4 or 9 it has to be remembered that the available sections within a particular W-designation do not comprise a continuous spectrum. Therefore, when a new λ is close enough, one must exit from the do-loop—that is, perform step 11. The allowable stress, F_a, can be obtained from the particular specifications followed.

If the buckling lengths around the x- and y-axes are different, with usually $L_x > L_y$, each time a new section is chosen it must be established which ratio is larger, KL_x/r_x or KL_y/r_y. The larger value is used as λ.

Some illustrative examples are given below to demonstrate the design procedure, first for solid section columns and next for built-up batten columns.

Example 5.1 Design a column using A36 steel and the AISC specification (ref. 5.14). The axial force is $P = 600$ kips (2.67 MN), and $L_x = L_y = 25$ ft (7.62 m). One end of the column is pinned, the other end built-in.

Solution From the AISC *Manual* (5.14) (p. 5-124), the recommended design value for K is taken as $K = 0.80$. Thus $KL = 25 \times 0.8 = 20$ ft (6.1 m). According to the load and the effective buckling length, first a W12 will be attempted,

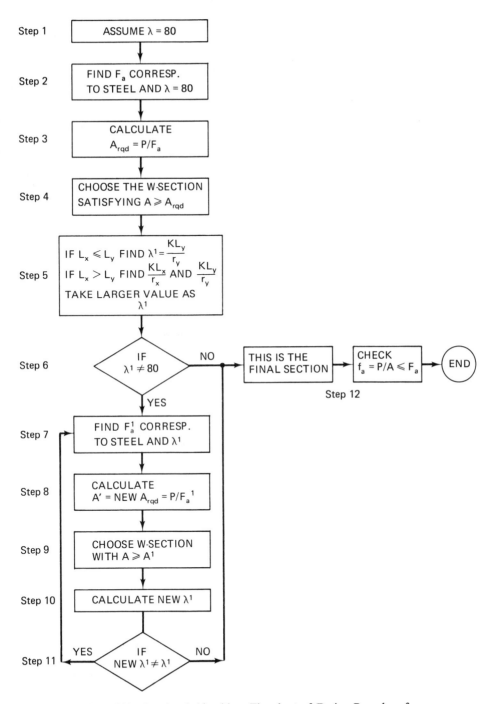

Figure 5.32 Longhand Algorithm. Flowchart of Design Procedure for Columns.

although this probably represents the limit of economical application for this series.

Step 1: Assume $\lambda = 80$.
Step 2: From the AISC *Manual* (p. 5-74) $F_a = 15.36$ ksi (105.90 MPa).
Step 3: $A_{rqd} = 600/15.36 = 39.06$ in.² (252 cm²).
Step 4: From the *Manual* (p. 1-24) W12 × 136 has $A = 39.9$ in. > 39.06 in.² and $r_y = 3.16$ in. (8.03 cm).
Step 5: Because $L_x = L_y$, $\lambda^1 = KL_y/r_y = (20 \times 12)/3.16 = 75.9$.
Step 6: $\lambda^1 = 75.9 < \lambda = 80$.
Step 7: From the *Manual* (p. 5-74, for $\lambda = 76$): $F_a^1 = 15.79$ ksi (108.87 MPa).
Step 8: $A^1 = 600/15.79 = 38.0$ in.² (245.16 cm²).
Step 9: From the *Manual* (p. 1-24) it is seen that there is no section other than W12 × 136 satisfying $A \geq A^1$, and therefore, proceed to
Step 11: New $\lambda^1 = \lambda^1$.
Step 12: As this is the final section, EXIT: $f_a = 600/39.9 = 15.04$ ksi < 15.79 ksi: OK

Next we try a W14 because the capacity of W12 was almost exhausted and the heaviest sections of any *W*-designation are never economical. In case such heavy sections are used, it is always advisable to try the next larger *W*-designation, which might produce a lighter section.

Steps 1, 2, 3: As before, $\lambda = 80$ and $F_a = 15.36$ ksi (105.90 MPa); $A_{rqd} = 39.06$ in.² (252 cm²).
Step 4: From the AISC *Manual* (5.14, p. 1-20) W14 × 145 (W36 cm × 2.11 kN/m) has $A = 42.7$ in. > 39.06 in. and $r_y = 3.98$ in. (10.11 cm).
Step 5: $\lambda^1 = (20 \times 12)/3.98 = 60.3$.
Step 6: $\lambda^1 = 60.3 < 80$.
Step 7: From the *Manual* (p. 5-74, for $\lambda^1 = 61$) $F_a^1 = 17.33$ ksi (118.49 MPa).
Step 8: $A^1 = 600/17.33 = 34.62$ in.² (227.16 cm²).
Step 9: From the *Manual* (p. 1-22) we get W14 × 120 (W36 cm × 1.75 kN/m).
Step 10: New $\lambda^1 = (20 \times 12)/3.74 = 64.2$
Step 11: New $\lambda^1 > \lambda^1$, therefore $F_a^1 = 17.02$ as $f_a = 600/35.3 = 17.00$ ksi < 17.02 ksi.   OK

It can be seen that a W14 section is better, as it is lighter by 16 lb/ft (234 N/m).

Example 5.2 Design for a truss a welded batten diagonal composed of two standard channels in A36 steel. The theoretical length of the diagonal is 22.25 ft (6.78 m) and the maximum compressive force is $P_{max} = 300$ kips (1.33 MN). Use AASHTO specifications (5.15). The force in the diagonal does not change sign and $P_{min} = 200$ kips. The number of cycles is 500,000 (bridge on a freeway with HS loading—see AASHTO (5.15, p. 151) table 1.7.2B-Stress Cycles, case II).

Solution The design of a bridge member must always be checked for fatigue. Therefore, the first part of the solution is to design this diagonal based on the instability criterion (see Section 1.2.6) and then to recheck the design considering the fatigue criterion (Section 1.2.7).

As pointed out under the heading "Columns as Truss Members" in Section 5.1.3 and under "Design Specifications" in Section 5.1.4, the K-factor is 0.85 for buckling in the plane of the truss and 1.00 for out-of-plane buckling. Thus 85% of the theoretical length of diagonals corresponds to the distance between the centers of their end connections. As the section is composed of two channels (Fig. 5.33) for buckling around the x-axis (in-plane), the effective buckling length

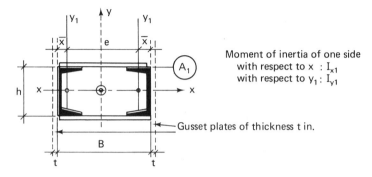

Figure 5.33 (Example 5.2).

is $K_x L_x = 0.85 \times 22.25 = 18.91$ ft (5.76 m) and for out-of-plane buckling $K_y L_y = 1.00 \times 22.25$ ft $= 22.25$ ft (6.78 m). The distance, h, normally depends upon the width, B. Let us suppose that we have freedom in determining this overall width, B. A good practice is to have this distance sufficient so that the moment of inertia about the y-axis (open axis) is at least 10% larger than that about the x-axis (material axis). In our special case this is still not sufficient due to different effective buckling lengths. Therefore, taking into consideration this difference in length one may write the following equation

$$I_y = 1.10 \left(\frac{K_y L_y}{K_x L_x} \right)^2 I_x \tag{5.53}$$

In our case, $I_y = 1.29 I_x$.

Using the notation given in Fig. 5.33 after substitution

$$I_y = 2 \left[I_{y1} + A_1 \left(\frac{e}{2} \right)^2 \right] \quad \text{and} \quad I_x = 2 I_{x1} \tag{5.54}$$

Equation (5.53) is written in the form

$$I_{y1} + \frac{A_1}{4} e^2 = 1.29 I_{x1}$$

Solving Eq. (5.54) for e we have

$$e_{\min} = 2\sqrt{1.29 r_{x1} - r_{y1}} \tag{5.55}$$

Of course this is only a minimum value for orientation, but any larger practical value can be adopted. The clear distance between the branches should be at least 4–5 in. (10–15 cm).

The "box" width, B, is then

$$B_{\min} = e_{\min} + 2\bar{x} \tag{5.56}$$

A practical width is also $L/15$, which in our case gives $B = 18$ in. (46 cm), and this value will be assumed for B.

If we now turn to the iteration process in steps 1, 2, and 3 (Fig. 5.32) and employ Fig. 5.25, it follows that $F_a = 13.5$ ksi for $\lambda = 80$ and that

$$A_{\text{rqd}} = \frac{300}{13.5} = 22.22 \text{ in.}^2 (143.4 \text{ cm}^2) \qquad A_{1\text{-rqd}} = 11.11 \text{ in.}^2 (71.7 \text{ cm}^2)$$

Step 4: From the AISC *Manual* (5.14, p. 1-36) 2C15 × 40 (2C38 cm × 584 N/m) has $A_1 = 11.80$ and $r_{x1} = 5.44$ in. (14.27 cm); $r_{y1} = 0.886$ in. (2.25 cm). From Eq. (5.55) $e_{\min} = 4.95$ in., $B_{\min} = 6.51$ in. < 18 in. Use 18 in.
Step 5: $\lambda^1 = K_x L_x / r_x = (18.9 \times 12)/5.44 = 41.7 \approx 42$.
Step 6: $\lambda^1 = 42 < 80$.
Step 7: From Fig. 5.25 [from AASHTO specifications (5.15), p. 455] for $\lambda^1 = 42$, $F_a^1 = 16.0$ ksi (110.3 MPa).
Step 8: $A^1 = 300/16 = 18.75$ in.^{2}; $A_1^1 = 9.37$ in.^{2} (60.48 cm^2).
Step 9: From the AISC *Manual* (p. 1-36) 2C15 × 33.9 (2C38 cm × 495 N/m) with $A_1 = 9.96$ in.^{2}, $r_{x1} = 5.62$ in.; $r_{y1} = 0.904$ in.; $e_{\min} = 5.04$ in.
Step 10: New $\lambda^1 = (18.91 \times 12)/5.62 = 40.4 \approx 41$.
Step 11: New $\lambda^1 = 41 < \lambda^1 = 42$.
Step 7a: From Fig. 5.25, for $\lambda^1 = 41$; $F_a^1 = 16.1$ ksi (111.0 MPa).
Step 8a: New $A^1 = 300/16.1 = 18.63$ in.^{2}; $A_1^1 = 9.32$ in.^{2} (60.11 cm^2).
Step 9a: From the AISC *Manual* (p. 1-36) 2C15 × 33.9 (the same section as in step 9).
Step 10a: New $\lambda^1 = 41$
Step 11a: New $\lambda^1 = \lambda^1 = 41$. END OF ITERATION PROCESS.
Step 12: EXIT: For $B = 18$ in.,

$$r_y = \sqrt{\frac{8.13 + 9.96(9.0 - 0.787)^2}{9.96}} = 8.26 \text{ in.} (20.98 \text{ cm})$$

$$\lambda_x = \frac{K_x L_x}{r_x} = \frac{18.91 \times 12}{5.62} = 40.4$$

$$\frac{K_y L_y}{r_y} = \frac{22.25 \times 12}{8.26} = 32.3 < 40.4$$

$$\therefore \quad F_a = 16.1 \text{ ksi } (111.0 \text{ MPa})$$

$$(f_a)_{\max} = \frac{300}{2 \times 9.96}$$

$$= 15.1 \text{ ksi } (104.1 \text{ MPa}) < 16.1 \text{ ksi } (11.0 \text{ MPa}): \qquad \underline{\text{OK}}$$

$$(f_a)_{\min} = \frac{200}{19.92} = 10.0 \text{ ksi } (68.9 \text{ MPa})$$

The stress range, $f_{sr} = 5.1$ ksi (35.2 MPa), should now be checked for fatigue. From table 1.7.2A1, stress category E, AASHTO specifications (5.15, p. 144) the stress range obtained, $F_{sr} = 12.5$ ksi (86.2 MPa), is larger than the calculated stress range of 5.1, for both the base and the weld metal. Therefore, no change is needed.

To design battens, their distribution along the diagonal, and their connection to the two sides (channels), let us follow the German DIN 4114 specifications for bridges (5.16).

From Eq. (5.46) the slenderness ratio, λ^1, for each side of the column must be less than or equal to $\lambda_x/2$. In our case this $\lambda^1 = \frac{1}{2}(40.4) = 20.2 < 50$: $\qquad \underline{\text{OK}}$

Therefore, the center-to-center distance between battens is
$$l = r_1\lambda_1 = r_{y1}\lambda_1 = 0.904 \times 20.2 = 18.3 \text{ in.:} \quad \text{use 18 in. (45 cm)}$$
According to Eq. (5.47) the ideal shear force is
$$V_i = 0.0125AF_a = 0.0125 \times 19.92 \times 16.1 = 4.0 \text{ kips (17.7 kN)}$$
As the distance, e, is $B - 2\bar{x} = 18.0 - 2(0.787) = 16.4$ in. $< 20r_{y1} = 18.1$ in., according to Eq. (5.49) no increase of V_i is required. Therefore,
$$V_i = 1.0 \times 4.0 = 4.00 \text{ kips (17.80 kN)}$$
The shear force in both battens is (Eq. 5.51)
$$2T = 4.00\frac{18}{16.4} = 4.39 \text{ kips}$$
or in one batten $T = 2.20$ kips (9.79 kN).

For trial dimensions of the batten we take the width as approximately $\frac{1}{2}B = 9$ in. and the thickness as $0.02B = 0.36$ in. $\approx \frac{3}{8}$ in. This section has to carry a shear force, T, and a moment of $8.5T$ (Fig. 5.34). The maximum shearing stress is
$$f_v = \frac{3}{2}\frac{T}{A} = \frac{3}{2}\cdot\frac{2.20}{\frac{3}{8}(9)} = 0.98 \text{ ksi} < 12.0 \text{ ksi}$$
The maximum bending stress is
$$f_b = \frac{Te}{I/c} = \frac{2.20(8.5)}{\frac{1}{6}\cdot\frac{3}{8}(9)^2} = 0.62 \text{ ksi} < 20.0 \text{ ksi}$$
Therefore, use for battens P_L-9 in. $\times \frac{3}{8}$ in.

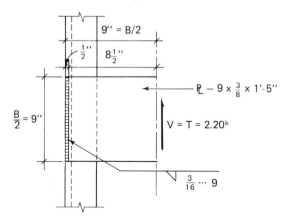

Figure 5.34 (Example 5.2).

For design of the fillet weld:

Maximum size: $\frac{3}{8} - \frac{1}{16} = \frac{5}{16}$ in. [5.15, section 1.7.21(C)]

Minimum size: $\frac{3}{16}$ in. (flange thickness $\frac{1}{2}$ in.) [5.15, section 1.7.21(B)]

At the end of the weld the maximum force per unit length of the weld according to Example 9.13 (Section 9.3.3, under "Elastic method") is
$$(T)_{max} = \sqrt{\left(\frac{H}{\Sigma l}+\frac{M_t}{J_p}y_{max}\right)^2 + \left(\frac{V}{\Sigma l}+\frac{M_t}{J_p}x_{max}\right)^2}$$

Buckling of Straight Prismatic Members   **173**

where, in our case, $H = 0$, $V = 2.20$ kips (9.79 kN), $M_t = Ve = 2.20(8.5) = 18.7$ kip-in. (211.28 kN-cm), $J_p = J_x = (9)^3 \cdot 1/12 = 60.75$ in.^{4} $y_{max} = 4.5$ in., and $x_{max} = 0$.

Thus:

$$(T)_{max} = \sqrt{\left[\frac{18.7}{60.75}(4.5)\right]^2 + \left(\frac{2.20}{9.0}\right)^2} = 1.41 \text{ kips/in. of weld}$$

According to AASHTO (ref. 5.15, p. 170), using an electrode E60 a weld of $\frac{3}{16}$ in. is needed (minimum size).

Example 5.3 Select a wide-flange section for the columns of a portal frame (Fig. 5.35) using A36 steel and AISC specification (5.14). The web of the rolled section lies in the frame plane. The horizontal beam member (girder) is a W12 × 79. Girts are attached at the midheight of the columns.

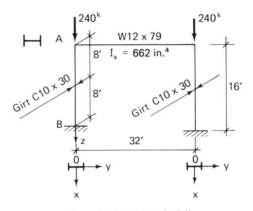

Figure 5.35 (Example 5.3).

Solution In this example the use of the chart in Fig. 5.18 [from the commentary on the AISC specification (5.12), p. 5-125] and AISC *Manual* tables for column design (5.14, part 3, pp. 3-1–97) will be illustrated.

The load of 240 kips (533.8 kN) is thought of as if transmitted by the horizontal element directly to the columns. For asymmetrical vertical loads or any horizontal loads, sidesway will occur, and therefore the effective buckling length of the columns is larger than 16 ft (for in-plane buckling: yoz-plane). For out-of-plane buckling (xoz-plane) there is no sidesway and the buckling length is 8 ft (conservatively).

To use the chart, the ratio

$$G = \frac{\sum I_c/L_c}{\sum I_g/L_g}$$

has to be found for the top and foot of the column for in-plane buckling. At the top,

$$G_A = \frac{I_c/16}{I_g/32} = 2I_c/I_g$$

Initially I_c will not be known. Therefore, the ratio I_c/I_g has to be assumed and later checked based on the obtained solution. If the discrepancy between the assumed and obtained value is large (say, at least 20%), another iteration cycle is needed in which the previously obtained section is usually assumed for a starting value.

In this example another approximation is made. In addition to axial forces the column is subjected to quite substantial bending moments caused by portal frame action. Therefore, these columns are so-called beam columns (see Chapter 8) and their design is different from column design. This effect will be disregarded here and the columns will be designed only with respect to the axial load. (See Example 8.4 for a redesign of this same problem).

First we assume the ratio I_g/I_c to be 0.5. Why? The load is not very large $(200 < P < 600)$ and the column height is moderate (12 ft $< L_c <$ 20 ft), and so most probably a below-average W12 will satisfy. The moment of inertia, I_x, of these sections is between 300 and 400. The beam has an I_x of 663 in.^{4}; therefore, it is logical to take as the first approximation a ratio $I_c/I_g \approx 0.5$. Therefore, $G_A = 2(0.5) = 1.00$. The coefficient at the column foot, G_B, is 1.0 for the fixed-end condition. The chart (Fig. 5.18) then gives $K_x = 1.32$ and $K_xL_x = 1.32 \times 16 = 21.12$ ft. For out-of-plane buckling $K_y = 1.00$ (actually less than 1.0 because no sidesway and no hinges in the 8-ft length!). Therefore, $K_yL_y = 8.0$ ft.

The AISC column tables (5.14) have been developed based on the assumption that

$$\frac{K_yL_y}{r_y} \geq \frac{K_xL_x}{r_x} \tag{5.57}$$

By multiplying both sides of this inequality by r_y it is seen that

$$K_yL_y \geq \frac{K_xL_x}{r_x/r_y} \tag{5.58}$$

This means that before using the AISC column tables one must always check this equation for each section. If this condition is satisfied, proceed with K_yL_y as the effective buckling length. If not, use the reduced value of

$$\frac{K_xL_x}{r_x/r_y}$$

For each section, this ratio of radii, r_x/r_y, is given in the tables. In our case, on the assumption that the inequality of Eq. (5.58) is satisfied—that is, for $K_yL_y = 8.0$ ft and $P = 240$ kips on p. 3-18 in the AISC Manual (5.14)—we see that a W12 × 45 can carry an axial load of 243 kips > 240 kips for an 8-ft effective buckling length. For that same section, $r_x/r_y = 2.65$ and Eq. (5.58) yields

$$8.0 > \frac{21.12}{2.65} = 7.97 \text{ ft } (2.43 \text{ m}).$$

That is, the equation is satisfied. The chosen section is OK. Now check the assumed ratio, I_c/I_g. In our case this ratio is

$$\frac{350}{662} = 0.529 \quad \text{(close enough to 0.5)}$$

There is no need to find a new K-coefficient and repeat the cycle.

Computer programming

If there is frequent need for the design of a large number of columns, a computer program for either a programmable desk calculator or a full-size electronic computer could be very useful.

First a program written in FORTRAN IV, level H, will be discussed, and then a program for a programmable desk calculator will be developed and explained. For both cases the cross section of the columns is a simple (solid) section.

FORTRAN IV program

The flowchart given in Fig. 5.32 can also be used for a computer program. Modifications have to be made to replace those steps the designer made when taking data out of existing tables, as too much core memory is needed to feed in complete tables. The only table of data to be fed in will concern wide-flange sections. The data which are needed are: wide-flange designation—number (WF) and weight (WEIGHT) in lb/ft; cross-sectional area (A) in in.^{2}; radius of gyration, r_x (RX), in inches, and r_y (RY), also in inches. If the sections are entered in ascending order—that is, starting with the smallest cross-sectional areas—the computer in searching for a satisfactory section will start from the first listed section up, until the restraining conditions are satisfied. In this way a section chosen by the computer will always be the best for the job (optimum) out of the sections supplied to the computer. The more nearly complete the list, the better the solution will be. In the illustrative problems given below only 50 sections out of all W12 and W14 wide-flange sections were entered as data.

For each problem the following additional data have to be supplied to the computer: actual buckling lengths with respect to the x-axis (LX) and y-axis (LY); corresponding factors of effective buckling length, K_x (KX) and K_y (KY); and the force (P). (The parenthetical letters are the computer terms for the variables.)

Once the computer has these data, it first takes the smallest section, calculates for this section slenderness ratios with respect to the x-axis [SX(1)] and the y-axis [SY(1)], and chooses the larger of these two and designates it SLEND(1). According to its magnitude with respect to C_c (limiting slenderness ratio; see under "Design Specifications" in Section 5.14) the computer has to calculate the corresponding allowable stress, F_a [FA(I)]. When now the force, P, is divided by this value, a new area [AREA2(I)] is obtained which is compared with the one chosen previously. If it is smaller than or equal to the previous value, the correct section is found and the process of going through the supplied table of sections is stopped. If the new area is larger than the one previously chosen, the computer will select the next section and repeat the process.

The flowchart for this procedure is shown in Fig. 5.36. The printout is shown on the two pages following Fig. 5.36. As illustration, one problem is discussed in the following example.

Example 5.4 Given: $L_x = 25$ ft, $L_y = 25$ ft, $K_x = K_y = 1.00$, and $P = 600$ kips. The iteration consisted of 27 choices (I = 27) and the solution was a W14 × 136 with an allowable stress of 15.40 ksi. In this particular case the results of each cycle operation are shown. Normally only final results are printed out.

Program for HP 9830A programmable desk calculator

Following the same idea, a program was developed for this type of computer with a microflowchart as shown in Fig. 5.37. The program is made more versatile by introducing data for the yield-point stress of the steel, F_y; that is, it can be used for

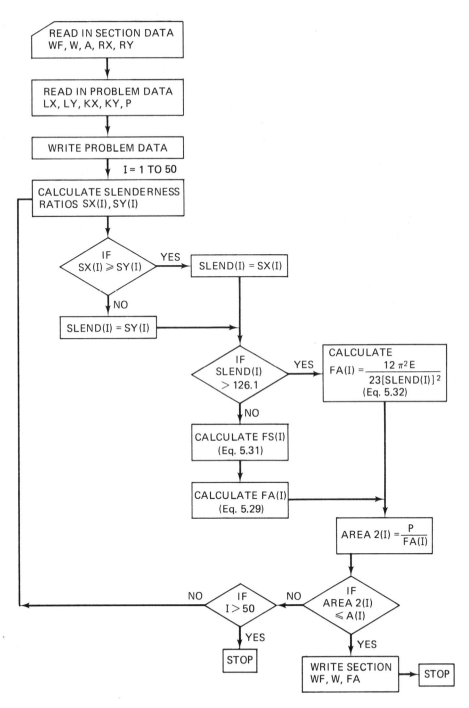

Figure 5.36 Computer Flowchart for Column Design Using FORTRAN IV.

```
54090 01  03-13-71  20,789     DESIGN OF STRAIGHT COLUMNS UNDER AXIAL LOAD, STEEL A36
                                PECS, 14 WF BEAMS AND 12 WF BEAMS

 1      C              DESIGN OF STRAIGHT COLUMNS UNDER AXIAL LOAD, STEEL A36,
 2      C       AISC SPECS, 14 WF BEAMS AND 12 WF BEAMS
 3      C       LX AND LY RESPECTIVE BUCKLING LENGTHS IN FT.
 4      C       KX AND KY RESPECTIVE COEFFICIENTS FOR THE EFFECTIVE BUCKLING LENGTH
 5      C
 6      C
 7              INTEGER WF, WEIGHT
 8              REAL LX, LY, KX, KY
 9              DIMENSION WF(50), WEIGHT(50), A(50), RX(50), RY(50), SLEND(50),
10             1 FA(50), FS(50), SX(50), SY(50), AREA2(50)
11      C       INFORMATION ABOUT 50 DIFFERENT SECTIONS OF 14 AND 12 WF BEAMS
12              READ(5,100) (WF(I), WEIGHT(I), A(I), RX(I), RY(I), I=1,50)             6
13       100    FORMAT(2I5, 3F10.2)                                                    6
14              WRITE(6,101)                                                           8
15       101    FORMAT(3X, 2HWF, 2X, 3HWEI, 9X, 1HA, 8X, 2HRX, 8X, 2HRY)
16              WRITE(6,100) (WF(I), WEIGHT(I), A(I), RX(I), RY(I), I=1,50)            8
17      C       INPUT DATA FOR THE PROBLEM
18        99    READ(5,102) LX, LY, KX, KY,                                           13
19       102    FORMAT(5F10.2)                                                        16
20              IF (P .EQ. 0.0) STOP                                                  16
21              WRITE(6,103) LX, LY, KX, KY, P                                        19
22       103    FORMAT(1H1, 4HLX =F10.2, 4H FT./ 5H LY =F10.2, 4H FT./                22
23             1 5H KX =F10.2/ 5H KY =F10.2/ 5H  P =F10.2, 5H KIP,)
24      C       FIRST SUPPOSE SLENDERNESS RATIO OF 60.0 WITH FA= 17.43 KSI
25              AREA1 = P/17.43                                                       22
26       106    FORMAT(5X, 7HAREA1 =F10.2)                                            23
27              DO 4 I= 1,50                                                          23
28              SX(I)= KX * LX * 12.0/RX(I)                                           24
29              SY(I)= KY * LY * 12.0/RY(I)                                           25
30              WRITE(6,107) I, SX(I), SY(I)                                          26
31       107    FORMAT(5X, 2HI=I5, 5X, 6HSX(I)=F10.2, 5X, 6HSY(I)=F10.2)              29
32              IF (SX(I) .GE. SY(I)) GO TO 10                                        29
33              SLEND(I) = SY(I)                                                      32
34              GO TO 11                                                              33
35        10    SLEND(I) = SX(I)                                                      34
36        11    IF(SLEND(I) .GT. 126.1) GO TO 2                                       35
37              FS(I)= 5.0/3.0 + (0.375 * SLEND(I))/126.1                             38
38             1 - (0.125 * SLEND(I)**3)/(126.1**3)
39              FA(I)= ((1.0 - (SLEND(I)**2)/(2.0 * (126.1**2))) * 36.0)/FS(I)        39
40              GO TO 3                                                               40
41         2    FA(I) = (12.0*9.8696*29000.0)/(23.0*((SLEND(I))**2))             39   41
42              FS(I) = 1.92                                                          42
43         3    AREA2(I) = P/FA(I)                                                    43
44              WRITE(6,108) AREA2(I), I, FA(I), FS(I)                                44
45       108    FORMAT(5X, 9HAREA2(I)=F10.2, 10X, 2HI=I5, 10X, 3HFA=F10.2,            47
46             1 10X, 3HFS=F10.2)
47              IF (AREA2(I) .LE. A(I)) GO TO 5                                       47
48         4    CONTINUE                                                              50
49              I = 50                                                                52
50              WRITE(6,104)                                                          53
```

```
54090 01  03-13-71  20,789     DESIGN OF STRAIGHT COLUMNS UNDER AXIAL LOAD, STEEL A3
                                PECS, 14 WF BEAMS AND 12 WF BEAMS

51       104    FORMAT(1H0, 42HNO SOLUTION WITH 14 WF SECTION IS POSSIBLE)            55
52              GO TO 99                                                              55
53         5    WRITE(6,105) I, WF(I), WEIGHT(I), FA(I)                               56
54       105    FORMAT(1H0, 3HI =I4, 10X, 4HWF =I4, 5X, 8HWEIGHT =I4,                 59
55             1 6H LB/FT, 5X, 4HFA =F10.2, 4H KSI)                               49
56              GO TO 99                                                              59
57              END                                                                   60

    23665 WORDS OF MEMORY USED BY THIS COMPILATION
```

PAGE 3

```
LX =    25.00 FT.
LY =    25.00 FT.
KX =     1.00
KY =     1.00
P  =   600.00 KIP.
```

I	SX(I)	I	SY(I)		FA		FS
1	52.36	1	212.77		3.30		1.92
AREA2(I)= 181.89							
2	51.46	2	205.48		3.54		1.92
AREA2(I)= 169.64							
3	51.11	3	201.34		3.68		1.92
AREA2(I)= 162.88							
4	51.55	4	158.73		5.93		1.92
AREA2(I)= 101.23							
5	51.19	5	157.07		6.05		1.92
AREA2(I)= 99.12							
6	50.85	6	156.25		6.12		1.92
AREA2(I)= 98.09							
7	50.17	7	122.45		9.93		1.92
AREA2(I)= 60.43							
8	56.82	8	99.34		13.06		1.90
AREA2(I)= 45.94							
9	49.83	9	121.95		10.00		1.92
AREA2(I)= 59.99							
10	56.50	10	98.68		13.14		1.90
AREA2(I)= 45.65							
11	49.59	11	120.97		10.14		1.92
AREA2(I)= 59.15							
12	49.26	12	100.00		12.98		1.90
AREA2(I)= 46.23							
13	56.18	13	98.36		13.18		1.90
AREA2(I)= 45.51							
14	48.94	14	99.34		13.06		1.90
AREA2(I)= 45.94							
15	55.76	15	97.72		13.26		1.90
AREA2(I)= 45.23							
16	48.78	16	81.08		15.23		1.87
AREA2(I)= 39.38							
17	55.56	17	97.40		13.30		1.90
AREA2(I)= 45.10							
18	48.62	18	80.86		15.26		1.87
AREA2(I)= 39.32							
19	55.25	19	97.09		13.34		1.90
AREA2(I)= 44.97							
20	48.31	20	80.65		15.28		1.87
AREA2(I)= 39.26							
21	54.95	21	96.46		13.42		1.90
AREA2(I)= 44.71							
22	48.15	22	80.65		15.28		1.87
AREA2(I)= 39.26							
23	47.92	23	80.00		15.36		1.87
AREA2(I)= 39.07							
24	54.45	24	95.85		13.50		1.90
AREA2(I)= 44.45							
25	47.69	25	79.79		15.38		1.87
AREA2(I)= 39.01							
26	53.67	26	94.34		13.61		1.90
AREA2(I)= 44.09							
27	47.54	27	79.58		15.40		1.87
AREA2(I)= 38.95							

PAGE 4

```
I =  27      WF =  14    WEIGHT = 136 LB/FT    FA =    15.40 KSI
```

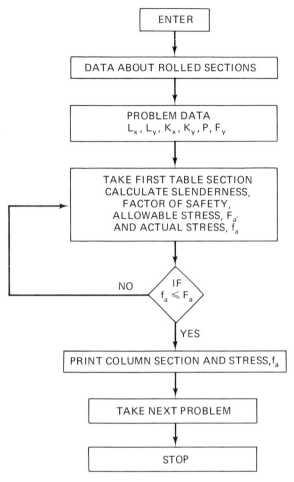

Figure 5.37 Micro Flowchart for Column Design on a Programmable Desk Calculator.

any type of steel. In the next two pages the complete listings of this program and the solution of Example 5.4 are given.

5.2 BUCKLING OF INITIALLY CURVED MEMBERS

If the centroidal axis of a column is not a straight line, the initial curve can be approximated by a sine curve (5.8) as

$$v_0(z) = (v_0)_{max} \sin \frac{\pi z}{L} \quad \text{(Fig. 5.38a)} \tag{5.59}$$

The bending moment now is (5.8)

$$M = -EI(v'' - v_0'') \tag{5.60}$$

180 DESIGN OF COMPRESSION MEMBERS

```
10 DIM A[50,5]
20 FIXED 2
30 PRINT TAB15"DESIGN OF STRAIGHT COLUMNS AXIALLY LOADED"
40 PRINT TAB15"STEEL FY=36 KSI OR 50 KSI, W12 OR W14 SECTIONS"
50 REM A(I,J) DATA MATRIX:J= 1 WIDE FLANGE,2 WEIGHT,3 AREA,4 RX,5 RY
60 REM FY=F1,LX=L1,LY=L2,KX=K1,KY=K2
70 REM F3=P/A, S1=SLEND.RATIO ABOUT X,S2=SL.RAT.ABOUT Y AXIS,S= MAX.SLEN
80 REM DERNESS,C=LIMITING SLEND.,,S3=SAFETY FAC.,
90 FOR I=1 TO 50
100 FOR J=1 TO 5
110 READ A[I,J]
120 NEXT J
130 NEXT I
140 DATA 12,14,4.16,4.62,0.75,12,16,4.71,4.67,0.77
150 DATA 12,19,5.57,4.82,0.82,12,22,6.48,4.91,0.85
160 DATA 14,22,6.49,5.54,1.04,12,26,7.65,5.17,1.51
170 DATA 14,26,7.69,5.65,1.08,12,30,8.79,5.21,1.52
180 DATA 14,30,8.85,5.73,1.49
190 DATA 14,34,10,5.83,1.53,12,35,10.3,5.25,1.54
200 DATA 14,38,11.2,5.87,1.55,12,40,11.8,5.13,1.93
210 DATA 14,43,12.6,5.02,1.09,12,45,13.2,5.15,1.94
220 DATA 14,43,12.6,5.82,1.89,14,48,14.1,5.85,1.91
230 DATA 12,50,14.7,5.18,1.96,12,53,15.6,5.23,2.48
240 DATA 14,53,15.6,5.89,1.92,12,58,17,5.28,2.51
250 DATA 14,61,17.9,5.98,2.45,12,65,19.1,5.28,3.02
260 DATA 14,68,20,6.01,2.46,12,72,21.1,5.31,3.04
270 DATA 14,74,21.8,6.04,2.48,12,79,23.2,5.34,3.05
280 DATA 14,82,24.1,6.05,2.48,12,87,25.6,5.38,3.07
290 DATA 14,90,26.5,6.14,3.7,12,96,28.2,5.44,3.09
300 DATA 14,99,29.1,6.17,3.71,12,106,31.2,5.47,3.11
310 DATA 12,106,31.2,5.47,3.11,14,109,32,6.22,3.73
320 DATA 12,120,35.3,5.51,3.13,14,120,35.3,6.24,3.74
330 DATA 14,132,38.8,6.28,3.76,12,136,39.9,5.58,3.16
340 DATA 14,145,42.7,6.33,3.98,12,152,44.7,5.66,3.19
350 DATA 14,159,46.7,6.38,4,12,170,50,5.74,3.22
360 DATA 14,176,51.8,6.43,4.02,12,190,55.8,5.82,3.25
370 DATA 14,193,56.8,6.5,4.05,12,210,61.8,5.89,3.28
380 DATA 14,211,62,6.55,4.07,12,230,67.7,5.97,3.31
390 DATA 14,233,68.5,6.63,4.1,12,252,74.1,6.06,3.34
400 DISP "FORCE P=";
410 INPUT P
420 DISP "STEEL FY=";
430 INPUT F1
440 DISP "BUCKLING LENGTH LX=";
450 INPUT L1
460 DISP "BUCKLING LENGTH LY=";
470 INPUT L2
480 DISP "K-FACTOR KX=";
490 INPUT K1
500 DISP "K-FACTOR KY";
510 INPUT K2
520 FOR I=1 TO 50
530 S1=K1*L1*12/A[I,4]
540 S2=K2*L2*12/A[I,5]
550 IF S2<S1 THEN 580
560 S=S2
```

```
570 GOTO 590
580 S=S1
590 F3=P/A[I,3]
600 C=SQR(2*(3.14↑2)*29000/F1)
610 IF S>C THEN 650
620 S3=(5/3)+(3*S/(8*C))-((S↑3)/(8*(C↑3)))
630 F2=(1-((S↑2)/(2*(C↑2))))*(F1/S3)
640 GOTO 660
650 F2=(12*(3.14↑2)*29000)/(23*(S↑2))
660 IF F3 <= (1.025*F2) THEN 700
670 NEXT I
680 PRINT "SOLUTION IMPOSSIBLE WITH GIVEN SECTIONS"
690 GOTO 800
700 PRINT
710 PRINT
720 PRINT TAB25"PROBLEM"
730 PRINT
740 PRINT TAB10" FORCE=";P;"STEEL=";F1;"LX=";L1;"LY=";L2;"KX=";K1;"KY=";K2;
750 PRINT
760 PRINT TAB25"SOLUTION"
770 PRINT
780 PRINT TAB15"WF SECTION"A[I,1];"X";A[I,2]
790 PRINT "NOMINAL STRESS=";F3;"ALLOWABLE STRESS=";F2;"MAX.SLEND.RATIO=";S
800 STOP
810 END
```

```
         DESIGN OF STRAIGHT COLUMNS AXIALLY LOADED
         STEEL FY=36 KSI OR 50 KSI, W12 OR W14 SECTIONS

                         PROBLEM

         FORCE= 600.00   STEEL= 36.00    LX= 25.00    LY= 25.00    KX=
1.00    KY= 1.00
                         SOLUTION
             WF SECTION 14.00    X 132.00
NOMINAL STRESS= 15.46   ALLOWABLE STRESS= 15.37    MAX.SLEND.RATIO= 79.79
```

If the column is hinged at its ends and there are no eccentricities, then the elastic line can be shown to be

$$v(z) = \frac{(v_0)_{\max}}{1 - P/P_e} \sin \frac{\pi z}{L} \tag{5.61}$$

where P_e is again the smallest Euler buckling load (Fig. 5.38b).

If the initial form has more than one-half–sine wave, $n \geq 2$ (Fig. 5.38c), and all other initial conditions are the same as before, then

$$v_0(z) = (v_{0n})_{\max} \sin \left(\frac{n\pi z}{L}\right) \tag{5.62}$$

and the complete solution is

$$v(z) = \frac{(v_{0n})_{\max}}{1 - P/P_{en}} \sin \frac{n\pi z}{L} + A \sin \alpha z + B \cos \alpha z + Cz + D \tag{5.63}$$

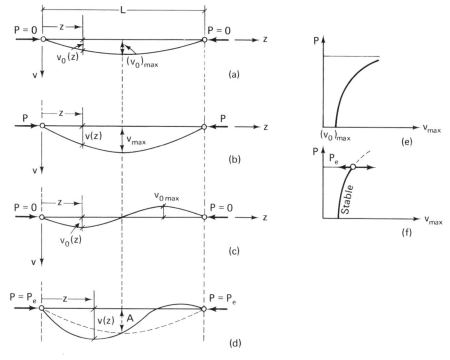

Figure 5.38 Buckling of Initially Curved Column.

where $P_{en} = (n\pi/L)^2 EI$ is the higher Euler load with n half-waves and $\alpha^2 = P/EI$ as before.

If the previous four boundary conditions are introduced, the elastic line is given by

$$v(z) = \frac{(v_{0n})_{max}}{1 - P/P_{en}} \sin \frac{n\pi z}{L} + A \sin \frac{m\pi z}{L} \qquad (5.64)$$

with $m = 1, 2, 3$. The constant A is zero when the system determinant is $\Delta \neq 0$, but it can have any value when $P = P_e$ or in general $P = P_{en}$.

For slowly increasing loads (Fig. 5.38e) again for $P < P_e$

$$v(z) = [v(P)]_{max} \sin \frac{n\pi z}{L} \qquad (5.65)$$

where

$$[v(P)]_{max} = \frac{(v_{0n})_{max}}{1 - P/P_{en}}$$

When the first Euler load, P_e, is reached, in spite of the initial curvature a true bifurcation point is reached. The corresponding diagram is shown in Fig. 5.38(f) and the corresponding elastic line in Fig. 5.38(d). For loads $P > P_e$, the elastic line is unstable.

5.3 BUCKLING OF THIN PLATES

Thin plates, which are stressed in their own plane (plane stress state), are often present in steel structures as webs of rolled beams and plate girders or columns and in folded and orthotropic plates.

As in Chapters 7, 8, and 10, the use of various buckling coefficients will be discussed in this and the next section (Section 5.4, "Postbuckling Strength of Plates"); a bare minimum of the underlying theory is presented in order to give the reader some feeling for these values.

When a plate is stressed in its plane under the action of external applied forces and a critical loading is reached, the plane state becomes unstable. For larger loads, after any disturbance a change in equilibrium results, the plate deflects laterally, and it becomes buckled. For a plate this does not necessarily imply failure, because the buckles are restrained in the transverse direction and the plate is usually capable of carrying loads beyond the first buckling load (see Section 5.4).

In practice, the buckling of rectangular plates is the most important (Fig. 5.39). If only one strip of a plate is isolated and considered (cross section $1 \cdot t$) and the Euler stress, F_e, is determined for the buckling length, b, of such a strip, one would have

$$F_e = \frac{\pi^2 EI}{b^2} \cdot \frac{1}{(1 \cdot t)} \tag{5.66}$$

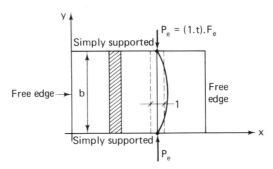

Figure 5.39 Euler's Buckling Stress, F_e.

where $I = 1 \cdot t^3/12$. When substituted in Eq. (5.66), the expression for the critical stress of an isolated strip is

$$F_e = \frac{\pi^2 E}{12} \cdot \left(\frac{t}{b}\right)^2 \tag{5.67}$$

Due to the fact that this strip in reality is not separated from the rest of the plate, instead of E, one must use $E/(1 - \mu^2)$. Using this value one obtains

$$F_e = \frac{\pi^2 E}{12(1 - \mu^2)} \cdot \left(\frac{t}{b}\right)^2 \tag{5.68}$$

With $E = 29{,}000$ ksi (200 MPa) and $\mu = 0.3$

$$F_e = 26{,}210 \left(\frac{t}{b}\right)^2 \tag{5.69}$$

The differential equation for plate buckling (f_x and f_y both compressive stresses) is given in reference 5.8 (p. 348)

$$E^1 \cdot \left(\frac{\partial^4 w}{\partial x^4} + 2\frac{\partial^4 w}{\partial x^2 \partial y^2} + \frac{\partial^4 w}{\partial y^4}\right) + t\left(f_x \frac{\partial^2 w}{\partial x^2} + 2f_v \frac{\partial^2 w}{\partial x \partial y} + f_y \frac{\partial^2 w}{\partial y^2}\right) = q_z \quad (5.70)$$

where $E^1 = Et^3/[12(1-\mu^2)] =$ flexural stiffness of the plate
$w(x, y) =$ deflection normal to the plate
$f_x, f_y, f_v =$ normal stresses in x- and y-directions and shear stress, respectively
$q_z =$ plate load per unit area in z-direction (normal to the plate).

If the stress, F_e, is introduced in Eq. (5.70), only the stress ratios appear and the critical, buckling stresses can be expressed for all loading cases in the form

$$F_{\text{crit}} = kF_e \quad \text{and} \quad F_{cv} = kF_e \quad (5.71)$$

or

$$F_{\text{crit}} = k\frac{\pi^2 E}{12(1-\mu^2)}\left(\frac{t}{b}\right)^2 \quad (5.72)$$

The buckling constant, k, depends upon the loading case, the aspect ratio, $\alpha = a/b$, and the support conditions along the plate edges (boundary conditions).

Case of uniform compression simply supported

The plate is loaded in the "longitudinal" direction, x, with uniform compressive stress, $f_x = f$; two other stress components, f_y and f_v, are zero. All four edges are simply supported (Fig. 5.40).

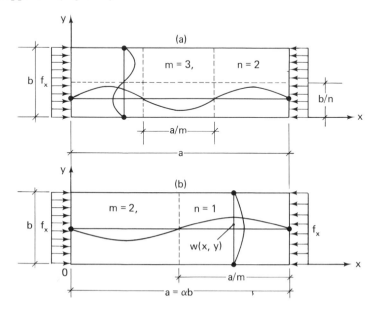

Figure 5.40 Plate with All Four Edges Hinged.

To solve Eq. (5.70) a product of functions is tried in the form

$$w(x, y) = A_{mn} \sin \frac{m\pi x}{a} \sin \frac{n\pi y}{b} \tag{5.73}$$

where m and n are integers. The buckled surface, $w(x, y)$, must satisfy the simply supported edge conditions. When $w(x, y)$ of Eq. (5.73) is substituted into Eq. (5.70), the buckling condition is obtained in the form

$$w(x, y) \cdot \left\{ \left[\left(\frac{m\pi}{a}\right)^2 + \left(\frac{n\pi}{b}\right)^2 \right]^2 - \left(\frac{\pi}{b}\right)^2 \frac{f}{F_e} \left(\frac{m\pi}{a}\right)^2 \right\} = 0 \tag{5.74}$$

This condition is satisfied either

1. For any compressive stress, f, when $w(x, y) = 0$,
2. For any (small) value of $w(x, y) \neq 0$, when the expression between brackets in Eq. (5.74) equals zero.

The undeflected equilibrium condition is stable for $f < F_{crit}$ and unstable for $f > F_{crit}$. For $F_{crit} = kF_e$, a bifurcation point of the elastic equilibrium is attained and the equilibrium is neutral for very small deflections, $w(x, y)$. If $\alpha = a/b$ is introduced, then the buckling coefficient, k, is (5.8)

$$k = \left(\frac{m}{\alpha} + n^2 \frac{\alpha}{m} \right)^2$$

where m and n are the numbers of sine half-waves in the x- and y-directions, respectively. For each pair of values of m and n there is a dfinite buckling stress and corresponding buckling form. The lines of interference ($w = 0$) form a rectangular mesh (Fig. 5.40a). In Fig. 5.40(b) the case of $m = 2$ is represented. The buckling constant, k, for $n = 1$ is

$$k = \left(\frac{m}{\alpha} + \frac{\alpha}{m} \right)^2 \tag{5.75}$$

The absolute smallest value for k follows from $dk/dm = 0$ and $(m)_{min} = \alpha$, which yields $k_{min} = 4$. This value is obtained when the plate length is an integer multiple of the width (because m and n are integers!). The buckles are complete squares, as $a/m_{min} = b$.

In Fig. 5.41 the dependence of the buckling coefficient, k, upon the aspect ratio, α, is represented. For each value of α a different value of number of waves, m, is controlling the minimum value. For the ratio, $\alpha = \sqrt{m(m+1)}$, the buckling stresses in case of m and $(m + 1)$ half-waves are equal to

$$F_{crit} = k \cdot m = k(m + 1)$$

This value occurs at the intersection points of the curves shown in Fig. 5.41.

For various edge conditions and types of loading, different buckling coefficients exist for elastic and inelastic behavior. These can be found in various manuals (5.22, 5.24, or 5.25).

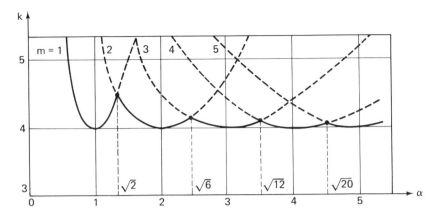

Figure 5.41 Buckling Constants for Cases in Fig. 5.40.

5.4 POSTBUCKLING STRENGTH OF PLATES

Equation (5.70) cannot be used for the investigation of postbuckling behavior of plates as the deflections of such a plate can no longer be considered small. Von Kármán et al. (5.23) in 1932 introduced the concept of the "effective" width, which greatly simplified the complicated mathematical expressions developed by him in 1910, for large-deflection theory of plates.

If we substitute a mesh of straight bars for a continuous plate (Fig. 5.42) and apply loads only in one direction, nonuniformly distributed perpendicular to their action, a physical model for plate postbuckling behavior is obtained. The intermediate horizontal bars, when the vertical ones are buckled, restrain them, which enable them to carry additional loads until the stress at the edges reaches yielding. The yielding zone spreads toward the center of the plate. Some small additional load is needed to achieve this spreading. The collapse is considered to occur at the moment of first yielding, and then the nonuniform stress distribution is substituted by a uniformly distributed stress over the "effective" width of the plate.

For simply supported plates this effective width, b_e, is (5.23)

$$b_e = b \sqrt{\frac{F_{crit}}{f_{max}}} \quad (5.76)$$

where F_{crit} is the buckling stress and f_{max} the edge stress of such a plate.

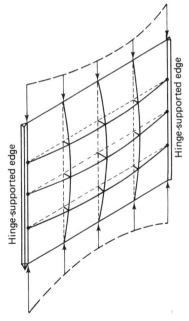

Figure 5.42 Model for Postbuckling Behavior of a Plate.

For structural steel design this formula is slightly modified to obtain a better agreement with experimental results.

From these considerations follow width-thickness ratios already discussed in some extent under "Design Specifications" in Section 5.1.4. To ensure that buckling will not occur, American specifications require that yielding be reached before buckling occurs—that is

$$(F_{crit})_{plate} \geq F_y \tag{5.77}$$

or

$$\frac{F_{crit}}{F_y} \geq 1$$

Substituting Eq. (5.72) in the above equation and taking $\mu = 0.3$, the following general equation for the limiting width-thickness ratios is obtained

$$\frac{b}{t} \leq 0.951 \sqrt{\frac{kE}{F_y}} = 162 \sqrt{\frac{k}{F_y}} \tag{5.78}$$

Some average k-values are:

$$\begin{aligned} &\text{Single angles (equal legs):} & k &= 0.425 \\ &\text{Flanges, web stiffeners:} & k &= 0.70 \\ &\text{Stems of tees:} & k &= 1.277 \\ &\text{Webs, cover plates:} & k &= 5.0 \end{aligned}$$

Substituting these k-values in Eq. (5.78) yields the values given by the AISC specification (ref. 5.14, p. 5-30) (Section 5.1.4, p. 5.17) by taking 0.7 of the theoretical value.

NOTATIONS

a	= length of plate	v'	= slope of tangent at any point of the elastic line
b	= width of rectangular section; width of a plate	w	= plate deflection, load intensity
b_e	= effective plate width		
c	= outer fiber distance from the centroid	α	= $\sqrt{P/EI}$, or plate aspect ratio, a/b
d	= depth of rectangular section	α_{em}	= $m\sqrt{P_e/EI}$
e	= end eccentricity in force application	α_e	= $\sqrt{P_e/EI}$
		β	= correction factor for shear stresses
k	= Musschenbroek's empirical factor; buckling constant	λ	= kL/r = slenderness ratio
n	= any integer	Δ	= determinant
r	= radius of gyration	ω	= buckling number
v_0, f	= deflection at midheight of the column	A, B, C, D	= constants of integration
		C	= Euler's "absolute elasticity"
$v(z)$	= deflections of the column elastic line	C_c	= slenderness ratio
		E	= Young's modulus

EI	= flexural stiffness of a beam	L_e	= effective buckling column height (length)
E^1	= $Et^3/[12(1-\mu^2)]$ = flexural stiffness of a plate	M	= bending moment
E_t	= tangential elastic modulus	P	= compressive force in a column
F_y	= yield point stress		
F_e	= critical Euler stress	P_e	= Euler critical (buckling) force
F_{crit}	= buckling stress for plates	P_u	= ultimate compressive force in a column
FS_e	= P/P_e = safety factor against Euler buckling	P_e'	= critical force of a built-up column
G	= ratio of column and girder stiffnesses	Q_s	= stress-reduction factor of unstiffened compression elements
I	= moment of inertia		
K	= coefficient of the effective buckling length	R	= radius of curvature
L	= column length, actual length	V	= shear force

PROBLEMS

5.1. Design a straight column using the most economical of available wide-flange sections in A36 steel, if the axial force is 600 kips. Column is hinged at both ends and 20-ft long. There are no lateral supports between the hinges. Use AISC specification (5.7).

5.2. An axially loaded column of 30-ft length, built in at both ends, has lateral supports every 6 ft to prevent buckling about the minor axis of the column. Use lightest wide-flange section using steel with $F_y = 50$ ksi if the force is 420 kips. Use AISC specification (5.7).

5.3. For a column with a cross section as shown in Fig. P-5.3 calculate the maximum allowable axial force this section can take for a buckling length of 12 ft. Take ends as hinged and no lateral supports in between. Use A36 steel and AISC specification (5.7).

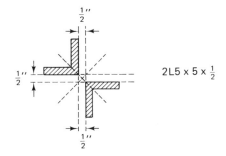

Figure P-5.3

5.4. If the angles are arranged as in Fig. P-5.4 and the loading and column conditions are as in Problem 5.3, what is the column capacity now? Use A36 steel and AISC specification (5.7).

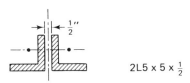

Figure P-5.4

5.5. For a column of 25-ft length with hinged ends, square structural tubing is used. Design this tubing if AISC specification (5.7) is used. Force $P = 100$ kip. (F_y for tubing is 46 ksi.)

5.6. Design a top chord of a roof truss to carry a compressive load of 150 kips. Choose for the member cross section two angles $\frac{1}{2}$ in. apart (for gusset plates). In the vertical truss plane, the member is braced at panel lengths of 6-ft intervals; for buckling out-of-plane member is braced at each second panel joint— that is, at 12-ft intervals. Use AISC specification (5.7) and A36 steel.

5.7. Design a welded batten column using two channels of A36 steel to carry a load of 142 kips. The effective length for the material axis is $K_x L_x = 29$ ft. No other lateral supports are provided. Use British BS 449 standards (5.18).

5.8. For the column designed in Problem 5.7 design the details of the battens and their connections to the channels. Use E70 electrodes, manual shielded-arc welding, and British BSS 449 specifications (5.18).

5.9. Select the lightest W-section to carry an axial compression load of 200 kips for a length of 24 ft, in a braced frame, with additional lateral supports in the weak direction (buckling out-of-frame plane) at 12 ft distance. Both column ends are hinged. Use high-strength, low-alloy steel with $F_y = 50$ ksi and AISC specification (5.7).

5.10. For a highway bridge a compressed diagonal of W-section carries forces of 58.5 kips and 78.0 kips due to dead and live load, respectively. If the theoretical length of this diagonal is 16 ft and the number of stress cycles less than 2×10^6, design the member using A36 steel, A325-F bolts, and AASHTO specification (5.15).

5.11. Select a W10 section of A36 steel to support an axial compressive load of 120 kips. The unsupported length of the column is 30 ft about both axes. Use AISC specification (5.7).

5.12. For the conditions as in Problem 5.11 choose a W10 section of steel with $F_y = 50$ ksi.

5.13. Using A36 steel and AISC column tables (5.14), determine the allowable axial load which the following columns can support:
a. A W8 × 58 with hinged ends and a length of 12 ft.
b. A W12 × 50 with fixed ends and a length of 16 ft.

5.14. Using AISC specification (5.7) and A36 steel, select the lightest available wide flange section if the axial load is 180 kips and $l_x = 18$ ft, $l_y = 6$ ft. Both column ends are fixed.

5.15. A 25-ft column is laterally supported in the weak direction at its mid-depth. Select the lightest W-section which can support an axial load of 320 kips. Use A36 steel and AISC specification (5.7). Both column ends are fixed.

5.16. Using a pair of A36 steel channels, at the distance of 10 in. back to back, design a built-up laced column 24 ft long with both ends pinned. The end tie plates have to be at 1 ft-6 in. from column ends. Use AISC specification (5.7).

5.17. Design a square structural tube for a pin-ended column of 8 ft and an axial load of 50 kips using AISC specification (5.7) ($F_y = 46$ ksi).

5.18. Design a hinged column, 48 ft in length, to carry a load of 400 kips. Use a cross section of four A36 steel corner angles with double lacing on all four sides. Use AISC specification (5.7).

5.19. Rework Problem 5.14 using steel with $F_y = 50$ ksi.

5.20. Rework Problem 5.15 using steel with $F_y = 50$ ksi.

REFERENCES

5.1. VAN MUSSCHENBROEK, P., *Physicae experimentales et geometricae*, Leiden, Netherlands, 1729.

5.2. EULER, L. *Methodus inveniendi lineas curvas ...*, Berlin, 1744; and "Sur la Force des Colonnes," in *Memoires de l' Academie Royale des Sciences et Belles Lettres*, Berlin, 1759, partially translated by J. A. Van den Broek (sections I to XVIII) and published as "Euler's Classic Paper 'On the Strength of Columns,'" *Am. J. Phys.*, vol. 15, no. 1, Jan.–Feb. 1947, pp. 309–18.

5.3. ENGESSER, F., *Zeitschrift für Architektur und Ingenieurwesen*, 1889, p. 455.

5.4. SHANLEY, F. R., "The Column Paradox," *J. Aeronaut. Sci.*, Dec. 1946, p. 678.

5.5. TIMOSHENKO, S. P., *History of Strength* (in Russian), Moscow, 1957, p. 47, fig. 22.

5.6. SCHLEICHER, F. *Taschenbuch für Bauingenieure*, 2nd ed., vol. I, Springer, Berlin, 1955.

5.7. AISC, *Specification for the Design, Fabrication, and Erection of Structural Steel for Buildings*, New York, 1978.

5.8. TIMOSHENKO, S. P., and J. M. GERE, *Theory of Elastic Stability*, 2nd ed., McGraw-Hill, New York, 1961, pp. 76–82.

5.9. SHANLEY, F. R., "Inelastic Column Theory," *J. Aeronaut. Sci.*, vol. 14, no. 5, May 1947, pp. 261–68.

5.10. HORSFALL, and SANDORFF, *Strain Distribution During Column Failure*, Lockheed Aircraft Corporation Report no. 5728, April 1946.

5.11. BLEICH, F., *Buckling Strength of Metal Structures*, McGraw-Hill, New York, 1952, p. 23.

5.12. AISC, *Commentary on the Specification for the Design, Fabrication, and Erection of Structural Steel for Buildings*, New York, 1969.

5.13. Structural Stability Research Council, *Guide to Stability Design Criteria for Metal Structures*, 3rd ed., Wiley, New York, 1976.

5.14. AISC, *Manual of Steel Construction*, 8th ed., Chicago, 1980.

5.15. AASHTO, *Standard Specifications for Highway Bridges*, 12th ed., Washington, DC, 1977. with 1978, 1979, 1980, and 1981 Interim Bridge Specifications).

5.16. Deutsche Industrie Normen, *Berechnungsgrundlagen für Stabilitätsfäle in Stahlbau* (*Knickung, Kippung, Beulung*), DIN 4114, Berlin, July 1952, February 1953.

5.17. BAES, L. "La poutre Vierendeel," *L'Ossature Métallique*, vol. 6, no. 3, 1937, pp. 125–52.

5.18. British Standards Institution, *Specification for use of structural steel in building*, (incorporating British Standard Code of Practice CP113), BS 449, London, reset and reprinted 1965.

5.19. L'institut Technique du Bâtiment et des Travaux Publics et le Centre Technique Industriel de la Construction Métallique, *Règles de calcul des constructions en acier*, Règles CM66, ed. Eyrolles, 1974.

5.20. Gosudarstvennyĭ Komitet Soveta Ministrov SSSR po delam Stroitelystva, Stroitelnye Normy i Pravila, Part II, Section V, Chapter 3: "Stalnye Konstruktsii," SN i P II-V, 3-62, Moscow, 1963.

5.21. TALL, L., ed., *Structural Steel Design*, 2nd ed., Ronald, New York, 1974, p. 288.

5.22. BROCKENBROUGH, R. L., and B. G. JOHNSTON, *United States Steel Design Manual*, 3rd printing, United States Steel Corporation, Pittsburgh, 1974.

5.23. VON KÁRMÁN, T., E. E. SECHLER, and L. H. DONNELL, "The Strength of Thin Plates in Compression," *Trans. Am. Soc. Mech. Engrs.*, vol. 54, APM-54-5, 1932, pp. 53–57.

5.24. KLÖPPEL, K. K., and J. SCHEER, *Beulwerte ausgesteifter Rechteckplatten*, vol. 1, Wilhelm Ernst, Berlin, 1960.

5.25. KLÖPPEL, K. K., and K. H. MÖLLER, *Beulwerte ausgesteifter Rechteckplatten*, vol. 2, Wilhelm Ernst, Berlin, 1968.

6

Torsion of Beams

6.1 GENERAL

Torsion is a phenomenon that can be caused by the application of an external torsional moment or as a secondary effect resulting from instability of a member that is loaded either axially or in bending.

6.1.1 Shear Center

An external torsional moment acts on a member when a torque is applied along its shear center axis (Fig. 6.1). This axis is an axis parallel to the centroidal axis through which loads applied perpendicular to it must pass so as to cause no twisting. Under the application of torque only, the shear center axis remains straight. For doubly symmetrical and doubly antisymmetrical sections the shear center coincides with the centroid. For other sections the shear center can be calculated or found from tables. If the plane in which transverse loading takes place does not pass through the shear center axis, torsion will result (Fig. 6.2). If the distance from the loading plane to the shear center axis is e and the load is P, the twisting moment is $T = Pe$. It is obvious that the channel shown in Fig. 6.2 must be restrained against overturning.

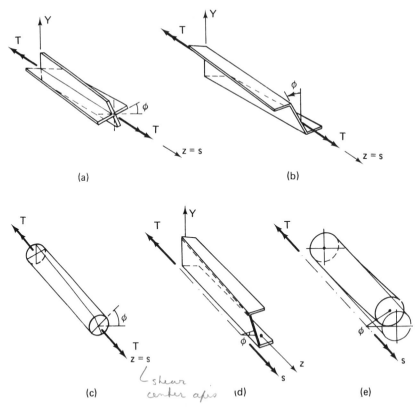

Figure 6.1 Torsion Due to Twisting Moment Acting Along Shear Center Axis (s), Free Ends: (a) Doubly Symmetrical; (b) Doubly Antisymmetrical; (c) Circular Section; (d) Singly Symmetrical; (e) Split Tube, Singly Symmetrical.

6.1.2 Pure and Restrained Torsion

The cross section of any member that is subjected to torsion, with the exception of closed circular sections, will exhibit warping out of its plane with possible distortion of the cross section. For example, in Fig. 6.1 all cases except case (c) will exhibit out-of-plane warping at the free ends. If there is no warping or no resistance to warping, we have pure torsion, also called St. Venant torque. If, however, there is warping and such warping is restrained, we have restrained or warping torsion causing additional cross-sectional distortion.

The total applied torque, T, is thus resisted by pure torsion, T_t, and the remainder, T_w, is carried by shears resulting from the bending resistance of the member elements.

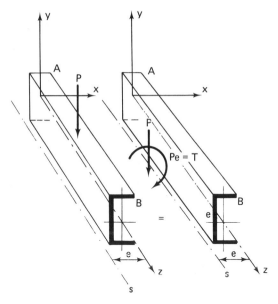

Figure 6.2 Torsion Due to Transverse Loading Not Passing Through Shear Center Axis.

6.2 PURE TORSION OF PRISMATIC MEMBERS

6.2.1 Introduction

It can be shown that for pure torsion

$$T_t = GJ\frac{d\phi}{dz} = C\frac{d\phi}{dz} \tag{6.1}$$

where T_t = applied torque, ϕ = angle of twist, G = shear modulus, and J = the torsional constant for a section which has the dimension of length to the fourth power (in.⁴). In the case of closed circular sections, J equals the polar moment of inertia.* Values of J are given in various manuals for various types of sections [e.g., AISC *Manual* (6.7), pp. 1-15 to 35 and 1-109 to 114].

6.2.2 Circular Closed Sections

For closed circular sections (Fig. 6.3) the rate of twist will be constant throughout the length and is equal to

$$\theta = \frac{d\phi}{dz} = \frac{T_t}{GJ} = \frac{T_t}{C} \tag{6.2}$$

It can also be shown that the shear stress varies linearly with the distance to the

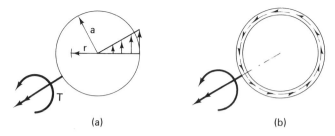

Figure 6.3 Pure Torsion of Circular Closed Sections: (a) Solid; (b) Tubular.

centroid and equals

$$f_v = \frac{T_t r}{J} \qquad (6.3)$$

where r is the distance to the centroid.

6.2.3 Tubular Closed Sections

For thin-walled tubular closed sections (Fig. 6.4) it can be shown that

$$f_v = \frac{T_t}{2At} \qquad (6.4)$$

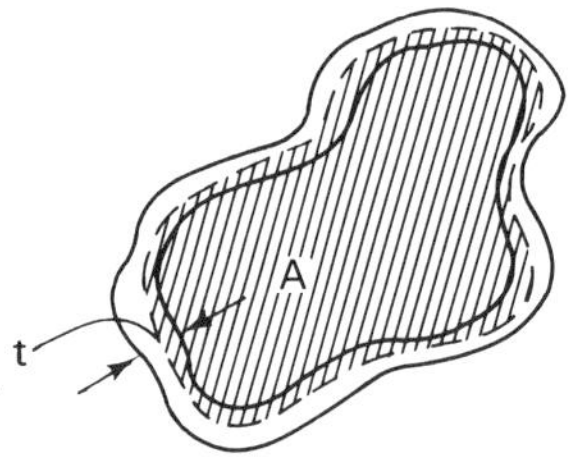

where A is the enclosed area. This equation is often referred to as Bredt's first formula. Also the rate or twist can be shown to be

$$\theta = \frac{T_t}{GJ} = \frac{1}{2GA} \oint_0^s f_v \, ds \qquad (6.5)$$

where ds is an element of length and the integration is around the entire periphery. For a constant wall thickness Eq. (6.5) becomes

Figure 6.4 Tubular Closed Section.

$$\theta = \frac{T_t S}{4GA^2 t} \quad \text{perimeter} \qquad (6.6)$$

where $S = \oint ds$. Often closed multicell shafts are used in fuselages and wings of aircraft to transmit torque. In civil engineering their application is mainly limited to concrete multicell box girders. In steel they are used far less, and the reader is referred to various textbooks in this area.

Example 6.1 As an example of the use of tubular sections subjected to torque consider the sections shown in Fig. 6.5. Each is subjected to a pure torque of 500 kip-in. (5,649 kN-cm). Calculate for each case the shear stress and the rate of twist; $G = 12 \times 10^3$ ksi (82.7×10^3 MPa)

Solution
(a) $A = 7.5 \times 7.5 = 56.25$ in.² (362.8 cm²); $\quad S = 30$ in. (76.2 cm) $4(7.5)$
(b) $A = 11.5 \times 5.75 = 66.125$ in.² (426.5 cm²); $\quad S = 34.5$ in. (87.6 cm)
(c) $A = \pi \times 4.25^2 = 56.745$ in.² (366 cm²); $\quad S = 26.70$ in. (67.8 cm)

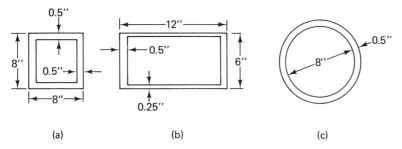

Figure 6.5 Closed Tubular (Box) Sections: (a) Square; (b) Rectangular; (c) Circular.

Next, according to Eq. (6.4) the shear stresses are as follows:

(a) $f_v = \dfrac{500}{2 \times 56.25 \times 0.5} = 8.89$ ksi (61.3 MPa)

(b) $f_v = \dfrac{500}{2 \times 66.125 \times 0.25} = 15.12$ ksi (104.3 MPa) (long side)

$f_v = \dfrac{500}{2 \times 66.125 \times 0.50} = 7.56$ ksi (52.2 MPa) (short side)

(c) $f_v = \dfrac{500}{2 \times 56.745 \times 0.50} = 8.81$ ksi (60.7 MPa)

The rates of twist according to Eqs. (6.5) and (6.6) are as follows:

(a) $\theta = \dfrac{500 \times 30}{4 \times 12 \times 10^3 \times 56.25^2 \times 0.5}$
$= 1.975 \times 10^{-4}$ rad/in. (7.78×10^{-5} rad/cm)

(b) $\theta = \dfrac{1}{2 \times 12 \times 10^3 \times 66.125}(7.56 \times 11.5 + 15.12 \times 23)$
$= 2.74 \times 10^{-4}$ rad/in. (10.78×10^{-5} rad/cm)

(c) $\theta = \dfrac{500 \times 26.70}{4 \times 12 \times 10^3 \times 56.475^2 \times 0.5}$
$= 1.73 \times 10^{-4}$ rad/in. (6.80×10^{-5} rad/cm)

6.2.4 Solid Rectangular Sections

The distribution of St. Venant shear stresses in a solid rectangular section is shown in Fig. 6.6. Except near the edges, the stress is nearly constant along the long sides, and its maximum value in the middle equals

$$f_v = \frac{T_t \gamma_t t}{J} \qquad (6.7)$$

where

$$J = \frac{bt^3}{3} - 2\psi t^4 \qquad (6.8)$$

and b and t are the long and the thin dimensions, respectively, and γ and ψ are numerical constants depending on the b-to-t ratio. Values for ψ and γ are given in

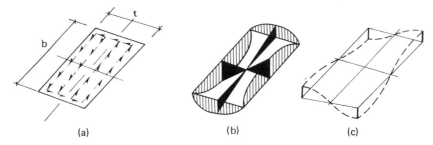

Figure 6.6 Direction and Distribution of Torsional Shear Stresses in a Rectangular Bar.

Appendix VI, Table 1. As b/t approaches 4.0, f_v and J approach

$$f_v = \frac{T_t t}{J} \tag{6.9}$$

$$J = \frac{bt^3}{3} - 0.21 t^4 \tag{6.10}$$

This last equation becomes, for thin rectangles,

$$J = \frac{bt^3}{3} \tag{6.11}$$

6.2.5 Open Sections

Open sections as shown in Fig. 6.7 behave as a series of rectangular elements and

$$J = \sum_i \frac{b_i t_i^3}{3} \tag{6.12}$$

The maximum shear stress in open shapes may be calculated from

$$f_v = \frac{T_t \delta t_f}{J} \tag{6.13}$$

where δ is a numerical factor called the stress coefficient, which includes the effect of stress increase due to the fillet. Figures 6.8 and 6.9 give values for δ.

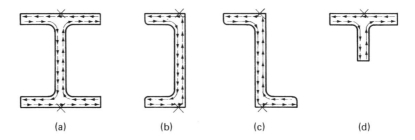

x indicates location of maximum shear stress

Figure 6.7 Open Structural Shapes: (a) I-section; (b) Channel; (c) Z-section; (d) T-section.

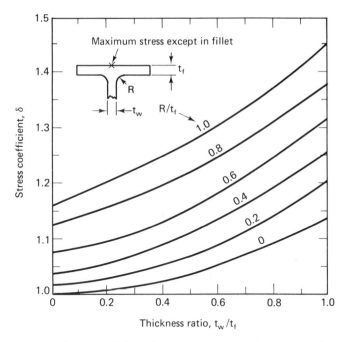

Figure 6.8 Stress Coefficients for T- or W-shapes with Parallel Flanges (6.3).

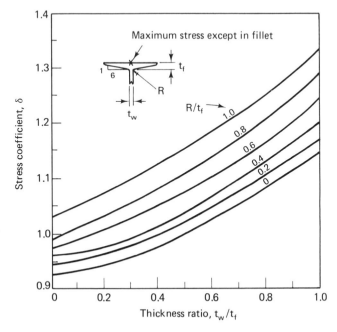

Figure 6.9 Stress Coefficients for T- or S-shapes with Sloping Inner Flanges (6.3).

199

6.3 RESTRAINED TORSION

6.3.1 Introduction

If warping of a section is resisted as shown in Fig. 6.10, warping out of plane at the support is impossible, causing forces, F, to develop. These are associated with bending of the flanges. This results in transverse shear forces, H.

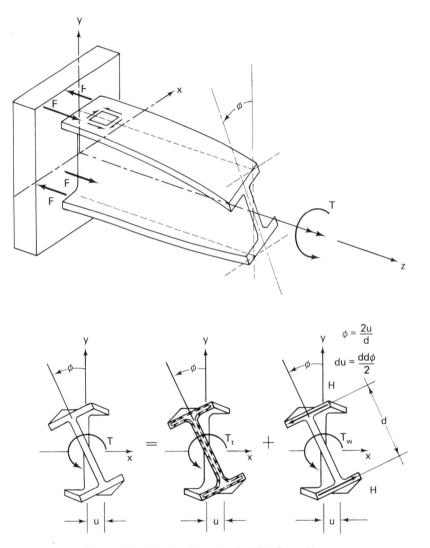

Figure 6.10 Warping Restraint in Wide-flange Beam.

6.3.2 I-Sections

In the case shown in Fig. 6.10 the flanges are the primary elements resisting lateral bending, and

$$T = T_t + T_w \tag{6.14}$$

The torsion bending resistance, T_w, neglecting web resistance to distortion, is (Fig. 6.10)

$$T_w = Hd \tag{6.15}$$

Also, if the bending moment in the flange is M_f, I_f the flange moment of inertia about the y-axis, and u the transverse deflection of the center of each flange at a section, then

$$M_f = -EI_f \frac{d^2u}{dz^2} = -EI_f \frac{d}{2} \frac{d^2\phi}{dz^2}$$

Now

$$H = V_f = \frac{dM}{dz} = -EI_f \frac{d^3u}{dz^3} = -EI_f \frac{d}{2} \cdot \frac{d^3\phi}{dz^3}$$

where $u = \phi(d/2)$, or $d\phi/dz = (2/d)\cdot(du/dz)$ and

$$T_w = -EI_f\left(\frac{d^2}{2}\right) \cdot \frac{d^3\phi}{dz^3} \tag{6.16}$$

The quantity $I_f(d^2/2)$, which has the dimension of length to the sixth power, is a measure of the torsional resistance of a wide-flange shape due to bending of its flanges about their major axes. It is called the warping constant, C_w. Values for C_w for various types of the sections are available in several manuals [e.g., AISC *Manual* (6.7), pp. 1-109 to 1-114]. Thus

$$T_w = -EC_w\frac{d^3\phi}{dz^3} = -C_1\frac{d^3\phi}{dz^3} \tag{6.17}$$

According to Eqs. (6.1), (6.14), and (6.17), now

$$T = C\frac{d\phi}{dz} - C_1\frac{d^3\phi}{dz^3} \tag{6.18}$$

For a variable torque, $T(z)$, this differential equation becomes

$$\frac{dT}{dz} = C\frac{d^2\phi}{dz^2} - C_1\frac{d^4\phi}{dz^4} \tag{6.19}$$

The general solution for Eq. (6.18) for the case of constant torque is

$$\phi = A_1 \sinh\left(\frac{z}{a}\right) + A_2 \cosh\left(\frac{z}{a}\right) + A_3 + \frac{Tz}{C} \tag{6.20}$$

where

$$a = \sqrt{\frac{C_1}{C}} = \sqrt{\frac{EC_w}{GJ}} \tag{6.21}$$

The arbitrary constants, A_1, A_2, and A_3, are to be determined from the boundary conditions of the member. The AISC *Manual* (6.7) gives values for a, C_w and J of rolled shapes (pp 1-109 to 114). Solutions of Eq. (6.20) for specific cases are given in Appendix VI, Table 2.

Various possible boundary conditions are

1. Pinned: no twist and free to warp, $\phi = 0$, $\phi'' = 0$
2. Free end: free to twist and to warp, $\phi'' = 0$
3. Fixed end: no twist and no warp, $\phi = \phi' = 0$

Example 6.2 For the beam shown in Fig. 6.10 find the stresses due to torsion.

Solution The boundary conditions are

$$\text{For } z = 0; \quad \phi = 0, \quad \frac{d\phi}{dz} = 0$$

$$\text{For } z = L: \quad \frac{d^2\phi}{dz^2} = 0 \quad (M_f = 0)$$

This yields (table 2 Appendix VI, Case 7b)

$$\phi = \frac{T \cdot a}{GJ}\left[\frac{z}{a} + \frac{\sinh(L-z)/a - \sinh L/a}{\cosh L/a}\right] \quad (6.22)$$

As a result of the torsional moment, T, the stresses can be classified into three categories:

1. Shearing stresses due to "pure" torsion, f_{vt}
2. Shearing stresses due to "warping" torsion, f_{vw}
3. Normal bending stresses due to restraining of warping, f_{bw}

These stresses for the beam shown in Fig. 6.10 can be calculated as follows:

$$f_{vt} = \frac{T_t t}{J} = Gt\frac{d\phi}{dz} \quad (6.23a)$$

$$f_{vw\text{-max}} = \frac{V_f Q_f}{I_f t_f} = \left(-EI_f \frac{d}{2}\frac{d^3\phi}{dz^3}\right)\frac{b^2 t_f}{8 I_f t_f} = -E\frac{b^2 d}{16}\frac{d^3\phi}{dz^3} \quad (6.23b)$$

$$\left(I_f = \frac{b^3 t_f}{12} \quad \text{and} \quad Q_f = \frac{b^2 t_f}{8}\right) \quad (6.23c)$$

$$f_{bw\text{-max}} = \frac{M(b/2)}{I_f} = \frac{Ed}{2}\frac{b}{2}\frac{d^2\phi}{dz^2} = \frac{Edb}{4}\frac{d^2\phi}{dz^2} \quad (6.23d)$$

and

$$f_v = f_{vt} + f_{vw} = GT\frac{d\phi}{dz} - E\frac{b^2 d}{16}\frac{d^3\phi}{dz^3} \quad (6.23e)$$

Example 6.3 Consider a 12W50 shape similar to the one shown in Fig. (6.10), 8-ft (2.44-m) long, subjected to a torque, T, applied at its free end. The base is fixed. Determine, in terms of T: (1)—(a) the angle of twist, $\phi(z)$; (b) $T_t(z)$; and (c) $T_w(z)$; (2) the extreme stresses at both ends. (T in lb-in.)

Solution From the AISC Manual (6.7, p. 1-113), $J = 1.78$ in.⁴ (74.5 cm⁴), $a = 52.3$ in. (132.8 cm), and $C_w = 1,880$ in.⁶ (504,847 cm⁶). Also:

$$G = \frac{E}{2(1+\mu)} = \frac{29 \times 10^6}{2(1+0.3)}$$

According to Eq. (6.22):

1. a. $\phi(z) = \dfrac{T(52.3)2.6}{29 \times 10^6 \times 1.78}$ *inches*

   $\times \left[\dfrac{z}{52.3} + \dfrac{\sinh(96-z)/52.3 - \sinh 96/52.3}{\cosh 96/52.3} \right]$ *RADIANS*

   $\phi = \dfrac{TL}{JG} = 4.84 \times 10^{-6} T$

   For $z = 0$: $\phi(0) = 0$ → B.C.   no twisting

   For $z = L = 96$ in. (2.44 m): $\phi(L) = 2.32 \times 10^{-6} T$

   b. $T_t(z) = GJ \dfrac{d\phi}{dz} = T\left[1 - \dfrac{\cosh(L-z)/a}{\cosh L/a} \right]$

   $= T\left[1 - \dfrac{\cosh(96-z)/52.3}{\cosh 96/52.3} \right]$

   For $z = 0$: $T_t = 0$ B.C.   no warping

   For $z = \dfrac{L}{2} = 48$ in. $= 1.22$ m: $T_t = 0.55T$

   For $z = L = 96$ in. $= 2.44$ m: $T_t = 0.69T$

   c. $T_w(z) = -EC_w \dfrac{d^3\phi}{dz^3} = T\left[\dfrac{\cosh(L-z)/a}{\cosh L/a} \right]$

   For $z = 0$: $T_w = T$ B.C.

   NOTE: $T = T_t(z) + T_w(z)$

   For $z = \dfrac{L}{2}$: $T_w = 0.45T$

   For $z = L$: $T_w = 0.31T$

   uniform torsion (moment) diagram

2. a. $f_{bw} = \pm \dfrac{Edb_f}{4}\left(\dfrac{d^2\phi}{dz^2}\right)$

   $= \pm 29 \times 10^6 \times \dfrac{(12.19 - 0.64)(8.08)}{4} \dfrac{Ta}{GJ}\left[\dfrac{1}{a^2} \dfrac{\sinh(L-z)/a}{\cosh L/a} \right]$

   For $z = 0$: $f_{bw} = \pm 0.62T$

   For $z = L$: $f_{bw} = 0$ → $\phi'' = 0$ $M_s = 0$

   b. $f_{vt} = \dfrac{T_t t_f}{J}$

   For $z = 0$ in the flange: $T_t = 0 \rightarrow f_{vt} = 0$

   For $z = L$ in the flange: $T_t = 0.69T \rightarrow f_{vt}^f = \dfrac{0.69 \times 0.640 T}{1.76}$

   max

   $= -0.247T$

   For $z = L$ in the web: $f_{vt}^w = \dfrac{0.37}{0.64} \times 0.247T = 0.143T$

   c. $f_{vw} = -E\dfrac{b^2 d}{16}\dfrac{d^3\phi}{dz^3} = \dfrac{ET}{a^2 GJ} \dfrac{b^2 d}{16}\left[\dfrac{\cosh(L-z)/a}{\cosh L/a} \right]$

   For $z = 0$: $f_{vw} = \dfrac{2.6T}{(52.3)^2} \dfrac{(8.08)^2(11.55)}{(16)(1.78)} = 0.025T$

   For $z = L$: $f_{vw} = \dfrac{0.025T}{3.21852} = 0.0078T$

The stress distribution is plotted in Fig. 6.11.

Restrained Torsion

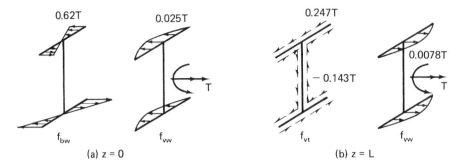

Figure 6.11 Stresses for Example 6.3: (a) $z = 0$; (b) $z = L$.

6.3.3 Channels

The twisting of channels occurs along their shear center axis and results in bending in the web (Fig. 6.12). The normal and shearing stresses in a channel are as follows:

1. Normal stresses due to warping

$$\text{In flange:} \quad f^f_{bw} = \frac{-Eeh}{2}\frac{d^2\phi}{dz^2} \quad (6.24a)$$

$$\text{In web:} \quad f^w_{bw} = \frac{-Eh}{2}(b-e)\frac{d^2\phi}{dz^2} \quad (6.24b)$$

2. Shearing stresses due to pure torsion

$$\text{In flange:} \quad f_{vf} = \frac{-Eh}{4}(b-e)^2\frac{d^3\phi}{dz^3} \quad (6.24c)$$

$$\text{In web at junction:} \quad f_{vw} = \frac{-Ebh}{2}\left(\frac{b}{2}-e\right)\frac{d^3\phi}{dz^3} \quad (6.24d)$$

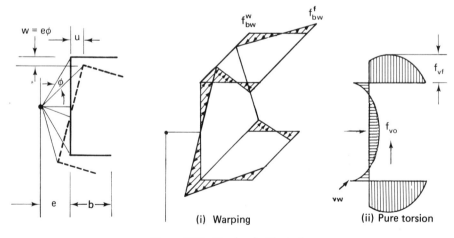

Figure 6.12 Torsion in Channels.

At midheight web: $$f_{vo} = \frac{-Eh}{2}\left(\frac{eh}{4} + be - \frac{b^2}{2}\right)\frac{d^3\phi}{dz^3} \tag{6.24e}$$

6.3.4 Box Sections

Restrained torsion in boxed beams as compared with I-beams and channels results in small normal stresses. In many practical cases it is allowable to neglect normal stresses in boxed beams and assume that the shearing stresses are the same as those for free torsion. To form some idea about the magnitude of the maximum normal stress in cantilever box beams, Fig. 6.13 shows the maximum values at the fixed end (6.6).

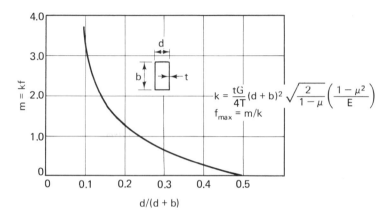

Figure 6.13 Maximum Normal Stress at the Fixed End of Cantilever Box Section (From T. Kármán and Chien Wei Zang, "Torsion with Variable Twist," *J. Aeronaut. Sci.*, vol. 13, no. 10, Oct. 1946).

6.4 COMBINED TORSION AND BENDING

Whenever a beam is loaded with a transverse bending load not passing through its shear center, we have combined torsion and bending. The simplified approach is given in Section 7.1.3, "Unsymmetrical Bending."

Example 6.4 Consider a wide-flange beam as shown in Fig. 6.14, assuming torsionally pinned-end conditions. Find the stress by the accurate procedure.
At A and B
$$\phi = \frac{d^2\phi}{dz^2} = 0$$

Solution According to table 2 Appendix VI, Case 4, the angle of twist is
$$\phi = \frac{ma^2}{GJ}\left[-\tanh\frac{L}{2a}\sinh\frac{z}{a} + \cosh\frac{z}{a} - \frac{z^2}{2a^2} + \frac{zL}{2a^2} - 1\right]$$

For this beam the same properties as those in Example 6.3 are given, while in addition

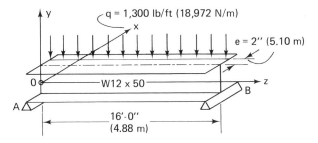

Figure 6.14 Eccentrically Loaded Wide-flange Beam.

$b = 8.08$ in. (20.5 cm); $\quad d = 12.19$ in. (30.96 cm);
$I_x = 394$ in.^{4} (16,441 cm^4) $\quad S_x = 64.7$ in.^{3} (1,060.2 cm^3);
$E = 29 \times 10^3$ ksi (199.96 $\times 10^3$ MPa);
$G = 11.154 \times 10^3$ ksi (76.9 $\times 10^3$ MPa)

1. *Simple Bending:* At $L/2$

$$M = \frac{qL^2}{8} = \frac{1.3 \times 16^2 \times 12}{8} = 499.2 \text{ kip-in. (5,640 kN-cm)}$$

$$f_b = \frac{M}{S_x} = \frac{499.2}{64.7} = 7.72 \text{ ksi (53.2 MPa)}$$

Shear at $z = 0$ is

Maximum shear force $V = \frac{qL}{2} = \frac{1.3 \times 16}{2} = 10.4$ kips (46.2 kN)

Shearing stresses at $z = 0$ at the midwidth of the flange due to bending are

$$f_v^f = \frac{10.4[(8.08/2) \times 0.64 \times 11.55/2]}{395 \times 0.64} = +0.614 \text{ ksi (4.2 MPa)}$$

and in the web

$$f_v^w = \frac{10.4[8.08 \times 0.64 \times 11.55/2 + 10.91/2 \times 0.37 \times 10.91/4]}{394 \times 0.37}$$

$$= 2.51 \text{ ksi (17.3 MPa)}$$

2. *Restrained Torsion:* The total torque per unit length is

$$m = q \cdot e = 1.3 \times 2 = 2.6 \text{ kip-in./ft}$$

$a = 52.3$ in. (132.7 cm) and $a^2 = 2,730$

$$\frac{ma^2}{GJ} = \frac{2.6 \times 2730}{19.96 \times 10^3 \times 12} = 0.0296$$

$$\frac{L}{a} = \frac{16 \times 12}{52.3} = 3.674; \quad \frac{L}{2a} = 1.837$$

$\sinh \frac{L}{2a} = 3.05922; \quad \cosh \frac{L}{2a} = 3.21852; \quad \tanh \frac{L}{2a} = 0.95051$

Thus

$$\phi = 0.0296\left(-0.95051 \sinh \frac{z}{a} + \cosh \frac{z}{a} - \frac{z^2}{2a^2} + \frac{zL}{2a^2} - 1\right)$$

and $\phi(0) = 0; \phi(L) = 0$

And so

$$\frac{d\phi}{dz} = 10^{-3}\left(-0.539 \cosh \frac{z}{a} + 0.567 \cosh \frac{z}{a} - 0.0109z + 1.043\right)$$

$$\frac{d\phi}{dz}(0) = 0.504 \times 10^{-3}; \quad \frac{d\phi}{dz}\left(\frac{L}{2}\right) = 0$$

$$\frac{d^2\phi}{dz^2} = 10^{-5}\left(-1.032 \sinh \frac{z}{a} + 1.085 \cosh \frac{z}{a} - 1.09\right)$$

$$\frac{d^2\phi}{dz^2}(0) = 0; \quad \frac{d^2\phi}{dz^2}\left(\frac{L}{2}\right) = -0.753 \times 10^{-5}$$

$$\frac{d^3\phi}{dz^3} = 10^{-7}\left(-1.974 \cosh \frac{z}{a} + 2.077 \sinh \frac{z}{a}\right)$$

$$\frac{d^3\phi}{dz^3}(0) = -1.974 \times 10^{-7}$$

Torsional shear at $z = 0$:

$$f_{vt}^f = Gt_f \frac{d\phi}{dz}(0) = 11.154 \times 10^3 \times 0.64 \times 0.504 \times 10^{-3}$$
$$= 3.60 \text{ ksi } (24.8 \text{ MPa})$$

$$f_{vt}^w = Gt_w \frac{d\phi}{dz}(0) = \frac{0.37}{0.64} \times 3.60 = 2.08 \text{ ksi } (14.3 \text{ MPa})$$

Warping shear at $z = 0$

$$f_{vw}^t = E\frac{b^2 d}{16}\frac{d^3\phi}{dx^3}(0) = 29 \times 10^3 \times \frac{8.08^2 \times 11.55}{16} \times (-1.974) \times 10^{-7}$$
$$= 0.284 \text{ ksi } (1.9 \text{ MPa})$$

$$f_{vw}^w = 0$$

All torsional and warping shear at $z = L/2$ is zero.

The warping normal stress at $z = L/2$ is

$$f_{bw} = \frac{Ebd}{4}\frac{d^2\phi}{dz^2}\left(\frac{L}{2}\right) = \frac{29 \times 10^3 \times 8.077 \times 12.19}{4}(-0.753 \times 10^{-5})$$
$$= -5.37 \text{ ksi } (37.0 \text{ MPa})$$

Now the combined stresses are as follows:

1. At $z = 0$:

   Shear stresses

   Flange: $\quad f_v^f = 0.614 + 3.60 + 0.284 = 4.50$ ksi (31.0 MPa)

   Web: $\quad f_v^w = 2.51 + 2.08 = 4.59$ ksi (31.6 MPa)

   Normal stresses none

2. At $z = L/2$

   Shear stresses: none

   Normal stresses: $f_b = 7.72 + 5.37 = 13.09$ ksi (90.3 MPa)

As can be seen in Section 7.1.3, Example 7.2 an approximate approach will yield slightly low values for the shear stresses and far too high values for the normal stresses.

6.5 DESIGN FOR TORSION

Torsion in general can be and should be avoided by the designer. Laterally loaded beams should be loaded through their shear center axis wherever possible. However, there are cases where torsion cannot be avoided. Examples are aircraft wings, curved beams, and beams in framed construction twisted as a result of bending of adjacent beams or, as in lintel beams, due to asymmetrical loading. For structures such as aircraft wings, torsion is a primary design consideration, as it may well be for curved beams. However, for most beams used in civil engineering structures, torsion is a secondary design consideration and very seldom determines the size of the beam. The important thing for a designer in such cases is to recognize torsion and to avoid it if possible or, if this cannot be done, to check its effect.

As will be shown in the next chapter, the torsional properties of beams are important, even if flexural loading does not directly cause torsion, in connection with their lateral stability.

NOTATIONS

a = $\sqrt{C_1/C} = \sqrt{EC_w/GJ}$
b = width of I-beam flange
d = depth of I-beam
f_v = shear stress
f_b = bending stress
f_t = torsional shear stress
f_w = warping shear stress
f^f = flange stress
f^w = web stress
u = displacement in x-direction
δ = stress coefficient
ϕ = angle of twist
μ = Poisson's ratio
A_1, A_2, A_3 = arbitrary constants
C = twisting constant

C_1 = EC_w
C_w = warping constant (in.^{6})
E = Young's modulus of elasticity
G = shearing modulus
I_f = moment of inertia of flange of I-beam (in.^{4})
J = torsional constant (in.^{4})
M_f = bending moment in flnage of I-beam due to warping restraint
T = twisting moment
T_t = St. Venant torque
T_w = warping torque
V_f = shearing forces in flange of I-beam

PROBLEMS

6.1. The cross section of a 6-ft simple beam consists of a channel C10 × 20; the total applied vertical load is 0.5 kip/ft and passes through the centroid of the channel (see Fig. P-6.1). Determine the stresses in this channel. (Assume section is torsionally and flexurally simply supported.) ($J = .37$ in.^{4}; $C_w = 57$ in.^{6}; $E = 29,000$ ksi; $\mu = 0.3$)

6.2. Analyze the channel of Problem 6.1 if instead of a uniformly applied vertical load a concentrated load of 3 kips is applied at midspan, passing through the centroid.

6.3. An ∠5 in. × 5 in. × ⅝ in. angle is used as a beam for a load (exclusive of own weight) of 2 kips acting at midspan (span length 8 ft) which passes through

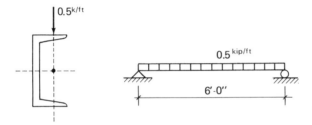

Figure P-6.1

the centroid (see Fig. P-6.3). Find the stresses in this angle if the beam ends are free to warp.

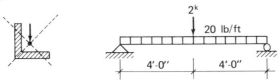

Figure P-6.3

6.4. Analyze the angle of Problem 6.3 if, instead of a concentrated load of 2 kips, in addition to the 20-lb/ft dead load a vertical uniform load of 0.25 kip/ft is acting through the centroid.

6.5. A crane girder as shown in Fig. P-6.5 has vertical and horizontal forces as the resultants of two crane wheel loads. Determine the maximum stresses in the crane girder taking into account torsion. The supports do prevent overturning of the beam, and the ends are free to warp. (Assume a cantilever 9 ft long uniformly loaded with 125 lb/ft and an end load of $12 + 9\,(.125) = 13.125$ kips and an end torque of $1.2 \times 20 = 24$ kips·in.)

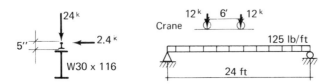

Figure P-6.5

6.6. A W16 × 40 cantilever beam supports an eccentric vertical force of 4 kips at the end of a span of 8 ft (see Fig. P-6.6). Calculate maximum stresses in this beam, neglecting its own weight.

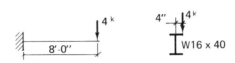

Figure P-6.6

6.7. Repeat Problem 6.6 if a 4-kip load is acting horizontally, instead of vertically, in the center plane of the top flange.

6.8. A concentrated vertical load of 10 kips is acting at midspan of a simply supported W-beam of A36 steel (see Fig. P-6.8). The ends are free to warp. Calculate the maximum stresses (normal and shear stresses) caused by this loading.

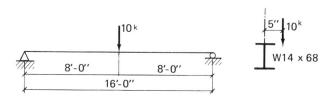

Figure P-6.8

6.9. Repeat Problem 6.8 if a 4-kip load is acting horizontally in the top plane of the upper flange with no vertical load.

6.10. Repeat Problem 6.8 if the 10-kip vertical load is replaced by a vertical uniformly distributed load of 1 kip/ft.

6.11. A 26-ft lintel beam has to support a wall. The superimposed load is equal to 1.9 kips per foot, acting eccentrically 6 in. from the web axis of the beam. Use a W-section of A36 steel to design this beam. The ends of the beam are torsional simple supports (free to warp but no rotation).

6.12. Design a beam to carry two symmetrically located concentrated loads of 22 kips each with an eccentricity of 4 in. with respect to the web plane (see Fig. P-6-12). Neglect the beam weight and assume for bending that this beam is behaving as a simple beam and for torsion as a torsionally fixed beam. Use a W-section of A36 steel and AISC specification (6.7). (Try W33 × 201.)

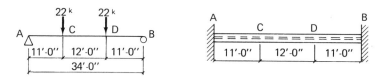

Figure P-6.12

6.13. Repeat Problem 6.12 if the two concentrated loads are 30 kips and have an eccentricity of 3 in.

6.14. Repeat Problem 6.12 if the eccentricities are on opposite sides of the web plane. (Try W30 × 116.)

6.15. A W10 × 33 beam of A36 steel spans 20 ft between two sloping girders (see Fig. P-6.15). There are no lateral bracings within the span length. The uniform load of 550 lb/ft (including beam weight) is vertical and acts at the center of the top flange. Determine the maximum stresses (normal and shear) if the beam ends are free to warp but torsionally fixed.

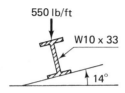

Figure P-6.15

6.16. Repeat Problem 6.15 using a C10 × 30 channel. ($J = 1.22$ in.^{4}; $C_w = 79.5$ in.^{6})

REFERENCES

6.1. TIMOSHENKO, S. P., and J. M. GERE, *Theory of Elastic Stability*, 2nd ed., McGraw-Hill, New York, 1961, pp. 212–24.

6.2. HEINS, C. P., JR., and P. A. SEABURG, *Torsion Analysis of Rolled Steel Sections*, Bethlehem Steel Corporation, Bethlehem, PA.

6.3. EL DARWISH, A., and B. G. JOHNSTON, "Torsion in Structural Shapes," *J. Struc. Div., Proc. ASCE*, no. ST1, 1965, pp. 203–27.

6.4. LYSE, I., and B. G. JOHNSTON, "Structural Beams in Torsion", *Trans. ASCE*, vol. 101, 1936, pp. 857–944.

6.5. TIMOSHENKO, S. P., *Strength of Materials*, Van Nostrand, Princeton, NJ, 1941, part 2, pp. 288–99.

6.6. VON KÁRMÁN, T., and CHIEN WEI-ZANG, "Torsion with Variable Twist," *J. Aeronaut. Sci.*, vol. 13, no. 10, Oct. 1946, pp. 503–10.

6.7. AISC, *Manual of Steel Construction*, 8th ed., Chicago, 1980.

7

Bending of Beams

7.1 STRAIGHT PRISMATIC BEAMS

A beam may be exposed to simple, biaxial, or unsymmetrical bending. As was pointed out in the previous chapter, if bending is combined with torsion, which will always happen if the plane of loading is not passing through the shear center, then it is called unsymmetrical bending. Simple bending occurs when the plane of loading coincides with one of the principal planes (containing the principal axis of all cross-sectional areas) for beam sections with two axes of symmetry (as I-sections), or when it is parallel to one of the principal planes for beams with open sections having one axis of symmetry (as C-sections) and is passing through the shear center. Biaxial bending is similar to simple bending, except that bending takes place about both principal axes without any torsion.

Initially it will be assumed that no lateral instability of the beam will occur (see Section 7.1.4, Lateral-Torsional Buckling) because the beam is laterally sufficiently supported so that no lateral-torsional buckling can take place.

7.1.1 Simple Bending

In simple bending the neutral axis (Fig. 7.1) always passes through the centroid of the cross-sectional area and is normal to the loading plane. Bending stresses are assumed to exist only in the longitudinal direction. Additional assumptions are as follows:

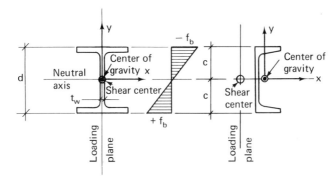

Figure 7.1 Simple Bending.

1. The material obeys Hooke's law, $\epsilon = f/E$ (is linearly elastic), with the same behavior in tension as in compression.
2. Plane cross sections normal to the beam axis remain plane and normal to the deformed axis (the elastic line) (Bernoulli's hypothesis).
3. The cross sections have at least one axis of symmetry.
4. The deformations are small as compared to the beam dimensions.

Maximum normal bending stresses are calculated from

$$f_b = \pm \frac{M}{S} \tag{7.1}$$

where S = elastic section modulus = I/c.

If the beam is exposed solely to uniform bending moments (no transverse forces), the elastic line of the beam is part of a circle, and sections remain really plane and the shear stresses are zero. If bending moments are due to any transverse loading on the beam, shear stresses are not zero and are calculated from the formula

$$f_v = V\frac{Q}{It} \tag{7.2}$$

where V = shear force, $Q = A\bar{y}_i$ is the statical (first) moment of that part of the area of cross section lying beyond the fiber under consideration, I = moment of inertia of the whole cross section, and t = width of the section.

For I-sections [see AISC *Manual* (7.1), section 1.5.1.2, p. 5-18] as the flanges take a very small amount of shear it is customary to use an "average" shear stress, $(f_v)_{av}$, and compare it with the allowable shear stress

$$(f_v)_{av} = \frac{V}{A_w} \leq F_v \tag{7.3}$$

where A_w = web area = dt_w and F_v = allowable shear stress.

In simple bending the deflections of a beam occur only in the plane of loading and can be calculated by integration of the following differential equation

$$\frac{d^2w}{dx^2} = \pm \frac{M}{EI} \tag{7.4}$$

where w = deflection at x and EI = flexural rigidity. The sign in Eq. (7.4) depends upon the convention defining the sign of the bending moments and the chosen orientation of the coordinate system, which determines the sign of the second derivative of the deflection with respect to the x-axis. If both (moment and second derivative) are of the same sign, a plus sign is used in Eq. (7.4); if they are of different signs, then a minus sign is chosen.

As pointed out earlier in this text (Section 1.2.4), beam deflection limitations often depend on the materials supported by that beam. For buildings, according to the AISC (7.1, section 1.13.1, Deflections, p. 5-42), beams and girders supporting plastered ceilings must be so proportioned that the maximum live load deflection does not exceed $\frac{1}{360}$ of the span. For highway bridges, according to the AASHTO (7.2, section 1.7.6, "Deflections," p. 152), members having simple or continuous spans must be designed so that the deflection due to live load plus impact shall not exceed $\frac{1}{800}$ of the span. For bridges in urban areas used in part by pedestrians, the ratio preferably shall not exceed $\frac{1}{1000}$. By comparison the German building specifications, DIN 1050 (7.3), state that usually the purpose of a building does limit the deflections: the deflection of floor beams and girders with a span larger than 16.5 ft (5 m) should be not greater than $\frac{1}{300}$ of the span and in the case of an overhang the deflection at the end should not exceed $\frac{1}{200}$ of the overhang length.

7.1.2 Biaxial Bending

Biaxial bending consists of simple bending about the two perpendicular principal directions, in the absence of torsion.

Sections with two axes of symmetry

When the loading plane (Fig. 7.2) passes through the centroid of a section with two axes of symmetry but does not coincide with either of the two principal axes, there is still only bending without torsion. The neutral line passes through the centroid but is no longer perpendicular to the loading plane. If the applied bending moment, represented by the vector, M, acting normal to the loading plane, is resolved into two components, M_x and M_y, then the extreme bending stresses are

$$f_b = \pm \frac{M_x}{S_x} \pm \frac{M_y}{S_y} \tag{7.5}$$

where S_x and S_y are elastic section moduli about the x- and y-axes, respectively.

Sections of arbitrary shape

If the beam section has an arbitrary shape composed of two or more different rolled sections—as, for example, the lintel beam shown in Fig. 7.3—the axes of reference (x and y in Fig. 7.3) to start with could be any axes. If they are principal axes, then Eq. (7.5) would apply. Thus

$$f_b = \pm \frac{M_1}{I_1} y_1 \pm \frac{M_2}{I_2} x_1 \tag{7.5a}$$

where M_1 and M_2 are vector component moments of the applied moment, M, in the directions of the principal axes, 1 and 2; I_1 and I_2 are the values of the moment of inertia of the cross section with respect to the principal axes; and x_1 and y_1 are coordinates of any particular point in the same coordinate system (1-2).

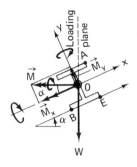

(a) Analytical method

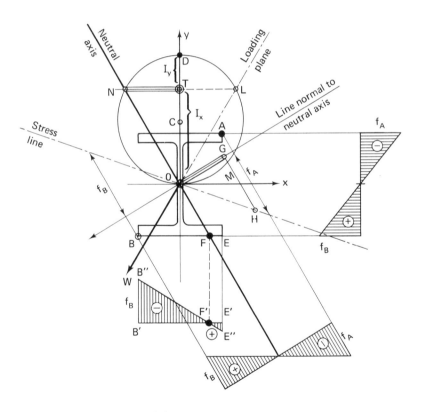

(b) Graphical method

Figure 7.2 Biaxial Bending with Double Symmetry.

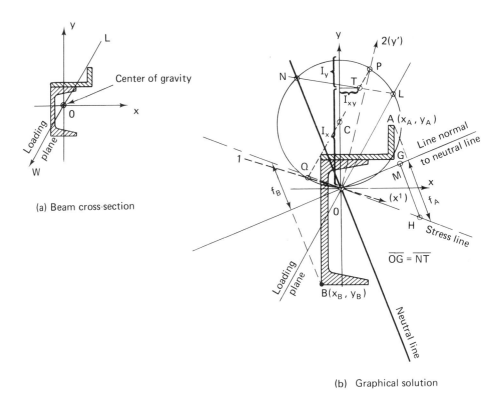

(a) Beam cross-section

(b) Graphical solution

Figure 7.3 Biaxial Bending of Any Shape.

If the initial coordinate system is not the principal one (x-y in Fig. 7.3), it often proves complicated to calculate the principal moments of inertia, I_1 and I_2, from the known values of I_x, I_y, and I_{xy} and especially to express the coordinates of characteristic points in that system, to be able to use Eq. (7.5a). To avoid such a calculation, the following well-known equation, given in any book on strength of materials, is used:

$$f_b = \frac{M_x I_y - M_y I_{xy}}{I_x I_y - I_{xy}^2} \cdot y + \frac{M_y I_x - M_x I_{xy}}{I_x I_y - I_{xy}^2} \cdot x \qquad (7.6)$$

where all the quantities refer to the arbitrary coordinate system (x-y).

Example 7.1 Design a purlin (Fig. 7.4) for a roof with a slope of 1 in 2. The purlins are supported by trusses 16 ft apart. Roof covering is $\frac{1}{4}$-in. corrugated asbestos roofing weighing 3.0 lb/ft² of roof surface with $\frac{1}{2}$-in. insulation board weighing $1\frac{1}{2}$ lb/ft². This board is attached to the underside of the purlins, so as to improve the inside appearance of the roof and reduce heat loss. For snow load assume that the building is located in the central latitudes of the United States and for wind load that the building is in an unexposed location protected by surrounding buildings.

Solution First the load analysis has to be performed so as to obtain the loading per linear foot of purlin. Next the structural analysis will be carried out to get extreme values of the bending moments and shear forces. To perform this analysis it must be decided beforehand how the purlin will be considered—as a simply supported or continuous beam. It should be noticed from Fig. 7.4(b) that for

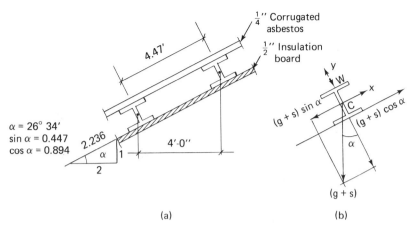

Figure 7.4 (Example 7.1).

simplicity the gravity and snow loads are both assumed to act through the section's center of gravity, and thus they have to be resolved into the x- and y-directions—that is, parallel and normal to the roof slope. Corrugated asbestos is capable of spanning 4 ft between purlins while also preventing the top flange from lateral-torsional buckling. However, it cannot carry the component of loads parallel to the roof slope to the ridge. Therefore, the purlin is exposed to biaxial bending. If no "sag" rods are used, the way to reduce moments about both axes is to choose a continuous-beam system for the purlin. Transportation and handling will limit the length of the purlins. Assume that in this case this is 50 ft. Therefore, it could be decided that the purlins will be continuous over three spans, 16 ft each, giving a total material length of 48 ft. Finally, when the extreme values of bending moments and shear forces are known from the structural analysis, then the purlins can be designed.

LOAD ANALYSIS

1. *Dead Load (per linear ft of purlin):*

Roof covering: 3.0×4.47 = 13.4 lb/ft
Insulation board: 1.5×4.47 = 6.7 lb/ft
Purlin: (Guess 50% of all other D.L.) = 10 lb/ft

Total: $g \approx 30$ lb/ft (vertical load)
(437.8 kN/m)

2. *Live Load:* For the central latitudes of the United States and a 6 in. in 12 in. slope, the snow load in lb/ft² of horizontal projection is 20 lb/ft²; therefore

Snow: $S = 20 \times 4.0 = 80$ lb/ft (1,167.5 kN/m)

The wind load for an unexposed location protected by surrounding buildings on a vertical surface is 20 lb/ft² (95.8 kN/m²). Pressure on a roof making an angle of $\alpha = 26°34'$ with the horizontal plane is

$$W_n = \frac{W\alpha}{45°} = \frac{20 \times 26.6}{45} = 11.8 \text{ lb/ft}^2 \text{ (56.6 kN/m}^2\text{) of roof surface}$$

The wind load per linear foot of purlin is

$11.8 \times 4.47 = 52.8$ lb/ft (770.6 kN/m) (perpendicular to the roof)

3. *Loadings to be Considered:* Two combinations of load must be considered:

    1. Dead load plus full snow load
    2. Dead load plus one-half snow load and full wind load

Maximum loading is then as follows:

    1. (D.L. + S) = 30 + 80 = 110 lb/ft (1.605 MN/m) (vertical)
    2. (D.L. + $\frac{1}{2}$S + W) = 30 + 40 = 70 lb/ft (1.022 MN/m) (vertical)
                  W = 52.8 lb/ft (0.771 MN/m) (normal to roof)

These loads will be resolved into components parallel to (x-component) and perpendicular to (y-component) the roof.

Loading Case 1:
    In x-direction: $110 \times 0.447 = 49.2 \approx 50$ lb/ft (730 N/m)
    In y-direction: $110 \times 0.894 = 98.3 \approx 100$ lb/ft (1,460 N/m)

Loading Case 2:
    In x-direction: $70 \times 0.447 = 31.3 \approx 30$ lb/ft (438 N/m)
    In y-direction: $70 \times 0.897 + 52.8 = 115.6 \approx 115$ lb/ft (1,678 N/m)

The AISC specification (7.7, section 1.5.6, p. 26) allows an increase in allowable stresses of one-third when produced by wind or seismic loading acting alone or in combination with the design dead and live loads. Therefore, it is customary to work with the same allowable stresses but to reduce the combined load in the same proportion. This procedure can be substantiated in the following manner.

The allowable stress in both cases should be equal to or larger than the calculated stress, which is some function ϕ of the loads

$$F_{\text{allowable}} \geq f_{\text{calculated}} = \phi(\text{D.L.} + \text{L.L.}) \tag{a}$$

and

$$\tfrac{4}{3} F_{\text{allowable}} \geq f_{\text{calculated}} = \phi(\text{D.L.} + \text{L.L.} + W) \tag{b}$$

or the same allowable stress in Eq. (b) is equal to

$$F_{\text{allowable}} \geq \tfrac{3}{4} f_{\text{calculated}} = \tfrac{3}{4}\phi_1(\text{D.L.} + \text{L.L.} + W) \tag{c}$$

As both components in loading case 2 are smaller than the corresponding components in case 1—

$$\tfrac{3}{4} \times 30 < 50 \text{ lb/ft}$$
$$\tfrac{3}{4} \times 115 = 86 < 100 \text{ lb/ft} \tag{d}$$

—only case 1 should be considered.

STRUCTURAL ANALYSIS. The purlins are, as already mentioned, continuous beams over three spans in both x- and y-planes. Using expressions for moments and shear forces from the AISC *Manual* (7.1, p. 2-126) the bending moment and shear force diagrams in the x-plane (Fig. 7.5) and in the y-plane (Fig. 7.6) are obtained.

Based on the plastic behavior of steel beams, the AISC specification (7.7, section 1.5.1.4, Bending) allows in our case for some moment redistribution—that is, the reduction of the minimum negative (support) moment by 10%, provided that the maximum positive moment is increased by 1/10 of the average negative moments (see Fig. 7.12).

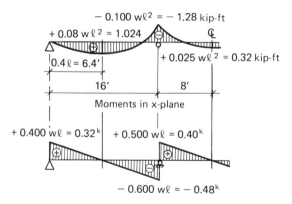

Figure 7.5 Shear Forces and Moments in x-plane.

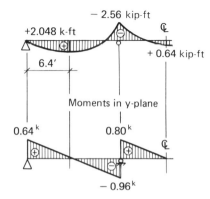

Figure 7.6 Shear Forces and Moments in y-plane.

This means that the "design" moments are

M_y in the x-plane (bending about weak axis) = Negative moment 9/10(−1.28)
= −1.152 kip-ft (−156.19 kN-cm). Positive moment 1.024 + (0.1/2)(1.28)
= +1.088 kip-ft (+147.51 kN-cm).

M_x in the y-plane (bending about strong axis) = Negative moment 9/10(−2.56)
= −2.304 kip-ft (−312.38 kN-cm). Positive moment 2.048 + (0.1/2)(2.56)
= +2.176 kip-ft (+295.03 kN-cm).

To start the design process it is customary to express the ratio of section moduli by an approximate coefficient, $S_n = S_x/S_y$, and to introduce it into Eq. (7.5):

$$f_b = \pm \frac{M_x}{S_x} \pm \frac{M_y}{S_x} \cdot S_n \leq F_b \tag{7.7}$$

From this, the required section modulus with respect to the x-axis is

$$(S_x)_{rqd} = \frac{1}{F_b}(\pm M_x \pm M_y \cdot S_n) = \frac{M^1}{F_b} \tag{7.8}$$

where $M^1 = M_x + S_n \cdot M_y$ and F_b = allowable bending stress, which can be

1. $0.66F_y$ for compact adequately braced rolled sections; or
2. $0.60F_y$ for sections with unbraced length, L_b, small enough to satisfy $L_b \leq L_u$ (see Section 7.1.4); or
3. Any value obtained from corresponding formulas given in Section 7.1.4 when $L_b > L_u$.

The coefficients, S_n, for wide-flange beams vary, but some average values can be taken to simplify the design process, such as the following

W6—$S_n = 3.5$
W8—$S_n = 7.0$
W10—Heavier sections (beyond 49 lb/ft): $S_n = 2.9$
 Lighter section (below 45 lb/ft): $S_n = 5.0$
W12—Heavier sections (beyond 53 lb/ft): $S_n = 3.0$
 Lighter sections (below 50 lb/ft): $S_n = 8.0$
W14—$S_n = 4$
W16—$S_n = 7$
W18, 24—$S_n = 6$
W27—$S_n = 8$
W30, 33—$S_n = 10$
W36—Heavier sections (beyond 230 lb/ft): $S_n = 7.0$
 Lighter sections (below 194 lb/ft): $S_n = 10.9$

An overall average value would be about $S_n = 6.0$. In our case the corrugated sheet diaphragm, which is very rigid in its own plane, protects the top flange when exposed to the positive beam moment for bending around the y-axis. In the vicinity of the supports, in the range of the negative moments, the bottom flange is protected only at the supports where the purlin is directly connected to the roof truss girder. At a distance of approximately $40r_y$ another lateral support should be provided by a simple diagonal bracing member if there is no special roof bracing already provided for other stability considerations.

Taking $S_n = 7$ (expecting some section in the range of W8) and using Eq. (7.8), we have, with $M = \pm 2.304$ kip-ft and ± 1.152 kip-ft,

$$(S_x)_{rqd} = \tfrac{1}{22}(2.304 \times 12 + 1.152 \times 12 \times 7) = 5.66 \text{ in.}^3 \text{ (92.8 cm}^3\text{)}$$

We take $F_b = 22$ ksi because flange support is provided by the corrugated sheeting.

The W8 × 10 has $S_x = 7.80$ in.^{3} and $S_y = 1.06$ in.^{3} with $S_n = 7.35$. The stress is

$$f_b = \frac{2.304 \times 12}{7.80} + \frac{1.152 \times 12}{1.06} = 3.54 + 13.04$$

$$= 16.59 \text{ ksi (114.38 MPa)} < 0.6F_y = 22 \text{ ksi (152 MPa)}$$

The total deflection of the purlin in its y-plane is [AISC *Manual* (7.1), p. 2-126]

$$(\Delta_{max}) = 0.0069 \frac{wl^4}{EI} = 0.0069 \frac{0.100 \times 16^4 \times 12^3}{29 \times 10^3 \times 30.8} = .088 \text{ in.} = \frac{L}{2182}$$

The deflection due to the live load only (snow load = $80 \times 0.894 = 71.5$ lb/ft) is

$$(\Delta_{max})_{L.L} = 1.05 \times \frac{0.0715}{0.100} = 0.75 \text{ in.} = \frac{L}{256}(1.9 \text{cm})$$

BENDING OF BEAMS

This deflection can be accepted as satisfactory because there are no deflection-sensitive materials connected with the roofing.

For biaxial bending the interaction equation for combined biaxial bending and axial loading could be used. This equation is given in the AISC specification (7.1, p. 5-26, equation 1.6-2) and is used for combined axial compression and bending when secondary moment effects are neglected. If the stress due to the axial compression is set equal to zero (because there is no such force acting in our case), then the formula is reduced to the following expression

$$\frac{f_{bx}}{0.60F_y} + \frac{f_{by}}{0.75F_y} \leq 1.0 \qquad (7.9)$$

The allowable stresses are now taken from the AISC specification (7.1, section 1.5.1.4.3). In our case

$$f_{bx} = 3.54 \text{ ksi}; \quad f_{by} = 13.04 \text{ ksi}; \quad 0.6F_y = 22 \text{ ksi};$$

and

$$0.75F_y = 27.0 \text{ ksi (186 MPa)}$$

Therefore, the interaction equation is

$$\frac{3.54}{22} + \frac{13.04}{27} = 0.161 + 0.483 = 0.644 < 1.00 \text{ ksi}: \qquad \underline{\text{OK}}$$

More about interaction equations can be found in this text in Section 8.2, Interaction Equations for Beam-Columns.

7.1.3 Unsymmetrical Bending

When bending takes place about one or both principal axes of a beam simultaneously with torsion, we speak about unsymmetrical bending. An overhead crane rail with both vertical and horizontal loads is loading its supporting crane girder in bending with accompanying torsion (Fig. 7.7)—that is, in unsymmetrical bending.

In Section 6.4 (Combined Torsion and Bending) the usual analytical treatment of the problem is given. Here, as a part of the practical designer's approach, the usual simplifications for unsymmetrical bending will be given. This simplified procedure is preferred to the more rigorous computation of bending and shear stresses resulting from the biaxial bending and torsion. This procedure is sufficiently accurate if the applied torsion is not excessive. A common practice is to substitute torsion by bending of horizontal flanges—either both of them or only the one which is closest to the applied horizontal load.

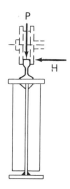

Figure 7.7 Unsymmetrical Bending.

If only a vertical eccentric load is applied (Fig. 7.8a), an equivalent couple is obtained from the following equilibrium equation:

$$Pe = Hd \qquad (7.10)$$

Apart from the vertical force acting on the section, both flanges are exposed to bending by the force, $H = Pe/d$, and bending and shearing stresses are calculated due to such bending occurring in the plane of flanges. The relatively thin web is disregarded in resisting these horizontal forces.

If torsion is caused by an applied eccentric horizontal force (Figs. 7.7 and 7.8b), then this horizontal force is assumed to cause bending only in the flange nearest to it.

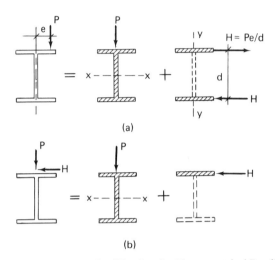

Figure 7.8 Design Simplification for Unsymmetrical Bending.

General experience is that this simplified design approach produces conservative values for the bending stresses as compared to those obtained by a more rigorous treatment and therefore is always on the safe side. Shearing stresses obtained in this way are smaller than the actual values and represent only a rough approximation of the true situation. Therefore, as mentioned above, if torsion is a substantial part of unsymmetrical bending, the simplified approach cannot safely be applied.

In the following example the same problem as in Example 6.4 will be used to illustrate this simplified procedure and allow a comparison of the results.

Example 7.2 A simply supported W12 × 50 beam of 16-ft (4.88-m) span is loaded by a uniform load of $w = 1{,}300$ lb/ft (18.97 kN/m) with an eccentricity of 2 in. (5.1 cm) (Fig. 7.9).

Solution Using Eq. (7.10), the equivalent horizontal uniform load is

$$w_h = 1{,}300 \times \frac{2}{11.55} = 225 \text{ lb/ft (3.28 kN/m)}$$

Maximum moment in the vertical plane is

$$_vM_{max} = \frac{wL^2}{8} = \frac{1.3(16)^2}{8} = 41.6 \text{ kip-ft (5,640.2 kN-cm)}$$

Maximum moment in the horizontal plane is

$$_hM_{max} = \frac{w_h L^2}{8} = \frac{0.225(16)^2}{8} = 7.2 \text{ kip-ft (976.2 kN-cm)}$$

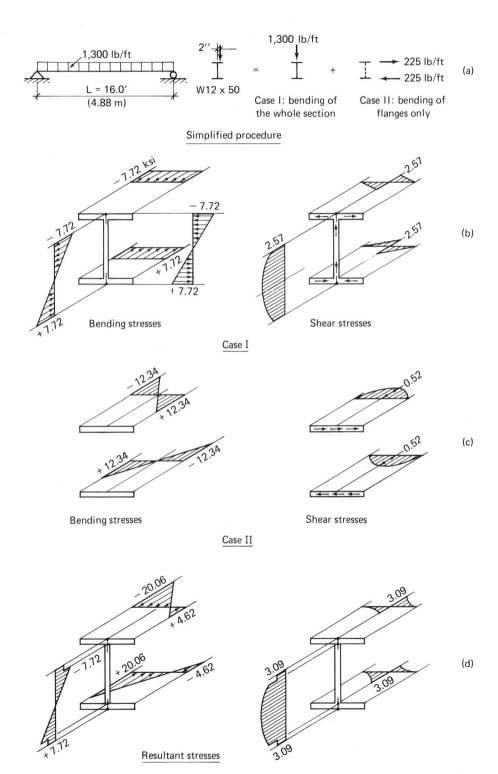

Figure 7.9 Simplified Unsymmetrical Bending.

Maximum shear force in the vertical plane is

$$v^{V_{max}} = 1.3 \times \tfrac{16}{2} = 10.4 \text{ kips (46.2 kN)}$$

Maximum shear force in the horizontal plane is

$$h^{V_{max}} = 0.225 \times \tfrac{16}{2} = 1.8 \text{ kips (8.0 kN)}$$

The section modulus about the x-axis is 64.7 in.³ (1,060.2 cm³). About the y-axis and for the whole section—that is, both flanges together—the value is 14.0 in.³ (229.4 cm³). For one flange only, $S_{y1} = 7.0$ in.³ (114.7 cm³). The web area is $A_w = 10.908 \times (0.371) = 4.05$ in.² (26.11 cm²), and for one flange, $A_f = 8.007 \times (0.641) = 5.18$ in.² (33.40 cm²).

Therefore, the resulting maximum bending stresses are, at midspan:

$$(f_b)_{max} = \frac{v^{M_{max}}}{S_x} + \frac{h^{M_{max}}}{S_y} = \frac{41.6 \times 12}{64.7} + \frac{7.2 \times 12}{7.0}$$
$$= 7.72 + 12.34$$
$$= 20.06 \text{ ksi (138.3 MPa)}$$

The resulting shear stresses are, at support:

$$(f_v)_{max} = \frac{v^{V_{max}}}{A_w} + \frac{3_h V_{max}}{2A_f} = \frac{10.4}{4.05} + \frac{3 \times 1.8}{2 \times 5.18} = 2.57 + 0.52$$
$$= 3.09 \text{ ksi (21.3 MPa)}$$

Using a more exact analysis for the same problem (see Example 6.4), the maximum combined stresses were obtained as follows:

At midspan $\left(z = \frac{L}{2}\right)$: $f_b = 13.09$ ksi (maximum bending stress)

At support $(z = 0)$: $f_v = 4.59$ ksi (maximum shear stress)

This means that the simplified method produced 31% higher bending stresses and a lower shear stress by 32.7%. As the shear stress usually does not control the design, the simplified approach is conservative. For heavier sections, these percentages are much smaller due to wider and thicker flanges.

7.1.4 Lateral-Torsional Buckling

It was earlier pointed out that only compressed members are subject to instability. As in any beam subject to bending, only one portion of the cross section is in compression, and this part may become unstable. Because this part is continuously connected through the web material to the tension part of the cross section, the stabilizing effect of the tension zone transforms free transverse buckling into a lateral-torsional buckling, causing lateral bending and twisting of the beam. Torsionally weak shapes which have a flexural rigidity in the plane of bending (normally the vertical plane and the web plane) that is significantly larger than their lateral rigidity will buckle more often than torsionally rigid shapes. Therefore, for instance, box-shaped sections are preferred for large beam spans or large unbraced lengths.

An exact analysis of lateral-torsional buckling for an arbitrarily shaped beam under different types of loading and boundary conditions is mathematically extremely complex (7.4). As in the majority of cases beam sections and those of plate girders possess double symmetry, the problem is greatly simplified by considering a beam of such a section exposed to pure bending in its web plane (7.5), in the elastic range and without distortion of the cross section. In that case the critical bending moment (producing lateral-torsional buckling) is

$$M_{\text{crit}} = \frac{\pi^2 E}{L}\left[I_y\left(\frac{C_w}{L^2} + \frac{GJ}{\pi^2 E}\right)\right]^{1/2} \tag{7.11}$$

where: C_w = warping torsional constant; J = St. Venant's (free) torsional constant; L = beam span or unbraced length; E, G = Young's and shear modulus, respectively; I_y = moment about weak axis of an I-section. This expression is usually simplified by introduction of the following approximations

$$\left.\begin{aligned} A &= 2b_f t_f \\ I_y &= 2\left(\frac{b_f^3 t_f}{12}\right) = \frac{b_f^3 t_f}{6} \\ J &= 2\left(\frac{b_f t_f^3}{3}\right) \\ C_w &= I_y \cdot \frac{d^2}{4} = \frac{b_f^3 t_f d^2}{24} \\ S_x &= \frac{I_x}{\frac{1}{2}d} = b_f t_f d \\ b_f &= 2r_y\sqrt{3} \end{aligned}\right\} \tag{7.12}$$

When the critical moment, M_{crit}, is divided by the section modulus, S_x, and after introducing the above simplifications, the critical stress is

$$F_{\text{crit}} = \sqrt{\left[\frac{\pi^2 E}{12(L/b_f)^2}\right]^2 + \left[\frac{\pi E}{\sqrt{18(1+\mu)}} \cdot \frac{A_f}{Ld}\right]^2} \tag{7.13}$$

The first term under the radical represents the contribution of torsional warping resistance and the second term St. Venant's torsional resistance.

For deep (slender) girders ($d \gg 0$) the value of the second term is much smaller than that of the first term and for practical purposes could be neglected. In this case only the warping term is taken instead of both terms—that is,

$$F'_{\text{crit}} = \frac{\pi^2 E}{12(L/b_f)^2} = \frac{\pi^2 E}{(L/r_y)^2} \tag{7.13a}$$

For shallow, stocky, thick-walled sections the opposite is valid and St. Venant's torsion prevails. Thus only the second term is kept—that is,

$$F''_{\text{crit}} = \frac{\pi E}{\sqrt{18(1+\mu)}} \frac{1}{L(d/A_f)} = \frac{0.65 E}{L(d/A_f)} \tag{7.13b}$$

with $\mu = 0.3$.

The AISC specification (7.7) uses the same approach and requires that both torsional resistances be investigated and the larger one used, provided its value is less than $0.6F_y$. Equations (7.13a) and (7.13b) are in terms of the critical stress. For design purposes both equations are first divided by the basic factor of safety of $\frac{5}{3}$ (see

Section 1.4.1). If for E is taken 29.0×10^3 ksi, then after dividing both equations by 1.67 the following expressions are obtained:

$$F'_b = \frac{171.388 \times 10^3}{(L/r_y)^2} \approx \frac{170 \times 10^3}{(L/r_T)^2} \qquad (7.13c)$$

$$F''_b = \frac{11.3 \times 10^3}{L(d/A_f)} \approx \frac{12 \times 10^3}{L(d/A_f)} \qquad (7.13d)$$

The first equation, Eq. (7.13c), is equation (1.5-6b) in the AISC specification (7.7).

The numerator is rounded off to 170×10^3, and r_y is replaced by r_T, the radius of gyration of a section comprising the compression flange plus $\frac{1}{3}$ the compression web area (for symmetrical sections actually $\frac{1}{6}$ the web area). This is done because it was found that part of the web acts together with the compression flange. The second equation, Eq. (7.13d), is equation (1.5-7) in the AISC specification (7.7, p. 22) but 11.3 is rounded off to 12. Both equations are multiplied by a correction factor, C_b, to compensate for the moment gradient—that is, the effect of the difference between the actual bending moment diagram within the unbraced length under consideration and a theoretical constant bending moment, as was assumed in the development of these formulas.

The modifier C_b is expressed by the polynomial

$$C_b = 1.75 + 1.05 \frac{M_1}{M_2} + 0.3 \left(\frac{M_1}{M_2}\right)^2 < 2.3 \qquad (7.14)$$

where M_1 is the numerically smaller of the two moments at sections where there are lateral bracings (Fig. 7.10). The sign convention for the moments, M_1 and M_2, is not the same as the "beam sign convention" according to which the plus and minus signs are shown in the bending moments diagrams of Fig. 7.10. If the moments within an unbraced length are producing single curvature, their ratio, M_1/M_2, is negative. If double curvature results, then M_1/M_2 is positive. Values of C_b are given in Table 7 of (7.7).

Equation (7.13d) (St. Venant's pure torsional resistance) is to be used either in the plastic or in the elastic range

$$F_b = \frac{(12 \times 10^3)C_b}{L(d/A_f)} \qquad (7.15)$$

The warping torsional resistance component as Eq. (7.13c)—

$$F_b = \frac{(170 \times 10^3)C_b}{(L/r_T)^2} \qquad (7.16)$$

—is applicable only in the elastic range—that is, when

$$\frac{L}{r_T} \geq \sqrt{\frac{(510 \times 10^3)C_b}{F_y}} \qquad (7.16a)$$

or when $F_b \leq \frac{1}{3}F_y$—that is, when F_b is only $\frac{1}{3}$ the yield stress or less.

For inelastic lateral-torsional buckling a parabola-type instead of a Euler-type equation is used:

$$F_b = \left[\frac{2}{3} - \frac{F_y(L/r_T)^2}{(1,530 \times 10^3)C_b}\right] F_y \qquad (7.17)$$

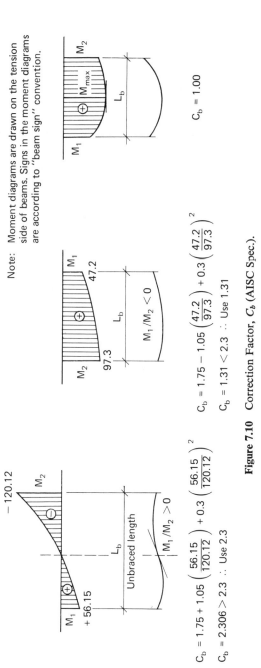

Figure 7.10 Correction Factor, C_b (AISC Spec.).

This is Eq. (1.5-6a) in AISC specification (7.7, p. 22). When the allowable stress, F_b, becomes equal to $0.6F_y$ there is no need to apply any of the above equations representing lateral instability. The limiting slenderness ratio, L/r_T, in that case is obtained from Eq. (7.17) substituting for $F_b = 0.6F_y$. Therefore, Eq. (7.17) is valid in the range

$$\sqrt{\frac{(102 \times 10^3)C_b}{F_y}} < \frac{L}{r_T} < \sqrt{\frac{(510 \times 10^3)C_b}{F_y}} \qquad (7.18)$$

The upper bound is obtained by equating Eq. (7.16) to Eq. (7.17).

All three formulas are only applicable in cases of bending of flexural members with at least one axis of symmetry being the weak axis and when loaded in that plane (web plane). For channels, when bent about their axis of symmetry the same equations apply. If the channel is bent about the principal axis perpendicular to the axis of symmetry, two critical moments are obtained, one positive and one negative, which are different in magnitude.

7.1.5 Design Considerations and Procedures for the Working Stress Method

In the present section the AISC and the AASHTO beam design specifications will be discussed briefly, including the necessary design aids, hints, and charts. Computer-aided designs for both a minicomputer and a full-size computer are introduced next.

Design considerations for beams are concerned with the strength, stability, and stiffness of the rolled sections chosen in the design process. Any of these three design criteria can control the design. For beams that are adequately laterally braced and for which the critical compressive stress is equal to or larger than the allowable stress in bending, the strength and not the stability criterion should be applied. Proportioning for stiffness involves giving due regard to the deflections so as to prevent excessive deformation under service conditions. The designer has to select a beam size which will satisfy all three conditions and at the same time is as light as possible.

The cheapest beam is always a standard rolled section. The dimensions of rolled wide-flange beams or standard I-sections are such that the shear in the web and web crippling (see Section 11.1, Beam Supports) almost never are governing factors in the design. Nevertheless, these have to be checked (see Example 7.3). Wide-flange sections are better than standard I-sections because for the same section modulus wide-flange shapes are lighter and have greater lateral stability, while it is also easier to connect their flanges to other members. The uniform thickness of the flanges makes connections to other structural members much simpler.

Additional shapes include miscellaneous sections as well as cold-formed steel structural members (7.6) for use in light construction (see Section 10.4, Light-Gage Steel Members). Castellated beams (Fig. 7.11) are obtained by flame-cutting the web of a section in a pattern as illustrated in Fig. 7.11a.

Open-web joists can also be used as another economical solution for roofs (see Section 10.2, Open-Web Joists). If they are used for floors, their vibrational characteristics must be investigated.

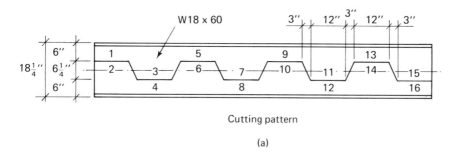

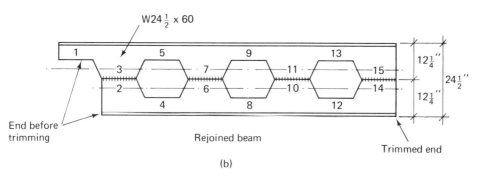

Figure 7.11 Castellated Beam.

Beam specifications

Buildings

The AISC specification (7.1, pp. 5.20–23) gives for compact hot-rolled sections which are symmetrical about and loaded in the plane of their minor axis, and which meet the requirements of compact and adequately braced sections, the allowable stress as

$$F_b = \pm 0.66 F_y \tag{7.19}$$

The requirements to be satisfied by a section in order to qualify as compact are specified so as to ensure the attainment of the full plastic moment (see Section 7.1.6, Plastic Design of Beams) before any local buckling will occur. This involves both the flanges and the web. Therefore, the width-thickness ratio of I-beam flanges, which is $\frac{1}{2}(b_f/t_f)$, must not exceed $65/\sqrt{F_y}$

$$\frac{1}{2}\frac{b_f}{t_f} \leq \frac{65}{\sqrt{F_y}} \tag{7.20}$$

Also, the depth-thickness ratio of the web, d/t_w, must not exceed the value $640/\sqrt{F_y}$

$$\frac{d}{t_w} \leq \frac{640}{\sqrt{F_y}} \tag{7.21}$$

This value is obtained from equation (1.5-4a) in the specification introducing $f_a = 0$. This is done because there is no applied axial compressive force present in the bending of beams. (For beam-columns the situation is different—see Chapter 8, Beam-Columns).

The requirements for adequate bracing of a compact section are such that no lateral instability can occur before the full plastic moment is reached. Therefore, the compression flange must be supported laterally at intervals, or the unbraced length, L_b, must satisfy the following:

$$L_b < \frac{76.0}{F_y} \cdot b_f \quad \text{or} \quad L_b < \frac{20,000}{(d/A_f) \cdot F_y} \tag{7.22}$$

If all these requirements are satisfied and the beam has negative moments acting at its supports (Fig. 7.12), another 10% increase in economy is allowed. Such beams may be proportioned for $\frac{9}{10}$ the negative moments produced by gravity loading which are maximum at the supports, provided that at the same time the maximum positive moment is increased by $\frac{1}{10}$ the average negative moments (Fig. 7.12).

If an I-beam is loaded so as to bend about its minor (web) axis, due to the fact that no lateral instability can occur before bending failure is reached the allowable stress is quite high and is given as

$$F_b = 0.75 F_y \tag{7.23}$$

This stress is also valid for solid round and square bars and for solid rectangular sections bent about their weaker axis.

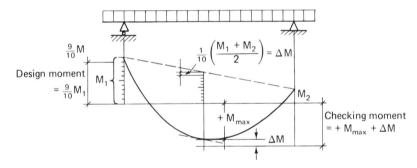

Figure 7.12 Moment Redistribution.

For noncompact or inadequately braced sections the discussion given in Section 7.1.4 and regarding Eq. (7.18) has to be followed.

Bridges

The 1977 AASHTO specifications as well as the interim specifications for bridges (7.2) give some explicit requirements for beam design. The most important ones are reviewed next.

DEPTH RATIOS [AASHTO (7.2–1977), section 1.7.6, p. 153]. Simple or continuous beams must be designed so that the deflection due to live load plus impact does not exceed $\frac{1}{800}$ the span, as already discussed in this text in Section 7.1.1.

MINIMUM THICKNESS OF METAL [AASHTO (7.2–1977), section 1.7.7, p. 153]. The web thickness of rolled beams must not be less than 0.23 in.

DIAPHRAGMS, CROSS-FRAMES, AND LATERAL BRACING [AASHTO (7.2—1977), section 1.7.17, p. 159]. Rolled beam spans must be provided with cross-frames or diaphragms at each end and with intermediate ones spaced at intervals not to exceed 25 ft. (7.6 m).

On spans with a concrete floor or other floor of equal rigidity adequately attached to the top flanges, one plane or system of lateral bracing must be provided near the bottom flange. If rolled beams are used as steel stringers in bridges with wooden floors, intermediate cross-frames must be placed between stringers more than 20-ft long.

ROLLED BEAMS, GENERAL [AASHTO (7.2—1977), section 1.7.42(A), p. 173]. Rolled beams, including those with welded cover plates, must be designed by the moment-of-inertia method.

The compression flanges of rolled beams supporting timber floors must not be considered laterally supported by the flooring unless the floor and fastenings are specially designed to provide adequate support.

BEARING STIFFENERS [AASHTO (7.2—1977) section 1.7.42(B), p. 173]. Suitable stiffeners must be provided to stiffen the webs of rolled beams at bearings when the unit shear in the web adjacent to the bearing exceeds 75% of the allowable shear for girder webs.

COVER PLATES [AASHTO (7.2—1977), section 1.7.12, p. 155]. The length of any cover plate added to a rolled beam must not be less than $(2D + 3)$ ft, where D is the depth of the beam in feet.

The maximum thickness of a single cover plate on a flange must not be greater than 2 times the thickness of the flange to which the cover plate is attached. Cover plates may be either wider or narrower than the flange to which they are attached. Cover plates wider than the flange must be provided with transverse end welds.

Any partial length welded cover plate must extend beyond the theoretical end by the terminal distance (2 times the nominal cover plate width for cover plates welded across their ends). All welds connecting cover plates to beam flanges must be continuous. Partial length welded cover plates shall not be used on flanges more than 0.8-in. (20-mm) thick for nonredundant load path structures subject to repetitive loadings which produce tension or reversal of stress in the member.

The allowable stresses due to bending in rolled beams are given in the AASHTO specifications (7.2—1977, p. 139): for the tension flange $f_b = 0.55F_y$; for the compression flange the same stress applies if the flange is supported laterally over its full length by embedment in concrete. If partially supported or if unsupported with $l/b_f < 36$ for structural steel (A36) or $22 < l/b_f < 30$ for high-strength, low-alloy structural steel (where l = length of unsupported flange, b_f = flange width), then an equation similar to Eq. (7.17) is to be used:

$$F_b = 0.55F_y \cdot \left[1 - \frac{(l/r')^2 F_y}{4\pi^2 E}\right] \qquad (7.24)$$

where $(r')^2 = b_f^2/12$ and l = length in inches of unsupported flange between lateral connections, knee braces, or other points of support. For continuous beams l may be taken as the distance from interior support to the point of dead load contraflexure if this distance is smaller than designated above.

The AASHTO specifications take into consideration the fact of moment redistribution in the plastic stage by allowing that continuous beams may be proportioned for negative moment at interior supports for an allowable unit stress 20% higher than permitted by the above formula, but in no case exceeding the allowable unit stress for compression flanges supported over their full length—that is, $F_b < 0.55 F_y$.

In the case of load factor design of simple and continuous beams of moderate length, the maximum moments and shears are calculated for factored loads by assuming elastic behavior of the structure. For compact sections—that is, satisfying the conditions (for F_y in lb/in.²)

$$\left.\begin{array}{ll} \text{(a)} \quad \dfrac{b_f}{2t} \leq \dfrac{1{,}600}{\sqrt{F_y}} & \text{(b)} \quad \dfrac{d}{t_w} \leq \dfrac{13{,}300}{\sqrt{F_y}} \\[1em] \text{(c)} \quad \dfrac{L_b}{r_y} \leq \dfrac{7{,}000}{\sqrt{F_y}} \quad \text{when} \quad M_2 \geq 0.7 M_1 \\[1em] \phantom{\text{(c)} \quad} \dfrac{L_b}{r_y} \leq \dfrac{12{,}000}{\sqrt{F_y}} \quad \text{when} \quad M_2 < 0.7 M_1 \end{array}\right\} \quad (7.25)$$

where L_b, M_1, and M_2 are the same as in Eqs. (7.14) and (7.15), and the maximum strength is computed as

$$M_u = F_y \cdot Z \tag{7.25a}$$

where Z is the plastic section modulus. If the section is not compact, but it satisfies the following conditions:

$$\left.\begin{array}{ll} \text{(a)} \quad \dfrac{b_f}{2t} \leq \dfrac{2{,}200}{\sqrt{F_y}} \\[1em] \text{(b)} \quad \dfrac{d - 2t_f}{t_w} \leq 150 \\[1em] \text{(c)} \quad L_b \leq \dfrac{20{,}000{,}000 A_f}{F_y d} \end{array}\right\} \quad (7.26)$$

where F_y is in lb/in.², then the maximum strength is computed as

$$M_u = F_y \cdot S \tag{7.26a}$$

where S is the elastic section modulus. If condition (c) is not satisfied, then the maximum strength shall be computed as

$$M_u = F_y \cdot S \left[1 - \dfrac{3 F_y}{4 \pi^2 E} \left(\dfrac{2 L_b}{b_f} \right)^2 \right] \tag{7.27}$$

When the ratio of two moments at both ends of the braced length, L_b, is less than 0.7, the maximum strength, M_u, as computed from Eq. (7.27) may be increased 20% but not to exceed $F_y \cdot S$.

Design aids

Longhand calculations

When the structural analysis for a beam is completed and the bending moment and shear diagrams obtained, the design process can start. Unfortunately, to perform this analysis the weight of the beam itself should be already known. To overcome this difficulty some reasonable estimate has to be made. For instance, for beams it was stated earlier that the ratio of depth to length of span preferably should not be less than $\frac{1}{25}$ (7.2—1977, section 1.7.4). If the beam is part of a continuous beam, then instead of the span the distance between the points of contraflexure under dead load is considered. If the beam is part of a frame, the ratio d/L can diminish to $\frac{1}{35}$ or $\frac{1}{40}$. Once this information about beam depth is obtained, then from the AISC *Manual* tables giving properties for the design of rolled sections (7.1, pp. 1-14 to 28) a beam section is selected matching the desired depth with an average weight for the sections of the same nominal depth. That weight is now used in the structural analysis as a first approximation. Once the design is performed and the selection of a beam section finished, the actual weight should be compared with the approximate value. If the difference is within 10% and on the safe side, no correction of the weight and structural analysis will be needed. If otherwise, a second iteration step is needed, after which a comparison again is made, etc.

Instead of this procedure, it is possible first to carry out the analysis omitting the dead weight of the beam and to design the beam on the basis of the results of such an analysis. Then this weight will be introduced in the analysis as a second step, the results of which are then checked and if necessary corrected and rechecked.

Sometimes it is possible to assume as a starting value a certain percentage of all dead load other than that of the beam itself—say, 3 to 6%. This is especially appropriate if, for example, the beam under consideration is supporting only a wall, thus carrying mainly a superimposed dead load. For heavier loads smaller percentages should be used. Another method refers to the total load W (7.17). Assumed beam weight, w, in pounds per foot is 1.25 times the total load, W, in kips.

Maximum moments for each section of beam span between cross-bracings have to be checked for stability so as to determine which moment governs the design. If, for this critical section, a compact, adequately braced rolled section is assumed, the corresponding allowable compressive bending stress, F_b, is used to calculate the required section modulus

$$S_{rqd} = \frac{M_{max}}{F_b} \qquad (7.28)$$

According to this value a section (normally a wide-flange section) is chosen and the assumption made about the section being compact and adequately braced is checked. If these conditions are not satisfied, a correction in the allowable stress is made and the adequacy of the previous section is checked. Once the section is found satisfactory, checking for the other span sections must follow.

If fatigue considerations are of importance, then this investigation has to follow. In buildings this is rarely the case, but in bridges very often.

One example is used to illustrate the outlined procedure.

Example 7.3 A two-span continuous beam is subjected to a uniformly distributed dead load of 3.0 kips/ft (43.8 kN/m) throughout its entire length of 60 ft (Fig. 7.13). The load is caused by a superimposed dead weight. Design this beam using A36 steel and AISC specification, if lateral bracings are (1) each 7.5 ft apart, and (2) 15 ft apart.

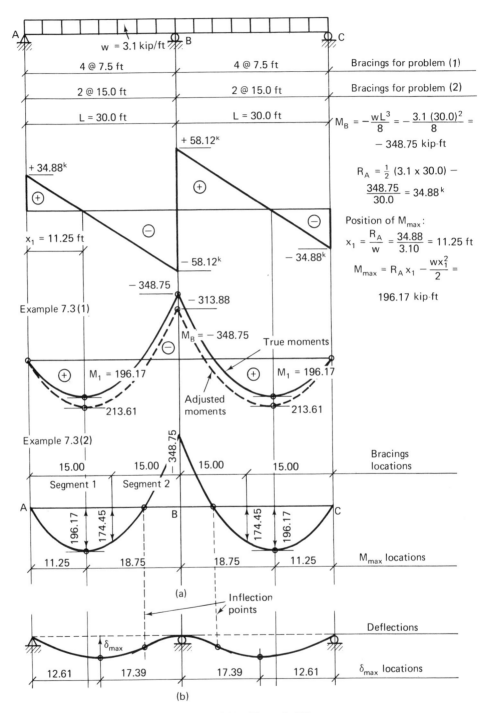

Figure 7.13 (Example 7.3).

Solution
1. BRACINGS 7.5 FT APART. As the given dead load of 3.0 kips/ft is on the heavy side for the beam weight estimate, use, say, 3%; thus $w = 90$ lb/ft (1.31 kN/m). If another approach is used based on the beam depth, d, then

$$d \approx \frac{0.9L}{20} = \frac{0.9(30 \times 12)}{20} = 16.2 \approx 18 \text{ in. (45.7 cm)}$$

where $0.9L$ represents the approximate distance of dead load contraflexure point from the left-hand support. From the AISC *Manual* (7.1, p. 1-18) the average weight of a W18 beam is $(35 + 119)\frac{1}{2} = 77.0$ lb. Therefore, a beam weight of about 80 lb/ft can be taken, which is close to the above 90 lb/ft. These two different weights can both be rounded to 100 lb/ft. Then the design load is 3.1 kips/ft instead of the given 3.0 kips/ft. Shear and moment diagrams are shown in Fig. 7.13.

Assuming that the selected beam will be a compact section adequately braced, the negative moment, $M_B = -348.75$ kip-ft, is first adjusted, as explained earlier, and the value of the design moment is

$$0.9 \times 348.75 = 313.88 \text{ kip-ft (425.56 kN-m)}$$

The required section modulus is therefore

$$S_{rqd} = \frac{313.88 \times 12}{24} = 156.94 \text{ in.}^3 \text{ (2,571.8cm}^3\text{)}$$

From the AISC *Manual's* allowable stress design selection table (7.1, p. 2-7) the section selected is a W24 × 76. Check for compact section:

Flange slenderness ratio: $\quad \dfrac{b_f}{2t_f} = \dfrac{8.985}{2 \times 0.682} = 6.59 < 10.8$

Web slenderness ratio: $\quad \dfrac{d}{t} = \dfrac{23.92}{0.44} = 54.4 < 107$

The section is compact. Check if the section is adequately braced [condition (5) in AISC specification (7.7), p. 20]:

$$12.7 b_f = 12.7 \times 8.985 = 114.11 \text{ in. (289.8 cm)} = 9.51 \text{ ft}$$

$$\frac{556}{d/A_f} = \frac{556}{3.90} = 142.56 \text{ in. (362.1 cm)} = 11.88 \text{ ft}$$

As the unbraced length, $l = 7.5$ ft $= 90$ in. (229 cm), is smaller than both of the above conditions, the W24 × 76 is adequately braced with supports 7.5 ft apart. Therefore, the allowable stress is as assumed

$$F_b = 0.66 F_y = 24 \text{ ksi}$$

The maximum field moment of 196.17 kip-ft (266 kN-m) has to be increased by $\frac{1}{10}$ the average of negative moments; that is

$$M_{pos} = M + \frac{1}{10}\left(\frac{M_A + M_B}{2}\right) = 196.17 + \frac{1}{10}\left(\frac{0 + 348.75}{2}\right)$$

$$= 213.61 \text{ kip-ft (289.62 kN-m)}$$

As the absolute value of this increased moment of 213.61 is still smaller than that of the adjusted negative moment (-313.88 kip-ft), no further check of bending stresses is required.

The shear stress in the web of the rolled section due to the maximum shear force of 58.12 kips (23.53 kN) is

$$f_v = \frac{V}{A_w} = \frac{58.12}{23.92 \times 0.44} = 5.52 \text{ ksi (38.06 MPa)} < 0.4F_y$$
$$= 14.5 \text{ ksi (100 MPa)}$$

Actual bending stress in W24 × 76 is

$$f_b = \frac{313.88 \times 12}{176} = 21.40 < 0.66F_y = 24 \text{ ksi (165.5 MPa)}$$

The last check is for stiffness—that is, the amount of deflection. Using the conjugate beam method (Fig. 7.13) the maximum deflection occurs at $x = 12.61$ ft (3.84 m) and equals

$$\delta_{max} = 0.38 \text{ in. (1 cm)} < \frac{L}{360} = 1.00 \text{ in. (2.5 cm)}$$

2. BRACINGS 15 FT APART. If the unbraced length is $l = 15$ ft $= 180$ in. $>$ 114.11 in., obviously the previously selected section W24 × 76 will be inadequately braced, though it will still be a compact section. No reduction of the negative moment is allowed. Therefore, the allowable stress needs to be reduced, and most probably this section will not be adequate. Assuming that the allowable stress will be $F_b = 20$ ksi (a drop of 20%), the required section modulus is

$$S_{rqd} = \frac{348.75 \times 12}{20} = 209.25 \text{ in.}^3 \text{ (3,429 cm}^3\text{)}$$

The best section from the same AISC design selection table (7.1, p. 2-7) would be a W27 × 84 with $S_x = 213$ in.^{3}. The bending stress is

$$f_b = \frac{348.75 \times 12}{213} = 19.65 \text{ ksi (135.5 MPa)}$$

To find the allowable compressive stress in the beam, two segments have to be considered. In segment 1, the bending moment is not an absolute maximum, but it occurs inside the segment. Therefore, $C_b = 1.00$. The radius of gyration for a W27 × 84 is $r_T = 2.49$ in. (7.1, p. 1-17). The slenderness ratio l/r_T is thus

$$\frac{l}{r_T} = \frac{15.0 \times 12}{2.99} = 72.3$$

This ratio is just between the two limits given in the AISC specification (7.7, p. 22); that is

$$53 < 72.3 < 119$$

Therefore, to obtain a value for the allowable stress, the larger value obtained from AISC equation (1.5-6a) [(7.7), warping torsional resistance] and AISC equation (1.5-7) [(7.7), St. Venant's torsional resistance: our Eq. 7.13d] is used, provided this value is not larger than $0.6F_y = 22$ ksi. Thus

$$F_b' = 24.0 - \frac{(72.3)^2}{1,181 \times 1.00} = 19.6 \text{ ksi (135.69 MPa)} \qquad \text{[AISC eq. (1.5-6a)]}$$

$$F_b'' = \frac{12 \times 10^3 \times 1.00}{15 \times 12 \times 4.19} = 15.9 \text{ ksi (109.21 MPa)} \qquad \text{[AISC eq. (1.5-7)]}$$

Because the larger value of the two, 19.6 ksi, is less than $F_b = 22$ ksi it can be used, and for segment 1 the allowable stress is

$$F_b = 19.6 \text{ ksi (135.83 MPa)}$$

The nominal bending stress in this segment is

$$F_b = \frac{196.17 \times 12}{212} = 11.10 \text{ ksi } (76.53 \text{ MPa}) < 19.6 \text{ ksi}: \quad \underline{\text{OK}}$$

In segment 2, the extreme moment occurs at the bracing (support B); therefore, the factor C_b(7.7, p. 5-22) is

$$C_b = 1.75 + 1.05\left(\frac{174.45}{348.75}\right) + 0.3\left(\frac{174.45}{348.75}\right)^2 = 2.35 > 2.3$$

use $C_b = 2.3$.

The lower limit of the slenderness ratio, l/r_T, for AISC equation (1.5-6a) is

$$53\sqrt{C_b} = 53\sqrt{2.3} = 80.4 > 72.3$$

Therefore, in segment 2 the allowable stress is

$$F_b = 0.60F_y = 22 \text{ ksi } (151.7 \text{ MPa})$$

The nominal stress is

$$F_b = \frac{348.75 \times 12}{213} = 19.6 \text{ ksi} < 22 \text{ ksi}: \quad \underline{\text{OK}}$$

The maximum deflection in this case will be

$$\delta_{\max} = 0.38\left(\frac{2,100}{2,850}\right) = 0.28 \text{ in.} < 1.00 \text{ in.}: \quad \underline{\text{OK}}$$

Beam tables and charts

In the AISC *Manual* (7.1), pp. 2-54–77 present charts showing allowable moments in beams (in kip-ft) for variable unbraced lengths (in ft) from 0 to 34 ft for moments between 0 and 2,220 kip-ft for A36 steel and between 0 and 3,050 kip-ft for steel qualities with $F_y = 50$ ksi. The unbraced lengths are between 0 and 34 ft. The moderating coefficient, C_b, is taken conservatively as 1.00. In the AISC *Manual's* beam tables (7.1, pp. 2-27–45) the uniform load constants are given for simply supported compact beams that are laterally supported—that is, with an unbraced length, $l \leq L_c$, and based on allowable bending stresses equal to 0.66F. Under the most recent AISC specification (7.7) practically all wide-flange sections are now compact, including those made of high-strength steels up to $F_y = 50$ ksi. These tables also give information about L_u, the maximum unbraced length of a compression flange (in ft) beyond which the allowable bending stress would be less than $0.6F_y$ when $C_b = 1$.

For quick design it is best to combine charts and tables and simply to select a section from the charts and check it using the tables. This is illustrated below using the alternatives of Example 7.3.

Solution

1. BRACINGS 7.5 FT APART. From the AISC charts (7.1, p. 2-57) for $M = 313.88$ kip-ft (425.56 kN-m) and $l = 7.5$ ft (2.29 m), the best choice is a W24 × 76 because it is the first section shown with solid lines to the right or above that point in the chart that has the coordinates (l, M). Next, the tables are entered for that section (7.1, p. 2-28) and from these

$$L_c = 9.5 \text{ ft } (2.90 \text{ m}) > l = 7.5 \text{ ft } (2.29 \text{ m})$$

which means that $F_b = 0.66F_y = 24$ ksi (165.5 MPa). Because our beam is not simply supported, the information given about the total allowable load of 94 kips (418 kN) means only a lower limit, not the true limit. In our special case of two equal continuous spans and symmetrical loading this is almost the true limit because the absolute value of middle support moment is numerically equal to that of the maximum simple beam moment. In our case the load is still larger than given, due to the adjusted moments.

2. BRACINGS 15 FT APART. In the case of $l = 15.0$ ft (4.57 m) the AISC charts (7.1, p. 2-57) will select a W27 × 84, as was obtained in the previous solution without charts. The AISC tables (7.1, p. 2-32) show that

$$L_u = 11.0 \text{ ft } (3.35 \text{ m}) < l = 15.0 \text{ ft } (4.57 \text{ m})$$

which means that lateral-torsional considerations will reduce the allowable stress below $0.6F_y$. The total load of 113 kips given in the table means only that our beam selection is on the safe side.

Computer-aided design

Program for a large computer system

This program (appearing in Appendix I-1), developed in FORTRAN IV, level H, is capable of determining the maximum moment and shear for a simply supported beam exposed to a uniformly distributed load and up to three concentrated forces of arbitrary intensities and arbitrary locations. Next it selects the beam section using A36 steel.

Therefore, the input data (see the complete listing of the program, card no. 19, Appendix I-1) are

span, L, in ft; unbraced length, LB, in ft; locations of three forces, X1, X2, and X3, in ft; uniformly distributed load, W, in kip/ft; and finally the intensities of the three forces, P1, P2, and P3, in kips.

For the maximum moment found in the analysis part of the program a section is selected from the information supplied to the computer at the beginning of the program under the heading "INPUT DATA FOR ROLLED WF-SECTIONS" (card no. 9). The information which has to be supplied to the computer depends upon how many sections are considered. In the program shown, a total of 50 wide-flange sections were considered. Input data for each section consists of

name, WF; weight, WEIGHT, in lb/ft; check, CHECK (1 if compact, 0 if not); section modulus, S, in in.^{3}; area of web, AWEB, in in.^{2}; limiting unbraced length for which the compact section can be considered adequately braced (L_c), LCOMP, in ft; and the limiting unbraced length beyond which the allowable stress is less than 0.6F (L_u), LU, in ft.

Sections are listed in ascending order according to their weight and section modulus. As the selection of a particular beam size starts from the lightest section, by comparing the required section modulus and the one provided by the section under consideration (card no. 70 or 85) the optimum solution—that is, the choice of the lightest beam—is ensured. In calculating the required section modulus, SREQRD, only two stresses are considered—namely, $0.66F_y$ for $l \leq L_c$ and $0.60F_y$ for $L_c < l \leq L_u$. This means that the section chosen by this computer program is assumed to be laterally supported. A flowchart is given in Fig. 7.14(a). Expanding this program to cover cases for which $l < L_u$ is easy. The only additional information needed is the radius of gyration (r_T), RT(I), in inches. The flowchart for this expanded program is given in Fig. 7.14(b).

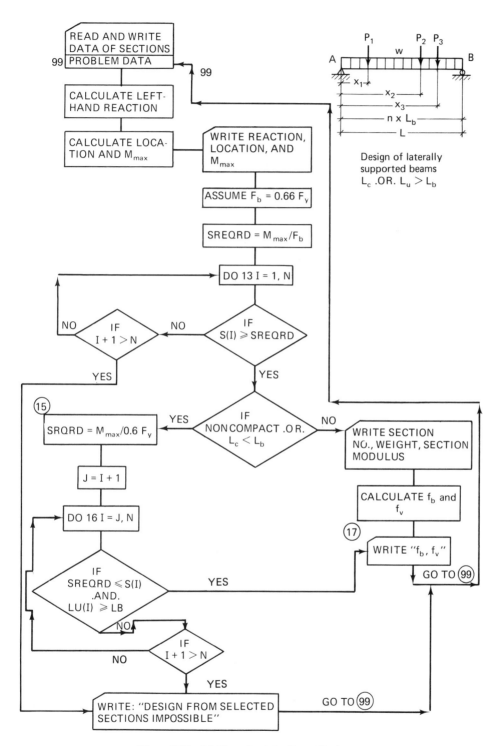

Figure 7.14 (a) Flow-chart for Beam Design.

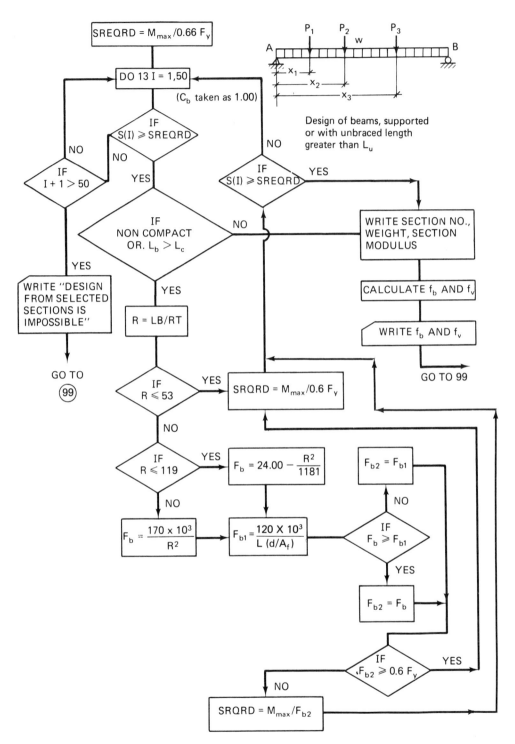

Figure 7.14 (b) Flow-chart for Beam Design.

240

Program for a minicomputer

The programs listed in Appendix I-2 are written for the Hewlett Packard 9820 minicomputer with option A (429 registers), and the Hewlett Packard 9830A.

Program for HP9820-A: To stay within the storage capacity of the RWM (read-write-memory), the whole design process of getting the results from a structural analysis and then selecting a section from a stored group of sections is divided into a 3-part program.

The first part of the program (Appendix I-2, Part I) will perform structural analysis of a simply supported beam loaded with a uniformly distributed load and up to three concentrated forces. The macroflowchart (Fig. 7.15) shows the major steps in this part of the program. The output of Part I yields the location of the absolute maximum moment and its magnitude as well as the maximum moment for each unbraced segment and its corresponding modifier, C_b.

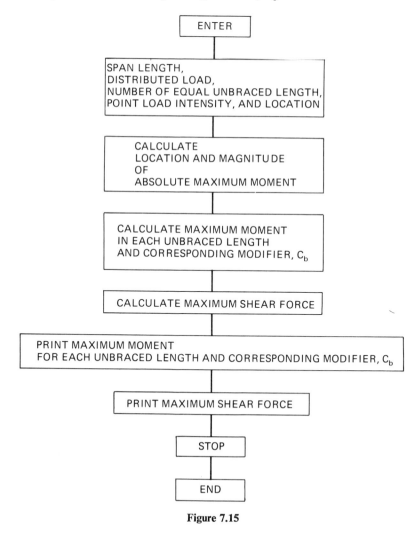

Figure 7.15

Straight Prismatic Beams **241**

The second part of the program (Appendix I-2, Part II) will select a rolled steel joist of minimum weight. The input is obtained from the output of Part I. First the absolute maximum moment and shear are entered, then the span length, the length of equal unbraced lengths, the maximum moment for the first unbraced segment, and the modifier, C_b, for this same section. This is repeated for each unbraced length. The macroflowchart is given in Fig. 7.16.

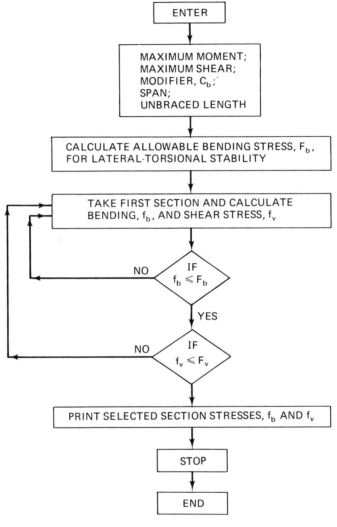

Figure 7.16

Part III of the program is used only once to store data about rolled sections in the computer. Thereafter data recorded on a magnetic card or cards are used. The flowchart of Part III of the program is shown in Fig. 7.17.

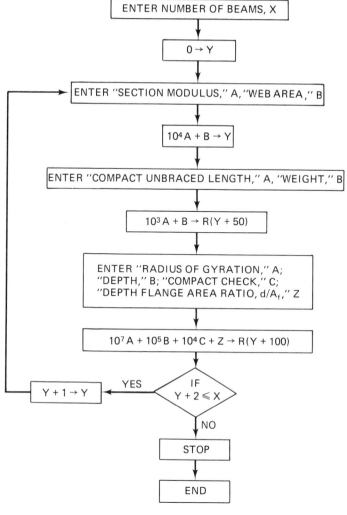

Figure 7.17

The input for Part III is composed of

number of sections wanted; section modulus in in.^{3}; web area in in.^{2}; maximum unbraced length, L_c, for a compact section in ft; weight in lb/ft; radius of gyration, r_T, in inches; section depth in inches; a check if a section is compact or not; and the ratio of the depth to the flange area, d/A_f, in in.$^{-1}$.

To be able to store all these data (8 different data per section) the technique of grouping data is used. In this way, instead of using $8 \times 50 = 400$ registers for 50 different beam sections, only $3 \times 50 = 150$ registers are needed.

As illustration, one example is given. Figure 7.18 shows the input data for Part I of the program as well as the output. There were five unbraced segments of $L_b = 10$ ft

Straight Prismatic Beams **243**

Figure 7.18 Program for Design of Simply Supported Beams, Part I: Moments, Shears, and Modifiers, C_b.

each. The output lists the location of the absolute maximum moment and then for each segment the maximum moment and the corresponding modifier, C_b. Figure 7.19a shows the input data for Part II

maximum shear force; span length; unbraced length, L_b; and for each segment the maximum moments and the corresponding modifiers, C_b.

244 BENDING OF BEAMS

```
            THIS PROGRAM
         DETERMINES THE
         BEST SECTION FOR
         THE GIVEN LOADS.




         MAX. SHEAR
                  106.26        MAX. MOMENT
                                         1426.02
         LENGTH
                   50.00         C
                                            1.21
         UNBRACED LENGTH
                   10.00


         MAX. MOMENT             MAX. MOMENT
                  937.59                   897.41

          C                       C
                    1.75                     1.75


                                 DEPTH
                                           36.00
         MAX. MOMENT
                 1454.04         WEIGHT
                                          230.00
          C
                    1.20         BENDING STRESS
                                           23.52

                                 SHEAR STRESS
                                            3.87
         MAX. MOMENT
                 1596.58

          C
                    1.00
```

Figure 7.19(a)

As output the final results are given as

depth of the chosen section; weight per foot; actual bending stress; and the shear stress.

In this example a W36 × 230 section was selected by the computer with actual stresses of

$$f_b = 23.52 \text{ ksi} \quad \text{and} \quad f_v = 3.87 \text{ ksi}$$

Program for HP9830A: The same problem of the analysis and design of a simple beam exposed to uniform load and up to three concentric loads is now solved using an HP9830A programmable desk calculator. Again, the program is divided into two parts—analysis and design, but using the overlaying techniques and common memory

DESIGN OF SIMPLE BEAMS OF A36 W.FLANGESTEEL SECTIONS

PART I:STRUCTURAL ANALYSIS
****

```
                          PROBLEM
L= 50    LB= 10    X1= 12.04    X2= 25    X3= 40.1    P1= 21.5    P2= 50.12
P3= 12    W= 2.5
```

SOLUTION OF THE PROBLEM

R-LEFT= 106.2588   KIP   R-RIGHT= 102.3612   KIP

LOCATION OF ABS.MAX.MOMENT X0 AND ITS MAGNITUDE M1
                X0= 25   M1= 1596.58

NUMBER OF UNBRACED SECTIONS IS 5

```
BRACING POINT X= 10              MOMENT M(X)= 937.588
SECTION NO.= 1     BETWEEN THE BRACINGS      MODIFIER CB= 1.75
BRACING POINT X= 20              MOMENT M(X)= 1454.036
SECTION NO.= 2     BETWEEN THE BRACINGS      MODIFIER CB= 1.197678443
MAX.MOMENT IS INBETWEEN THE BRACING POINTS M1= 1596.58

SECTION NO.= 3     BETWEEN THE BRACINGS      MODIFIER CB= 1
BRACING POINT X= 40              MOMENT M(X)= 897.412
SECTION NO.= 4     BETWEEN THE BRACINGS      MODIFIER CB= 1.254593359
BRACING POINT X= 50              MOMENT M(X)= 0
SECTION NO.= 5     BETWEEN THE BRACINGS      MODIFIER CB= 1.75
X1= 12.04    M(P1)= 1098.153952    X2= 25    M(P2)= 1596.58    X3= 40.1
M(P3)= 890.86338
```

Figure 7.19(b)

PART II:STRUCTURAL DESIGN
****

SOLUTION OF THE DESIGN PROBLEM PART II

```
UNBRACED BEAM SEGMENT NO. 1    AL.BEND.FB= 24
WF 36   X 230   ITERATION'S CYCLE 40
NOM.BEND.STRESS IS 13.44212186    NOM.SHEAR STRESS IS 3.894546254

UNBRACED BEAM SEGMENT NO. 2    AL.BEND.FB= 24
WF 36   X 230   ITERATION'S CYCLE 40
NOM.BEND.STRESS IS 20.84639427    NOM.SHEAR STRESS IS 3.894546254

UNBRACED BEAM SEGMENT NO. 3    AL.BEND.FB= 24
WF 36   X 230   ITERATION'S CYCLE 40
NOM.BEND.STRESS IS 22.89003584    NOM.SHEAR STRESS IS 3.894546254

UNBRACED BEAM SEGMENT NO. 4    AL.BEND.FB= 24
WF 36   X 230   ITERATION'S CYCLE 40
NOM.BEND.STRESS IS 12.86612186    NOM.SHEAR STRESS IS 3.894546254

UNBRACED BEAM SEGMENT NO. 5    AL.BEND.FB= 24
WF 36   X 230   ITERATION'S CYCLE 40
NOM.BEND.STRESS IS 0     NOM.SHEAR STRESS IS 3.894546254
```

FINAL SELECTED WF SECTION IS
                            36   X 230

Figure 7.19(c)

after the input the program proceeds automatically and gives the final result. The input is: span, unbraced length, abscissas, magnitudes of point loads, and the intensity of uniform load. The program is written in BASIC and a full listing is given in the Appendices I-3.1 and 3.2. As an illustration, the same problem shown in Fig. 7.18 is solved and the output is shown in Figs. 7.19b and c.

7.1.6 Plastic Design of Beams

In Chapter 1 the plastic behavior of mild steel was briefly discussed and the three accompanying phenomena described. The simple plastic theory for steel has as a basic assumption deformational response as represented by the ideal or Kist stress-strain diagram (Fig. 7.20). Therefore, no work-hardening is taken into consideration and the formation of ideal plastic hinges is possible. It is also assumed that the stress adaptation is occurring at one cross section without the spreading of the hinge in length along the elastoplastic boundary of the beam. Such ideal plastic hinges when occurring in statically indeterminate structures make possible the "equalization" of moments and thus coaction between different beam sections. It was pointed out earlier that this second phenomenon actually results in economy; therefore, only statically indeterminate structures are designed on the basis of plastic behavior. The AISC specification (part 2) allows plastic design of simple or continuous beams, braced or unbraced planar rigid frames. In bridge design this is not allowed. Therefore, here the plastic design of continuous beams in buildings only will be considered.

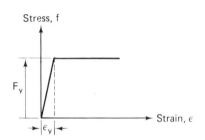

Figure 7.20 Kist Diagram.

Design of continuous beams

Although the first plastic phenomenon of stress adaptation within one and the same section makes even a simply supported beam statically indeterminate to the first degree, the simple plastic theory allows the consideration of continuous beams as statically determinate structures at the collapse stage. Each formed plastic hinge reduces the number of redundancies by one. When $(r + 1)$ hinges have developed (where r is the degree of redundancy), the structure as a whole becomes a kinematically movable mechanism which will collapse. Next to "complete" collapse, useful service life is also terminated when a local or partial collapse occurs within a structure. For a continuous beam this is actually the normal collapse mode because simultaneous collapse of all spans of a continuous beam very seldom occurs. A beam is unstable when three hinges develop along a single straight line. Therefore, a simple beam fails when a single hinge develops at any place along its span. A span of a continuous beam will fail when three or two hinges (for interior and exterior spans, respectively) are formed. It is assumed that each span may have different cross section or moment capacity, but that this moment, M_p, is constant within each span length.

Consider the continuous beam shown in Fig. 7.21a. Mainly two methods are applied when analyzing continuous beams: the mechanism and the statical method. For illustration purposes, both methods will be discussed.

There are three conditions which have to be satisfied by the bending moment diagram if it is to represent the true collapse state:

1. *Mechanism Condition:* The moment diagram must show a sufficient number of plastic hinges so that a collapse mechanism is developed.
2. *Equilibrium Condition:* The moment diagram must represent the state of equilibrium of all forces (including the reactions) acting on the collapse mechanism.
3. *Yielding Condition:* The moment diagram cannot show at any location a bending moment greater than the full plastic moment of that structural member.

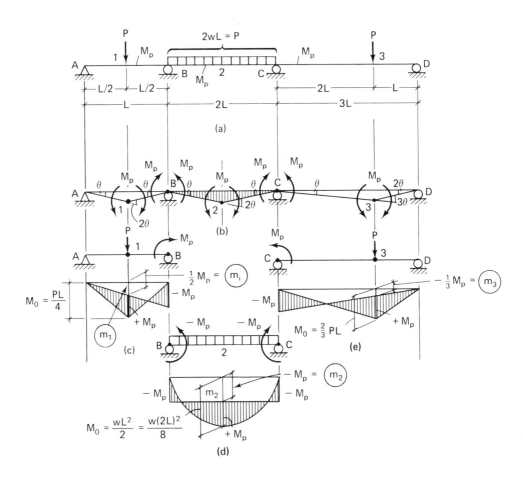

Figure 7.21

Solution by mechanism method

The mechanism method starts with the assumed collapse mechanism and the solution is then checked by the two other conditions. In a continuous beam each span has to be investigated; places of possible hinges are always at the locations of extreme values of the bending moment.

SPAN AB. At A in Fig. 7.21b a hinge is already existing, which means that only two additional hinges have to be formed—totaling three hinges in a row—for collapse to occur. The location of the two potential hinges is obvious: under the load P at 1 and at support B. A virtual work equation is written relating the external and internal work, which must be equal to one another to satisfy equilibrium. Relative rotations are shown in Fig. 7.21b. Note that only relative rotations are important; therefore, beam segments are drawn as straight lines. Once M_p is reached at 1, moments between A and B are constant. As the curvature, M/EI, equals a constant during the next virtual movement, no change of curvature occurs. Therefore, the relative rotation is actually occurring, as if in a straight beam.

$$\text{External work} = \sum (\text{force} \cdot \text{displacement})$$
$$= P \cdot \frac{L}{2}\theta$$

$$\text{Internal work} = \sum (\text{moment} \cdot \text{rotation})$$
$$= M_p \cdot (2\theta + \theta)$$

Equating external work and internal work gives

$$\tfrac{1}{2}PL\theta = 3M_p \cdot \theta$$

which yields the collapse load, P, as

$$P = \frac{6M_p}{L} \tag{a}$$

Checking the yielding condition (see the sketch in Fig. 7.21c), the moment at 1 is

Simple beam moment: $M_0 = \dfrac{PL}{4} = \dfrac{6M_p}{L} \cdot \dfrac{L}{4} = \dfrac{3}{2}M_p$

Redundant moment: $m_1 = -\tfrac{1}{2}M_p$

Final moment: $M_1 = (\tfrac{3}{2} - \tfrac{1}{2})M_p = M_p:$ OK

As all three conditions are now satisfied, this is the true collapse mechanism for span AB, and the collapse load is expressed in terms of the beam capacity, M_p.

SPAN BC. Hinges must develop at three locations to have a collapse of this span. These locations are at both supports, and due to the symmetry at midspan

$$\text{External work} = 2\int_0^L (x\theta)(w\,dx) = 2w\theta\left(\frac{x^2}{2}\right)_0^L = wL^2\theta$$

Internal work = $M_p \cdot (\theta + 2\theta + \theta)$

Equating external and internal work yields
$$wL^2\theta = 4M_p \cdot \theta$$
The amount of the external work, $wL^2\theta$, can also be interpreted as
$$w[\tfrac{1}{2} \cdot (L\theta) \cdot 2L]$$
where the expression between brackets represents the area of the triangle shaded in Fig. 7.21b. This means that the work done by the uniform load is the product of the load intensity and the area of the geometrical figure delineated by the collapsed beam shape. From the previous equality of work one obtains
$$w = 4\frac{M_p}{L^2} = \frac{P}{2L}$$
Therefore,
$$P = 8\frac{M_p}{L} \qquad (b)$$

Checking the yielding condition (see Fig. 7.21d), the moment at 2 is

Simple beam moment: $\quad \dfrac{wL^2}{2} = \dfrac{4M_p}{L^2} \cdot \dfrac{L^2}{2} = 2M_p$

Redundant moment: $\qquad m_2 = -M_p$

Final moment: $\qquad\qquad\qquad\qquad M_p \quad$ OK

SPAN CD. Again, as in the first span, AB, only two hinges have to be developed—namely, at support C and at 3.

$$\text{External work} = P \cdot 2L\theta$$
$$\text{Internal work} = M_p \cdot (\theta + 3\theta)$$
$$2PL\theta = 4M_p\theta$$
$$P = 2\frac{M_p}{L} \qquad (c)$$

Checking the yielding condition, the moment at 3 [Fig. 7.21(e)] is

Simple beam moment: $\quad \dfrac{2}{3}PL = \dfrac{2L}{3} \cdot 2\dfrac{M_p}{L} = \dfrac{4}{3}M_p$

Redundant moment: $\qquad m_3 = -\dfrac{1}{3}M_p$

Final moment: $\qquad\qquad\qquad\qquad M_p \quad$ OK

Comparing the results of collapse load values given in Eqs. (a), (b), and (c), the smallest collapse load is
$$P = 2\frac{M_p}{L} \qquad (d)$$

Therefore, when P reaches this value, span CD will collapse. If the load were given and we were designing the beam—that is, trying to find the required M_p—then the largest M_p would be the design plastic moment. For instance, when we solve all three equations for M_p we have

From Eq. (a): $M_p = \frac{1}{6}PL$

From Eq. (b): $M_p = \frac{1}{8}PL$

From Eq. (c): $M_p = \frac{1}{2}PL$ controls

Therefore, the design moment is

$$M_p = \tfrac{1}{2}PL \tag{e}$$

Solution by statical method

In this method a bending moment diagram satisfying the equilibrium and yielding conditions is drawn and then is checked for the mechanism condition. The same bending moment diagrams as in Fig. 7.21(c, d, e) are drawn, and from the geometry of each diagram the following three equations are obtained, one for each span:

$$\frac{PL}{4} = M_p + \frac{1}{2}M_p \quad \text{or} \quad P = 6\frac{M_p}{L} \tag{a}$$

$$\frac{wL^2}{2} = M_p + M_p \quad \text{or} \quad P = 8\frac{M_p}{L} \tag{b}$$

$$\frac{2}{3}PL = M_p + \frac{1}{3}M_p \quad \text{or} \quad P = 2\frac{M_p}{L} \tag{c}$$

As can be easily seen, the collapse loads are exactly the same as before.

A check for mechanism condition shows that each moment diagram satisfies this also.

The scope of this book precludes further detailed discussion of plastic design, but the interested reader is referred to existing excellent books dealing solely with this subject (7.8, 7.9, 7.10, 7.11, and 7.12).

As a further illustration, the same problem as in Example 7.3 can now be solved using plastic analysis and design.

Example 7.4 A two-span continuous beam is subjected to a uniformly distributed dead load of 3.0 kips/ft (43.78 kN/m) (at service level) throughout its entire length of 60 ft (9.14 m) (Fig. 7.22a). Design this beam using A36 steel and AISC specification (7.7), part 2, if lateral bracings can be provided at places required by the analysis.

Solution

ULTIMATE LOADS. According to section 2.1 of the AISC specification, a factored load is equal to 1.7 times the sum of the given live load and dead load. In our case this means

$$w_u = 1.7 \times 3.1 = 5.27 \text{ kips/ft (76.91 kN/m)}$$

MECHANISM METHOD. The plastic hinges in the field will no longer develop in the beam centers because the moments at supports A and C are zero and the hinges will be pushed by the negative support moment at B toward the supports A and C. If this hinge distance is denoted as x, then from the geometrical relations shown in Fig. 7.22(b) the external work is

$$2w_u[\tfrac{1}{2}\theta x \cdot L] = w_u L \theta x \tag{a}$$

The internal work is

$$2M_p\theta\left[\frac{L}{L-x}+\frac{x}{L-x}\right]=2\left(\frac{L+x}{L-x}\right)M_p\theta \qquad (b)$$

Equating expressions (a) and (b), rearranging, and solving for M_p we have

$$M_p=\frac{1}{2}(w_uL)\left(\frac{L-x}{L+x}\right)\cdot x \qquad (c)$$

For an extreme value of M_p,

$$\frac{dM_p}{dx}=0 \qquad (d)$$

or

$$x^2+2Lx-L^2=0 \qquad (e)$$

From this Eq. (e) the value for x is

$$x=(-1+\sqrt{2})L=0.414L \qquad (f)$$

In our case:

$$x=(0.414)(30)=12.426 \text{ ft } (3.787 \text{ m}) \qquad (g)$$

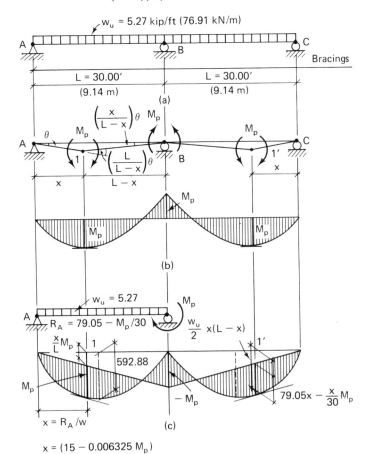

Figure 7.22 (Example 7.4).

and from Eq. (c):

$$M_p = \frac{1}{2}(5.27)30\frac{30 - 12.426}{30 + 12.426}(12.426)$$

$$= 406.88 \text{ kip-ft} = 4{,}883 \text{ kip-in. (551.705 kN-m)} \qquad \text{(h)}$$

From the AISC *Manual*'s plastic design selection table (7.1, p. 2-13) a W24 × 55 section (A36 steel) has $M_p = 402$ kips/ft (or 1.2% less than required). In our working stress design (Example 7.3, Section 7.1.5) a W24 × 76 section was selected. An economy of 27.6% resulted from plastic design.

SOLUTION BY STATICAL METHOD. In Fig. 7.22c it is shown that the simple beam moment is

$$M_0 = M_p\left(1 + \frac{x}{L}\right) = \frac{w_u}{2}x(L - x) \qquad \text{(a)}$$

When solved for M_p, the same expression as Eq. (c) above is obtained—that is

$$M_p = \frac{1}{2}(w_u L)\left(\frac{L - x}{L + x}\right)x \qquad \text{(b)}$$

The solution, then, can be obtained as was already done in this example according to the mechanism method. Instead of differentiating the expression for M_p and equating it to zero, as was done in the mechanism method, the distance x will be found from the condition that the shear force there must be zero. When this value of x is substituted in Eq. (b), the value of M_p is finally obtained as before:

$$M_p = 407 \text{ kip-ft (551.82 kN-m)} \qquad \text{(c)}$$

and the design will follow the same path as in Eq. (h).

7.2 CURVED BEAMS

7.2.1 Beams of Large Curvature

First it is assumed that the curved beam axis, the loads, and the reaction forces at the supports are all in one and the same plane. Secondly, it is also assumed that the initial radius of curvature is small, less than about 6 times the beam depth. In that case the stresses are no longer linearly distributed and they are not zero at the centroidal axis; that is, the neutral axis does not pass through the centroid of the cross section, but at a distance, $r = R - u$, from the center of curvature (Fig. 7.23b). Using notations from that figure, stresses at distances $(r + \eta)$ and $(r - \eta)$ are

$$f = +\frac{M\eta}{Au(r + \eta)} \quad \text{and} \quad f = -\frac{M\eta}{Au(r - \eta)} \qquad (7.29)$$

The edge stresses are

$$f_e = +\frac{Mx_e}{Au(r + x_e)} \quad \text{and} \quad f_i = -\frac{Mx_i}{Au(r - x_i)} \qquad (7.30)$$

where A is the area of the cross section.

To calculate the distance u—that is, the distance between the centroid and the neutral axis—an auxiliary value, $\bar{u}$, has to be calculated from

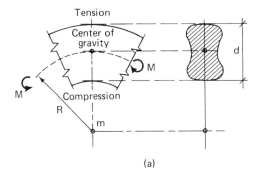

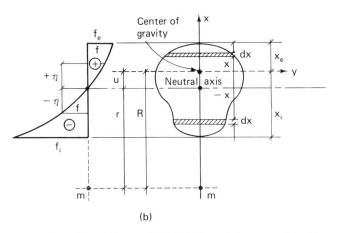

Figure 7.23 Curved Beam of Small Radius of Curvature $R < 6d$.

$$\bar{\mu} = \frac{1}{AR}\left(\int_{-x_i}^{+x_e} x\, dA + \frac{1}{R}\int_{-x_i}^{+x_e} x^2\, dA + \frac{1}{R^2}\int_{-x_i}^{+x_e} x^3\, dA + \cdots + \frac{1}{R^{n-1}}\int_{-x_i}^{+x_e} x^n\, dA\right) \quad (7.31)$$

The distance u is now equal to

$$u = R\frac{\bar{\mu}}{1+\bar{\mu}} \quad (7.32)$$

If the radius of curvature of the neutral line, r, is increasing, the curvilinear distribution of stresses becomes less and less pronounced until for $r \to \infty$ it is the same as for a straight beam. For a radius, $R \geq 10 \times$ (beam depth), the straight-line stress distribution formula of a straight beam can be applied.

I-beams with large curvature (7.13)

For I-beams with large curvature the following relation (see Fig. 7.24) is valid:

$$r = \frac{A}{b_1 l_n(r_{01}/r_{12}) + b_2 l_n(r_{12}/r_{23}) + \cdots} \quad (7.33)$$

or

$$r = \frac{A/2}{b_1(v_1 + v_1^3/3 + v_1^5/5) + b_2(v_2 + v_2^3/3 + v_2^5/5) + \cdots} \qquad (7.34)$$

where

$$v_1 = \frac{r_{01} - r_{12}}{r_{01} + r_{12}}; \qquad v_2 = \frac{r_{12} - r_{23}}{r_{12} + r_{23}}$$

One numerical example (see Fig. 7.25) will be presented to give the reader some feeling for the magnitudes involved.

Using Eq. (7.33) the radius of the neutral axis is

$$r = \frac{33.0}{12 ln\frac{35}{34} + 0.5 ln\frac{34}{16} + 12 ln\frac{16}{15}}$$

$$= 22.0 \text{ in. } (55.9 \text{ cm})$$

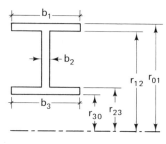

Figure 7.24 Curved Beam of I-section.

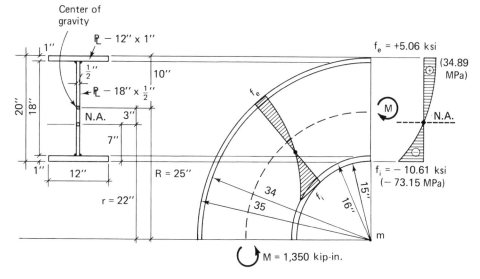

Figure 7.25 Stresses in Curved Beams.

If instead Eq. (7.34) is used, then the result is

$$r = \frac{33.00/2}{12[0.014493 + (0.014493)^3/3] + 0.5[0.360000 + (0.360000)^3/3] + 12[0.032258 + (0.032258)^3/3]}$$

$$= 22.03 \text{ in. } (56 \text{ cm}): \quad \underline{\text{OK}}$$

where

$$v_1 = \frac{1.0}{69} = 0.014493; \qquad v_2 = \frac{34 - 16}{34 + 16} = 0.360000; \qquad v_3 = \frac{16 - 15}{16 + 15} = 0.032258$$

If a moment, $M = 1{,}350$ kip-in. (152.53 kN-m), is supposed to act on this section, the extreme stresses are

$$f_e = \frac{1{,}350 \times 13}{33 \times 3(22+13)} = 5.06 \text{ ksi } (34.89 \text{ MPa})$$

$$f_i = \frac{-1{,}350 \times 7}{33 \times 3(22-13)} = -10.61 \text{ ksi } (-73.15 \text{ MPa})$$

If the beam is treated as a straight beam, the stresses are ± 10.64 ksi.

Deformation of flanges

In the web and flanges of a curved beam exposed to bending moment radial compressive forces are created due to the curvature and the tensile or compressive forces in the flanges (Fig. 7.26a). These deviator forces, $p = f_b t/r$, are distributed across the flanges (Fig. 7.26b), causing them to bend. This actually means that the outside tips of the flanges will be in a region of lower stresses; that is, they will take a smaller load and due to this nonuniform stress distribution the stress in the web is increased.

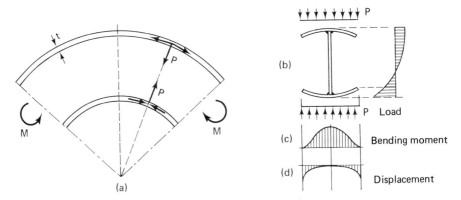

Figure 7.26 Flange Deformation of a Curved Beam in Bending.

Further, the forces, p, are causing bending stresses in flanges, and the corresponding bending moment diagram is approximately as in Fig. 7.26c, with the corresponding bending deformation as in Fig. 7.26d. A differential equation of the fourth order results when the problem is analyzed. Stüssi (7.14) used a statical approach and finally arrived at the following expression:

$$f' = \gamma \cdot f \tag{7.35}$$

where γ is a proportionality factor, f is the nominal bending stress obtained from Eq. (7.30) (the larger of the two), and f' is the bending stress in the flange normal to the beam axis. The γ-factor depends upon a ϕ-value, and in the region of about $1.0 < \phi < 3.0$ with $\phi = b^2/t \cdot r$ (where b is flange overhang—i.e., equal to about half the flange width minus the web thickness) γ is 1.8. Next an equivalent stress is calculated for this two-dimensional stress state in the flanges, using the expression

$$f_{\text{equiv}} \approx \sqrt{f_i + (f')^2 + f_i f'} \tag{7.36}$$

This stress should be less than $\frac{4}{3} f_{b\text{-allowable}}$. In our numerical example (Fig. 7.25) we have

$$\phi = \frac{b^2}{t \cdot r} = \frac{(5.5)^2}{1 \times 22} = 1.375 \quad \text{and} \quad \gamma = 1.8$$

Therefore

$$f' = 1.8(10.61) = 19.10 \text{ ksi} \, (131.69 \text{ MPa})$$

$$f_{\text{equiv}} = \sqrt{(10.61)^2 + (19.10)^2 + 10.61(19.10)} = 26.10 \text{ ksi} \, (179.95 \text{ MPa})$$

For A36 steel, in general $f_{b\text{-allowable}} = 0.6 F_y = 22$ ksi (151.7 MPa). Thus

$$\tfrac{4}{3} f_b = 29.3 \text{ ksi} \, (202.02 \text{ MPa}) > 26.1 \text{ ksi}: \quad \text{OK}$$

7.3 TAPERED BEAMS

For efficiency and/or architectural reasons beams are often tapered in depth. For moderate taper, when the taper angle 2α is less than 40° (Fig. 7.27) (usually the case in structural design), the simple bending formulas for bending stress are sufficiently accurate, but the effect of taper on shear stress can be large (7.15).

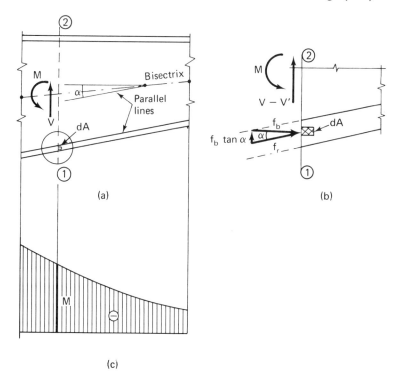

Figure 7.27 Tapered Beams.

Considering the area element, dA, in the flange at section 1-2 (Fig. 7.27b), the bending stress, f_b, which is obtained by the usual formula, (M/Iy), has to be increased to

$$f_r = \frac{f_b}{\cos \alpha} \tag{7.37}$$

because bending stress in the thin flange must be parallel to the flange boundary surfaces. The vertical stress component, $f_b \tan \alpha$, when multiplied by the element of area, dA, and integrated over the whole flange cross section, will give a vertical force, V', which will provide for the web a slight relief from the shear force, V:

$$V' = \tan \alpha \iint_{A_f} f_b \, dA \qquad (7.38)$$

Practically, an average bending stress for the flange is found and V' is approximated by

$$V' = f_{av} A_f \cdot \tan \alpha \qquad (7.39)$$

If the bending moment increases with beam height, as it normally does (Fig. 7.27c), then the shear force, V, is reduced by this component, V', which is actually the result of bending and tapering.

Timoshenko has shown (7.16) that for a wedge-shaped beam the maximum shearing stress is about 3 times the average shearing stress and occurs at points most remote from the neutral axis, which is in direct opposition to the results obtained for prismatic bars.

NOTATIONS

b	= ½ flange width minus web thickness	H	= horizontal force
b_f	= flange width	L_b	= unbraced length
d	= depth of a wide-flange section	M	= bending moment
f_b	= calculated (nominal) bending stress	M_p	= full plastic moment
		P	= vertical force
f_v	= calculated shear stress	Q	= statical moment
$(f_v)_{av}$	= average shear stress	R	= radius of curvature of curved beams
l	= unbraced length	S_n	= S_x/S_y = section moduli coefficient
r	= radius of curvature of neutral axis in a curved beam		
r_T	= radius of gyration for rolled beam sections	S_x, S_y	= elastic section modulus with respect to x- and y-axes, respectively
t	= width of section	S_{y1}	= section modulus about y-axis of a single flange
t_f	= flange thickness		
t_w	= web thickness	V	= shear force
u	= displacement of neutral axis in curved beams	μ	= Poisson's ratio or coefficient
		ϕ	= proportionality factor, angle rotation
A_f	= flange area		
A_w	= web area	γ	= factor in tapered beams, proportionality factor in curved beams
C_b	= moment gradient coefficient		
EI	= flexural rigidity	δ	= deflection
F_b	= allowable bending stress	$\bar{\mu}$	= auxiliary value in curved beams
F_v	= allowable shear stress		

PROBLEMS

7.1. A floor frame system has stringers at 8-ft center-to-center with a span of 30 ft (see Fig. P-7.1). The stringers are simply supported by the floor beams. If the floor deck consists of a reinforced concrete "one-way" slab 4-in. thick directly supported by stringers that carry a live load of 125 lb/ft², design the stringers without counting on composite action between the slab and the steel beam, but assuming that lateral buckling of the stringers is prevented. Check deflection.

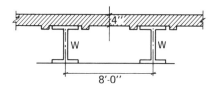

Figure P-7.1

7.2. Design a lintel made up of two A36 steel angles to carry an 8-in. brick wall (130 lb/ft³) spanning a 5-ft-8-in. opening with 4-in. bearing at both ends (see Fig. P-7.2). A triangular wall loading may be assumed with height equal to half the span. Angles are not connected to one another; the masonry wall does not provide lateral support.

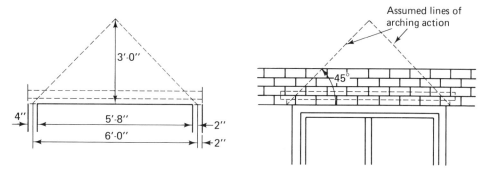

Figure P-7.2

7.3. Design T-subpurlins of A36 steel supporting precast reinforced-concrete slabs of $1\frac{1}{2}$ in. thickness (concrete weighs 150 lb/ft³) and a 2-in. coating of "nailing concrete" (poured on top of precast slabs) (see Fig. P-7.3). The distance between the subpurlins is 30 in. The main purlins are spaced at a 6-ft horizontal distance on a roof slope of 1:2. Assume a snow load of 20 lb/ft² (horizontal area). Consider subpurlins as beams continuous over four spans.

7.4. Design main purlins for the same loading conditions as in Problem 7.3 (again see Fig. P-7.3). The distance between roof trusses (the span length for main purlins) is 20 ft. Analyze the purlins as simple beams. The subpurlins

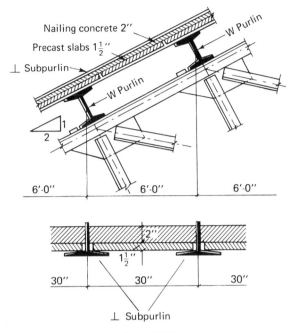

Figure P-7.3

laterally support the compression flanges of the main purlins at intervals of 2 ft–6 in.

7.5. A two-span continuous beam is subjected to a uniformly distributed load of 3.2 kips/ft (including beam weight) over its entire length of 2 × 35 ft (see Fig. P-7.5). Lateral bracing is provided at 7-ft intervals. Select a W-member without cover plates using A36 steel and AISC specification.

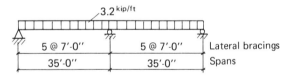

Figure P-7.5

7.6. Design a beam of A36 steel for a simple span of 30 ft–0 in. with the loading as shown in Fig. P-7.6. Lateral bracings are provided at beam ends and at midspan.

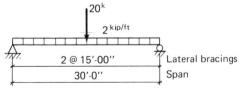

Figure P-7.6

7.7. The beam shown in Fig. P-7.7 is a W12 × 50 of A50 steel. The uniformly distributed load, $w = 1.0$ kip/ft, is acting through the centroid of the section but at an angle of 45° with the vertical beam axis. Find the maximum bending and shearing stresses. Lateral bracings are provided only at beam ends. Is this beam safe according to AISC specification?

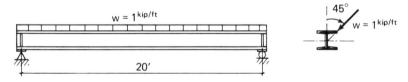

Figure P-7.7

7.8. Design a beam subjected to the loading shown in Fig. P-7.8. Use a wide-flange shape of A36 steel and assume that lateral bracing is provided at the beam ends only.

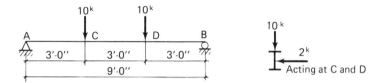

Figure P-7.8

7.9. A beam with overhangs is exposed to the loading system shown in Fig. P-7.9. Design this beam using A36 steel and AISC specification subject to the requirement that the maximum deflection has to be smaller than 0.5 in. at beam ends and smaller than 1 in. at midspan.

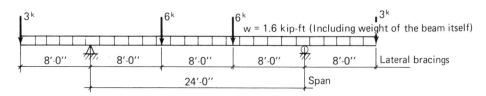

Figure P-7.9

7.10. Design a continuous beam using a W-shape of A36 steel; strengthen the midsupport by using cover plates (see Fig. P-7.10). Lateral bracings are located at 10-ft intervals. Use AISC specification.

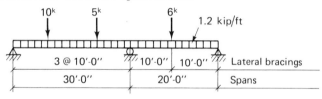

Figure P-7.10

7.11. Design the most economical beam section of A36 steel using AISC specification, if the beam is fully laterally supported and simply supported. Loading is a uniform load of 3 kips/ft (not including beam weight) with three loads at $\frac{1}{4}$ points of 25 kips each. The span is 24 ft.

7.12. Same as Problem 7.11, but with three sections of an unbraced length of 8 ft.

7.13. Same as Problem 7.11, but without any intermittent lateral support.

7.14. A W-section is to be used in a span of 12 ft to carry a midspan concentrated load of 250 kips. Use steel with $F_y = 50$ ksi and AISC specification. The beam is simply supported and is braced only at the supports.

7.15. Obtain the lightest W-section for a simply supported beam of 42 ft with two 20-kip symmetrical concentrated loads at 14 ft from each end. Lateral supports are available at the ends and concentrated loads. Use A36 steel and AISC specification.

7.16. Same as Problem 7.15, but with lateral supports at the ends and at midspan.

7.17. Same as Problem 7.15, but without any intermittent lateral supports between the ends.

7.18. A W16 × 40 of A36 steel is simply supported and has an effective span of 18 ft. It carries a uniform load of 1.2 kips/ft (including beam weight) and a midspan load of 12 kips. The beam is laterally supported every 6 ft. Find (1) maximum bending stress and (2) maximum shearing stress, and compare them with the allowable stress according to AISC specification.

7.19. Same as Problem 7.18, but unbraced lengths are 9 ft.

7.20. Same as Problem 7.18, but lateral supports are provided only at beam ends.

REFERENCES

7.1. AISC, *Manual of Steel Construction*, 8th ed., Chicago, 1980.

7.2. AASHTO, *Standard Specifications for Highway Bridges*, 12th ed., Washington, DC, 1977; (with 1978, 1979, 1980, and 1981 Interim Bridge Specifications).

7.3. Deustche Industrie Normen, *Stahl im Hochbau*, DIN 1050, Berlin, 1968.

7.4. BRESLER, B., T. Y. LIN, and J. B. SCALZI, *Design of Steel Structures*, 2nd ed., J. Wiley, New York, 1968, p. 386.

7.5. TIMOSHENKO, S., "Theory of Bending, Torsion, and Buckling of Thin-walled Members of Open Cross-Sections," *J. Franklin Inst.*, vol. 239, nos. 3, 4, 5, 1945.

7.6. American Iron and Steel Institute, *Specification for the Design of Cold-formed Steel Structural Members*, Washington, DC, Sept. 3, 1980.

7.7. AISC, *Specification for the Design, Fabrication and Erection of Structural Steel for Buildings*, 1978.

7.8. NEAL, B. G., *The Plastic Methods of Structural Analysis*, 3rd ed., Chapman & Hall, London, 1977.

7.9. HEYMAN, J., *Plastic Design of Frames*, Cambridge University Press, Cambridge, 1971.

7.10. Massonnet, C. E., and M. A. Save, *Plastic Analysis and Design*, vol. 1: *Beams and Frames* (1965); vol. 2: *Plates, Shells, and Disks* (1972); North-Holland, New York.
7.11. Disque, R. O., *Applied Plastic Design in Steel*, Krieger, New York, 1971.
7.12. Tall, L., ed., *Structural Steel Design*, 2nd ed., Ronald, New York, 1974.
7.13. Deutcher Stahlbau-Verband, *Stahlbau*, vol. 1, 2nd ed., Cologne, 1971, pp. 148–50.
7.14. Stüssi, F., *Entwurf und Berechnung von Stahlbauten*, vol. 1: Grundlagen, Springer-Verlag, 1958, p. 501.
7.15. Shanley, F. R., *Mechanics of Materials*, McGraw-Hill, New York, 1967, p. 191.
7.16. Timoshenko, S., *Strength of Materials*, 3rd ed., Van Nostrand Reinhold, New York, 1956, part 2, p. 62.
7.17. Perez, A. J., "Estimating Beam Size," *Civil Engineering—ASCE*, Dec. 1971, p. 67.

8

Beam-Columns

8.1 INTRODUCTION

Almost all members in a structure are subjected to both bending and axial load—in either tension or compression. When the axial load in a member causes compression and the amounts of compression and bending are both significant, we usually classify such a member as a beam-column. Beam-column behavior can be caused by an axial load acting simultaneously with end moments, by transverse loading, or simply because the load, though parallel to the column axis, is acting eccentrically (Fig. 8.1a, b, c, and d).

Present design procedures mostly make use of interaction equations which are either (1) empirically determined on the basis of experiments and related to the working stress approach; or (2) semiempirical, based on theoretically developed ultimate strength interaction curves. Both design philosophies are based on the ultimate strength concept regardless of whether the working stress or plastic design method is used.

8.1.1 Ultimate Strength of Beam-Columns

If we consider a wide-flange beam which is bent about its major axis by end moments and simultaneously by an axial load, P (Fig. 8.2), we observe several effects. First we have a loading range where the behavior is linear elastic. Due to the presence

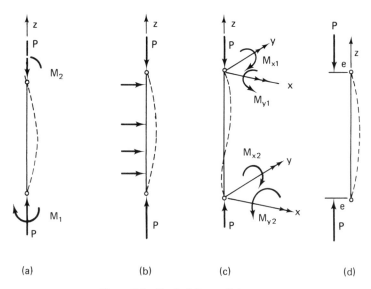

Figure 8.1 Typical Beam-Columns.

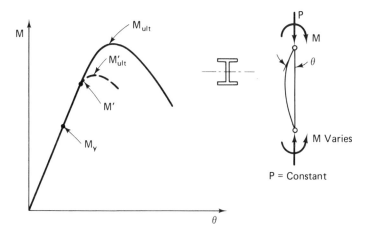

Figure 8.2 M-θ Relationship for Wide-flange Beam-column with Varying End Moments and Constant Axial Load.

of residual stresses, first yielding occurs as soon as the compressive yield stress is reached in any section. We shall designate this moment at first yield M_y. As the moment increases the behavior becomes increasingly nonlinear until no additional moment can be supported and the maximum or ultimate moment (M_{ult}) is reached. Failure as described in this case is assumed to take place in the plane of bending. This will automatically be the case if a member is bent about its weak axis. If, however, the member is bent about its strong axis, as in this example, failure in the plane of bending is possible only if the member is adequately braced against lateral-torsional buckling. If no such bracing is present, the ultimate moment capacity will be reduced considerably to a critical value, M'_{ult}. In both modes of failure (bending or lateral-torsional) it is assumed that local buckling is prevented.

To determine the ultimate strength of beam-columns failing in the plane of bending several methods can be used. All are based on the relationship between the moment, M, the axial load, P, and the curvature, ϕ. Typical M-P-ϕ curves for wide-flange beams are shown in Fig. 8.3. In some methods the ultimate strength is determined on the basis of an assumed deflection shape (8.1, 8.2), while others make use of numerical integration (8.3, 8.4, 8.5, 8.6, 8.7). At the same time the M-θ curve can often be constructed. When carrying out ultimate strength calculations it is best to represent the results in the form of P-M-L/r ultimate strength interaction curves. A set of these is shown in Fig. 8.4 in the presence of a linear residual stress of $0.3F_y$. Such sets of interaction curves can be drawn for a particular shape (wide flange), different end moment ratios, and different residual stress levels (8.14). For comparison, different end moment ratios are shown in Fig. 8.5 for an L/r_x value of 60. As can be seen, the most severe case is that with two equal end moments causing single curvature, while the least severe is that of two equal end moments causing double curvature. The effect of residual stress levels is comparatively small.

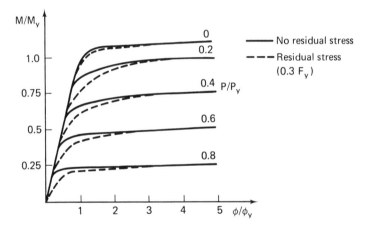

Figure 8.3 M-P-ϕ Curves for Wide-flange Beam (From R. L. Ketter, E. L. Kaminsky, and L. S. Beedle, "Plastic Deformation of Wide-flange Beam-columns," *Trans. ASCE*, vol. 120, 1955, pp. 1028–69).

8.1.2 Differential Equation for Beam-Columns (Fig. 8.6)

Assuming failure in the plane of bending, we have (8.9)

$$M_z = M + Py = -EI\frac{d^2y}{dz^2} \tag{8.1}$$

where M is the primary moment and Py the secondary moment. If EI is constant, dividing by EI and differentiating twice yields

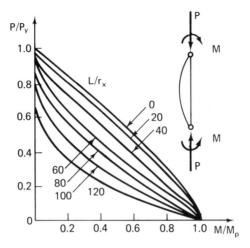

Figure 8.4 Ultimate Strength Interaction Curves for a Wide-flange Beam with Equal End Moments, Including Residual Stress of $0.3F_y$. (From R. L. Ketter and T. V. Galambos, "Columns under Combined Bending and Thrust, *Trans. ASCE*, vol. 126 (I), 1961, p. 1).

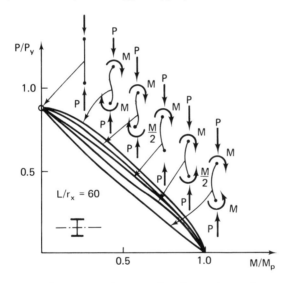

Figure 8.5 Ultimate Strength Interaction Curves for Various End Moment Ratios.

$$\frac{d^4y}{dz^4} + k^2\frac{d^2y}{dz^2} = -\frac{1}{EI}\frac{d^2M}{dz^2} \tag{8.2}$$

where $k^2 = P/EI$. Substituting d^2y/dz^2 for $-M_z/EI$ yields

$$\frac{d^2M_z}{dz^2} + k^2 M_z = \frac{d^2M}{dz^2} \tag{8.3}$$

The solution is

$$M_z = A \sin kz + B \cos kz + g(z) \tag{8.4}$$

Introduction   267

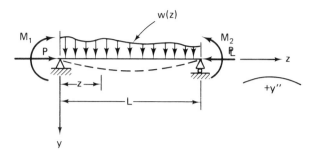

Figure 8.6 Beam-column.

where $g(z)$ is any value of M satisfying Eq. (8.3). Several cases are shown in Appendix VII.

8.1.3 Moment Magnification Factor

In the case of beam-columns the expression for the maximum moment can be written as follows:

$$M_{z\text{-max}} = M\left(\frac{C_m}{1-\alpha}\right) \tag{8.5}$$

where $\alpha = P/P_e$; C_m is a moment gradient coefficient that takes into account the variation of the moment between the supports and can be written

$$C_m = 0.6 - 0.4\beta \tag{8.6}$$

where β is the algebraic ratio of the numerically smaller to the larger end moment, which may vary from -1 to $+1$. The term $C_m/(1-\alpha)$ is considered the magnification factor.

8.2 INTERACTION EQUATIONS FOR BEAM-COLUMNS

The use of an analytical approach based on the solution of a differential equation is not very practical and has almost entirely been replaced by empirically determined beam-column interaction equations.

8.2.1 Zero-Length Members

The ultimate strength interaction relationship for strong axis bending is shown in Fig. 8.7. The "analytical" curve is derived from analytically determined ultimate strength interaction curves (Figs. 8.4 and 8.5). The solid line represents the approximation

$$\frac{M}{M_p} = 1.18\left(1 - \frac{P}{P_y}\right) \tag{8.7}$$

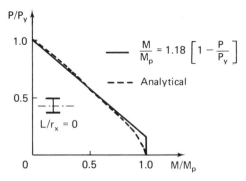

Figure 8.7 Interaction Curves for Zero-length Members (From ASCE-WRC Committee, "Commentary on Plastic Design in Steel," Chapter 7, Compression Members, *ASCE Manual of Engineering Practice* no. 41, 1961).

8.2.2 Instability in the Plane of Bending

The two dashed curves in Fig. 8.8 represent the analytical ultimate strength curves for $L/r_x = 40$ and 120, while the solid lines represent the equation

$$\frac{P}{P_{\text{crit}}} + \frac{M}{M_p} = 1.0 \tag{8.8}$$

The two dot-dash lines represent the equation

$$\frac{P}{P_{\text{crit}}} + \frac{M}{M_p(1-\alpha)} = 1.0 \tag{8.9}$$

where P_{crit} is the axial force which would cause failure if no moment were present, which is equal to

$$\frac{\pi^2 E_t}{(KL/r)^2} \quad \text{[Eq. (5.16)]}$$

As can be seen, the amplification factor, $1/(1-\alpha)$, has very little effect at small L/r_x ratios ($L/r_x \leq 40$).

For loading cases different than that with two equal end moments Eqs (8.8) and (8.9) are too conservative and so the additional correction factor. C_m, should be introduced (see Eq. 8.7), as shown in Fig. 8.9 and represented by the expression

$$\frac{P}{P_{\text{crit}}} + \frac{M}{M_p}\left(\frac{C_m}{1-\alpha}\right) = 1.0 \tag{8.10}$$

or, substituting the values for α and C_m,

$$\frac{P}{P_{\text{crit}}} + \frac{M}{M_p}\left[\frac{1}{(1-P/P_e)}\right](0.6 - 0.4\beta) = 1.0 \tag{8.11}$$

The solid curves shown in Fig. 8.9 do not pass through the point $M/M_p = 1$ when $P = 0$ except for the case when $\beta = +1.0$. Thus when Eq. (8.11) falls below the line given by Eq. (8.8), this latter equation (Eq. 8.11) is used.

Interaction Equations for Beam-Columns

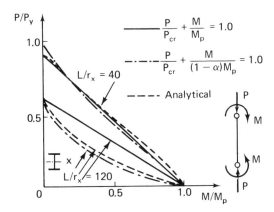

Figure 8.8 Interaction Formula With and Without Use of the Amplification Factor (From R. L. Ketter, "Further Studies in the Strength of Beam-columns," *Proc. ASCE*, vol. 87, no. ST-6, August, 1961, pp. 135–52).

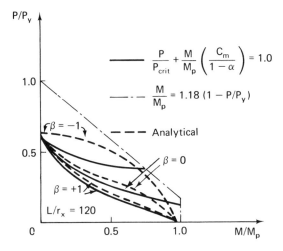

Figure 8.9 Interaction Formula for Unequal End Moments (From R. L. Ketter, "Further Studies on the Strength of Beam-columns," *Proc. ASCE*, vol. 87, no. ST-6, Aug, 1961; and L. Tall, ed., *Structural Steel Design*, 2nd ed., Ronald Press, New York, 1974).

8.2.3 Instability by Lateral-Torsional Buckling

To include the possibility of lateral-torsional buckling we can replace M_p by M_m, where M_m now is the maximum bending moment the member can carry when bending about its strong axis without lateral buckling, in the case where P equals zero. The revised ultimate strength interaction equations are thus modified to

$$\frac{P}{P_{\text{crit}}} + \frac{MC_m}{M_m(1 - P/P_e)} \leq 1.0 \tag{8.12}$$

$$\frac{M}{M_p} = 1.18\left[1 - \frac{P}{P_y}\right] \leq 1.0 \tag{8.13}$$

where $M < M_p$.

8.3 WORKING STRESS DESIGN METHOD FOR BEAM-COLUMNS

Instead of expressing the interaction equation in terms of ultimate loads, the working stress method makes use of allowable stresses. The AISC, AASHTO and AREA specifications all use the working stress method. Both the AISC and AASHTO specifications in addition use the plastic design (load factor) design approach. Only the AISC and the AASHTO working stress methods are discussed as the AREA specifications are very similar.

8.3.1 AISC Working Stress Method

According to the AISC *Manual* (8.11) beam-columns must be proportioned to saitsfy the following requirements:

$$\frac{f_a}{F_a} + \frac{C_{mx}f_{bx}}{(1 - f_a/F'_{ex})F_{bx}} + \frac{C_{my}f_{by}}{(1 - f_a/F'_{ey})F_{by}} \leq 1.0 \tag{9.14}$$
[AISC eq. 1.6-1a]

$$\frac{f_a}{0.60F_y} + \frac{f_{bx}}{F_{bx}} + \frac{f_{by}}{F_{by}} \leq 1.0 \tag{8.15}$$
[AISC eq. 1.6-1b]

When $f_a/F_a \leq 0.15$, the following equation may be used in lieu of Eqs. (8.14) and (8.15):

$$\frac{f_a}{F_a} + \frac{f_{bx}}{F_{bx}} + \frac{f_{by}}{F_{by}} \leq 1.0 \tag{8.16}$$
[AISC eq. 1.6-2]

In the above equations the subscripts x and y indicate the axis of bending. For C_m the reader is referred to pp 5-118–120 and appendix A, table 8 of the AISC specifications (8.11). The AISC specifications give values for F_a (appendix A, table 3) and for F'_e (appendix A, table 9) for different slenderness ratios.

8.3.2 AASHTO Working Stress Method

In the 1978/1979/1980/1981 interim specifications (8.12) AASHTO uses practically the same equations as the AISC's and so they are not repeated here; only the differences will be discussed briefly. The AASHTO specifications do not allow the use of Eq. (8.16) while in Eq. (8.15) instead of $0.60F_y$ in the first denominator they use $0.472F_y$, which means that instead of a factor of safety of 1.67 AASHTO uses 2.12. The same is true for F'_e, which is calculated with a factor of safety of 2.12 instead of the AISC value of $\frac{23}{12} = 1.917$. Also, AASHTO uses M_1/M_2 as positive when the curvature is single and negative when in reverse curvature.

8.3.3 Examples

Because of the similarity between the AISC and AASHTO specifications only the AISC ones will be used in the following three examples. Design can be carried out by selecting a trial section and checking the design equations or by using the AISC column tables for axial loads (8.11, pp. 3-15–97).

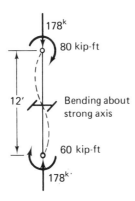

Figure 8.10 Beam-column Braced at Midpoint Against Buckling in the Weak Direction; Examples 8.1 and 8.2.

Example 8.1 Design a 12-ft (3.7-m) beam-column subjected to loads as shown in Fig. 8.10. Initially assume that lateral buckling is prevented. Use A36 steel and the AISC specifications (8.11). The column is braced at its midpoint against buckling in the weak direction.

Solution The effective length is

$$K_x = 1.0 \rightarrow K_x L = 12 \text{ ft } (3.7 \text{ m})$$
$$K_y = 0.5 \rightarrow K_y L = 6 \text{ ft } (1.85 \text{ m})$$

If no moment is acting, it can be seen from the AISC column tables that a W8 × 31 will carry 178 kips (791.7 kN) and a W10 × 33 will carry 189 kips (840.7 kN). Because of the reverse curvature the moment magnification effect will be small and we shall try a W10 × 45. Another approach to arrive at a trial section is to assume an allowable stress on the basis of the slenderness:

$$\frac{K_x L}{r_x} \approx \frac{12 \times 12}{4} = 36$$

$$\frac{K_y L}{r_y} \approx \frac{6 \times 12}{2} = 36$$

According to AISC table 3-36 (8.11, p. 5-74) $F_a = 19.50$ ksi (134.5 MPa); therefore, assume $F_a = 18$ ksi (124 MPa) and that $A_{req} = 178/18$ is 10 in.² (64.5 cm²).

A W10 × 39 = 11.5 in.² (74.2 cm²); therefore, choose a W10 × 45 with $L_c = 8.5$ ft > 6.0 ft. Also

$$\frac{K_x L}{r_x} = \frac{12 \times 12}{4.32} = 33.3; \qquad \frac{K_y L}{r_y} = \frac{6 \times 12}{2.01} = 36 \qquad \text{Use 36:}$$

$$F_a = 19.50 \text{ ksi } (134.4 \text{ MPa}); \qquad f_a = \frac{178}{13.3} = 13.48 \text{ ksi } (92.9 \text{ MPa})$$

$$\frac{M_1}{M_2} = 0.75 \rightarrow C_m = 0.6 - 0.4(0.75) = 0.3 < 0.4 \qquad \text{Use 0.4:}$$

$$\frac{K_x L}{r_x} = 33.3 \rightarrow F'_e = 134.74 \text{ ksi } (929.0 \text{ MPa}) \qquad \text{[AISC table 9]}$$

$$F_b = 0.66 F_y \qquad \text{[AISC } \textit{Manual}\text{, section 1.5.1.4]} = 24 \text{ ksi } (141.5 \text{ MPa})$$

$$f_b = \frac{M_{max}}{S} = \frac{80 \times 12}{49.1} = 19.55 \text{ ksi } (134.8 \text{ MPa})$$

Check the interaction equation (Eq. 8.14)

$$\frac{13.48}{19.50} + \frac{0.4 \times 19.55}{(1 - 13.48/134.74)24} = 0.691 + 0.362 = 1.05 > 1$$

From Eq. (8.15):

$$\frac{13.48}{22.00} + \frac{19.55}{24.00} = 0.612 + 0.815 = 1.427 > 1$$

Select a W12 × 50: $A = 14.7$ in.^{2}; $S = 64.7$ in.^{3} (1,060 cm^3); $r_x = 5.18$ in. (13.2 cm); $r_y = 1.96$ in. (5.0 cm); $L_c = 8.5$ ft > 6.0 ft:

$$\frac{K_x L}{r_x} = \frac{144}{5.18} = 27.80; \quad \frac{K_y L}{r_y} = \frac{72}{1.96} = 36.7$$

Use 36.7 → $F_a = 19.44$ ksi (134.0 MPa); $f_a = 178/14.7 = 12.10$ ksi (83.4 MPa): $F'_e = 193.3$ ksi (1,332 MPa)

$$F_b = 0.66 F_y = 24 \text{ ksi } (141.5 \text{ MPa}); \quad f_b = \frac{80 \times 12}{64.7} = 14.84 \text{ ksi } (102.3 \text{ MPa})$$

First check Eq. (8.15):

$$\frac{12.10}{22.00} + \frac{14.84}{24.00} = 0.550 + 0.618 = 1.168: \quad \underline{\text{N.G.}}$$

After having tried a W14 × 53, we next try a W14 × 61: $A = 17.90$ in.^{2} (115.5 cm^2); $S = 92.2$ in.^{3} (1511 cm^3); $r_x = 5.98$ in. (15.2 cm); $r_y = 2.45$ in. (6.2 cm); $L_c = 10.6$ ft > 6.0 ft. Also

$$\frac{K_x L}{r_x} = \frac{144}{5.98} = 24.1; \quad \frac{K_y L}{r_y} = \frac{72}{2.45} = 29.4; \quad F_a = 19.98 \text{ ksi } (137.8 \text{ MPa})$$

$$f_a = \frac{178}{17.90} = 9.94 \text{ ksi } (68.5 \text{ MPa}); \quad F'_e = 257.3 \text{ ksi } (1,774 \text{ MPa})$$

$$F_b = 24 \text{ ksi } (141.5 \text{ MPa}); \quad f_b = \frac{80 \times 12}{92.2} = 10.41 \text{ ksi } (71.8 \text{ MPa})$$

Equation (8.15) yields

$$\frac{9.94}{22.00} + \frac{10.41}{24.00} = 0.452 + 0.434 = 0.886 < 1: \quad \underline{\text{OK}}$$

Equation (8.14) now gives

$$\frac{9.94}{19.98} + \frac{0.4 \times 10.41}{(1 - 9.94/257.3)24} = 0.497 + 0.180 = 0.677 < 1: \quad \underline{\text{OK}}$$

The W14 × 53 was only 4% overstressed. The next most economical section is a W14 × 61.

The same column can now be redesigned using the column tables. According to the AISC *Manual* (pp. 3-9–12) we can determine an equivalent required tabular load, P_{tab}, by rewriting Eqs. (8.14), (8.15), and (8.16) as follows:

$$\begin{aligned} P_{\text{tab}} &= P + P'_x + P'_y \\ &= P + \left[B_x M_x C_{mx} \left(\frac{F_a}{F_{bx}} \right) \left(\frac{\alpha_x}{\alpha_x - P(KL)^2} \right) \right] \\ &\quad + \left[B_y M_y C_{my} \left(\frac{F_a}{F_{by}} \right) \left(\frac{\alpha_y}{\alpha_y - P(KL)^2} \right) \right] \end{aligned} \quad \begin{array}{r} (8.17) \\ \text{[modified Eq. (8.14)]} \end{array}$$

$$P_{tab} = P + P'_x + P'_y$$

$$= P\left(\frac{F_a}{0.6F_y}\right) + \left[B_xM_x\left(\frac{F_a}{F_{bx}}\right)\right] + \left[B_yM_y\left(\frac{F_a}{F_{by}}\right)\right] \quad \begin{array}{c}(8.18)\\ \text{[modified Eq. (8.15)]}\end{array}$$

when $f_a/F_a \leq 0.15$ finally

$$P_{tab} = P + P'_x + P'_y = P + \left[B_xM_x\left(\frac{F_a}{F_{bx}}\right)\right] + \left[B_yM_y\left(\frac{F_a}{F_{by}}\right)\right] \quad \begin{array}{c}(8.19)\\ \text{[modified Eq. (8.16)]}\end{array}$$

where B_x and B_y are tabulated bending factors, while α_x and α_y are components also shown in the column tables (AISC *Manual*, pp 3-15–45). To illustrate the use of these modified interaction equations in conjunction with the tables we shall rework the Example 8.1.

Example 8.2 Design a column for the same problem as in example 8.1 (Fig. 8.10) using the AISC column tables (8.11).

Solution From column table I (8.11, p. 3-21) an average value of B_x of 0.195 appears a good estimate. As a first estimate we use Eq. (8.19) and assume F_a/F_{bx} to be unity; thus

$$P_{tab} = 178 + 0.195 \times 80 \times 12 = 178 + 187 = 365 \text{ kips } (1{,}624 \text{ kN})$$

which would indicate a W14 × 68 [400 kips (1,779 kN)] or a W14 × 61 [358 kips (1,592 kN)]. If we select the latter section, we find in the table that $A = 17.9$ in.^{2} (115.5 cm^2); $r_x/r_y = 2.44$; $r_y = 2.45$ in. (6.2 cm); $L_c = 10.6$ ft (3.23 m); $L_u = 21.5$ ft (6.55 m); $F_{bx} = 24$ ksi (141.5 MPa); $B_x = 0.194$; $\alpha_x = 95.4 \times 10^6$; $C_{mx} = 0.4$; Now; $K_yL/r_y = (6 \times 12)/2.45 = 29.4$; $K_xL/(r_x/r_y) = 12/2.45 = 4.89 <$ 6 ft. So use K_yL.

From AISC table 3.36, $F_a = 19.98$ ksi:

$$P(KL)^2 = 178(72)^2 = 0.923 \times 10^6$$

Equation (8.17) now yields

$$P_{tab} = 178 + \left[0.195 \times 80 \times 12 \times 0.4\left(\frac{19.98}{24.00}\right)\left(\frac{95.4}{95.4 - 0.923}\right)\right] = 178 + 62.9$$

$$= 240.9 \text{ kips } (1{,}072 \text{ kN}) < 358 \text{ kips } (1{,}592 \text{ kN}): \quad \underline{\text{OK}}$$

Equation (8.18) gives

$$PF_a/(0.6F_y) = 178 \times 19.98/22.0 = 161.6$$

$$P_{tab} = 161.6 + 0.195 \times 80 \times 12\left(\frac{19.98}{24.00}\right) = 161.6 + 155.8$$

$$= 317.4 \text{ kips } (1{,}420 \text{ kN}) < 358 \text{ kips } (1{,}592 \text{ kN}): \quad \underline{\text{OK}}$$

If we had selected a W14 × 53, we would have $A = 15.6$ in.^{2} (100.7 cm^2); $r_x/r_y = 3.07$; $r_y = 1.92$ in. (4.9 cm); $L_c = 8.5$ ft (2.2 m); $L_u = 17.7$ ft (5.4 m); $F_{bx} = 24$ ksi (141.5 MPa); $B_x = 0.201$; $\alpha_x = 80.6 \times 10^6$; $C_{mx} = 0.4$; $KL_y/r_y = (6 \times 12)/1.92 = 37.5$; $L_x/(r_x/r_y) = 12/2.45 = 4.89$ ft < 6 ft; so use L_y. From AISC table 3.36, $F_a = 19.38$ ksi (133.6 MPa)

$$P(KL)^2 = 0.923 \times 10^6$$

We shall only check Eq. (8.18) which yields: $PF_a/(0.6F_y) = 178 \times 19.38/22 = 156.8$

$$P_{tab} = 156.8 + 0.201 \times 80 \times 12\left(\frac{19.38}{24.00}\right)$$

$$= 312.6 \text{ kips } (1{,}390 \text{ kN}) > 302 \text{ kips } (1{,}343 \text{ kN}): \quad \underline{\text{N.G.}}$$

To arrive at an initial column size we can often use the rather simple approximate expression

$$P_{eq} = P + 2.1\frac{M_x}{d} \tag{8.19a}$$

where P_{eq} is the equivalent required tabular load and d the depth of the member in its plane of bending.

Example 8.3 If we apply this approximation to the previous example we get

$$P_{eq} = 178 + 2.1\left(\frac{80 \times 12}{14}\right) = 322 \text{ kips}$$

assuming a W14 beam. According to the column table on page 3-21 of the AISC Manual for $K_y L = 6$ ft we need a W14 × 61, which is exactly the section we finished up with in Examples 8.1 and 8.2.

As a fourth example we shall discuss the same problem solved in Chapter 5, Example 5.3 (Fig. 5.35).

Example 8.4 Select a wide-flange section for the columns of a portal frame (Fig. 8.11) using A36 steel and AISC specification (8.11). The web of the rolled section lies in the frame plane. The horizontal beam member (girder) is a W12 × 65. Girts are attached at the midheight of the columns. At the top of the column a moment of 100 kip-ft and at the bottom one of 50 kip-ft are acting.

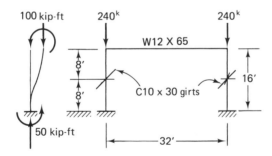

Figure 8.11 Portal Frame, Example 8.4—Webs in Plane of Frame.

Solution As in Example 5.3 we find that $G = 2I/I_{cg}$ and assume $K_y L_y = 8.0$ ft (2.4 m) and $K_x L_x = 21.12$ ft (6.4 m); $P = 240$ kips (1,067.5 kN), $M_{top} = 100$ kip-ft (135.6 kNm), and $M_{base} = 50$ kip-ft (67.8 kNm). From the AISC column tables (8.11, p. 3-24) an average value of B_x is about 0.22. Again as a first estimate we use Eq. (8.18) and assume F_a/F_{bx} to be unity. Thus

$$P_{tab} = 240 + 0.22 \times 100 \times 12 = 504 \text{ kips } (2{,}242 \text{ kN})$$

On the basis of this preliminary estimate we select a W14 × 82 with a capacity of 465 kips (2,068 kN).

Now from the AISC table: $A = 24.1$ in.^{2} (156 cm^2); $I_x = 882$ in.^{4} (36716 cm^4); $r_x/r_y = 2.44$; $r_y = 2.48$ in. (7.7 cm); $L_c = 10.7$ ft (3.26 m); $L_u = 28.1$ ft (8.6 m)—that is, $F_{bx} = 24$ ksi (141.5 MPa); $B_x = 0.196$; $\alpha_x = 131.4 \times 10^6$; $C_{mx} = 0.85$ (because of sidesway). Now

$$\frac{K_y L_y}{r_y} = \frac{8 \times 12}{2.48} = 38.7; \qquad \frac{I_c}{I_g} = \frac{882}{663} = 1.33$$

$$G_A = \frac{I_c/16}{I_g/32} = 1.33 \times 2 = 2.66; \qquad G_B = 1.0 \text{ (fixed base)}$$

Accoridng to Fig. 5.18, $K_x = 1.52$, and so

$$\frac{K_x L_x}{r_x/r_y} = \frac{1.52 \times 16}{2.44} = 9.97 \text{ ft (3 m)}$$

$$K_y L_y = 8 \text{ ft (2.4 m)} \quad \text{and thus} \quad K_y L_y < \frac{K_x L_x}{r_x/r_y}$$

Therefore, use 9.97 ft (3.0 m)

$$\frac{K_x L_x}{r_x} = \frac{1.52 \times 16 \times 12}{6.05} = 48.23; \qquad F_a = 18.51 \text{ ksi (127.6 MPa)}$$

$$P(KL)^2 = P(K_x L_x)^2 = 250(1.52 \times 16 \times 12)^2 = 21.86 \times 10^6$$

Equation (8.17) now yields

$$P_{tab} = 240 + 0.196 \times 100 \times 12 \times 0.85 \left(\frac{18.51}{24.00}\right)\left(\frac{131.4}{131.4 - 21.86}\right) = 240 + 185$$

$$= 425 \text{ kips (1,890 kN)} < 447 \text{ kips (1,988 kN)}; \qquad \underline{\text{OK}}$$

[Tabular capacity for 9.97 ft (3.0 m) = 447 kips (1,988 kN) (8.11, p. 3-21).]
Equation (8.18) yields

$$P_{tab} = 202 + 0.196 \times 1{,}200 \left(\frac{18.51}{24.00}\right) = 202 + 181.4$$

$$= 383.4 \text{ kips (1,705 kN)} < 447 \text{ kips (1,988 kN)}: \qquad \underline{\text{OK}}$$

When comparing the preceding with Example 5.3 it can be seen that the addition of bending moments increases the section size from a W12 × 50 to a W14 × 82.

8.4 PLASTIC (ULTIMATE) DESIGN METHOD FOR BEAM-COLUMNS

The plastic design method, also referred to as load factor design, is based directly on Eqs. (8.12) and (8.13) and makes use of appropriate load factors.

8.4.1 AISC Plastic Design Method

In part 2 of the AISC specifications (8.11) the equations that control the design of beam columns are identical to Eqs. (8.12) and (8.13) and are therefore not repeated here. In these, $P_e = \frac{23}{12} A F'_e$, where F'_e is defined as before; thus

$$P_e = \frac{23}{12} \cdot A \cdot \frac{12 \, \pi^2 E}{23(Kl_b/r_b)^2} = \frac{\pi^2 EA}{(Kl_b/r_b)^2} \tag{8.20}$$

Also:
$$P_{crit} = 1.7AF_a \tag{8.21}$$

For columns braced in the weak direction M_m is defined as follows
$$M_m = M_p \tag{8.22}$$

And for columns unbraced in the weak direction
$$M_m = \left[1.07 - \frac{(l/r_y)\sqrt{F_y(\text{ksi})}}{3{,}160}\right]M_p \leq M_p \tag{8.23}$$

8.4.2 AASHTO Load Factor Design Method

The AASHTO specifications (8.12) use the following equations
$$\frac{P}{0.85A_sF_{crit}} + \frac{MC}{M_u(1 - P/A_sF_e)} \leq 1.0 \tag{8.24}$$

$$\frac{P}{0.85A_sF_y} + \frac{M}{M_p} \leq 1.0 \tag{8.25}$$

where
$$F_{crit} = F_y\left[1 - \frac{F_y}{4\pi^2 E}\left(\frac{KL_c}{r}\right)^2\right] \tag{8.26}$$

if
$$\frac{KL_c}{r} \leq \sqrt{\frac{2\pi^2 E}{F_y}}$$

and
$$F_{crit} = \frac{\pi^2}{(KL_c/r)^2} \tag{8.27}$$

for
$$\frac{KL_c}{r} > \sqrt{\frac{2\pi^2 E}{F_y}}$$

The C-factor is same as C_m defined before in Eq. (8.7) except for the different sign convention of M_1/M_2, and KL_c/r is the effective slenderness ratio in the plane of bending, while M_u is the maximum bending strength in kip-feet as descussed in Chapter 5.

8.4.3 Examples

Example 8.5 Design the beam-columns in the frame shown in Fig. 8.12 using AISC specifications (8.11) and A36 steel. Assume no sidesway.

Solution We have $q_u = 1.70 \times 2 = 3.4$ kips/ft (49.6 kN/m). According to plastic analysis, the moment diagram is as shown in Fig. 8.12b.

BEAM BC. We have $M = 340$ kip-ft (461 kN-m); assume full lateral support.

$$Z_{rqd} = \frac{340 \times 12}{36} = 113.3 \text{ in.}^3 \text{ (1,857 cm}^3\text{)}$$

Try W24 × 55:

$$Z = 134 \text{ in.}^3 \text{ (2,196 cm}^3\text{)} > 113.3 \text{ in.}^3 \text{ (1,857 cm}^3\text{)}: \quad \underline{\text{OK}}$$

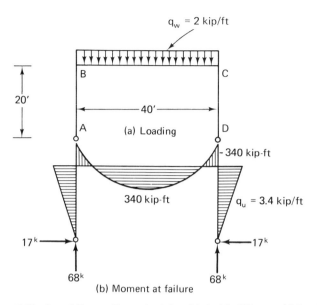

Figure 8.12 Portal Frame, Examples 8.5 and 8.6—No Sidesway (a) Loading; (b) Moments at Failure.

Check web thickness for local buckling:

$$\frac{d}{t_w} \leq \frac{412}{\sqrt{F_y}}\left(1 - 1.4\frac{P}{P_y}\right)$$

$$P_y = AF_y = 16.2 \times 36 = 583.2 \text{ kips } (2{,}594 \text{ kN})$$

$$P = 17 \text{ kips } (76 \text{ kN}); \qquad \frac{P}{P_y} = 0.0290 < 0.27$$

$$\frac{d}{t_w} \leq \frac{412}{\sqrt{36}}[1 - 1.4(0.0290)] = 65.9$$

From AISC p. 1-17

$$\frac{d}{t_w} = 59.7 < 65.9: \qquad \underline{\text{OK}}$$

Check flange thickness:

$$\frac{b_f}{2t_f} = 6.9 < 8.5: \qquad \underline{\text{OK}}$$

Use W21 × 62.

COLUMN AB. We assume the column to be fully braced in the weak direction, and therefore $M_m = M_p$. Also

$$M = 340 \times 12 = 4{,}080 \text{ kip-in. } (460 \text{ kN-m}); \qquad P = 68 \text{ kips } (302 \text{ kN});$$

$$L = 20 \text{ ft} = 240 \text{ in. } (6.1 \text{ m})$$

Try W14 × 68:

$$A = 20.0 \text{ in.}^2; \qquad \frac{d}{t_w} = 33.8: \qquad \underline{\text{OK}}$$

$$\frac{412}{\sqrt{36}}\left[1 - 1.4\left(\frac{68}{20 \times 36}\right)\right] = 59.6 > \frac{d}{t_w} = 33.8: \quad \text{OK}$$

$$\frac{b}{2t_f} = 7.0 < 8.5: \quad \text{OK}$$

$$M_p = 36 \times 115 = 4{,}140 \text{ kip-in. (468 kN-m)}$$

$$\frac{M}{M_p} = \frac{4{,}080}{4{,}140} = 0.985 < 1.00: \quad \text{OK}$$

Equation (8.15) yields

$$\frac{P}{P_y} + \frac{M}{1.18 M_p} = \frac{68}{36 \times 20} + \frac{4{,}080}{1.18(4{,}140)}$$

$$= 0.094 + 0.835 = 0.929 < 1.00: \quad \text{OK}$$

Check Eq. (8.12):

$$P = 68 \text{ kips (302 kN)}; \quad P_{\text{crit}} = 1.70 \times 20 \times F_a;$$

$$M_m = M_p = 4{,}140 \text{ kip-in.}; \quad \frac{L}{r_x} = \frac{20 \times 12}{6.01} = 39.9 \rightarrow F_a = 19.2 \text{ ksi}$$

$$F_e' = 93.8 \text{ ksi}; \quad P_{\text{crit}} = 34 \times 19.2 = 652.8 \text{ kips (2,904 kN)}; \quad C_m = 0.6$$

$$P_e = \tfrac{23}{12} A F_e' = \tfrac{23}{12} \times 20 \times 93.8 = 3{,}595.7 \text{ kips (15,994 kN)}$$

and

$$\frac{P}{P_{\text{crit}}} + \frac{C_m M}{(1 - P/P_e) M_m} = \frac{68}{652.8} + \frac{0.6(4{,}080)}{(1 - 68/652.8)4{,}140} = 0.104 + 0.660$$

$$= 0.764 < 1.00: \quad \text{OK}$$

Example 8.6 Design the same column of the frame as in Example 8.5 (Fig. 8.12) using the AASHTO load factor equation.

Solution Assume that $\tfrac{1}{3}$ the 2 kips/ft is live load. Then:

$$q_u = 1.30[\tfrac{4}{3} + \tfrac{5}{3}(\tfrac{2}{3})] = 3.18 \text{ kips/ft (4.3 kN/m)}$$

$$\frac{KL_c}{r} = 39.9 \leq \sqrt{\frac{2\pi^2 E}{F_y}}$$

Thus:

$$F_{\text{crit}} = F_y\left[1 - \frac{F_y}{4\pi^2 E}\left(\frac{KL_c}{r}\right)^2\right] = 36\left[1 - \frac{36}{4\pi^2 \times 29 \times 10^3} \times 39.9^2\right]$$

$$= 34.2 \text{ ksi (236 MPa)}$$

We have $C = 0.6$, and

$$F_e = \frac{\pi^2 E}{(KL_c/r)^2} = \frac{\pi^2 \times 29 \times 10^3}{39.9^2} = 179.8 \text{ ksi (1,240 MPa)}$$

$$P = 68 \times \frac{3.18}{3.40} = 63.6 \text{ kips (283 kN)};$$

$$M = 4{,}080 \times \frac{3.18}{3.40} = 3{,}816 \text{ kip-in. (431 kN-m)}$$

$$M_u = 36 \times 115 = 4{,}140 \text{ kip-in. (468 kN-m)}$$

Also ($b'/t = 7.0 < 8.4$ and $d/t = 33.8 < 70$)

Equation (8.24) yields

$$\frac{63.6}{0.85 \times 20 \times 34.2} + \frac{3{,}816}{4{,}140[1 - 68/(20 \times 179.8)]} = 0.109 + 0.939$$

$$= 1.048: \quad \underline{\text{N.G.}}$$

This is slightly underdesigned, and so we would have to use the next larger section—that is, a W14 × 74.

8.5 FOREIGN DESIGN PRACTICES FOR BEAM-COLUMNS

The German DIN 4114 specifications (8.21) use working stress interaction eqautions similar to the U.S. specifications except that a distinction is made between:

1. Bending about an axis of symmetry, or bending such that the distance to the compression fiber is larger than the distance to the tension fiber, in which case

$$\omega \frac{P}{A} + 0.9 \frac{M}{S_c} \leq F_{\text{all}} \tag{8.28}$$

2. Bending about an axis such that the distance to the compression fiber is smaller than the distance to the tension fiber, in which case the following two equations must be satisfied:

$$\omega \frac{P}{A} + 0.9 \frac{M}{S_c} \leq F_{\text{all}} \tag{8.29}$$

$$\omega \frac{P}{A} + \frac{300 + 2\lambda}{1{,}000} \frac{M}{S_t} \leq F_{\text{all}} \tag{8.30}$$

where ω and λ are as discussed in Chapter 5 and F_{all} is the specified allowable compressive stress (DIN 1073, 1050, and 120).

French specifications CM66 (8.22), similar to the German specifications, also distinguish between the case when the bending axis is an axis of symmetry or the distance to the compression fiber is smaller than the distance to the tension fiber.

British specifications BS 153 (8.23) and BS 449 (8.24) for bridges and buildings, respectively, use the simple working stress interaction equation

$$\frac{f_a}{F_a} + \frac{f_b}{F_a} < 1 \tag{8.31}$$

8.6 COMBINED BENDING AND AXIAL TENSION

In the case of combined bending and tension working stress interaction equations are mostly used. The AISC specification (8.11) uses Eq. (8.15), which is repeated here

$$\frac{f_a}{0.60F_y} + \frac{f_{bx}}{F_{bx}} + \frac{f_{by}}{F_{by}} \leq 1.0 \tag{8.15}$$

[AISC eq. (1.6.16)]

The only additional stipulation is that the computed bending compressive stress taken alone must not exceed the allowable value of F_b for beams (8.11), section 1.5.1.4; see also Chapter 7 in this text). For plastic design the extreme fiber stress nowhere should exceed F_y.

NOTATIONS

d	= web depth	F'_e	$= \dfrac{12\pi^2 E}{23(Kl_b/r_b)}$
f	= actual stress		
f_a	= actual axial stress	F_y	= yield stress
f_b	= actual bending stress	G	= ratio of stiffnesses
k^2	$= P/EI$	I	= moment of inertia
l_b	= actual unbraced length in plane of bending	I_c	= column moment of inertia
		I_g	= girder moment of inertia
r	= radius of gyration	K_x, K_y, K	= effective length factors
r_x, r_y, r_b	= radius of gyration about x-axis, about y-axis, and in plane of bending, respectively	L	= length of columns
		M_z	= total moment
		M_1	= smallest numerical moment
t_f	= flange thickness	M_2	= largest numerical moment
t_w	= web thickness	$M_{\max}$	= maximum moment
α	$= P/P_e$	M_p	$= F_y Z$ = plastic moment capacity
α_x, α_y	= components		
β	= end moment ratio coefficient	M'_m	$= M_u$ = ultimate bending moment capacity
θ	= rotational angle at support		
ϕ	= curvature	M	= primary moment
A	= area	M_{ult}	= ultimate or maximum moment
B_x, B_y	= bending factors	P'_x, P'_y	= equivalent axial loads
C_m, C	= moment correction factor	P_e	$= P_{\text{crit}}$ = critical Euler load
E	= Young's modulus of elasticity	P'_e	$= \tfrac{23}{12} A F'_e$
F	= allowable stress	P_{tab}	= tabulated concentric column load
F_{crit}	= ultimate axial column stress capacity		
		S	= section modulus
F_a	= permissible axial stress if only axial force	Z	= plastic modulus
F_b	= permissible bending stress if only bending moment		

PROBLEMS

8.1. A floor beam acting as a frame member is loaded as shown in Fig. P-8.1. If no translation of the joints can take place and lateral bracing is provided at the beam ends only, design this beam using A36 steel and AISC specification (8.11).

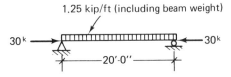

Figure P-8.1

8.2. A beam-column with 300 kips of axial load is loaded by a horizontal girt at midheight by a force of 10 kips acting in the column web plane (plane of bending)(see Fig. P-8.2). The column is hinged at both ends without sidesway and has a height of 24 ft. At midheight, movement in the weak direction (out of plane of bending) is prevented. Design this beam-column selecting a W12 of A36 steel and using AISC specification (8.11).

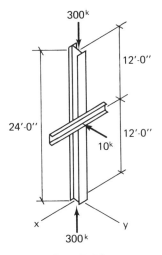

Figure P-8.2

8.3. A column is part of a braced frame and carries an eccentric force of $P = 280$ kips acting with an eccentricity of 12 in. about its strong axis (see Fig. P-8.3).

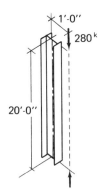

Figure P-8.3

Taking the column height of 20 ft as the effective buckling length, select the lightest W14 in A36 steel according to the AISC specification (8.11).

8.4. In a braced frame a W27 × 84 of A36 steel is used as the horizontal member of 40 ft span, transmitting to the top of the column an axial force of 35 kips and a moment, producing tension in the outside column flange, of 325 kip-ft (see Fig. P-8.4). Design the column if A36 is used and the column is hinged at its base. The column height is 21 ft. The column is braced in the weak direction by horizontal members every 7 ft. Use AISC specification (8.11).

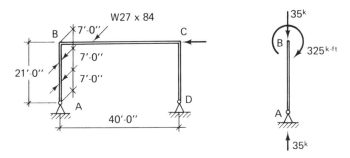

Figure P-8.4

8.5. A beam-column as part of an unbraced frame is loaded by a horizontal beam (W24 × 84) with an axial load of 175 kips and a moment of 365 kip-ft acting about its strong column axis (see Fig. P-8.5). In the weak direction there

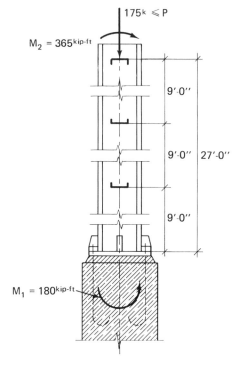

Figure P-8.5

Problems 283

is no sidesway and there is bracing at the third points. If the height of the column (distance between the axis of the horizontal frame beam and column-bearing plate on the foundation) is 27 ft, design this column using steel with $F_y = 50$ ksi and AISC specification (8.11). The moment at the base is 180 kip-ft (counter-clockwise). The frame span is 60 ft.

8.6. A compressed chord of a roof truss is composed of two angles continuous over four truss panels. Roofing is directly supported by the chord, and therefore this chord is acting as a beam-column, with lateral buckling prevented. If the loading is as given in Fig. P-8.6, design this chord by selecting single or double equal-leg angles of A36 steel and using the AISC specification (8.11).

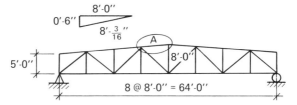

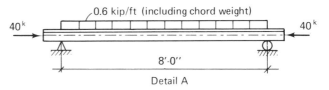

Detail A

Figure P-8.6

8.7. Redesign the column in Problem 8.4 using the plastic design method [part 2 of the AISC specification (8.11)]. Use load factors of 1.7 to get force and moment at ultimate level.

8.8. A horizontal beam in a floor framing system is loaded according to Fig. P-8.8. Assume lateral bracing at midspan and beam ends. Select the lightest W-shape in A36 steel, using AISC specification (8.11).

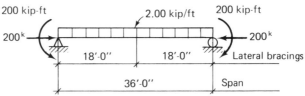

Figure P-8.8

8.9. Redesign the beam-column in Problem 8.2 suing high-strength, low-alloy A441 steel.

8.10. Redesign the beam-column in Problem 8.5 using ASTM A36.

8.11. Repeat Problem 8.1 if instead of longitudinal load of 30 kips a longitudinal load of 40 kips is applied and in addition a vertical concentrated load of 10 kips at the midpoint as well as an estimated dead load of 0.05 kip per foot. Use A36 steel, AISC specification, and Eq. (8.19b).

8.12. Repeat Problem 8.2 for an axial load of 420 kips and a midheight force of 12 kips. Select a W14 of A36 steel and use AISC specification (8.11).

8.13. Repeat Problem 8.3 for $P = 200$ kips and an eccentricity of 3 in. in both the x and y directions.

8.14. Repeat Problem 8.3 for a column 14-ft long, a load of 160 kips, and an eccentricity of 6 in. in both the x and y directions.

8.15. Repeat Problem 8.5 for an axial load of 280 kips, a top moment of 250 kip-ft and a base moment of 125 kip-ft (all other data are the same).

8.16. The top chord of a roof truss shown in Fig. P-8.16 is composed of two angles continuous over three truss panels. The roofing is directly supported by this chord with lateral buckling prevented. If the roof loading in the top chord, including its own weight, is 0.5 kip/ft, design the chord using A36 steel and AISC specification (8.11).

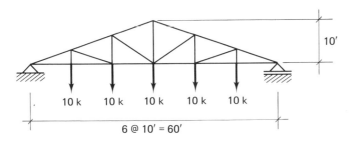

Figure P-8.16

8.17. Redesign the column of Problem 8.13 using A441 steel.

8.18. Repeat Problem 8.8 for a uniform load of 2.2 kips/ft, an axial load of 140 kips, and an end moment of 180 kip-ft using A441 steel.

8.19. Redesign the column of Problem 8.12 using A441 steel.

8.20. Redesign the top chord of Problem 8.16 for a chord load, including its own weight, of 1.0 kip/ft using A441 steel and AISC specification (8.11).

REFERENCES

8.1. BLEICH, F., *Buckling Strength of Metal Structures*, McGraw-Hill, New York, 1952.

8.2. KETTER, R. L., "Stability of Beam-columns Above the Elastic Limit," *Proc. ASCE*, vol. 81, separate no. 692 May 1955.

8.3. NEWMARK, N. M., "Numerical Procedure for Computing Deflections, Moments, and Buckling Loads," *Trans. ASCE*, vol. 108, 1943, pp. 1161–234.

8.4. OJALVO, M., and Y. FUKOMOTO, *Nomographs for the Solution of Beam-column Problems*, Bulletin no. 18, WRC, New York, 1962.

8.5. KETTER, R. L., and T. V. GALAMBOS, "Columns Under Combined Bending and Thrust," *Trans. ASCE*, vol. 126, (I) p. 1–25, 1961.

8.6. KETTER, R. L., "Further Studies on the Strength of Beam-columns," *Proc. ASCE*, vol. 87, pp. 135–52 ST-6, Aug. 1961.

8.7. OJALVO, M., "Restrained Columns," *Proc. ASCE*, vol. 86, no. EM-5, pp. 1–12, Oct. 1960.

8.8. KETTER, R. L., E. L. KAMINSKY, and L. S. BEEDLE, "Plastic Deformation of Wide-flange Beam-columns," *Trans. ASCE*, vol. 120, pp. 1028–69, 1955.

8.9. TIMOSHENKO, S. P., and J. M. GERE, *Theory of Elastic Stability*, 2nd ed., McGraw-Hill, New York, 1961.

8.10. YU, C. K., and L. TALL, "A514 Steel Beam-columns," *Publ. International Association for Bridge and Structural Engineering* (31-II), 1971, pp. 185–213.

8.11. AISC, *Manual of Steel Construction*, 8th ed., Chicago, 1980.

8.12. AASHTO, *Standard Specifications for Highway Bridges*, 12th ed., Washington, DC, 1977 (with 1978, 1979, 1980, and 1981 Interim Bridge Specifications).

8.13. HARPER, I., "Design of Beam Columns," *Eng. J. AISC*, vol. 4, no. 2, April 1967, pp. 41–61.

8.14. TALL, L., ed., *Structural Steel Design*, 2nd ed., Ronald, New York, 1974.

8.15. ASCE-WRC Committee, "Commentary on Plastic Design in Steel, Chapter 7, Compression Members, *ASCE Manual of Engineering Practice*, no. 41, 1961.

8.16. MASSONNET, C., "Stability Considerations in the Design of Steel Columns," *J. Struc. Div., Proc. ASCE*, vol. 85, no. ST-7, Sept. 1959, pp. 75–111.

8.17. HORNE, M. R., "The Stanchion Problem in Frame Structures Designed According to Ultimate Carrying Capacity," *Proc. Inst. Civ. Engrs.*, vol. 5, no. 1, part 3, April 1956, pp. 105–46.

8.18. AUSTIN, W. J., "Strength and Design of Metal Beam-Columns," *Proc. ASCE*, vol. 87, no. ST-4, April 1961, p. 1.

8.19. GALAMBOS, T. V., *Structural Members and Frames*, Prentice-Hall, Englewood Cliffs, NJ, 1968, chapter 5.

8.20. VAN KUREN, R. C., and T. V. GALAMBOS, "Beam-column Experiments," *Proc. ASCE*, vol. 90, no. ST-2, pp. 223–256, April 1964.

8.21. *Stahl im Hochbau*, 13th ed., Verlag Stahleisen, Dusseldorf, 1969, section 7.3.2.1, pp. 649–50.

8.22. L'Institut Technique du Batiment et des Travaux Public and Le Centre Technique Industriel de la Construction Métallique, *Règles de Calcul des Constructions en Acier*, 5th ed., Règles CM66–1974, ed. Eyrolles, Paris, 1974, p. 71.

8.23. British Standards Institution, *Specification for Steel Girder Bridges*, BS 153, London, 1966, Parts 3B and 4.

8.24. British Standards Institution, *Specification for the Use of Structural Steel in Buildings*, BS 449, London, 1965.

9

Connections

9.1 INTRODUCTION

A structure is assembled from individual structural members by means of connections. Connections, if not designed properly, can represent weak links in a structure, diminish its serviceability because of large deflections, and, from a practical point of view, raise the cost of the fabrication and erection of a structure significantly. The design and detailing of connections often is as important as the design of the members themselves. Many recent structural failures have been caused by failure of connections, and it is therefore extremely important that the engineer carefully examine the detailing and design of all connections as well as their construction.

A good connection must be practical, cheap, and safe. Unfortunately, this is not easily achieved, owing to the very complex behavior of connections. Most connections are indeterminate to a high degree. Mathematical models used in their analysis are only very rough approximations of the real situation, and practical knowledge of the actual deformational response of joints is essential. Loads are applied through bolts, rivets, pins, or welds and are concentrated on small portions of the joints. They flex, and because of the substantial shear and moment that must often be transmitted the application of simple technical bending theory is not possible. The distribution of stresses usually is highly irregular, as it is affected by the deformation of the fasteners and the connected material itself, making an exact theoretical analysis impracticable. The safe and economical design in proportioning of connections therefore is semiempirical, as the analysis is controlled by experience, past practice, and the results of well-conducted experiments. For this reason both actual behavior and design methods will be discussed in this chapter. Section 9.2 deals with performance of connections, while Section 9.3 discusses their design when loaded statically. The behavior and design of connections under dynamic loading and their fatigue strength also are considered, but to a lesser extent. The main connecting devices used at present are bolts and welds. In the United States rivets are now obsolete and pins are at present used solely for supports or other special joints. Trusses with pinned joints have largely disappeared. The increased strength of bolts due to the use of high-strength steels and the omission of hardened washers has helped bolting to replace riveting. The cost of high-strength A325 bolts is about three times that of rivets. The reduced number of bolts in a joint as compared to the number of rivets, the omission of washers, and the reduced cost for installing bolts as compared to riveting, however, have offset the higher material cost. It is expected that the present trend toward welding and bolting will continue.

Historically, the first fasteners were ordinary bolts, because the material used in the first iron structures was cast iron, which was too brittle for riveting. In the 1840s, with the replacement of cast iron by wrought iron as the main structural material, riveting became predominant. The reason was its superiority over ordinary bolts because the connected parts were kept tight while also considerable friction between faying surfaces existed, which prevented slip under moderate shearing loads. Ordinary bolts, or "unfinished bolts," as they are now called, could not produce such a clamping force. Deformations of joints using ordinary bolts are larger and more bolts are needed per joint than when using rivets. High-strength bolts, which exhibit a much higher pretensioning force, reversed this situation. The first applications of high-strength bolts occurred when replacing rivets which had worked loose in the joints of an ore bridge. As these bolts remained tight after several years of operation and their increased clamping force was beneficial for the fatigue strength of joints, high-strength bolts started to be used on their own merit for connections in general, and especially for field connections where bolting is done on the erection site. Shop connections were still riveted, but by about 1970, welding was used for nearly all shop fabrication. As the use of rivets is almost obsolete, their design and performance will not be discussed in this book.

The design of pins is discussed in Chapter 11.

9.1.1 Types of Connections

The AISC specification (9.1) defines the following three basic types of constructions:

> *Type 1.* Commonly designated as *rigid-frame* (continuous frame), assumes that beam-to-column connections have sufficient rigidity to hold virtually unchanged the original angles between intersecting members.
> *Type 2.* Commonly designated as *simple* framing (unrestrained, free-ended), assumes that, insofar as gravity loading is concerned, the ends of beams and

girders are connected for shear only, and are free to rotate under gravity load.
Type 3. Commonly designated as *semirigid framing* (partially restrained), assumes that the connections of beams and girder possess a dependable and known moment capacity intermediate in degree between the rigidity of Type 1 and the flexibility of Type 2.

Although these definitions describe primarily beam and column framing, they indirectly specify the design of all connections, which have to be consistent with the assumptions as to type of construction. Type 1 construction is unconditionally permitted. Type 2 construction is permitted but connections should have adequate capacity to resist wind moments, and adequate rotation capacity to avoid overstress of the fasteners or welds under combined gravity and wind loading. Type 3 (semirigid) construction is also permitted, but only upon evidence that the connections to be used are capable of furnishing, as a minimum, a predictable proportion of full end restraint.

According to the relative position of the connected members and the corresponding force actions, which have to be transferred by that connection, connections are subject to

1. Axial shear, when the load P is applied along a line of action which coincides with the axis of the structural member (usually a truss member or beam flange) and the connection (Fig. 9.1a, b).
2. Eccentric shear (torsion and shear) when the force action line does not pass through the center of a bolt or weld group, such as is the case for a bracket attachment (Fig. 9.2), a framed beam to column connection, or a beam-web splice (Fig. 9.3).
3. Tension in bolted connections when the force or moment is producing normal tensile stresses in the shank of the bolts. The simplest case of fasteners in tension is the T-hanger connection (Fig. 9.4a) or in the case of bending and/or eccentric tensile forces of a beam-to-column connection in a frame (Fig. 9.4b).
4. Combined axial tension and shear as in end connections of bracing diagonals to columns (Fig. 9.5), or in the connection of a bracket to a column web or in a beam-to-column connection (Fig. 9.4b).

9.1.2 Types of Fasteners

As already mentioned, bolting and welding are now the main connecting techniques. Pins were widely used in the past for truss joints when it was believed that joints should behave like ideal joints, i.e., no transfer of moments to the truss members and no moment transfers across the joints. The main-chord members of such pinned trusses were eye bars. However, it was soon realized that secondary stresses, produced by moments which were transferred to the joints by more or less rigid riveted end connections of truss elements, are minor and therefore of no big concern. On the contrary, the deflections of riveted trusses were much smaller relative to pinned connections and the safety of the whole truss was increased by creating multiple redundancy. In pinned trusses, failure of one truss member (eye bar) meant failure of

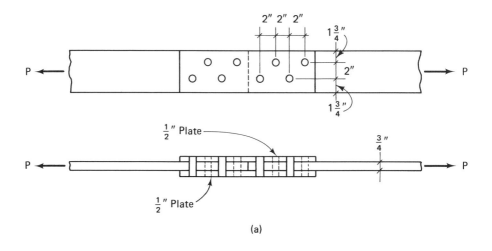

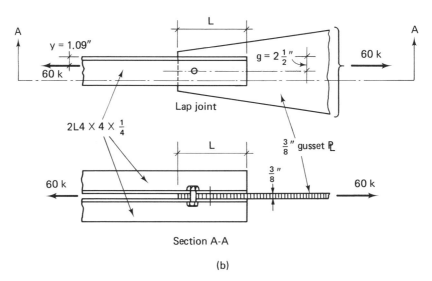

Figure 9.1 (Part (a) from Crawley and Dillon, *Steel Buildings*, 2nd Ed., John Wiley & Sons, New York, 1977, Fig. 7.24, 7.26, 7.27.)

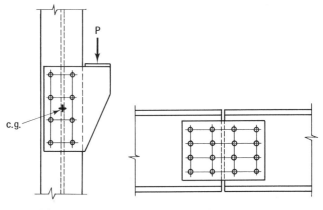

Figure 9.2   Figure 9.3

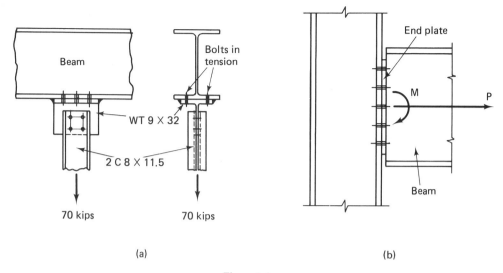

Figure 9.4

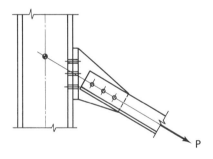

Figure 9.5

the whole structure, which proved not to be true for riveted trusses. For that reason, both eye bars as truss members and pins as fasteners are now obsolete. Pins are now used only for special joints on cantilevered trusses or multispan girders, and for supports. Therefore, pins are discussed in this text, in Chapter 11, but eye bars are not.

Bolts will be discussed in this section, and welds in the next one.

According to the AISC specification (9.1) steel bolts must conform to one of the following standard specifications (latest edition):

Low-Carbon Steel Externally and Internally Threaded Standard Fasteners, ASTM A307 (unfinished bolts)
High Strength Bolts for Structural Steel Joints, Including Suitable Nuts and Plain Hardened Washers, ASTM A325
Quenched and Tempered Steel Bolts and Studs, ASTM A449
Quenched and Tempered Alloy Steel Bolts for Structural Steel Joints, ASTM A490

In connections, A449 bolts may be used only in bearing-type connections requiring bolt diameters greater than $1\frac{1}{2}$ in. A449 bolt material is also acceptable for high-strength anchor bolts and threaded rods of any diameter. Therefore, in the following text only unfinished bolts of A307 and high-strength bolts of A325 and A490 will be discussed. Turned bolts and ribbed bolts are practically obsolete, and will not be discussed.

9.1.2.1 Unfinished (Rough) Bolts (ASTM A307)

For temporary connections or for connections involving only the transmission of small static forces or moments in building construction (light-weight structures), unfinished, rough bolts may be used. The limited use of these bolts stems from large deformations experienced with such connections and their low fatigue strength. The steel used for rough bolts (ASTM A307) has a minimum tensile stress of $F_u = 60$ ksi (414 MPa), so the clamping force produced by tightening of nuts is insufficient to produce a friction type of connection. Upon load application friction will be overcome and permanent slip will occur. Such slip will be followed first by elastic and then plastic deformations of both the bolt shank and the hole surface. Consequently, the deformations of the bolted connection even under working loads are rather large.

Unlike rivets, the strength of bolts is specified in terms of a tensile test of the complete threaded fastener. The weakest section of any bolt is its threaded portion, and the strength of the bolt is usually computed by using the "stress area," an average area based on the nominal and root diameters [AISC *Manual* (9.2), p.4-141]. In Table 9.1 the properties of A307, A325, and A490 structural bolts are given.

Unfinished bolts are forged from rolled steel rods and are allowed rather large tolerances in shank and thread dimensions. Consequently holes are punched or drilled $\frac{1}{16}$ in. larger than the nominal bolt diameter.

Table 9.1 PROPERTIES OF STRUCTURAL BOLTS

ASTM (9.3) Designation	Type Name	Bolt Diameter, D		Tensile Strength Stress Area*		Proof Load Stress Area*	
		in.	(mm)	ksi	(MPa)	ksi	(MPa)
A307	Low-carbon steel	all	all	55	(380)	—	—
A325	High-strength steel bolts	$\frac{1}{2}-\frac{3}{4}$	(13–19)	120	(830)	85	(587)
		$\frac{7}{8}-1$	(22–25)	115	(793)	78	(538)
		$1\frac{1}{8}-1\frac{1}{2}$	(29–38)	105	(723)	74	(511)
A490	High-strength alloy steel bolts	$\frac{1}{2}-4$	(13–100)	150	(1,036)	120	(828)

*Stress area $= 0.785[D - 0.9743/n)]^2$, where $D =$ nominal bolt size and $n =$ number of threads per inch [See AISC *Manual* (9.2), p. 4–141].

To assure proper functioning of bolted connections under load, the parts must be tightly clamped between the bolt head and the nut. To prevent damage of the basic connection material, washers are used. Washers also serve to distribute the clamping force evenly over the bolted member and to prevent the threaded portion of the bolt from bearing on the connecting piece.

9.1.2.2 *High-Strength Bolts*

History of high-strength bolts

The high-strength bolt is not new, but its extensive application to structural joints is of recent origin (9.4). The bolt is made of accurately controlled, quenched and drawn medium-carbon steel. When applying a high-tensile bolt to structures, it is torqued to a high tension and supported by hardened washers under the bolt head and the nut. This high tension in the bolt will produce a high clamping force with corresponding high friction between the faying surfaces, of an intensity sufficient to transmit the force without shearing or bearing on the bolt.

The first tests on high-strength bolts in the United States were made in 1938 (9.5). The bolt tensions were low in comparison to modern practice. The next step came in October 1947 when K. H. Lenzen (9.6) conducted tests of joints of balanced design but with various washers, to avoid losing grip in reverse loading.

The American Society for the Testing of Metals has developed specifications (9.3) to control their quality under the ASTM designation A325, and the Research Council on Riveted and Bolted Structural Joints has also prepared a set of specifications (9.6).

Since the 1950s the A325 high-strength bolt has become the prime field fastener of structural steel in the United States.

Bolt dimensions and quality

The latest specifications call for a heavy hexagonal-head structural bolt, a heavy, semifinished hexagonal nut, and either one washer or no washers, depending upon the tightening method used.

Tension is induced in the bolt by turning the nut. This operation sets up a combined tension and torsional shear that results in development of a lower maximum tension in the bolt (see the discussion under "Combined Shear and Tension," p. 335, in Section 9.2.1.7). Tests have shown that the ultimate shearing stress of A325 bolts is about 70% of the bolt ultimate tensile stress.

In addition, A490 bolts are made of quenched and tempered alloy steel for use with high-strength steel members. These bolts are marked on the head with "A490" and on the nut with "2H" or "DH."

To assemble a bolted joint (9.7), the bolts must be smaller than the bolt holes by $\frac{1}{16}$ in. (2 mm); as bolts do not expand and fill the holes like rivets, it is essential that the bolts be drawn extremely tight to protect the joint against slipping. Thus, the load is transferred across the joint by friction between the connected parts rather than by shear on the fasteners, except in case of large static loads.

In general, bolt tension has little effect on the ultimate strength of the joints because loads less than the ultimate load cause the plates to slip sufficiently to produce bearing on the bolts. Therefore, surface preparation and bolt tension have little or no effect on the ultimate strength of either a bolted or riveted joint. They do have, however, an important effect on the load-slip characteristics of the joint (Fig. 9.6).

The most recent research (9.8) has shown that the value of the slip factor does not depend decisively on the friction between the contact surfaces, but on the shearing of

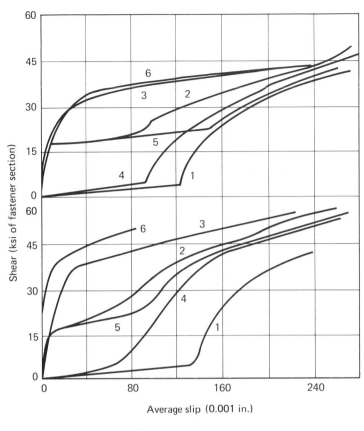

Figure 9.6 Relationship Between Load and Slip. (From 9.9).

the protruding parts of the two surfaces—that is, their roughness and strength.

The installation of high-strength bolts must be adequate to tighten each bolt to its proof load i.e., to at least 70% of the minimum required tensile strength. This can be done by either of two methods, the turn-of-nut-tightening method and bolt tightening by use of a direct-tension indicator. Both A325 and A490 bolts tightened by the calibrated-wrench method (i.e., by torque control) must have a hardened washer under the element (nut or bolt head) that is being turned in tightening. Two hardened washers should be used with all A490 bolts used to connect material having a specified minimum yield point less than 40 ksi (276 MPa) (9.10).

Turn-of-nut tightening

When this method is used to provide the specified bolt tension, first enough bolts must be brought to a "snug-tight" condition to ensure that the parts of the joint are brought into good contact with one another. *Snug-tight* is defined as the tightness attained by a few impacts of an impact wrench or the full effort of a person using an oridinary spud wrench. According to table 4 of the AISC specification for A325 and A490 bolts (9.2), all bolts in the joint must then be tightened additionally.

Direct tension indicator

This English method uses the *coronet* load indicator. This is a hardened washer with a series of protrusions on one face (9.28). The washer is inserted between the bolt head and the gripped material with the protrusions bearing against the underside of the bolt, leaving a gap. Upon tightening, the protrusions are flattened and the gap is reduced. Bolt tension is evaluated from measurements of the residual gap, Fig. 9.7. This method yields significant cost savings over both turn-of-nut and calibrated wrench installations. In order to achieve the minimum required bolt tension, the manufacturer recommends tightening until the gap is reduced to 0.015 in. (0.38 mm) or less.

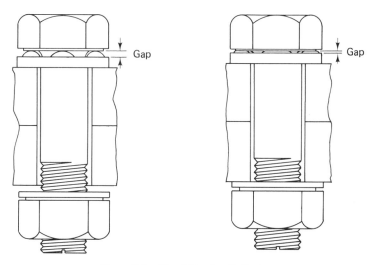

Figure 9.7 Direct Tension Indicator.

High-strength bolts are applied in friction-type and bearing-type connections. In friction-type (designated F) connections the load is carried solely by the friction between the connected parts. In a bearing-type connection, after overcoming the friction, load is transmitted by shear and bearing. For shearing capacity it is important to determine whether the threads are included in the shear plane (designation N) or whether they are excluded (designation X).

9.1.3 Electric-Arc Welding

Welding is an ancient art (9.11) in spite of the fact that welding engineering is a relatively new field. Prehistoric bracelets have been discovered that were made by forge welding. The art of working and hardening steel, an advanced stage in metalworking that doubtless took centuries to reach, was commonly practiced 30 centuries ago in Greece and is mentioned by Homer.

Welding consists of joining two pieces of metal by establishing between them a metallurgical atom-to-atom bond, as distinguished from previously described mechanical connections held together by friction or mechanical interlocking.

There are many welding processes (about 35 of them), but at present for civil engineering structures the most important is electric-arc welding, which is one of the fusion-welding processes. Due to limited space, only this form of welding will be discussed in this text.

The material discussed in the following sections is mainly based on AISC specifications (9.1) and (9.2) and material published by the American Welding Society (9.18, 9.20).

History

Although the history of arc welding could be considered to have started with the discovery of the electric arc in 1801 by H. Davy, it really did not because no attempts were made for its practical use until 1881 (9.12). Six years later a patent covering rights on a carbon-arc-welding process was issued (9.13). In 1888, the carbon electrode was replaced with a metallic one which upon striking an arc gradually melted and added fused drops of metal to the weld. Patents were issued on the metallic-arc process in the United States in 1889.

As an industrial process, arc welding was introduced around the turn of the century, but it was not received with a great deal of acclaim since the mechanical properties of the joints produced were of a dubious nature. Unquestionably the greatest contribution to the advancement of arc welding was the addition of a coating to the electrode wire for general improvement of its mechanical properties and performance. In 1910 there was produced a heavily fluxed or coated electrode, making possible the "shielded arc" characteristic of modern welding. After World War I welding techniques gradually gained ground and several important achievements were recorded.

Processes

In arc welding, the intense heat required to reduce metal to a liquid state is created by an electric arc which is formed between the work to be welded and metal wire called an electrode. The electrode, held in a suitable holder, is brought close to the metal to be welded, causing an arc to be formed between the tip of the electrode and the work. The instant the arc is formed, the temperature of the work at the point under the tip of the electrode jumps to approximately 6,500°F (3,450°C)(Fig. 9.8a). Much of what goes on in and about the arc is only imperfectly understood.

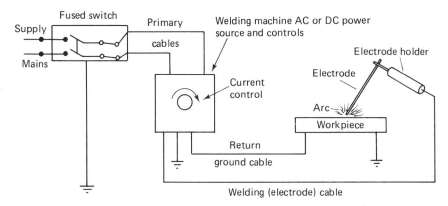

Figure 9.8a Arc-welding Circuit.

The shielded arc

One of the difficulties when welding with a bare wire electrode is the fact that when molten metal in the arc stream as well as that in the molten crater in the workpiece are exposed to the atmosphere their oxidation takes place very rapidly. Thus coated electrodes are used to protect the molten metal from the air and to stabilize the arc and make more effective use of the arc energy (Fig. 9.8b). This shielded metal-arc welding is the most usual type of welding.

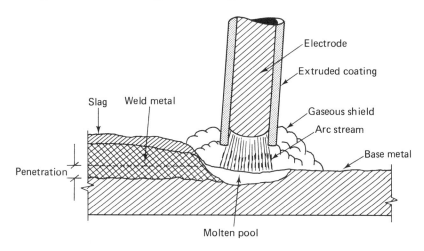

Figure 9.8b Elements of the Shielded Metal-arc-welding Process.

Submerged (hidden) arc welding

This method is used for automatic or semiautomatic welding. The arc is completely hidden from view at all times, covered by a mound of granular agglomerated flux (Fig. 9.9b). A bare electrode wire coiled on a reel is fed by mechanically powered drive rolls continuously into the arc. Current is fed to the wire through contact jaws

Introduction   **297**

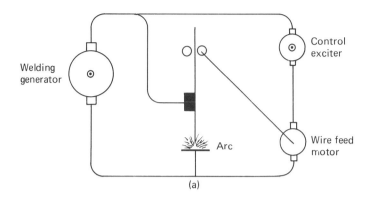

(a)

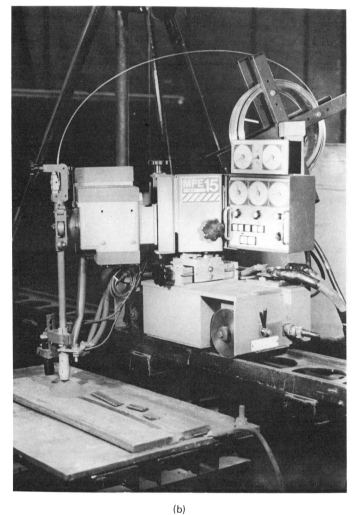

(b)

Figure 9.9 Submerged Arc Process.

between which the wire passes. These jaws are close to the arc area, so that the current has a relatively short distance to travel as compared with work with electrodes (Fig 9.9a).

Mechanical properties of the weld metal deposited by this process are consistently of a high quality, and higher welding speeds can be employed than in any other welding process (9.14).

Electroslag

This automatic process is used primarily (9.15) for the butt welding of mild steel and medium- and high-tensile structural steels in a vertical position, Equipment has also been devised to enable the process to be used for welding fillet welds. The process can be used for butt welding mild steel plates up to 18 in. (45.6 cm) in thickness and plates of high-strength steels between 1 and 4 in. (25 to 102 mm) in a vertical position.

The electroslag process (Fig. 9.10) is fundamentally different from arc welding. The electrode wire is fed into the space between the edges of the two plates to be joined, and an arc is established under a blanket of flux powder. When a molten slag bath is formed, the arc is automatically extinguished, though the current continues to flow because the specially compounded flux produces a molten slag which is highly conductive. The flow of current heats up the slag bath to a temperature above the melting point of steel. The slag in turn heats up the edges of the parent metal to the

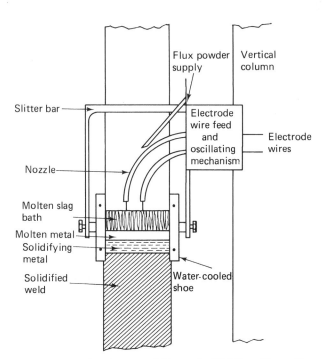

Figure 9.10 Schematic Diagram of Electroslag Welding. (From *Manual, Semi-Automatic and Automatic Arc Welding*, by Eric Flintham, The British Oxygen Comp. Ltd., Hammersmith House, London, 1966.)

fusion point and melts off the electrode wire as it is fed into the slag bath. Thus a column of deposited metal is built up, welding the two plates together.

In building construction electroslag welding is used (9.16) for heavy-section welding with success.

Equipment

The arc-welding process can be (1) wholly manual, (2) semiautomatic, or (3) automatic. The equipment primarily depends upon the process used and is varied to satisfy the specific needs of each type.

Manual metal-arc welding

Electricity supplied from public mains is usually at too high a voltage for arc welding. Therefore, the voltage of alternating-current mains is reduced to between 50 and 100 volts by means of a static transformer for ac welding or by means of a transformer-rectifier or motor-generator for dc welding. If ac supply is available, the choice of ac or dc equipment depends upon several factors (9.17).

The latest development in the evolution of electrode coatings has been the addition of iron powder. In many cases, because of their higher deposition rates, the iron powder–type electrodes prove to be the fastest type of electrode. Iron powder electrodes increase welding speed up to 50%. One such electrode, for instance, is the E-6015 or E-6024 designated by the American Welding Society (9.18), specifications A5.1 and A.5.5). The "E" classification designates electrodes of mild steel and low-alloy steel. The first two digits (here 60) indicate the approximate tensile strength in ksi. The last digit indicates a group of welding technique variables, such as current supply and application. The next-to-last digit indicates a welding position number (see Positions of Welding,"p. 9.24).

The choice of the proper welding rod can become quite involved with special steels. The best admonition is to choose the steel you want to use and then seek the assistance of a welding expert to choose the rod to be used and specify the conditions for the welding process (9.19).

Semiautomatic welding

The early methods of semiautomatic welding were adaptations of the existing automatic welding processes in which the operator took over some of the functions previously controlled electrically or mechanically. These semiautomatic welding processes are useful for repetitive fabrications where the location or the shortness of the weld lengths makes the capital expense and/or setup costs of automatic welding equipment uneconomical, but they have largely been superseded by the more attractive gas-shielded metal-arc processes such as gas-shielded metal-arc welding with either bare or flux-cored electrodes.

Transformer-rectifier sets are used for the semiautomatic process. Wire for semiautomatic welding equipment is normally supplied (Fig. 9.11) wound on 4-in. (102-mm) or 12-in. (305-mm) diameter reels with plastic or hardboard centers. Flux-cored electrode is usually supplied in coils wound on steel formers, the nominal weights of electrode being about 25 lb (111 N) or 50 lb (222 N).

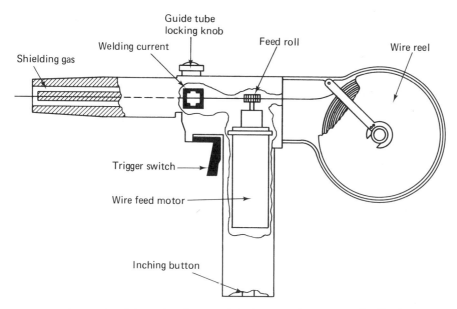

Figure 9.11 Schematic Diagram of Reel-on-gun Arrangement for Semi-automatic Welding (From 9.14).

Automatic welding

Basically the equipment for all automatic arc-welding processes consists of a power source with a contactor and auxiliaries; a welding head complete with electrode wire feed unit, welding nozzle assembly, powder supply or gas supply, and cooling water (if used); and a control unit and means of moving the welding head or workpiece at the required welding speed.

For submerged arc welding (Fig. 9.9.b) with one electrode a 1,200A (ampere) set will cover most requirements. For electroslag welding with wire it is customary to use currents of about 600 A on each electrode.

Welding terms and definitions

Types of welds

In most civil engineering structures only two types of welds are used: groove (or butt) and fillet welds. Most of the welds (about 80%) are fillet welds. Butt welds comprise about 15%, and the remaining 5% are made up of the slot, plug, and other special welds. Therefore, in this book only groove and fillet welds are treated.

A groove weld is a weld made in the groove between two members to be joined (Fig. 9.12)(9.20). According to the thickness of the joining plates, the edge preparation has to be done properly. Actually, success in welding depends about 75% on weld preparation and only 25% on the welder.

A fillet weld is a weld of approximately triangular cross section (Fig. 9.13). The theoretical throat, t_e, is used in design calculations, but the size of the fillet weld is always given as the nominal size of the leg, a—usually in sixteenths of an inch (approx.

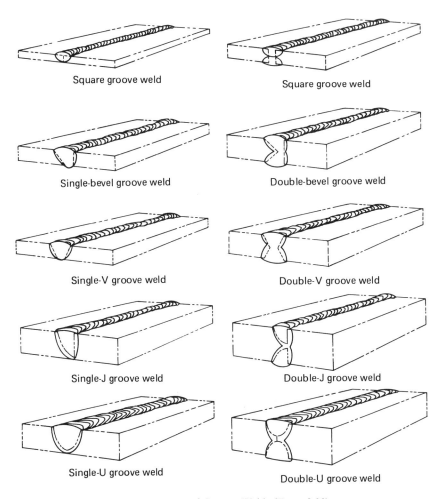

Figure 9.12 Types of Groove Welds (From 9.20).

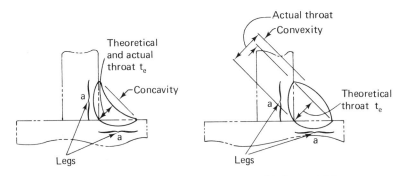

Figure 9.13 Convex and Concave Fillet Welds (From 9.20).

2 mm). The effective throat thickness of a fillet weld, t_e, is the shortest distance from the root to the face of the diagrammatic weld. For example, for an equal-leg fillet

$$t_e = 0.707a \tag{9.1}$$

where a is the leg size. For fillet welds made by the submerged arc process the effective throat thickness is equal to the leg size for $\frac{3}{8}$-in. (9.5-mm) and smaller fillet welds and equal to the theoretical throat plus 0.11 in. (2.8 mm) for fillet welds over $\frac{3}{8}$ in. (9.5 mm).

The size of a groove weld is equal to the thickness of the welded parts if these parts are of the same thickness (Fig. 9.14a). If the thicknesses are different, then the weld size, a, is equal to the smaller thickness of the parts to be joined (Fig. 9.14b).

For fillet welds the AISC specification (9.1) gives the minimum size of fillet welds as determined by the thickness of the two parts joined, except that the weld size need not exceed the thickness of the thinner part joined unless a larger size is required by calculated stress.

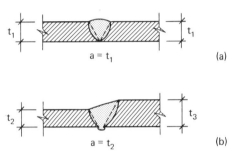

Figure 9.14 Size of Groove Welds.

In case of a T-joint, one rule of thumb (9.21) says that the fillet size should equal $\frac{3}{4}$ the plate thickness required to develop full plate strength. Using this method (Fig. 9-15) a $\frac{3}{8}$-in. (10-mm) fillet weld on a $\frac{1}{2}$-in. (13-mm) plate should "beat the plate." Actually, in the study cited so did $\frac{11}{32}$-in. (9-mm) and $\frac{5}{16}$-in. (8-mm) fillets. Only when the fillet size was reduced to $\frac{1}{4}$ in. (6 mm) did weld failure occur at a stress of 12.3 kips per linear inch (2,154.1 KN/m)—more than 3.8 times the AWS allowable. The German specifications for welded buildings, DIN 4100 (9.22), give the following formula for the calculation of the recommended (best-suited) throat of a fillet weld (measured in millimeters)

$$t_e \geqq \sqrt{t} - 0.5 \geqq 3 \text{ mm} \tag{9.2}$$

where t is the thicker of the two parts joined. If this formula is applied to the case of Fig. 9.15, where $t = \frac{1}{2}$ in. (12.7 mm), the leg size is found to be

$$a = 1.41 t_e = 4.3 \text{ mm} = 0.1704 \text{ in.} \approx \tfrac{3}{16} \text{ in.} (9.3) \tag{9.3}$$

The AISC specification (9.1) gives $a = \frac{3}{16}$ in. as a minimum, not as the best size.

The same German specifications state (Fig. 9.16) that the thickness, t, of an element with fillet welds on both sides of it must be larger than $\frac{1}{4}$ in. (6 mm) to prevent the cutting of this element by the weld penetration.

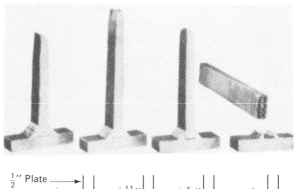

Figure 9.15 (From 9.21).

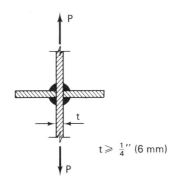

Figure 9.16 Welds on Both Sides of a Plate.

The AISC specification (section 1.17.3) gives the maximum size of a fillet weld used along edges of connected parts (Fig. 9.17) as equal to the thickness of the material if it is less than $\frac{1}{4}$ in. (6 mm) thick. For thickness of $\frac{1}{4}$ in. (6 mm) or more the maximum size is $\frac{1}{16}$ in. (2 mm) less than the thickness of the material. For the minimum effective length of strength fillet weld these specifications give a length not smaller than 4 times the nominal size.

When the strength required is less than that developed by a continuous fillet weld of the smallest permitted size, in order to economize sometimes intermittent fillet welds are used (Fig. 9.18). The effective length of any segment

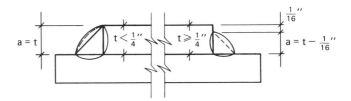

Figure 9.17 Maximum of Fillet Weld along Edges of Connected Parts.

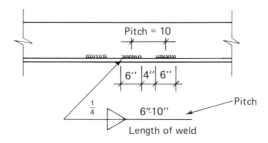

Figure 9.18 Intermittent Fillet Weld.

of intermittent fillet welding must be not less than 4 times the weld size, with a minimum of $1\frac{1}{2}$ in. (38 mm) (see 9.2.2.4 about their adverse effect on fatigue).

Welding symbols

A welding symbol may consist of as many as eight elements (Fig. 9.19): reference line, arrow, basic weld symbols, dimensions, supplementary symbols, finish symbols, tail and specification, process, or other reference.

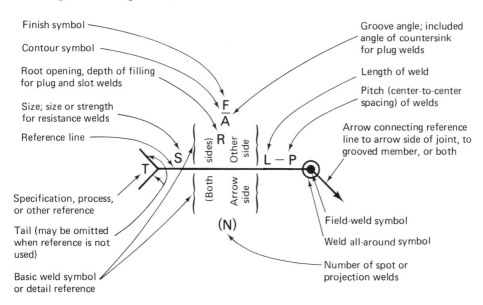

Figure 9.19 Standard Location of Elements of a Welding Symbol (From 9.2, p. 4–148).

Types of joints

According to the relative position of the parts to be welded—that is, whether they are in the same, parallel, or different planes—basically there are five types of joints

Introduction   **305**

1. *Butt Joint:* both parts in the same plane (Fig. 9.12).
2. *Lap Joint:* parts in parallel planes (Fig. 9.20a).
3. *Edge Joint:* parts in parallel planes (Fig. 9.20b).

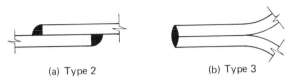

(a) Type 2   (b) Type 3

Figure 9.20 Lap and Edge Joint.

4. *T-joint:* parts perpendicular to one another (Fig. 9.21a).
5. *Corner Joint:* parts perpendicular to one another (Fig. 9.21b).

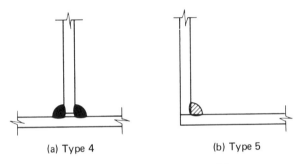

(a) Type 4   (b) Type 5

Figure 9.21 T- and Corner Joint.

Position of welding

In Fig. 9.22 are shown four different positions of welding depending on the location of the weld axis. The easiest position for work—that is, without need for a special electrode and with maximum ease and speed—is the flat position. Therefore, in steel workshops jigs and fixtures are used. A welding jig is a device capable of holding the component parts to be welded in the proper relative location and fit-up (Fig. 9.23). A fixture is similar to a jig, except that it permits changing the position of the work during welding so as to place the joint in the most convenient position.

9.1.3.1 Welding Procedures for Lowest Costs

For every welding job there is one best design and one best procedure. These two put together will result in the lowest possible cost.

Choice of steel

Although in Section 9.1.3.2 (Weldability of Steels) the choice of steel will be discussed more fully, for now the effect of steel composition on welding speed is stressed. The steel composition ranges best suited for maximum welding speeds (9.13) are shown in Table 9.2.

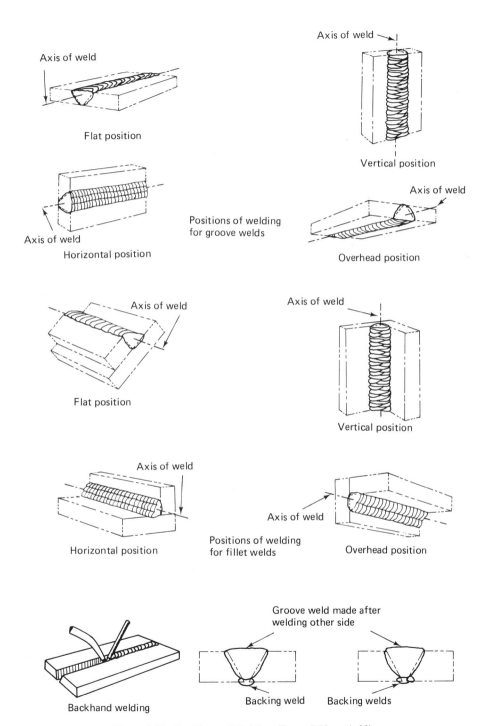

Figure 9.22 Positions of Welding (From 9.13, p. 1–33).

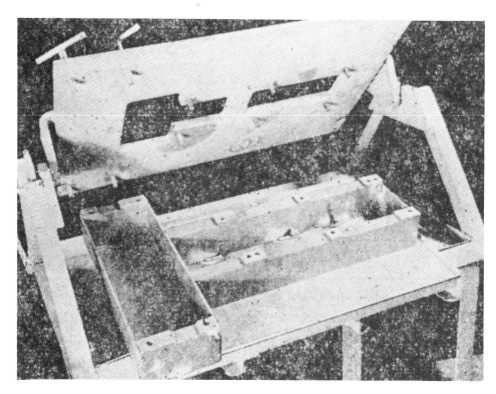

Figure 9.23 Simple Welding Jig (From 9.13, p. 2–8).

Table 9.2 STEEL COMPOSITION FOR WELDING PURPOSES

Component	Low	Preferred	High
Carbon	.10%	.13–.20%	.25%
Manganese	.30	.40–.60	.90
Silicon	—	$\leq .10$	.15 (max.)
Sulfur	—	$\leq .035$	.05 (max.)
Phosphorus	—	$\leq .03$	.04 (max.)

Type of weld

For civil engineering steel structures, as already explained, mainly two types of welds are used: fillet and groove welds. In Fig. 9.12 different groove welds are shown according to edge preparation. Basically the type of weld with a flame-cutting edge preparation is to be preferred. Single-V welds are easiest to fabricate. Still, they can be used only for relatively thin plates—$\frac{3}{8}$–$\frac{5}{8}$ in. (10–16 mm)—because the volume of the weld increases rapidly with the thickness. More welding means slower speeds, larger distortions, residual stresses (see Section 9.1.3.4), and more costly welds. The single-V weld is preferred when welding can be done from one side (9.23). Double-V

308 CONNECTIONS

welds need only one-half the welding metal for the same plate thickness and the same opening angle as single-V welds, but they require that welding from the back side be feasible. With larger plate thicknesses the double-V weld will require too much additional metal, and therefore a U-weld is used with slopes of 20° or smaller.

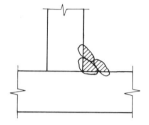

A weld must have a good penetration (see Fig. 9.8b) in the parent or base material along all contact surfaces and must be free from cracks, porosity, and slag inclusions. To get a fine-grain structure of the deposited metal several passes (Fig. 9.24), especially for groove welds, are preferred. Special attention must be given to the first pass at the root of the weld. In case of smaller thicknesses this first pass has to be done with thin wire $\frac{1}{8}$ in. (3 or 3.25 mm) in diameter.

It should be noted that for buildings the practical minimum size of fillet welds is $\frac{1}{8}$ in. (3 mm) (for bridges $\frac{3}{16}$ in.) and that the most economical welds are those made in a single pass, which are therefore preferable in spite of the fact that many passes generally improve weld metal quality. The maximum size of a fillet weld made in one pass is

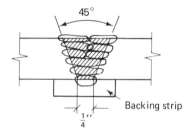

Figure 9.24 Welds with Several Passes.

1. $\frac{3}{8}$ in. (10 mm) in the flat position
2. $\frac{5}{16}$ in. (8 mm) in horizontal or overhead positions; or
3. $\frac{1}{2}$ in. (13 mm) in the vertical position

Electrode

For every job type the electrode, its size, the type and amount of current, and arc length and speed have to be chosen. Most of this information can best be obtained from appropriate handbooks (9.13, 9.20).

When considering procedure selection, for fast-to-fill joints submerged arc welding should also be considered. Although the deposition rates of hand welding are advancing, at present the maximum approaches 0.25 lb/min. The submerged arc-welding process's rate of deposition starts where hand welding stops. Also, the maximum welding current for manual welding corresponds to the minimum current for submerged arc welding.

9.1.3.2 Weldability of Steels

By "weldability" is understood ability to produce economical welds that will be sound and crack-free and will meet satisfactorily the engineering requirements of the joint. The chemistry and structure of the base metal and the weld metal are of

Introduction 309

prime concern here. The effects of the heating and cooling cycles associated with fusion welding are mainly localized in the weld metal and the heat-affected zone (HAZ). The HAZ—that is, the parent steel surrounding the weld metal and the weld itself—will exhibit various hardness distribution across a weld (perpendicular to the welding direction). The hardness in steel depends upon how rapidly the steel is cooled. The increase in hardness as the fusion zone is approached is explained by the fact that the maximum temperature becomes higher as the fusion zone is approached. The situation is made worse by the fact that this region, heated to the higher temperature, will generally have a faster rate of cooling than the regions not heated so high. Hardness means embrittlement and cracks in the HAZ or in the weld. A crack may occur during or after welding in a weld itself or in other parts of the HAZ. Good design and proper welding procedure, possibly with preheating of thicker parts, will prevent these cracking problems. Several factors affect weld cracking during the welding process, such as (9.24) joint restraint that causes high stresses in the weld, bead shape (convex or concave), carbon and alloy content of the base metal hydrogen and nitrogen pickup, and rapid cooling rate. The cracks in the HAZ depend mainly upon high carbon content, hydrogen embrittlement, and rate of cooling. Most steels can be welded at the average plate thickness without worrying about weld cracking. As plate thickness increases, however, weld cracks and underbead cracks (Fig. 9.25) may become problems and require special precautions.

9.1.3.3 *Weld Defects*

Deciding whether a weld is "good" or "bad" depends on which defects have been discovered by inspection and if such defects will affect weld performance in service. Inspection has to start with checking edge preparation, continue through the welding process, and finish with inspection of the final appearance of the weld. There are certain telltale signs which will reveal considerable information to a qualified inspector after the welding is done. Besides inspection, testing of test coupons (Fig. 9.26) in destructive tests is performed to establish the suitability of a proposed structural metal and to qualify the welding process or operator. Nondestructive testing by X-ray or gamma-ray radiography, magnetic flux, or ultrasonic inspection can detect welding defects such as cracks, lack of fusion, voids, and slag inclusion.

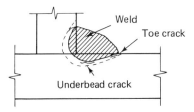

Figure 9.25 Underbead Cracking.

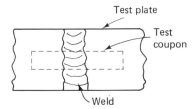

Figure 9.26 Test Coupon of a Weld.

9.1.3.4 *Distortion and Residual Stresses Due to the Welding*

Besides the residual stresses created by rolling processes in the steel mill, the heat cycle in welding will introduce either distortions or, if distortions are prevented, residual stresses (Fig. 9.27). It is not possible to restrain all parts of a fabrication, and some distortion is inevitable.

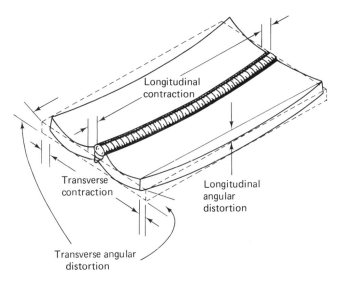

Figure 9.27 Effect of Single-side Welding of an Unrestrained Butt Joint (From 9.14).

Longitudinal contraction and distortion

Along the weld axis the weld tends to contract more than the parent metal, causing some distortion and residual stresses. The contractional stress is partially absorbed by the resistance of the parent metal, causing stretching of the weld metal, but the ultimate shrinkage may be of the order of $\frac{1}{80}$ in. per foot of weld (1 mm/m).

Transverse contraction

If a weld is made between two plates which are free to move, the plates will be drawn together by the transverse contraction of the weld metal. It will be seen that the gap between the plates will narrow and close up in advance of the arc. If the plates are thin and long enough, they eventually will "scissor" or overlap one another. The amount of drawing together depends upon the speed of welding: the greater the speed, the less the amount of movement of the plates. The shrinkage caused in the width of an assembly will be of the order of $\frac{1}{16}$ in. (1.6 mm) for a standard butt joint preparation.

Transverse angular distortion

As mentioned, the first run of deposited metal in a butt joint pulls the plates toward one another. The next run of deposited metal, on top of this, tries to do the same but is prevented by the first run, which would have to be compressed before the plates could come closer together at the root. As the second run contracts transversely it creates a "pull" at the weld face, and the resistance of the first run to compression creates a "push" at the root of the joint, causing the plates to lift out of alignment with one another (Fig. 9.28). Angular distortion will be increased (Fig. 9.29) in proportion to the number of runs deposited in the weld (approximately 1° per run). Balanced heat input (Fig. 9.30)—that is, balanced welding—will prevent angular distortion.

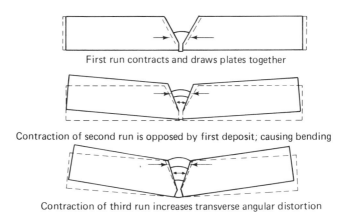

Figure 9.28 Angular Distortion in a Butt Weld Due to the Contraction of Weld Metal (From 9.14).

Figure 9.29 Angular Distortion Caused by Fillet Weld, Showing Increased Distortion with Multiruns with a Smaller Electrode to Produce the Same Size of Weld (From 9.14).

Lamellar tearing (see also Section 2.6)

Lamellar tearing of plates in excess of 1.5 in. (38 mm) in welded joints with tension directed through the thickness, as already discussed, has become quite a problem in the United States since the 1960s (9.25). In 1972 the AWS for the first time included a section on lamellar tearing in its code [AWS D1.1-72 (9.26)].

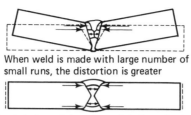

When weld is made with large number of small runs, the distortion is greater

When welding is balanced, heat input is balanced, eliminating transverse angular distortion and producing a slight contraction

Figure 9.30 Effects of Excessive Number of Runs and of Balanced Welding (From 9.14).

Lamellar tearing (Fig. 9.31) is due to the variation in strength in steel in different directions. The German specifications [DIN 4100 (9.22)] require that welded structural elements exposed to tension or bending tension in through-thickness be avoided unless sufficient strength in the direction of material thickness is demonstrated. The same specifications limit to 30 mm (1.18 in.) the thickness of flange plates directly welded to the web of plate girders and to 50 mm (1.97 in.) the thickness of any additional flange plate.

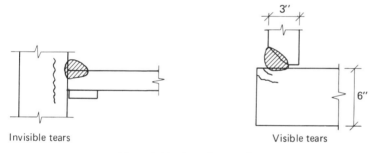

Invisible tears   Visible tears

Figure 9.31 Lamellar Tearing.

Welding sequence

Joint preparation, size of electrode used, welding current setting, run length per electrode, and sequence of welding a joint or number of joints in an assembly are factors which play a part in achieving a satisfactory welded fabrication. Good designers will specify the smallest-size welds that fufill the statical requirements, keeping in mind "balanced" welding.

It is always desirable to weld away from a point of restraint, and the welding of a joint should be done from the center line outward to each free end. The application of this principle and the "doubling-up" method to eliminate transfer angular distortion of a T-joints is shown in Fig. 9.32 for the deposition of a single-run fillet weld either side of the vertical member.

The "planned wandering" method for welding long butt joints with two operators is shown in Fig. 9.33. The other alternative is the "stepback method" (Fig. 9.34), which again should be evenly balanced about the center line of a joint.

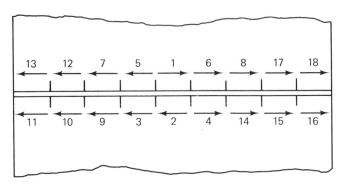

Figure 9.32 Sequence of Welding Fillet Welds on Either Side of a Vertical Member, One Operator, Doubling-up Method (From 9.14).

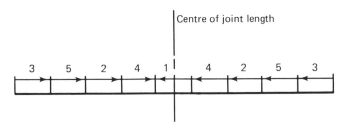

Figure 9.33 Planned Wandering Method for Welding Long Butt Joint with Two Operators; Welding Simultaneously after First Deposit (From 9.14).

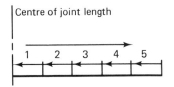

Figure 9.34 Stepback Method for Welding Long Butt Joint with Two Operators; One Either Side of Center (From 9.14).

Residual stresses can be measured mechanically by removing part of an internally stressed body, thus destroying the equilibrium of residual stresses. By measuring relative strains the corresponding stresses can be obtained. In Fig. 9.35 (9.27) the shrinkage stresses in a frame knee joint are shown.

Correction of distortions

Due to many still uncertain factors involved, distortion may occur even if the welding has been done in a planned sequence. Rectification of distortion usually can be achieved by mechanical means and the application of local heating, but only in cases in which distortion is not too excessive.

Stress relieving

When a welded structure is exposed to statical loading high residual stresses combined with service stresses may very soon bring weld zones to local plastic yielding. These plastic deformations cause the diminishing of weld stresses, and consequently

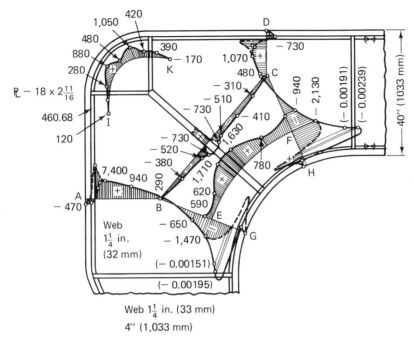

Figure 9.35 Shrinkage Stresses in kg/cm² in a Frame Knee Joint (From 9.27).

these stresses may completely vanish. The conditions for plastic yielding, however, vary depending upon particular geometry.

Although it is impossible to judge welding stresses with respect to their effect upon the statical strength without considering the structure itself, it is still possible on the basis of past experience to conclude that, in general, residual stresses due to welding have little or no effect on the statical strength provided the material has good mechanical properties.

If a welded structure is exposed to dynamic loading, the available experience is that in structures made of mild structural steel a large decrease in the fatigue strength due to welding stresses is not to be expected.

9.2 CONNECTION PERFORMANCE

9.2.1 Bolted Connections

As was pointed out earlier in this text, the performance of a bolted connection, where a group of bolts is transferring a certain force (force in general terms), is highly redundant and very complicated. Therefore, we shall first consider a single bolt exposed to shear and bearing and then extend this to a group of bolts. We shall consider a single bolt, although a single bolt is seldom used. Most specifications require a minimum of two or three bolts. The intent of this requirement is to avoid the possibility of a faulty bolt and subsequent complete connection failure and to avoid possible relative rotation of the members connected by a single faulty bolt, which will act as as a loose pin.

In this discussion loading will be assumed to be static, i.e., the force is applied slowly to reach its full magnitude without exhibiting any dynamic effects.

9.2.1.1 Single Bolt in Concentric Shear-Bearing–Type Behavior

A single bolt under a concentric shear would behave somewhat as a pin, except for differences introduced by possible initial tension of the bolt which would produce friction between the connected parts. If the load is sufficiently large to overcome such friction and produce failure, the failure can be of one of the following four types: (a) shearing of the bolt; (b) bearing failure in the zone of contact; (c) tension failure of the connected plate; or (d) tear-out of the plate behind the bolt (Fig. 9.36). Which type of failure will actually take place depends upon the dimensions and the relative strength of the bolt materials and those of the connected parts.

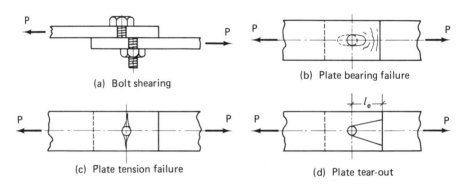

Figure 9.36 Four Types of Failure.

Shearing of the bolt: The shearing capacity of the bolt is determined by assuming a uniform stress distribution over the cross section of the bolt. Assuming P is the load to be carried by one single bolt, then the uniform shear stress is

$$f_v = \frac{P}{m(\pi d^2/4)} \tag{9.4}$$

where P = load carried by bolt
d = nominal diameter of bolt
m = number of active shear planes (usually one for *single shear* and two for *double shear*)

The AISC specification (9.1) in Table 1.5.2.1 (p.5-24) gives the following allowable shear stresses in bearing-type connections: for A307 bolts, 10.0 ksi (69.0 MPa); for A325 bolts with threads included in shear planes (N-A325), 21.0 ksi (144.8 MPa); and when threads are excluded from shear planes (X-A325), 30.0 ksi (206.8 MPa). For A490 bolts the comparable values are 28.0 ksi (193.1 MPa) and 40 ksi (275.8 MPa) respectively. In friction-type connections with standard size holes (hole diameter = $d + \frac{1}{16}$ in., where d is the nominal bolt diameter) the allowable shear stress for A325 bolts is 17.5 ksi (120.7 MPa), while for A490 it is 22.0 ksi (151.7 MPa). These provisions for the shearing stresses are largely based on empirical rules. Unfinished A307 bolts cannot be used in friction-type connections.

All stresses in high-strength bolts, either computed, given, or allowable, are *nominal* and not *real* stresses. Because of the high clamping force, the service load is transmitted by friction and not by bolt shearing. This is illustrated by the following evaluation of the friction forces. Although friction coefficients varying from about 0.20 to 0.60 have been observed in tests of joints having unpainted faying surfaces and a tight mill-scale covering (9.29), a value of 0.35 may be taken as typical for a conservative estimate of the limit of frictional resistance. Using this value according to the AISC *Manual* (9.2), table 3 (p. 5-214) for a 1-in. diameter A325 bolts, the minimum fastener tension is 51 kips and a friction force of $0.35 \times 51 = 17.8$ kips is obtained. The allowable shear force (9.2, p. 4-5) for 1-in. diameter F-A325 bolts is 13.7 kips, and 16.5 kips for N-A325 bolts. The computed value based on friction is larger than the force based on nominal shearing stress.

Bearing of the bolt: Bearing or contact stresses are nominal stresses computed by dividing the load on the bolt by the product of its diameter and the thickness of the connected parts. The distribution and magnitude of the true bearing stresses (Fig. 9.37) show a high local compressive stress, f_{cm}. On the basis of elastic theory, H. Hertz (9.30) gives the following expression for this stress at the center of two cylindrical contact surfaces with radii $d_1/2$ and $d_2/2$ (9.31) and the length of the cylinders assumed as infinite:

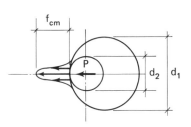

Figure 9.37 Distribution of Bearing Stresses.

$$f_{cm} = 0.789 \sqrt{\frac{PF}{2(1-\mu^2)} \cdot \frac{d_1 - d_2}{d_1 d_2}} \tag{9.5}$$

where P is the load per unit length. For $E = 29{,}000$ ksi (Young's modulus), $\mu = 0.3$ (Poisson's ratio), and d_1 and d_2 (diameters of the hole and the bolt, respectively), the stress f_{cm} in ksi becomes

$$f_{cm} = 100 \sqrt{P\left(\frac{d_1 - d_2}{d_1 d_2}\right)} \tag{9.6}$$

where P is in kips/inch and d_1 and d_2 are in inches. This means that the bearing pressure increases as the clearance between the hole and bolt increases.

If $d_1 = d_2 + 1/16$ in. then Eq. (9.6) can be written as

$$F_{cm} = 100 \sqrt{\frac{P}{16d^2 + d}} \tag{9.7}$$

For an A325 bolt of $d = 1$ in. and a material thickness of 1 in., the AISC *Manual* (9.2, p. 4-6) gives a force $P = 72.5$ kips for a normal distance of $3d$ between bolts. With this value, the maximum stress f_{cm} becomes

$$f_{cm} = 100\sqrt{\frac{72.5}{16+1}} = 206.5 \text{ ksi} \qquad (9.8)$$

Equation (9.5) was developed for infinitely long cylinders and unlimited elasticity of the material, and therefore should be modified before applying it to cylinders with a finite length and real material. Still, the high value obtained in Eq. (9.8) indicates that local contact (bearing) pressures lie in the yielding range. The accompanying shearing stresses acting on oblique planes and producing yielding in shear are also very important. Because of this yielding and consequent plastic deformation, the contact zone is widened and the clearance between hole and bolt is largely reduced. Consequently, the bearing pressure will drop. Therefore, after initial yielding, equilibrium will be established, with moderate contact (bearing) pressure stresses. The bolt material is confined in the hole and cannot fracture. Consequently, bearing failure happens in the material in which the hole is made (parent or basic material) when deformation due to crushing becomes excessive. In a connection transmitting a tensile force the tensile strength is not impaired when the computed bearing stress at working load is as much as 2.25 times the allowable tensile stress of the connected material. For this reason the previous AISC specification (Feb. 1969) allowed bearing stress of $1.35F_y$, which is approximately 2.25 times $0.6F_y$, the allowable tensile stress. Tests have shown (9.32) that a linear relationship exists between the ratio of critical bearing stress to tensile strength of the connected material and the ratio of fastener spacing (in the line of force) to fastener diameter. The following equation affords a good lower bound to published test data for single-fastener connections with standard holes, and is conservative for adequately spaced multifastener connections

$$\frac{F_{p,\text{cr}}}{F_u} = \frac{l_e}{d} \qquad (9.9)$$

where $F_{p,\text{cr}}$ = critical bearing stress, F_u = tensile strength of the connected material, l_e = distance, along a line of transmitted force, from the center of a fastener to the nearest edge of an adjacent fastener, or to the force edge of a connected part (in the direction of stress), and d = diameter of a fastener. For $l_e = 3d$ the critical bearing stress is $3F_u$. With a factor of safety of 2.0, the allowable bearing stress is $F_p = 1.5F_u$. the value given in the 1978 AISC specification. Section 1.5.1.5.3 of this specification (9.1) establishes this value for the maximum allowable bearing stress

$$F_p = 1.5F_u \qquad (9.10)$$

Spacing and/or edge distance may be increased to provide the required bearing stress, or the bearing force may be reduced to satisfy a spacing and/or edge distance limitation. Thus, the specification provides for adjustment of spacing and/or edge distance in the direction of stress in terms of calculated bearing stress or vice versa, rather then providing a single criterion. A safety factor of 2.0 is always used. Accordingly, the AISC *Manual* (9.2) on pp. 4-6 to 4-9 gives the bearing capacities related to four different fastener spacings: 3 in., minimum spacing to obtain full bearing capacity, preferred spacing of $3d$, and absolute minimum spacing of $2\tfrac{2}{3}d$.

Tension of the connected plate: This is discussed in Chapter 4, Design of Tension Members.

Tear-out of the plate: The end distance l_e in the direction of stress (Fig. 9.38) required to prevent the tear-out of the plate may be obtained by consideration of the plate shear strength and the load transmitted by the end bolt. If we take (for simplicity

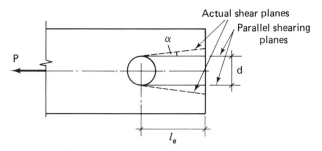

Figure 9.38 Plate Tear-out.

and greater security) two parallel planes with $\alpha = 0$ as the shearing planes and assume that the plate shearing strength, F_{us} equals

$$F_{us} = \frac{F_u}{\sqrt{3}} \approx 0.6 F_u \tag{9.11}$$

(although this relation is only valid for yielding in a stress state of pure shear, obtained for Huber-Hencky-Mises yield condition), the necessary plate tear-out force for a given plate thickness t becomes

$$P = 2t\left(l_e - \frac{d}{2}\right)0.6 F_u \tag{9.12}$$

The force has to be equal to or larger than the bearing strength of the bolt

$$P \geq F_p dt \tag{9.13}$$

i.e.,

$$F_p \leq \frac{P}{dt} \tag{9.14}$$

Solving for l_e/d from Eq. (9.12) and Eq. (9.13) we have

$$l_e/d \geq 0.5 + 0.83(F_p/F_u) \tag{9.15}$$

Eq. (9.15) can be approximated by the following simpler expression (as a lower bound for any $l_e/d \geq 2$):

$$\frac{l_e}{d} \geq \frac{F_p}{F_u} \tag{9.16}$$

the same as the expression given in Eq. (9.9). Introducing for F_p the value from Eq. (9.14) and taking 2.0 as a factor of safety, the final relation becomes

$$l_e \geq \frac{2P}{F_u t} \tag{9.17}$$

where l_e is the distance between the center of a bolt and the nearest edge of an adjacent fastener or the edge of the connected plate in the direction of the force.

The AISC specification, section 1.16-4 gives the above expression as equations (1.16-1) and (1.16-2). To limit excessive hole deformation, AISC also gives the limit of the nominal bearing stress as $1.5F_u$ [see Eq. (9.10)]. When this value is substituted into Eq. (9.16) increased by a factor of safety of 2.0, the upper boundary of the distance l_e becomes $3.0d$. Without the factor of safety, or the lower bound (the minimum distance) $l_e = 1.5d$, the traditional minimum distance in steel design practice.

Friction-type behavior

Figure 9.39a shows a single high-strength bolt connection transmitting a tensile force P and Fig. 9.39b shows its free-body diagram. By pretensioning the bolt up to the proof load T specified by the AISC specification (9.2, table 3, p. 5-214) by one of the three methods explained earlier in Section 9.1.2.2 of this text, a clamping compressive force C of equal magnitude is created. Owing to this compression of the

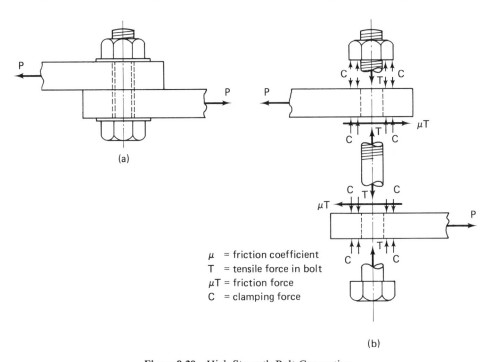

μ = friction coefficient
T = tensile force in bolt
μT = friction force
C = clamping force

Figure 9.39 High Strength Bolt Connection.

plates in their faying surfaces and depending upon the friction coefficient μ, a corresponding friction force μT is present. The friction coefficient μ depends on the surface condition of the connected plates, as already discussed on page 317, with a conservative value of 0.35. To avoid any slip, we must have

$$\mu T \geq P \qquad (9.18)$$

Taking the nominal shear stress in friction type A325 bolts in standard holes ($d + \frac{1}{16}$ in.) as $F_v = 17.5$ ksi (120.7 MPa) and the proof load stress (for diameters of $\frac{1}{2}$ to 1 in.) as 65 ksi (448.2 MPa) the friction stress is 22.8 ksi (157.2 MPa). Therefore, the factor of safety against slip is

$$\frac{22.8}{17.5} = 1.3 \tag{9.19}$$

which is considered sufficient in view of the conservative friction coefficient μ.

9.2.1.2 Bolt Groups in Concentric Shear

A bolt group is in concentric (axial) shear if the line of action of the resultant force, which has to be transmitted, passes through the center of gravity of such a group.

If the friction between the faying surfaces of connected parts is never overcome by the force (this is the design criterion for the friction type connection) then each bolt of the group will carry an equal share and the connection will behave quite elastically. If the friction is overcome and permanent slip occurs, the bolt shank and the surface of the hole come into contact and the shearing and bearing stresses become real stresses and are no longer nominal stresses. In that case the first and the last bolts of the group carry a much larger share. An additional 65% is carried by these bolts when the number of bolts in one line is 6 or more. Hertwing and Peterman (9.33) have measured the force distribution of riveted connections, which behave similarly to bearing-type bolt connections. Figure 9.40 shows their results for different numbers of rivets in one row for the normal case that the cross-sectional area of the splice is equal to the area of the connected parts. For 6 rivets in one row the end rivets carry half of the load. There have been many attempts to determine analytically the distribution of forces in individual fasteners, mostly rivets, and De Jonge (9.34) summarizes these efforts. The inelastic slip and local yielding of both the connected parts and the bolt shanks or rivets make any mathematical analysis based on elasticity impossible. Extensive experimental work has established that inelastic slipping and local yielding tend to equalize the forces so that at failure practically equal participation of fasteners exists. This observation serves as the basis for the design of bolt groups exposed to concentric shear.

9.2.1.3 Bolt Groups in Torsion and Eccentric Shear

When the resultant force acting on a bolt group does not pass through the center of that group and imposes only shear or bearing stresses in bolts (no tension in bolt shanks), the eccentric load may be replaced by an equivalent force and a couple acting at the centroid of the bolt group. This case is demonstrated by a typical bracket connection to the flanges of a column (Fig. 9.41), transmitting crane girder reactions P and H, or by beam-web splices (Fig. 9.42) transmitting a bending moment M and a shear force V acting at the web joint.

The *classical (elastic) approach* assumes rigid bracket or splice plates and separately considers the action of the shear forces (vertical and/or horizontal) and the moment. The shear forces are divided equally among the bolts producing vectorial

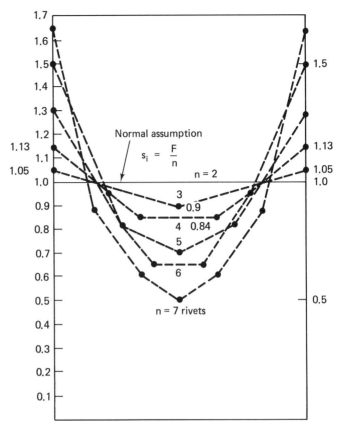

Figure 9.40 Forces in Rivets When $A_{\text{splice}} = A_{\text{bar}}$ (From 9.33).

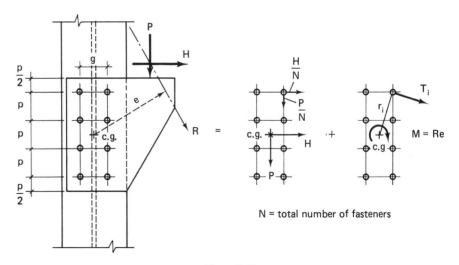

Figure 9.41

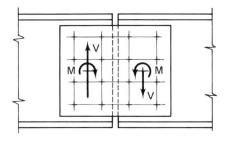

Figure 9.42

horizontal and vertical components. The moment is considered as a torsional moment tending to rotate the plate and the opposing bolts around the center of gravity of the bolt connection. The forces in the bolts therefore are proportional to the bolt distances r_i as measured from the centriod acting normal to their radial distances. As the bolts and plate are elastic and not rigid, as assumed, they will have translatory and rotational displacements. Ignoring these translations and fully imposing only rotational displacements produces a very conservative assessment of the final forces in the bolts when the force components are summed vectorially.

Research during the last decade (9.32, 9.35, and 9.36) introduced *ultimate strength analysis* based on a factor of safety of 2.5, consistent with that used in other types of connection design. In the ultimate strength method the translation and rotation displacements are reduced to a pure rotation about a point called the instantaneous center of rotation (Fig. 9.43). For two-dimensional equilibrium

$$\sum_{i=1}^{N} R_i \sin \theta_i = 0$$

$$P - \sum_{i=1}^{N} R_i \cos \theta_i = 0 \qquad (9.20)$$

$$P(l + r_0) - \sum_{i=1}^{N} R_i \times r_i = 0$$

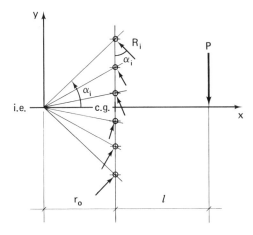

Figure 9.43

The deformations Δ_i of bolts are still taken proportional to the distances r_i from the instantaneous center of rotation. This center has to be found by trial-and-error until Eqs. (9.20) are approximately (within 2–3%) satisfied. The correlation between the load on the bolt and its deformation is expressed as (9.37)

$$R_i = R_{\text{ult}}(1 - e^{-10\Delta_i})^{0.55} \tag{9.21}$$

where R_i = shear force in a single fastener at any given deformation (Fig. 9.44)
   R_{ult} = ultimate shear load of a single fastener
   Δ_i = total deformation of a fastener, including shearing, bending, and bearing deformation, and local bearing deformation of the plate
   e = base of natural logarithms = 2.718...

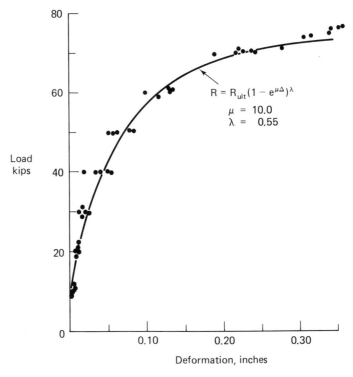

Figure 9.44 Load-Deformation Response of Bolts (From Crawford and Kulak, *Journal of the Structural Division*, American Society of Civil Engineers, Vol. 97, No. ST 3, March 1971, p. 772).

The maximum deformation $\Delta_{\max}$ was experimentally obtained as 0.34 in. By applying this deformation to the fasteners most remote from the instantaneous center, ($r_{\max}$), the maximum shear force $R_{\max}$ for these fasteners can be computed. For fasteners in less remote locations the deformations Δ_i are computed from

$$\Delta_i = 0.34 \frac{r_i}{r_{\max}} \tag{9.22}$$

while the corresponding forces R_i can be found from Eq. (9.21). The ultimate shear load for A325 fasteners is obtained using the ultimate bolt shear stress $(f_v)_{max}$ and the corresponding bolt cross section. $(f_v)_{max}$ is taken as about

$$(f_v)_{max} = 0.62 F_u = 0.62(120) = 74.4 \text{ ksi} \tag{9.23}$$

The elastic method can always be applied for a safe but conservative design giving a factor of safety ranging from 2.5 to 3 with 1978 AISC bolt stresses. The modified elastic method cannot be used with the 1978 AISC stresses allowed in the specifications because it will give a factor of safety of only about 2.0. The ultimate method results in a factor of safety of 2.5. In the AISC *Manual* (9.2), Tables X to XVIII have been prepared based upon using the instantaneous center concept for each fastener pattern and each eccentric condition. The nondimensional coefficient C was obtained by dividing the ultimate load P by R_{ult}. The derived values can be safely used with any fastener diameter and are conservative when used with ASTM A490 bolts. By use of these tables, margins of safety are provided equivalent to those obtained for bolts used in joints less than 50 in. long, subject to shear produced by concentric load in both friction-type and bearing-type connections (9.2, p. 4-60).

9.2.1.4 Single Bolt in Tension

To understand the effect of an applied axial tension on a pretensioned high-strength bolt, first consider the simplest possible arrangement, a single bolt clamping two plates tightly together. In Fig. 9.45 a single bolt with the contributary portion of the connected plates is shown. The joined plate pieces have an initial thickness t and an effective area of contact between the plates of A_p. The cross section of the bolt is A_b. Before tightening the bolt to the pretensioned force T_i (when the bolt is in a snug-tight position) the increase in bolt length and decrease in the plate thickness of $2t$ are zero (Fig. 9.45a). When the pretensioning force T_i is introduced by any of the three methods discussed previously (Section 9.1.2.2) the shank of the bolt is increased by ΔL_{bi} (Fig. 9.45b) which is proportional to the introduced force T_i and

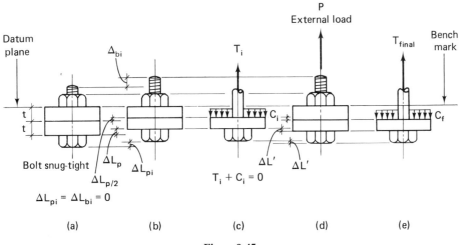

Figure 9.45

the bolt length between the nut and the bolt head, $2t$. If K_b is the spring constant for the bolt (the force necessary to produce a unit extension of the bolt) then

$$T_i = K_b \cdot \Delta L_{bi} \qquad (9.24)$$

Due to this force T_i the plates will be squeezed together an amount ΔL_{pi} which, by experience, is much less than ΔL_{bi}.

If k_p is the plate's spring constant then

$$C_i = -k_p L_{pi} \qquad (9.25)$$

As there are no other forces acting, force equilibrium requires that (see the free-body diagram, Fig. 9.45c)

$$T_i + C_i = 0 \qquad (9.26)$$

or

$$k_b \Delta L_{bi} - k_p \Delta L_{pi} = 0 \qquad (9.27)$$

The force C_i is acting on a much larger effective area of plates (a circular area of about 3 bolt diameters) than the bolt force T_i (a circular area of one bolt diameter); therefore, even for the same elastic properties of plate and bolt materials, as the forces are opposite and equal

$$k_p \gg k_b \quad \text{and} \quad \Delta L_{pi} \ll \Delta L_{bi} \qquad (9.28)$$

In Fig. 9.46 the plots of Eqs. (9.24) and (9.25) are shown as two straight lines OA and OB, respectively, with $|\Delta L_{pi}| \ll \Delta L_{bi}$. If now an external force P, tending to separate the plates, is applied two different situations may arise:

1. $P \geq T_i$. In this case the plates will separate, $C_{\text{final}} = 0$, and the bolt will carry the whole external force P in addition to its initial tensile force T_i. This definitely will produce yielding of the bolt and collapse of the connection.
2. $P < T_i$. The force will only reduce the pressure between the plates and the bolt shank, and total plate thickness will increase by the same amount $\Delta L'$, because they remain in contact. This means that the compression in the plates will decrease by an amount $\Delta C = k_p \Delta L'$ (positive quantity) while the tensile force in the bolt will increase by $\Delta T = k_b \Delta L'$, or

$$\Delta L' = \frac{\Delta T}{k_b} = \frac{\Delta C}{k_p} \qquad (9.29)$$

From this it follows that

$$\Delta C = \frac{k_p}{k_b} \cdot \Delta T \qquad (9.30)$$

Equilibrium at this final stage (see Fig. 9.46) requires that

$$P = \Delta T + \Delta C = \left(1 + \frac{k_p}{k_b}\right) \Delta T \qquad (9.31)$$

Therefore

$$\Delta T = \frac{P}{1 + k_p/k_b} \qquad (9.32)$$

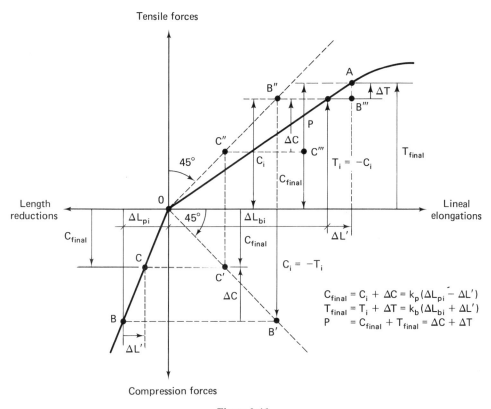

Figure 9.46

For the same material the spring constants are proportional to the corresponding effective areas: A_p of the plates and A_b of the bolt. In this case the final expression for increase in the bolt tensile force is

$$\Delta T = \frac{P}{1 + A_p/A_b} \tag{9.33}$$

If the ratio A_p/A_b is taken approximately as 9, then

$$\Delta T \approx 0.10 P \tag{9.34}$$

Thus until the applied load, P, exceeds the bolt pretensioned force, T_i, the increase in bolt tension is rather small, only about one-tenth the applied tensile load. For high-strength bolts the initial tension force is quite large, so they can be loaded in direct tension and still perform satisfactorily. Unfinished ASTM A307 bolts have a small pretensioned force which is normally less than the force P caused by service loading and the plates will lose compression. Therefore, such a connection will behave quite differently from one using high-strength bolts and can be applied only if the loads are small, i.e., in so-called light steel construction.

9.2.1.5 Concentric Tension on Bolt Groups

In the introduction of this chapter (Section 9.1) and in Fig. 9.4a the T-hanger connection was given as an example of a connection subjected to concentric tension. If the inner part of the flange of a T-section (Fig. 9.47a) is flexurally very stiff, its deformation due to the applied load of $4T$ relative to the bolt extension is very small and no prying action or compression at the flange tips will occur. On the contrary, a very flexible flange (Fig. 9.48a) due to the deformation caused by bending will press the support with its outer portions and develop contact stresses resulting in two prying forces Q acting somewhere near the flange tips. Once established, these forces can be

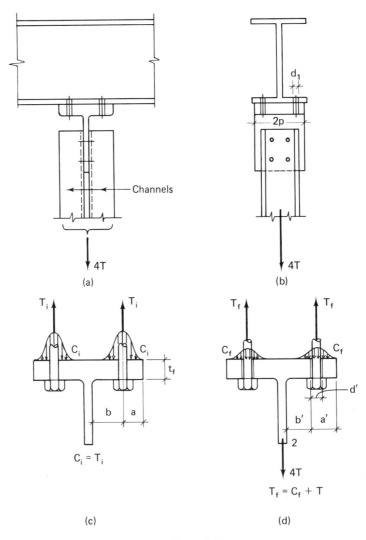

Figure 9.47

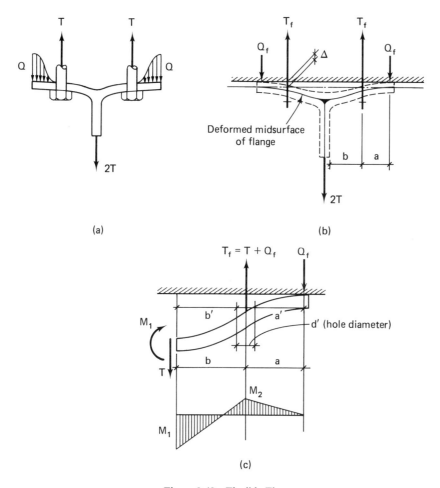

Figure 9.48 Flexible Flange.

active until the collapse occurs. As the bolt forces must be the equilibrating forces resisting the action of the applied load of $4T$ and the prying forces $2Q$, it is obvious that prying forces reduce the bolt capacity and, therefore, must be taken into account. Let us now consider the behavior of a T-hanger connection for both cases, i.e., when the flange is very stiff and when it is very flexible.

Very stiff flange: We consider a T-connection with four high-strength bolts (Fig. 9.47a, b) connecting a T-section to a beam. The flange of this T-section is very stiff. When the bolts are tightened to provide at least the minimum tension T_i as prescribed by the specification (9.38), the localized contact pressure C_i, between the flange and the support surfaces, symmetrically distributed around the bolt, balances the pretension force T_i, as shown in the free-body diagram in Fig. 9.47c. When the pulling force $4T$ is now applied, the contact pressure C_i will decrease, lose its symmetry, but remain still localized as shown in Fig. 9.47d. The flange thickness originally was reduced by a small amount Δ_i equal to

$$\Delta_i = \frac{C_i}{k_p} \tag{9.35}$$

where k_p is the spring constant of the flange. This reduction will also decrease, i.e., the flange thickness will increase, after the pulling force is applied. The flexural deformation of the inner part of the flange will be extremely small, practically remaining straight, so that its tips will not be pressed against the support. If the applied force $4P$ is sufficiently large the compression forces C_f may decrease to zero and the original flange thickness t_f will be maintained. Separation between the upper surface of flange and the support will occur because of bolt extension. In that case it is valid to assume a statically determinate distribution of the external force $4T$ to all four bolts and the final force T_f in each bolt is

$$T_f = T \tag{9.36}$$

Figure 9.47d shows an intermediate stage when the compression force C_f is not equal to zero but

$$C_f \ll C_i \tag{9.37}$$

Very flexible flange: A much more complicated behavior will take place if the flange is very flexible. Using a simple model, part of this complex problem is qualitatively explained. The connection now consists of only two bolts. In the initial stage of zero external load we have as before

$$T_i = C_i \tag{9.38}$$

When an external load of $2T$ is applied, the flange is bent (Fig. 9.48c). If the outer flange portions $2a$ remain pressed against the supporting surface, prying forces Q develop. The distribution of these pressures is not known, as they are greatly affected by the local surface roughness and localized yielding. Tests have shown (9.39) that the contact is concentrated near the flange tips, at least in cases of thin flanges where the prying action is of significance. The force in bolt $T_f > T_i$ must be equal to

$$T_f = T + Q_f \tag{9.39}$$

when separation between the flange and support surface at bolt location occurs. At that moment, the thickness of flange t_f will have been completely recovered and the bolt elongation will be equal to the deflection of the midsurface of the flange. The bolt elongation now is

$$\Delta = \frac{T + Q_f - T_i}{k_b} \tag{9.40}$$

where k_b is the spring constant of the bolt. If the prying force Q_f and the applied load T are uniformly distributed along the line distance p (Fig. 9.47b) and the flange behaves as an ordinary beam, the deflection of its midsurface at the bolt line can be found. The elimination of deflection Δ from that expression and Eq. (9.40) yields

$$Q_F = \left[\frac{\frac{1}{2} - \frac{Ept_f^3}{12ab^2 k_b}\left(1 - \frac{T_i}{T}\right)}{\left(\frac{a}{b}\right)\left(\frac{a}{3b} + 1\right) + \frac{Ept_f^3}{12ab^2 k_b}} \right] \cdot T \tag{9.41}$$

This equation shows at least qualitatively those factors which control the magnitude of Q_F. The prying force Q_F cannot be negative. Consequently, when the second term in the numerator becomes larger than one-half, no prying force will be developed. Prying decreases with the increase of the flexural stiffness of the inner part of the flange as well as with an increasse in the outer span a (increase in moment M_2, Fig. 9.48c). This beneficial effect of an increase in a is limited because for relatively large values of a the contact zone is no longer concentrated at the flange tips and the prying force cannot be treated as a line load. The initial bolt tension T_i has an influence as well as the bolt spring constant k_b.

As the model used represents a very simplified prying process, empirical modifications of Eq. (9.41) are necessary before it can be used in practical design. The AISC Manual (9.2) on p. 4-89 gives the following equation for the force per bolt including prying action in kips, B_c, when the applied tension load per bolt is T:

$$B_c = T\left[1 + \frac{\delta\alpha}{(1+\delta\alpha)}(b'/a')\right] \qquad (9.42)$$

where $\delta = 1 - d'/p$, $\alpha = (Tb'/M - 1)\delta$, and $M = M_p/2 = pt_f^2 F_y/8$.

The notation used is explained in Figs. 9.47 and 9.48. The following assumptions were made in deriving the above equation. The factor of safety for bolts in tension and tee flanges in bending is ≥ 2.0. Required full plastic moment in bending of the flange is $pt_f^2 F_y/4$. The final force in the bolts shifts from the bolt center line to the stem side of the bolts by a distance of $d'/2$. Moment M_1 (Fig. 9.48c) is always larger than moment M_2, and the outer span $a < 1.25b$. An application of this method is given in Example 9.12.

9.2.1.6 Bending and Eccentric Tensile Forces on Bolt Groups

In beam-to-column connections (Fig. 9.49) the bolts are exposed to combined shear and tension; this topic will be discussed in the next section (9.2.1.7). Here, we want to consider only the action of a bending moment. A similar situation exists in the bracket connection to a column web shown in Fig. 9.50. The top bolts in the

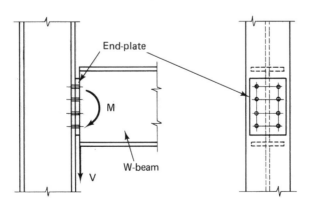

Figure 9.49 Beam-to-Column End-plate Connection.

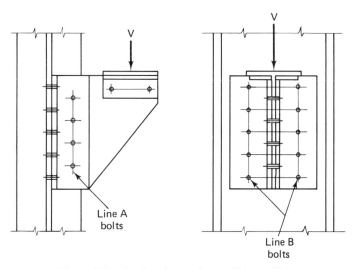

Figure 9.50 Bracket Connection to Column Web.

column web (bolts along lines *B*) under the action of a bending moment will have axial tension of varying intensities while the bottom part of the connection material (not only bolts but also the legs of the angle) will be in compression. The distribution of these stresses in bolts and contact surfaces will depend upon the initial compressive stress in the contact area between the web and the legs of the angle created by pretensioning of the bolts. If the bolts used for this connection are high-strength bolts, either A325 or A490, then this precompression can be computed more or less accurately from the initial bolt tension T_i. In the unloaded condition, each bolt is tensioned to T_i producing the equilibrating contact stresses represented by their resultant force, $C_i = T_i$ as shown in the free-body diagram (Fig. 9.51a). When the moment is applied the change of stresses will depend mainly upon the plate stiffness in Fig. 9.49 or the angle stiffness in Fig. 9.50. Neglecting local deformations of either one due to bending and applying the concept of plane sections, the center of rotation will be at the centroid of the group of bolts. This rotation will change both the forces in the bolts and in the contact area. Having in mind the behavior of a single bolt in tension (Section 9.2.1.4) and the diagram (Fig. 9.46) representing this behavior, we know that the change in plate compression forces (ΔC on the diagram) will be much larger than the change in the bolt tension, T. As long as the force introduced by moment action is smaller than T_i there will be no separation and this statement is true. Therefore, ΔT can be disregarded and it can be assumed that the change occurs only in the compression forces. For a clockwise applied moment, the maximum reduction in the compression force C_i will take place at the top bolt, C_m. If the total number of bolts is N, then the expression

$$I = \sum_{i=1}^{N} (d_i)^2 \tag{9.43}$$

where d_i is distance of any bolt from the centroidal axis, has the meaning of the

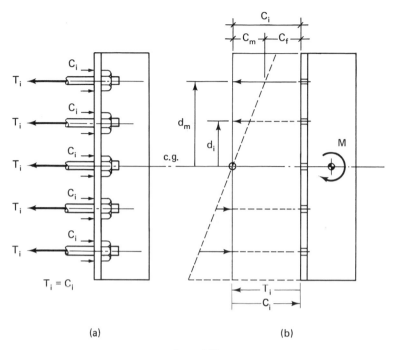

Figure 9.51

moment of inertia of bolt cross sections having a unit area. The maximum loss of compression around the top bolts, C_m, is

$$C_m = \frac{M}{I} \cdot d_m \qquad (9.44)$$

If $C_m \leq C_i$ no separation of contact surfaces will occur, the connection remains tight, and the corresponding increase in the bolt tension force ΔT is relatively small. The final force in the bolt will be only slightly greater than the initial tension.

The method is conservative in assuming the plane section will remain plane. Either the end plate or the legs of the angles will be deformed, at least during the transmission of the bending moment to the plates. While the beam of Fig. 9.49 is still elastic, the stress will vary linearly from mid-depth. Due to the flanges, there will be concentrations of force at the top and bottom of the beam. The compressive forces in the flange and portion of the web may be transmitted by bearing without affecting the bolts. The tensile forces must, however, be transmitted by the bolts. Since the tension flange is below the bolt group, the force in the flange must be carried to the bolts by the butt plate, causing it to bend out of its plane (9.29, p. 848). This definitely will affect the force distribution to the bolts. However, at low to moderate loads acting on pretensioned bolt groups, the assumption of invariable tension in the bolts and linear variation of contact stress from the centroid of the group is probably a fair approximation of actual behavior (9.29, p. 850).

When the bolts used in the connection are A307, unfinished bolts, the pretensioning force is relatively small and of variable, uncontrolled intensity. In that case, separation of planes will occur under moderate loads and the force distribution in bolts will be different from that occurring when using high-strength bolts. The resisting moment is formed by tensile bolt forces and compressive contact forces (Fig. 9.52). When the moment is applied, it is assumed that stresses vary linearly from the centriod of the effective areas. As a result the upper bolts will be in tension and the lower portion of the plate in compression. The maximum tensile force in bolts under the working load moment is computed from the stresses and the bolt's contributary areas. If this force does not exceed the allowable bolt capacity, the connection is properly designed.

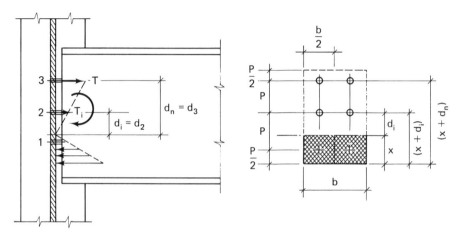

Figure 9.52

The neutral axis is found by taking moments about the lower edge of the compression zone

$$\frac{bx^2}{2} = \sum_{i=k}^{n} A_i(x + d_i) \tag{9.45}$$

where $n =$ number of bolts in one vertical line,
$k =$ the number of the first row of bolts above the neutral axis

(in Fig. 9.52 $n = 3, k = 2$). The A's are the total bolt areas in each row. The location of the neutral axis, i.e., between which bolt rows, has to be determined by a trial-and-error procedure. Once x is found, the moment of inertia of the effective area is

$$I = \frac{bx^3}{3} + \sum_{i=k}^{n} A_i(d_i)^2 \tag{9.46}$$

and the maximum force in a bolt is given by

$$T_{max} = \frac{M}{I} A_n \cdot d_n \tag{9.47}$$

To find the *ultimate resistance* of a high-strength bolted moment connection (Fig. 9.53) a tensile stress in all the bolts no higher than their elastic proof load is assumed, and the yield point stress, F_y, is on the compression area. Equilibrium of horizontal forces requires that

$$F_y bx = 2 \sum_{i=k}^{n} T_i \tag{9.48}$$

where T_i = elastic proof load, and k and n as before for two bolts in each horizontal row. Knowing T_i, F_y, and b the position of the neutral axis is easily found, after a proper estimate for k.

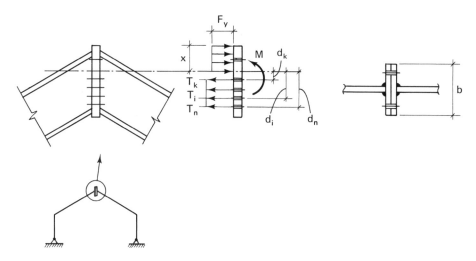

Figure 9.53

The external applied ultimate moment M_u must equal

$$M_u = \frac{F_y bx^2}{2} + \sum_{i=k}^{n} T_i d_i \tag{9.49}$$

If the material which transmits the moment to the bolts is rigid enough, M_u obtained from Eq. (9.49) will yield a conservative estimate. If the material which transmits the moment to the bolts is not sufficiently rigid, more stressing of the outside bolts than indicated by the assumed linear strain variation will occur and M_u, calculated from Eq. (9.49), will become unconservative. In the case of extreme flexibility of the butt plates, only the outermost bolts will actually oppose the moment at high loads. If the strain in these bolts is more than the material can carry, bolts will crack. Next the adjacent pair of bolts will pick up the load, which is now larger than the previous one (due to the reduction in $d_{n-1} < d_n$), and this pair of bolts will crack, resulting in progressive bolt collapse until the whole connection is destroyed.

9.2.1.7 Combined Shear and Tension on Bolts

In many types of connections such as the end connections of a bracing diagonal (Fig. 9.54) or beam-to-column connection (Fig. 9.49), bolts are exposed to combined tension and shear. Tests have shown (9.40) that the ultimate strength of a rivet (or bolt) could be presented by an elliptical interaction equation

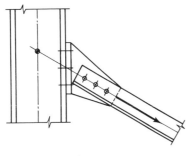

Figure 9.54

$$\left(\frac{f_v}{F_{vu}}\right)^2 + \left(\frac{f_t}{F_u}\right)^2 = 1 \qquad (9.50)$$

where f_v and f_t are the shearing and normal components of stress at failure under the combined action of shear and tension and F_{vu} and F_u are the failure stresses under pure shear and tension, respectively. On the basis of the above ultimate stress relation both the AISC and the AASHTO specifications give working stress versions for different types of rivets and bolts. The AISC specification gives (for bearing-type connections) the allowable tension stress F_t as reproduced in Table 9.3.

Table 9.3 ALLOWABLE TENSION STRESS (F_t) FOR FASTENERS IN BEARING-TYPE CONNECTIONS

Description of Fastener	Threads *Not* Excluded from Shear Planes	Threads Excluded from Shear Planes
A325 bolts	$55 - 1.8 f_v \leq 44$	$55 - 1.4 f_v \leq 44$
A490 bolts	$68 - 1.8 f_v \leq 54$	$68 - 1.4 f_v \leq 54$
A307 bolts	$26 - 1.8 f_v \leq 20$	

For A325 and A490 bolts used in friction-type connections, the maximum shear stress allowed in table 1.5.2.1 (9.2, p. 5-28) shall be multiplied by the reduction factor

$$1 - f_t(A_b/T_b) \qquad (9.51)$$

where $f_t =$ the average tensile stress due to a direct load applied to all of the bolts in a connection and T_b is the specified pretension load of the bolt (9.2, table 1.23.5, p. 5-59).

The AASHTO specifications (9.41) require use of high-strength bolts for fasteners subject to tension or combined shear and tension. In bearing-type connections the shear stress should not exceed

$$F_{vc} \leq \sqrt{F_v^2 - (0.6 f_t)^2} \qquad (9.51a)$$

where $F_v =$ shear strength of the bolt, ϕF, as given in table 1.7.71A (9.41, p. 236)
$\quad f_t =$ tensile stress due to the applied load

The relation [Eq. (9.51a)] is a working stress version of Eq. (9.50).

9.2.1.8 Fatigue Strength of Bolted Connections

The fatigue strength of bolted connections is considerably reduced by the mere existence of holes. Bars with open holes exposed to fatigue testing in tension break through cross sections weakened by holes, and the crack starts at the hole edges (Fig. 9.55). In Section 2.4 the effect of stress concentrations and the corresponding reduction in fatigue strength was discussed. The maximum tensile stress in a bar of rectangular cross section with a single hole (Fig. 9.56) according to A. Henning (9.42) is

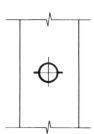

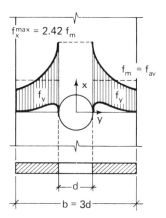

Figure 9.55 Bar Failure in Fatigue Tension Test.

Figure 9.56 Axial and Perpendicular Stresses in Bars with Holes in Tension (From 9.42).

$$f_{x\text{-max}} = f_{av}\left[3 - 2\left(\frac{d}{b}\right) + 0.8\left(\frac{d}{b}\right)^2\right] \quad (9.52)$$

where d = hole diameter, b = plate width, and f_{av} = average tensile stress.

In the limiting case where $d/b = 0$, the stress concentration factor, $f_{x\text{-max}}/f_{av} = f_k = 3.0$. Due to this stress concentration even if the hole is "open" (no fasteners inside), the fatigue strength is reduced in pulsating tension by 35%; under alternating loading (stress reversal) this reduction is 42% (Fig. 9.57) (9.43).

Besides the effect of stress concentrations due to the presence of a hole a or sudden change of shape, when a load is applied to the plate through the bearing of a fastener on the material around the hole, the combination of the tangential stress (Fig. 9.58(9.44)) and the effect of strain concentrations produces an f_k approximately equal to 6, or twice as large as before (Fig. 9.59). The maximum stress at the side of the hole occurs at a distance equal to about one-sixth the diameter from the center line A-A on the bearing side (Fig. 9.58).

This new stress and strain concentration will further reduce the fatigue strength, even more than in the case of an "open" hole. When the bearing stress is raised to 5 times the tensile stress, for instance, the fatigue strength of double-lap joints is decreased to even less than the value for single-lap joints with bolts in bearing and no clamping.

The clamping force of either rivets or unfinished bolts usually is not high enough to warrant taking its effect into account. However, when the bearing stress is reduced or even eliminated because of high-strength bolts exerting large clamping forces, the fatigue strength will be increased considerably. High clamping forces of bolts, with bearing eliminated, will produce quite a different stress distribution. In one study (9.44), with no axial load on the specimen the stress distribution at the side of the hole consisted of tangential compression across the thickness of the plate. The addition of an axial force, equal to 40% of the average compressive stress, exerted by the washer on the plate reduced this tangential stress at the surface of the plate to nearly zero and changed the stress at the center of the plate to a tension approximately equal

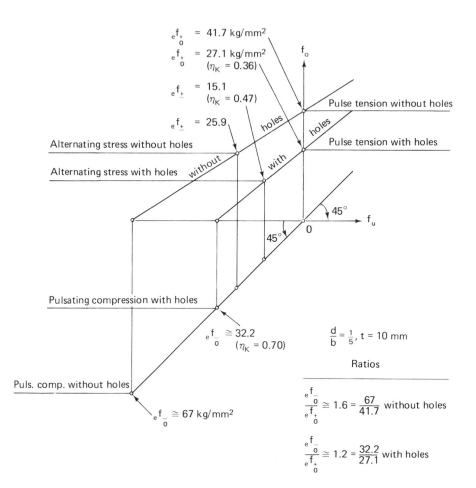

Figure 9.57 Goodman Diagram of a Bar Without and With Holes (From 9.43).

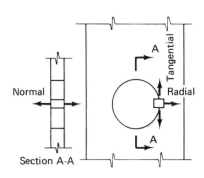

Figure 9.58 Designation of Stresses and Strains at the Side of a Hole (From 9.44).

to the axial stress. For bolts in direct tension, the current Research Council on Riveted and Bolted Structural Joints specifications (9.6) permit a stress of 44 ksi (303 MPa) for A325 bolts on the nominal bolt area. This stress is approximately two-thirds the proof load of the bolt. Thus if the total load (direct load plus prying forces) actually supported by the bolts does not produce stresses that exceed 44 ksi (303 MPa), there should be little or no variation of stress in the bolt during loading cycles and fatigue should not be a problem.

It could be said that the high clamping force of a high-strength bolt is so effective in reducing the stress concentrations adjacent to a bolt hole that other locations of stress concen-

338 CONNECTIONS

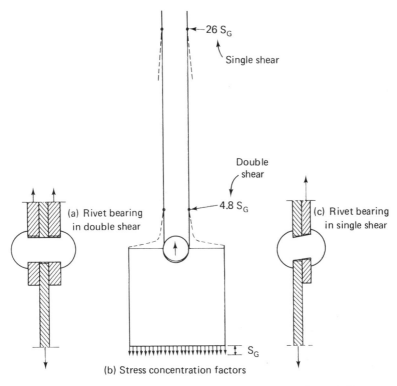

Figure 9.59 Stress and Strain Concentration at the Sides of a Hole Produced by Rivet Bearing (From 9.44, p. 1372).

tration may become critical in fatigue. For example, poor surface conditions resulting from rolling or fabrication may cause stress concentrations of sufficient magnitude to initiate failure. Thus a fatigue crack will not always appear at the bolt hole (Fig. 9.55) but often in front of the hole (9.45). If very high alternating stresses are applied, bearing will occur and the crack will again pass through the hole.

In conclusion, it could be stated that high-strength bolts show far superior performance in fatigue than rivets or unfinished bolts. Unfinished bolts are to be used only in light structures with prevailing statical loading or for temporary structures. It is essential that unfinished bolts be tight to obtain benefits similar to those obtained when using high-strength bolts (9.46). The clamping force is the most important variable governing the fatigue strength of the joints tested. Joints connected with high-strength bolts show much higher fatigue strength, sometimes even approaching the yield point of the steel (9.47), as compared to those connected with rivets.

9.2.2 Welded Connections

A successful welded connection cannot be just an imitation of a good bolted connection where bolts are simply substituted by welds. What is proper in a bolted connection does not necessarily work for a welded connection. If typical bolt details were used for welds (9.48), weld and plate cracking would be likely to occur. The reason is the basic difference between bolts and welds, mechanical connectors and metallurgical ones: the first being discontinuous, elastic and working by friction, while the second is continuous in nature, rigid, and with an atomic bond. The first type can act as "crack arrester," whereas the second type facilitates crack propagation. Therefore, a good designer, besides being completely up-to-date on modern welding technology (Note: Anything done 5 years ago in welding design or fabrication could be obsolete!), has to give full attention during detailing of a connection to the force or stress path. This path must be provided in a least disturbing way, avoiding fatigue-sensitive configurations. A stress coming into an element has to go out of this element. Failure to provide such a path is the main reason for structural design problems. Figure 9.60 shows a double-vee butt weld with reinforcements. The flow of stresses through the plate, before reaching the weld, is parallel to the plate surfaces and relatively uniformly distributed. Because of weld reinforcement, stress flow lines start to deviate, trying to remain tangential to the surfaces of reinforcement. After passing through the weld they cannot abruptly restore the initial flow existing before the weld, and a concentration of stress in section 1-1 will result. If the reinforcement was ground to a smooth transition this disturbance would not occur. This is an example demonstrating that a stronger weld does not necessarily mean a stronger or better connection. A stronger-than-needed weld may mean higher heat input, higher distortions and residual stresses, more fabrication time and weight, and finally more cost, with a negative result in the connection quality. The weld reinforcement actually acted here as a stress raiser.

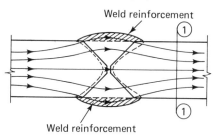

Figure 9.60

A similar situation arises in case of a cover plate fillet-welded to a rolled beam (Fig. 9.61a). If the connection is made by a small transverse fillet (Fig. 9.61b) a high concentration of stresses in the otherwise adequate cover plate will occur. A larger fillet will distribute the stresses more uniformly (Fig. 9.61c). It is best, though, to reduce the thickness of the cover plate by a transition in thickness and to make a connection with a small fillet weld (Fig. 9.61d) without high stress concentration.

Figure 9.62a shows the simplest and most efficient way of welding a lug to a rolled section. A path must be provided to transfer the force to that part of the section that lies parallel to the force action, in this case the web (shaded in the figure). If the rolled section is in a different position (Fig. 9.62b), the parts of the section parallel to the force are now the flanges. Therefore, the lug has to be properly welded to the flanges and not to the web at all.

Similarly, when a floor beam frames into a main girder on a bridge, the flanges of

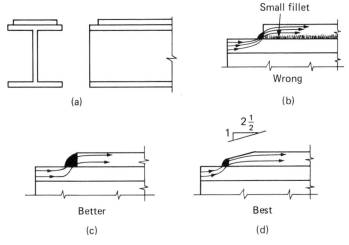

Figure 9.61

the girder will take the moment action due to the floor beam deflection and end rotation, and consequently have to be connected with the floor beam (Fig. 9.62c). A bracket is welded to the top of the floor beam and bolted together with the beam-web to the transverse stiffener of the girder. The stiffener is welded only to the web, although the flanges (shaded area in figure) are parallel to the force action. Tension force in the top flange of the floor beam will enter through the welded bracket and the bolted connection into the stiffener. The weld of the stiffener to the girder web will transfer this force to the web. The force action out-of-plane of the web will produce bending stresses in the web and soon fatigue cracks in the web will develop at the top of the stiffener and progress on both sides of it almost parallel to the girder axis. The stiffener actually had to stretch over the whole web depth and should be welded to the top flange to be able to transmit the force through this weld directly to the flange without locally bending the web.

Another important consideration is the rigidity of a welded connection that is much higher than the rigidity of a similar bolted one. When the design of the connection is poor, a bolted connection is more "forgiving" than a welded one, mainly because of its lesser rigidity. Therefore, the consideration of strain compatibility in welded connections is extremely important. As an example, consider a stringer supported by and running through the web of a box girder bent (Fig. 9.62d). The web of the stringer is cut out for the box, while only the top and the bottom flanges run continuously. A slot, flame-cut and ground, has been cut into the web of the box girder to receive the lower flange of the stringer. If the bottom flange were only supported by both webs of the box girder, i.e., if it were left to pass unattached through the web, only crippling of the web and bearing pressure on the contact areas would have to be checked. But if the bottom flange were welded to the box girder webs, weld cracking at the tips of the flange would be imminent. For even quite a moderate bending stress and corresponding strain existing in the girder webs in the region of the slot, a flange welded to that web should accommodate this same strain, i.e., it should transversely elongate for the total displacement of the web equal to the product

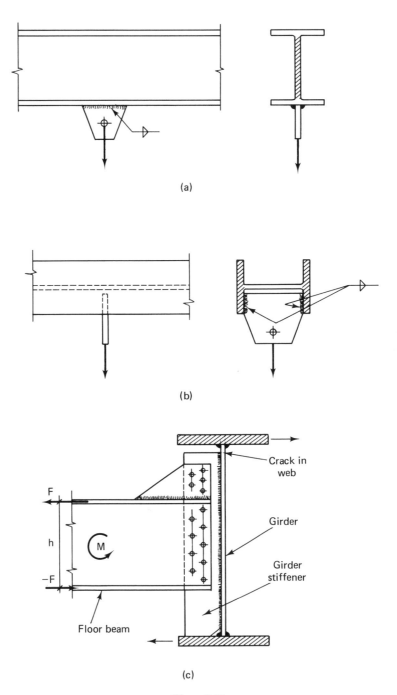

Figure 9.62

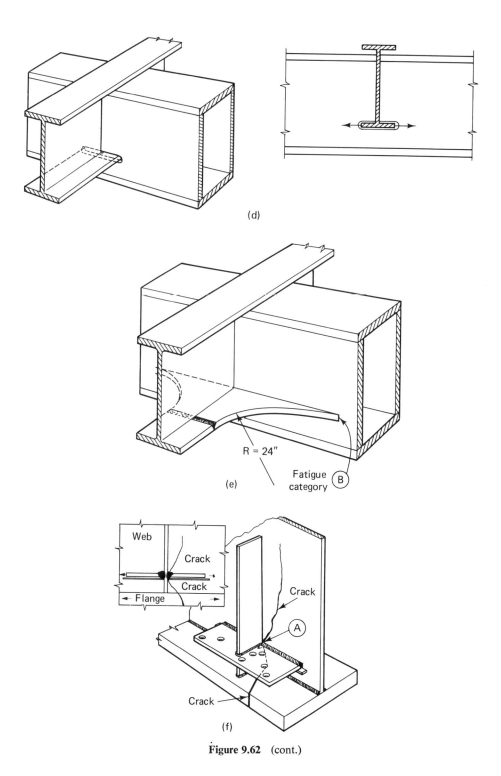

Figure 9.62 (cont.)

of the web strain and the flange width. Having in mind the large flange area, obviously a very large force would be required and no weld of reasonable size could transfer such a force. Weld cracking in most cases would trigger fatigue cracking of the webs and gradual failure, provided the circumstances bar an even worse event, namely brittle failure. Instead of slotting the webs, it would be better to stop the web and bottom flange of the stringer in front of the box webs and use an attachment, as shown in Fig. 9.62e. A wide plate cut to a large radius (2 ft) is welded between the box web and the girder bottom flange.

In a welded joint there are no secondary or less important elements. Even an interrupted backing bar can cause main members to crack. The *Structural Welding Code* (AWS D1.1-80, 9.49) in section 3.13 under Groove Weld Backing, requires that steel backing shall be made continuous for the full length of a weld. All necessary joints in steel backing shall be complete joint penetration butt welds. In a 345-kV transmission line some poles failed just because of the discontinuity of backing bars. The pole shaft, having a dodecagon cross section, had two longitudinal groove welds made over backing bars, which were discontinuous. The poles failed 1 ft from the foundation because of cracking exactly at the backing bar discontinuity located on the tension side of the pole. This discontinuity had a severe notch effect and it proved to be wrong to think that a backing bar is used just to keep weld metal from spilling through. A similar situation may arise in attaching a gusset plate of a horizontal bracing to a girder web close to the tension flange (Fig. 9.62f) and at a place where an intermediate transverse stiffener is positioned. In the given detail, the gusset plate was slotted to let the stiffener pass. The stiffener was also welded transversely at the bottom to the tension flange. The backing bars for horizontal groove weld joining the gusset and web plates were discontinuous. Being so close to the bottom flange (5 in) there was no space for overhead fillet-welding the gusset plate to the girder web. Therefore, a groove weld using a backing bar was chosen. The stress in the web at the attachment weld was high. The notch, created by the interrupted backing bar increased that stress to a much higher level and under repeated loading of this bridge structure fatigue cracking occurred first in the weld and then through the web it propagated downwards to the flange and through the flange. The web crack also propagated upwards through the web, where in the region of low web bending stresses it finally stopped.

Figure 9.62f illustrates several errors of the detail shown: the concentration of welds in a small area with the crossing of welds at point A and with transverse fillet welds of the stiffener cutting across the tension flange. Enough room should be left so that the web can adjust itself to new stress conditions. A concentration of welds will introduce highly restrained welds with high level locked-in stresses which aggravate the whole problem. The attachment plate should not have had a slot, thus providing a gap in the flow of forces from horizontal bracing members that are supposed to be in equilibrium at the gusset plate. Because of this gap, forces instead will flow through a part of the web and produce a high stress in it. The web portion below this plate does not require a stabilizing stiffener, being in tension. Therefore, a far better design (Fig. 9.63) is to stop the stiffener at the attachment and cope it to avoid crossing of welds. The attachment as shown in Fig. 9.63 is a wide plate cut to a radius of 12

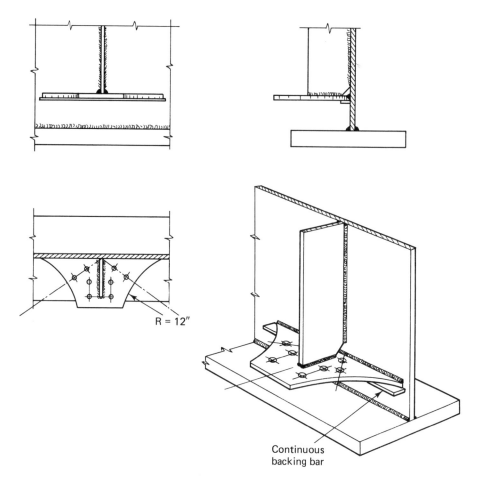

Figure 9.63

in. (or smaller, but larger than 6 in. to stay in fatigue category C, the same as for the web-to-stiffener fillet weld).

In summary, welded connections are quite different from bolted connections. The force path and compatibility of strains have to be examined carefully. The force can be taken only by areas parallel to their line of action. Fatigue and lamellar tearing considerations are important and not secondary in the design process. The proper size and shape of welds is most important and backing bars, when used, must be continuous (or groove welded together by complete joint penetration butt welds).

9.2.2.1 *Concentric Forces on Welds*

In building and bridge construction only two types of welds are used: butt and fillet welds. Plug or slot welds have exhibited poor fatigue performance (Category F), mainly due to discrete, high concentrated fields of residual stresses from fabrication. Therefore, only the performance and behavior of butt and fillet welds will be discussed.

The static strength of a weld is proportional to its type, size, fabrication technique (hand or machine made, one-run versus multirun welds, sequence of welding, preheating), and the type of electrode used. A designer has to decide on the type and size of weld, but is less concerned about the required edge preparations and the type of electrode to be used. Normally, the fabricator, on the basis of experience and expertise, will choose the particular type of weld according to the size of plate, position of weld, available equipment, and preparation of weld edges. The fabricator also chooses the matching electrode best suited for the job (when matching weld metal is required). AWS D1.1-80 *Structural Welding Code* in table 4.1.1 (9.49 p. 44) gives matching filler metal requirements. For ASTM A36 steel with a minimum yield point of 36 ksi (250 MPa) and tensile strength of 58 ksi (345 MPa) the matching electrode of the 60 series has a minimum yield point of 50 ksi (345 MPa) and a minimum tensile strength of 67 ksi (460 MPa). Electrodes of the 70 series have corresponding values of 57 ksi (395 MPa) and 70 ksi (485 MPa), respectively. Therefore, the mechanical properties of the filler metal normally enable it to outperform the basic metal.

A full penetration groove weld fabricated with matching electrode in a statically loaded configuration is therefore stronger than the basic material and need not be checked by any calculation. Only the French specifications (9.50) say that no calculation is required for full penetration butt welds of matching electrodes. The AISC specification (9.1, table 1.5.3) gives for tension or compression normal to or parallel to the axis of weld the same allowable stress as for the base metal.

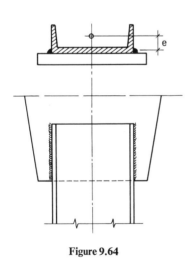

Figure 9.64

Fillet welds may run parallel or normal to the force action, or under an angle. The capacity of a fillet weld depends upon the direction of the loading and the length of the weld itself. Transverse welds have more strength than parallel welds. This is mainly due to the nonuniform distribution of stresses in parallel welds and the eccentricity of the applied force which normally occurs in end connections (Fig. 9.64). For such joints the force is acting in a plane parallel to the weld plane, producing therefore additional, nonuniform bending stresses. Extensive tests carried out in Germany (9.51) in 1965–1966 showed a sharp increase in shear stresses near the ends of parallel welds, which depended upon the length of the weld compared to its size. For a length-to-size ratio of 105, the maximum shear stress was 2.9 times the average. Therefore, the German specifications for welded building structures (9.52) limit the length to weld-size ratio to a minimum of 15 and a maximum of 60.

More recent research (9.53) found also that transverse welds are about 44% stronger than longitudinal welds (see Fig. 9.65). They also show a decrease in deformation capacity, in that longitudinal welds are nearly four times more ductile than transverse welds. American specifications do not restrict the weld length and design the weld by shear stress in the plane of weld throat irrespective of the angle between the force and the weld. This procedure seems to be a good practical decision which greatly simplifies the design procedure.

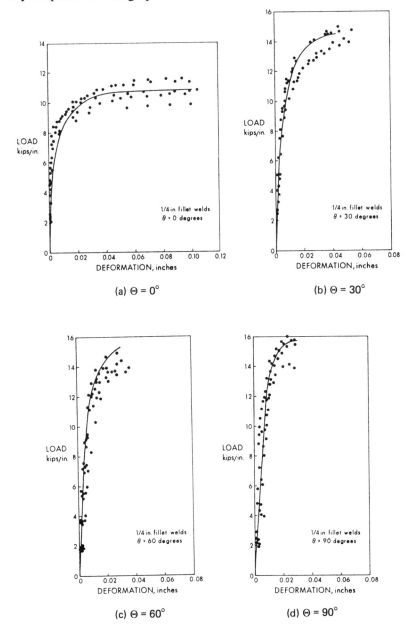

Figure 9.65 Load versus Deformation (From 9.53).

Connection Performance 347

9.2.2.2 Torsion and Eccentric Shear Forces on Welds

As mentioned earlier (Sections 9.1 Introduction and 9.2.1.3 Bolt Groups in Torsion and Eccentric Shear) there are many situations wherein eccentric forces must be accommodated. We shall limit our discussion to the case in which the eccentric load acts in the same plane as the weld group. This type of connection occurs frequently in practice—for example, when using brackets supporting beams or girders which cannot be centered on their supporting columns, Fig. 9.66a; girder web splices, Fig. 9.66b; and double angle connections, Fig. 9.66c.

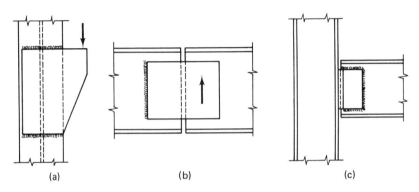

Figure 9.66

Owing to the action of an eccentric force, a weld group is subject to a combination of direct shear and moment. The previous method for the design of eccentrically loaded welded connections was similar in many ways to the one used for bolted connections. The weld must transmit a direct shear stress and a shearing stress due to torsion. For purposes of allowable stress design, the usual direct stress formula and the torsion formula are used in vector combination (9.53). The superposition implies elastic response of the welds. It is known that such an approach produces an over-conservative design in comparison with the ultimate strength of the weld. As connection design is basically ultimate design with appropriate factors of safety (between 2 and 3), a new and better method of design has been developed (9.54), similar to the approach used for determining the ultimate load on eccentrically loaded bolted connections (see Section 9.2.1.3). The continuous weld is divided into elemental lengths and the resisting force on each element of weld is assumed to act at its center (Fig. 9.67). The continuous weld is then similar to a line of bolts in a bolt group, each element of weld being analogous to a bolt, except that in the case of weld the maximum strength and deformation will depend upon the angle of the resisting force. It is assumed that the overall strength of the weld group will be the sum of the individual capacities.

When the element under consideration is in a horizontal section of the weld leg, the angle which the resultant force, R_i, on that section makes with the axis of the weld is given by

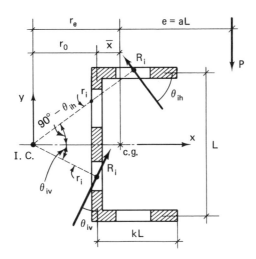

Figure 9.67

$$\theta_{ih} = \tan^{-1}\left(\frac{x_i}{y_i}\right) \tag{9.53}$$

For an element in a vertical portion, this angle is given by

$$\theta_{iv} = \tan^{-1}\left(\frac{y_n}{r_0}\right) \tag{9.54}$$

The radius of rotation of the element, r_i, is expressed as

$$r_i = (x_i^2 + y_i^2)^{1/2} \tag{9.55}$$

The following empirical relationship is now used to obtain the ultimate deformation of each element, $\Delta_{\max}$, in inches,

$$\Delta_{\max} = 0.225\,(\theta + 5)^{-0.47} \tag{9.56}$$

in which θ is expressed in degrees. The element of weld which first reaches its ultimate deformation is located next by finding the element for which the ratio of $\Delta_{\max}/r_{\max}$, i.e., of ultimate deformation to radius of rotation, is a minimum. The deformation of any other element is given by

$$\Delta_i = \left(\frac{r_i}{r_{\max}}\right) \cdot \Delta_{\max} \tag{9.57}$$

in which $r_{\max}$ = the radius of rotation for that element of weld which first reaches its ultimate deformation.

The resisting force, R_i, acting at the center of the ith element and at an angle θ_i, in degrees, is found from

$$R_i = R_{\text{ult}}\,(1 - e^{-\mu \Delta_i})^\lambda \tag{9.58}$$

in which

$$R_{\text{ult}} = \frac{10 + \theta}{0.92 + 0.0603\theta} \tag{9.59}$$

$$\mu = 75 e^{0.0114\theta} \tag{9.60}$$

$$\lambda = 0.4 e^{0.0146\theta} \tag{9.61}$$

The resisting forces are next resolved into two components, and the equation of equilibrium, as applied for the instaneous center, I.C., can now be checked

$$\sum F_x = 0 \qquad (9.62)$$

$$\sum F_y = 0 \qquad (9.63)$$

$$\sum M_{\text{I.C.}} = 0 \qquad (9.64)$$

From Eq. 9.64, the external applied load, P, can be found when written for the I.C.

$$P(e + r_e) - \sum_{i=1}^{n} (r_i \times R_i) = 0 \qquad (9.65)$$

The true instantaneous center (I.C.) is found if the sum of vertical components of the weld resistance forces equals the force P found from Eq. (9.65). If it is not equal, a new trial location of the I.C. must be chosen, and the procedure repeated until Eq. (9.65) is satisfied. The value of P so obtained is the ultimate load which the weld group can sustain.

The empirical equations (9.58) to (9.61) are based on 23 tests on $\frac{1}{4}$-in. (6.35 mm) fillet welds made by E-60 electrodes. Using a minimum factor of safety of 3.33 the load level is reduced from ultimate to the service load. The practical design of welds will be explained later in Section 9.3 under Connection Design.

9.2.2.3 Bending of Welds

If an eccentric load does not lie in the same plane as the weld group, the welds will be exposed to shear and moment (Fig. 9.68). The fillet weld in the compression zone cannot deform under the action of the rotational movement and therefore the connected plate is in bearing over part of its depth (9.55). The usual design approach is based on the assumption of the elastic response of the weld to the load. The welds are designed to transmit a direct shear stress and a moment-induced shear stress based on ordinary elastic beam analysis. Because of the restricted rotation of a part of the weld, this assumption is not correct and the elastic procedure results in a high factor of safety based on the ultimate strength of the weld.

Tests in the past having large ratios of eccentricity of load to weld length (9.56) have shown a factor of safety on the order of 5, when based on the ultimate strength of the weld according to the 1969 AISC specification. In more recent tests (9.57) which considered the combined effects of shear force and bending moment on two parallel fillet welds used in the connection of a column bracket, the ratio of eccentricity to weld length was small—varying from 0 to 1.3. The factor of safety was a function of that ratio, varying between 2.7 and 5.7. In calculating these factors of safety the effect

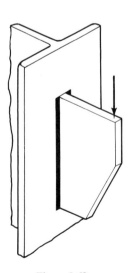

Figure 9.68

of the variation in strength of an element of weld with the angle of applied load was not considered. An ultimate strength analysis, which takes into consideration the true load-deformation response of individual weld elements, is discussed in (9.53).

Let us consider the case shown in Fig. 9.69a. The welds in the tension zone are divided into individual elements of equal length and because of the simultaneous shearing and bending each one in the top part will rotate around the same instantaneous center (I.C.,) with its resisting force, R_i induced by the load, acting through the center of gravity of the element and normal to the radius of rotation, r_i. The deformation of each weld element in the tension zone varies linearly with its distance from the instantaneous center and in the direction of the resisting force. The connecting plate in the compression zone of the connection is in direct bearing at the time the ultimate load is reached. At that moment, a critical weld element, usually but not necessarily the farthest from the instantaneous center of rotation, reaches its ultimate deformation, Δ_{max}, as given by Eq. (9.56). (note that Eqs. (9.58) to (9.61) are also valid.) Next it is assumed that the compression zone stress distribution in the plate is triangular with the maximum stress equal to the yield point of the plate material.

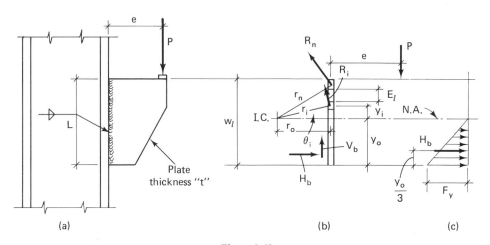

Figure 9.69

Their resultant force, H_b, is now given by

$$H_b = \frac{y_0 \sigma_y t}{2} \quad (9.66)$$

The length of the weld in the compression zone is assumed to resist a vertical component of force only, and it will have a uniform vertical deformation over the whole length. The vertical shear force, V_b, resisted by the length of weld below the neutral axis can therefore be expressed by

$$V_b = \frac{y_0}{(w_l - y_0)} \sum (R_i)_v \quad (9.67)$$

in which $(R_i)_v$ = vertical resisting force component of a weld element = $(r_0/r_i) \times R_i$.
To find the ultimate load, P, only one initial value is assumed for r_0. All the

forces, R_i, in the weld elements are thereby defined, V_b, the vertical shear force of the weld, and H_b, the bearing force. In order for the connection to be in equilibrium we have

$$\Sigma F_x = 0; \quad H_b - \sum_{i=1}^{n}(R_i)_h = 0 \quad (9.68)$$

in which
$(R_i)_h$ = Horizontal force component = $y_i/r_i \times R_i$.

$$\Sigma F_y = 0; \quad \Sigma(R_i)_v + V_b - P = 0 \quad (9.69)$$

$$\Sigma M_{\text{I.C.}} = 0; \quad P(e + r_0) - \sum_{i=1}^{n}(R_i r_i) - r_0 V_b - \left(\frac{2}{3}\right)y_0 H_b = 0 \quad (9.70)$$

By eliminating the unknown ultimate load P from the last two equations, the following equation is obtained, in which the only unknowns are y_0 and r_0

$$\sum_{i=1}^{n}(R_i)_v + V_b - \frac{1}{(e+r_0)}\sum_{i=1}^{n}[(R_i r_i) + r_0 V_b + \left(\frac{2}{3}\right)y_0 H_b] = 0 \quad (9.71)$$

Equation (9.71) gives the other unknown value of y_0, once an initial r_0 is assumed. If Eq. (9.68) is also satisfied, the ultimate load, P, has been found and its value is given by Eq. (9.69).

Comparative tests have shown the validity of this approach, which would be attractive to designers provided ultimate load tables, such as Table 9.4, are available.

Table 9.4 TYPICAL ULTIMATE LOAD TABLE

L, in.	\multicolumn{8}{c}{e, in.}
	2	4	6	8	10	12	14	16
2.00	13.26	6.64	4.47	3.35	2.66	2.22	1.95	1.70
4.00	51.88	26.37	17.61	13.23	10.54	8.82	7.56	6.62
6.00	110.70	58.90	39.47	29.68	23.73	19.79	16.86	14.86
8.00	173.80	103.51	69.92	52.57	42.11	35.11	30.16	26.34
10.00	231.93	158.05	108.54	81.99	65.69	54.81	46.98	41.10
12.00	286.75	218.57	154.93	117.57	94.38	78.76	67.57	59.17
14.00	339.36	280.78	207.71	159.16	128.16	107.12	91.93	80.48
16.00	390.20	342.28	265.36	206.30	166.77	139.59	119.90	105.03
18.00	439.84	402.11	325.67	258.14	210.06	176.15	151.49	132.73
20.00	488.59	459.80	387.36	313.69	257.57	216.72	186.57	163.69

Note: The ultimate loads tabulated, in kips, are for connections of the type shown below.

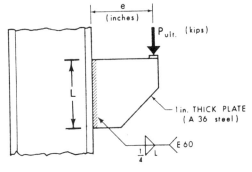

Weld Arrangement in Table 9.4 (From J.L. Dawe and G.L. Kulak, "Eccentrically Loaded Welded Connections," *Journal of the Structural Division*, Proc. ASCE, Paper No. 10457, Vol. 100, No. ST4, April 1974.)

9.2.2.4 Fatigue Strength of Welded Joints

All welding processes intoduce some discontinuities in or near the welds apart from creating distortions, strong residual stress fields, and possible sharp stress raisers. Good welding techniques can only minimize flaws and other unwanted consequences caused by large heat input; they cannot completely eliminate them. The fatigue strength of welded joints for the same stress range depends very much upon the type and form of detail and the internal structure of the welds. It is very important to reduce as much as possible stress concentrations introduced by change of cross section and possible weld defects that are not permitted by ASTM and AWS. It is very irrational to cut out those parts of weldments containing discontinuities and weld them again. This will result in a worse condition than the original one.

In Table 9.5 some of the characteristic stress concentration factors for welds

Table 9.5 STRESS CONCENTRATION FACTORS, K_a

Type of Joint	Sketch	K_a
Butt weld with smooth transition		1.0
Butt weld with abrupt transition		1.6
Butt weld with corrections		1.0
Joint with fillet welds of 45°		2.2
Joint with fillet welds of 1:2		2.8
Joint with butt welds		1.6
Joint with butt welds and smooth transition		1.0
Web with two stiffeners fillet-connected		1.5
Web with the pair of stiffeners with machined weld reinforcement		1.0
Pair of stiffeners with one fillet weld each		2.0
Specimen with asymmetrical stiffener connected by two fillet welds		1.3
Web with one side stiffener connected with one fillet weld		1.9

Connection Performance

are given. The details involved are plate joints and stiffener-to-web attachments.

The type of a detail and therefore the magnitude of the stress concentration factor will limit the allowable stress range, F_{sr}. Only details that provide the highest fatigue strength should be considered.

At weld toes and weld ends of both fillet- and groove-welded details, high stress concentrations occur. Consequently, at these points we have the smallest allowable stress. Therefore, intermittant fillet welds show poor fatigue strength. Details where fatigue crack growth is possible from points of stress concentrations provide the lowest allowable stress range (categories E and F). Details which involve failure from internal discontinuities, such as porosity, slag inclusion, cold laps, and other comparable conditions, will have a high allowable stress range (9.58). This is primarily due to the fact that there is no geometrical stress concentration at such discontinuities other than the effect of the discontinuity itself.

A wide class of groove and fillet-welded details are covered by the next worst Category E. However, there are alternate details for which a higher stress range is allowed. Connections with fillet welds usually show a smaller fatigue strength than those with butt welds.

The actual fatigue design of welded connections in buildings and bridges is discussed in Section 9.3 under Connection Design.

9.3 CONNECTION DESIGN

In the preceding section, the behavior of individual fasteners as well as groups of fasteners or complete connections was discussed. In this section the design of connections in beam framing, column connections, and truss connections will be treated, using available techniques and design aids.

9.3.1 Flexible (Shear) Beam Framing Connections

In Section 9.1.1, Types of Connections, it was pointed out that in Type 2 connections the ends of beams are connected to columns or girders for shear only and that the rotational flexibility of such a connection, where critical, must be sufficient to accommodate the rotation of simple beam ends under the action of gravitational load.

Basically, two types of simple shear connections are mostly used: web framing and seated beam. In web-framing connections, the web of a beam is connected by means of two web angles (Fig. 9.70), one-sided or single angle (Fig. 9.71), or a single plate (Fig. 9.72).

For many years, "standard double-angle" connections have been most commonly used, with rivets in the past and with bolts and welds more recently. The search for a somewhat simpler connection, easier to handle in the field, resulted in the single angle (one-sided) web connection (9.59). The most desirable form is the one in which the angle is welded to the column, usually in the fabricating shop, and the beam is bolted to it in the field. The single plate framing connection has been considered to

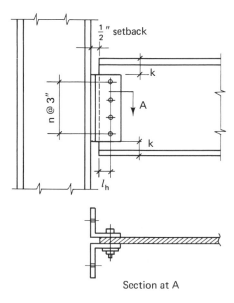

Figure 9.70 Double Angle Connection.

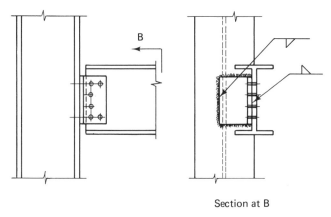

Figure 9.71 Single Angle Connection.

be a flexible connection that is economical in both material and fabrication requirements in the erection of steel buildings (9.60).

Double-angle framing connection: The conventional practice in designing high-strength bolted connections is to consider the shear as uniformly distributed among the bolts, neglecting the eccentricities e_1 and e_2 in both legs of the framing angles (Fig. 9.73).

The thickness of the angle legs should not be too excessive, because the flexibility of the connection is mainly due to the bending and twisting of the angle. A leg thickness of $\frac{3}{8}$ to a maximum of $\frac{5}{8}$ in. will not offer a large resistance to the beam-end rotation, and this will assure flexibility.

To illustrate the design procedure the following examples are presented.

Connection Design 355

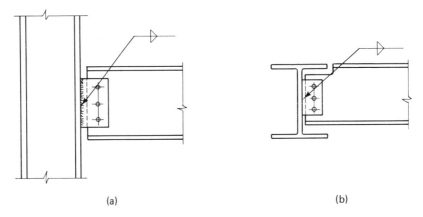

Figure 9.72 Single Plate Framing.

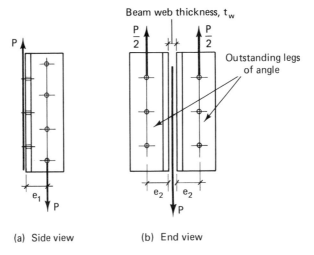

Figure 9.73 Framing Angles.

Example 9.1 A W30 × 116 beam of A36 material has to be connected to a W12 × 190 column, also of A36 material, using A325-N bolts. The reaction force to be transmitted by the shear connection is 140 kips (623 kN). The beam is uncoped. Use AISC specification.

Solution The first problem facing a designer is the choosing of the bolt diameter, d. Generally speaking, in one and the same structure only one or at the most two different diameter sizes are used. For buildings, the basic size is $\frac{3}{4}$ in. For light structures $\frac{5}{8}$ in., and for heavier structures $\frac{7}{8}$ in. are used. For bridges the AASHTO specifications (9.41) in section 1.7.22(B), gives two basic sizes, $\frac{3}{4}$ in. and $\frac{7}{8}$ in. (19.05 and 22.22 mm). It also states that $\frac{5}{8}$ in. (15.88-mm) diameter fasteners shall not be used in members carrying calculated stress, with some exceptions. The diameter of fasteners in angles carrying calculated stresses shall not exceed one-fourth the width of the leg in which they are placed.

The diameter is also influenced by the thickness of the members connected together by the bolt. German construction experience (9.61), gives the following expression as a good practical measure for the diameter, d

$$d_{in} \approx \sqrt{2t_{in}} - 0.1_{in} \qquad (d_{cm} \approx \sqrt{5t_{cm}} - 0.2_{cm}) \qquad (9.72)$$

in which t is the thinnest of plate thicknesses fastened together. Figure 9.74 gives the recommended diameters as a function the thinnest part in the grip.

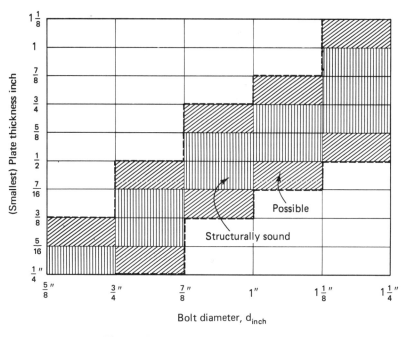

Figure 9.74 Recommended Bolt Diameters.

The beam-web thickness of a W30 × 116 is $\frac{9}{16}$-in. (see AISC *Manual*, 9.2, p. 1-14). As we have two web angles, the thickness of one should be one-half of $\frac{9}{16}$, rounded to the next higher $\frac{1}{16}$-in., i.e., $\frac{5}{16}$ in. Applying Eq. 9.72 we have

$$d \approx \sqrt{2 \times \tfrac{5}{16}} - 0.1 = 0.69 \text{ in.} \approx \tfrac{3}{4} \text{ in.}$$

From Fig. 9.74 we see that a thickness of $\frac{1}{4}$ in. can accommodate bolts of $\frac{5}{8}$ in. or $\frac{3}{4}$ in. in diameter and if required even $\frac{7}{8}$ in. in size. Therefore, we choose the average value of $\frac{3}{4}$ in.

The allowable capacity of a framed beam connection is the smallest one obtained considering bolt shear, bolt bearing on connecting material, beam-web tear-out (block shear if beam coped—see the next example, 9.2), shear on the net area of the connection angles or connection plates, and local bending stresses (normally disregarded).

Clearance for assembly is essential in all cases. The angle thickness was chosen considering bearing and shear capacity of the web of the beam. The leg size of the angles is dictated by the clearance for assembly. In the section at A, Fig. 9.75b, the required assembling clearances are introduced, using data from the AISC *Manual* (9.2), tables for assembling clearances and stagger for tightening, p. 4-132. Next we determine the number of bolts considering shear.

Connection Design

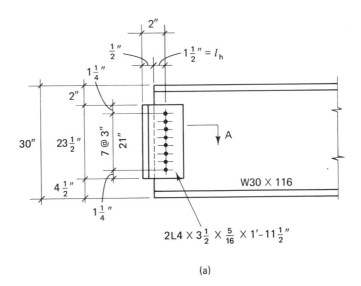

(a)

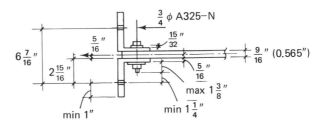

(b) Section at A

Figure 9.75 Example 9.1.

From table 1-D, p. 4-5 of the AISC *Manual* we see that the capacity of A325-N bolts in double shear is 18.6 kips. From table 1-E and material strength $F_u = 58$ ksi, on p. 4-6 of the AISC *Manual* we have a bearing capacity of $\frac{9}{16}$ in. of material thickness and a usual spacing of 3 in. 36.7 kips > 18.6 kips. Using the smaller capacity the number of bolts, *n*, is given by

$$n = \frac{18.6}{140} = 7.5 \approx 8 \text{ bolts of } \tfrac{3}{4}\text{-in diameter}$$

For an end distance of $1\tfrac{1}{4}$ in. (see table 1.16.5.1, p. 5-51 of the AISC *Manual*) the total length of web angle, *l*, will be

$$l = 2 \times (1\tfrac{1}{4}) + 7 \times 3.0 = 23\tfrac{1}{2} \text{ in.}$$

The net shear capacity of two angle legs is (allowable stress $0.3F_u = 17.4$ ksi for A36 angles)

$$R = 2 \times \{17.4 \times \tfrac{5}{16} \times [23.5 - 8(\tfrac{3}{4} + \tfrac{1}{16})]\} = 2 \times 92.4$$
$$= 184.97 > 142.0 \qquad \text{OK}$$

Since this beam is not coped, beam-web tear-out does not require checking. The minimum horizontal end distance, l_h, has to be checked according to section 1.16.5.3 of the AISC specification (9.1).

$$l_h \geq 2P_R/F_u t \qquad (9.73)$$

in which P_R = beam reaction in kips divided by the number of bolts
= 140/8 = 17.5 kips/bolt
F_u = 58 ksi tensile strength of the beam-web material
t = thickness of the beam-web = 0.565 in.

After substitution of the corresponding values in Eq. (9.73) we have

$$l_h \geq 2 \times 17.5/58 \times (0.565) = 1.07 \text{ in.} < 1.5 \text{ in.} \qquad \underline{\text{OK}}$$

The connection as sketched is adequate. If for some reason a smaller gage is required on the outstanding legs of the connection angles, the bolts must be staggered with those in the beam web.

The same problem could be solved using data from the *Manual* table 1-F (for l_h) and tables 11-A, 11-B, and 11-C (pp. 4-10, 25, and 27, respectively).

From table 1-F (p. 4-10) for $l_h = 1\frac{1}{2}$ in. and A36 material, i.e., $F_u = 58$ ksi, we see that one fastener and 1-in. thick material can carry 43.5 kips. As our thickness is $t_w = 0.565$ in. and we have 8 bolts, the beam reaction force that can be transmitted by the chosen group of bolts is sufficiently big, i.e.,

$$R_{max} = 8 \times (.565) \times 43.5 = 196.6 \text{ kips} < 140 \text{ kips:} \qquad \underline{\text{OK}}$$

From table 11-A (p. 4-25) for bolts in bearing-type connections of $\frac{3}{4}$-in. diameter bolts A325-N we read for an angle thickness $t = \frac{5}{16}$ in. a capacity of 148 kips > 140 kips for an angle length of $23\frac{1}{2}$ in. and 8 bolts. From table 11-C (p. 4-27) we see that the allowable shear force in the connection angles for A36 material and 8 bolts in a length of $23\frac{1}{2}$ in. is 185 kips > 140 kips: $\underline{\text{OK}}$ All the values are the same as previously obtained.

Example 9.2 A W21 × 57 stringer beam is shear connected to a W36 × 150 floor beam transmitting a shear force of 90 kips. Use A325-F bolts, AISC specification, and A36 material.

Solution Assume $\frac{3}{4}$-in. diameter for A325-F bolts. Web thickness of the stringer is $\frac{3}{8}$ in. (AISC *Manual*, p. 1-16). The connection angles are $\frac{1}{4}$-in. thick (AISC *Manual*, table II-A, p. 4-24) for $\frac{3}{4}$-in. bolt diameter. Therefore, the thinnest part in the grip, Fig. 9.76d, is $\frac{1}{4}$ in. and from Fig. 9.74 we see that a $\frac{3}{4}$-in. diameter for the bolts is satisfactory. As this coincides with the initially assumed diameter of $\frac{3}{4}$-in. we do not need to change the bolt size. From the same table II-A we read that 92.8 kips < 90 kips requires 6 bolts and an angle length of $17\frac{1}{2}$ in. We have to cope the stringer for a length of $5\frac{3}{4}$ in. and depth of 2 in., Fig. 9.76c, and therefore, web tear-out (block shear) has to be checked. When values for bolt forces are used according to the 1978 AISC specification, which are much higher than before, and we have a relatively thin web, failure can occur by a combination of shear along the vertical plane through the fasteners plus tension along a horizontal plane through the bottom fastener (Fig. 9.77). The resistance to block shear in kips is (AISC *Manual* p. 4-11)

$$R_{BS} = 0.30 A_v F_u + 0.50 A_t F_u \qquad (9.74)$$

Vertical web area, A_v, is

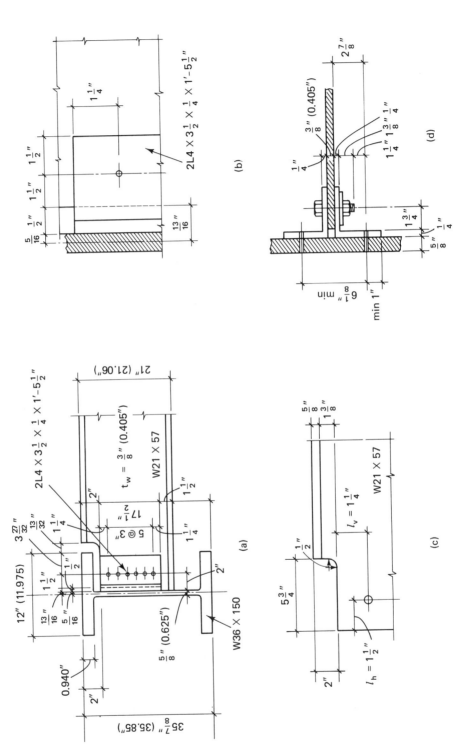

Figure 9.76 Example 9.2.

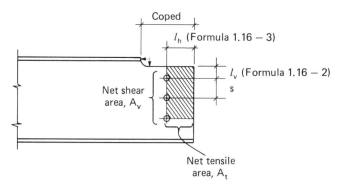

Figure 9.77 Block Shear.

$$A_v = t_w[l_v - d_h/2 + (n-1)(S - d_h)] \tag{9.75}$$

in which t_w = web thickness
l_v = vertical distance from the top hole to the web edge in the direction of shear force
d_h = diameter of hole = $d_{bolt} + \frac{1}{16}$ in.
n = number of bolts in vertical row
s = spacing of holes

Horizontal tear-off area, A_t is

$$A_t = t_w\left(l_h - \frac{d_h}{2}\right) \tag{9.76}$$

in which l_h = horizontal distance from the center of the holes to the web end. Substituting our values into Eqs. (9.75) and (9.76) we get

$A_v = 0.405\{1.25 - \frac{1}{2}(\frac{3}{4} + \frac{1}{16}) + (6-1)[3 - (\frac{3}{4} + \frac{1}{16})]\} = 4.77$ in.^{2}
$A_t = 0.405[1.50 - \frac{1}{2}(\frac{3}{4} + \frac{1}{16})] = 0.44$ in.^{2}

Therefore, resistance to block shear, R_{BS}, from Eq. (9.74) is

$R_{BS} = 0.30(4.77)58 + 0.50(0.44)58 = 95.8$ kips > 90 kips: OK

The same checking could be performed using the AISC *Manual*, table I-G, p. 4-11. Coefficient C_1 for $l_v = 1.25$ in. and $l_h = 1.50$ in. is 1.13. Coefficient C_2 for $n = 6$ and a $\frac{3}{4}$-in. bolt is 2.96. Then, resistance to block shear in kips is

$R_{BS} = (C_1 + C_2)F_u t_w = (1.13 + 2.96)58(0.405) = 96.1$ kips

which is close to 95.8 kips, obtained by our previous calculation. The net shear in the connection angles is checked using table II-C, p. 4-27. For a bolt diameter of $\frac{3}{4}$ in., the length of connection angles of $17\frac{1}{2}$ in. and for an angle thickness of $\frac{1}{4}$ in. the force is 110 kips. This is larger than 90 kips.   OK

Since the connection could slip into bearing, although for friction bolts this is a remote possibility, the beam-web and connecting material should be checked for bearing stress.

The edge distance $l_v = 1\frac{1}{4}$ in. has to satisfy eq. (1.16-2) of the AISC *Manual*, p. 5-50, i.e.

$$l_v \geq 2P/F_u t_w \tag{9.77}$$

Connection Design

in which P = force transmitted by one fastener to the critical connected part, i.e.,

$$P = \frac{90}{6} = 15 \text{ kips.}$$

When this value and the values $F_u = 58$ ksi and $t_w = 0.405$ in. are substituted into Eq. 9.77 we see that

$$l_v = 1.25 \approx \frac{2(15)}{58(0.405)} = 1.28 \text{ in.}$$

The same could be found by using the AISC *Manual*. From table I-F at $F_u = 58$ (p. 4-10) we read for $l_v = 1.25$ in. a coefficient of 36.3 per inch of web thickness. Then the permissible bearing is $R = 36.3 \times 0.405 \times 6$ bolts $= 88.2$ kips ≈ 90 kips.   OK

The clearances are checked in Fig. 9.76d similar to Example 9.1.

Single angle one-sided framing connection: The most desirable form of the single angle connection is one in which the angle is welded to the supporting member, usually in the fabricating shop, and the beam is bolted to it in the field. The study given in (9.59) investigated failure behavior of such connections. The moment due to the eccentricity of the bolts of weld group in the outstanding angle leg is taken into account. The study showed that the factor of safety available for the bolts and welds is greater than 2.5. Rotations on the order of 0.024 rad (1°22′30″) can be achieved in all sizes of connections from 2 to 12 bolts and the bulk of this flexibility is available in the rotation of the supported beam (due to the slip of the beam with respect to the connection angle, rather than to the deformation of the angle itself).

The design of this type of connection is illustrated in the next example (9.3). The table on page 4-84 of the AISC *Manual* is used.

Example 9.3 Design a one-sided connection for a W16 × 36, $F_y = 36$ ksi to transmit an end reaction of 43 kips to a column flange of a W14 × 90 ($t_f = 0.710$ in.), Fig. 9.78. Use ASTM A325-X bolts in the beam-web leg and outstanding leg.

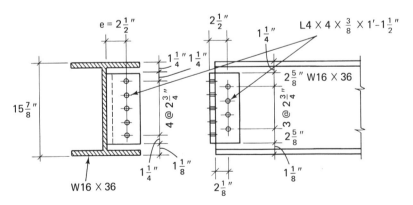

Figure 9.78 One-sided Bolted Connection, Example 9.3.

Solution As the beam-web thickness is $\frac{5}{16}$ in. (see *Manual*, p. 1-18), the angle thickness should be at least the same or a bit stronger, say $\frac{3}{8}$ in. Using Eq. (9.72) the bolt diameter will be:

$$d = \sqrt{2 \times \tfrac{3}{8}} - 0.1 = 0.77 \text{ in.} \qquad \text{use } \tfrac{3}{4} \text{ in.}$$

The shear capacity of one $\frac{3}{4}$-in. A325-X bolt in single shear is $r_v = 13.3$ kips (*Manual*, p. 4-5, table I-D). The bearing capacity on beam-web even for minimum spacing is $r_p = 14.7$ kips > 13.3 kips (*Manual*, table I-E, p. 4-6). The required number of bolts in the beam-web therefore is

$$n = \frac{43}{13.3} = 3.2 \qquad \text{use } 4$$

The maximum angle length is $13\frac{5}{8}$ in. (*Manual*, p. 1-18, distance T). If we round this length off to $13\frac{1}{2}$ in. and use bolt end distances of $1\frac{1}{4}$ in., the remaining length of 11 in. can be divided into four parts of $2\frac{3}{4}$ in. each. This will give 5 bolts in the outstanding leg. This number has to be larger than the number in the web leg, because of the twisting moment. The eccentricity, e, is $2\frac{1}{2}$ in. (Fig. 9.78). To arrive at the number of bolts quickly, we shall use coefficients from the table given in *Manual*, p. 4-84. This coefficient, C, is equal to the ratio of force to bolt capacity, which is $43/13.3 = 3.23$. For one row of bolts, as in our case, the coefficient in the table closest to 3.23 is 4.02, and this coefficient requires 5 bolts. The assembly clearances require a stagger of $1\frac{3}{8}$ in. for $\frac{3}{4}$-in. bolts for a clear distance $F = 2\frac{1}{2} - \frac{3}{8} = 2\frac{1}{8}$ in. (*Manual*, p. 4-132, Stagger for Tightening). The connection is sketched in Fig. 9.78. If the outstanding leg were fillet welded with E-70XX instead of being bolted (Fig. 9.79) table III, given in the *Manual* on p. 4-31, could be used. First, the weld size has to be determined. Table 1.17.2A (*Manual*, p. 5-52) gives for a flange thickness of 0.710 in., a minimum size of $\frac{1}{4}$ in. Section 1.17.3 (*Manual*, p. 5-53) gives the maximum size of fillet welds along edges of connected parts as

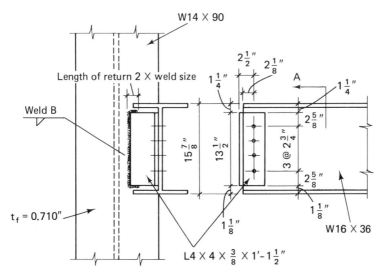

Figure 9.79 One Leg Welded, Example 9.3.

$$\tfrac{3}{8} \text{ in.} - \tfrac{1}{16} \text{ in.} = \tfrac{5}{16} \text{ in.}$$

Since the capacities shown in table III are for two angles, it will be convenient to double the given reaction, i.e., $2 \times 43 = 86$ kips. We shall try both weld sizes for the given angle length of $13\tfrac{1}{2}$ in. From table III weld B, by interpolation, we see that a $\tfrac{1}{4}$-in. weld has a capacity of 68.9 kips < 86 N.G., and a $\tfrac{5}{16}$ in. weld a capacity of 86.1 kips ≈ 86 kips.   OK Therefore, we chose a weld size of $\tfrac{5}{16}$ in.

Single plate connection: As already mentioned, this connection comprises a single plate, with prepunched bolt holes, that is shop welded to the supporting member. During erection, the beam with prepunched holes is brought into position and field bolted to the framing plate (9.62).

The standard design procedure for a single plate framing connection is to assume that each bolt carries an equal portion of the total shear load and, in agreement with the simple support assumption, that relatively free rotation occurs between the end of the beam and the supporting member. Because of this, this connection is often called the "shear tab." This connection derives its limited ductility from bolt deformation in shear and slippage until in bearing, deformation of holes either in beam and/or plate, and out-of-plane bending of the plate. Tests have demonstrated, though, that the single plate connection can develop a significant end moment. The magnitude of it depends upon the stiffnesses of (a) the bolt group (number, size, and configuration), (b) the plate, and (c) the beam web, and on the flexural stiffness of the beam itself and its supporting structural elements (column or girder). When the connection is used on both sides of a supporting structure, with resulting symmetry, the connection may be considered attached to a rigid support. On the basis of research (9.62) the following design procedure is suggested. First, select a plate thickness, t, $\pm \tfrac{1}{16}$ in. of the supported beam web, Fig. 9.80. Then, compute number of bolts required from beam reaction and bolt capacity in shear and bearing. Insure ductility by providing: (a) the bolt diameter, d, to plate, t (or beam—web, t_w) thickness ratio, d/t, according to Table 9.6, and (b) a distance of the center of bolt holes to the plate edge of $2d$, in which $d =$ bolt diameter in inches. Enter the design curve (Fig. 9.81) with the beam span to depth ratio, L/D, and read $(e/h)_{\text{ref}}$. Next use the following equation to obtain (e/h):

$$e/h = (e/h)_{\text{ref}} \times \left(\frac{n}{N}\right) \times \left(\frac{S_{\text{ref}}}{S}\right)^{0.4} \tag{9.78}$$

in which

    $n =$ number of bolts
    $N = 5$ for $\tfrac{3}{4}$-in. and $\tfrac{7}{8}$-in. bolts and 7 for 1-in. bolts
    $S_{\text{ref}} = 100$ for $\tfrac{3}{4}$-in. bolts, 175 for $\tfrac{7}{8}$-in. bolts, and 450 for 1-in. bolts
    $s =$ section modulus of beam

Compute h:

$$h = (n - 1)p \tag{9.79}$$

in which $n =$ number of bolts

    $p =$ pitch

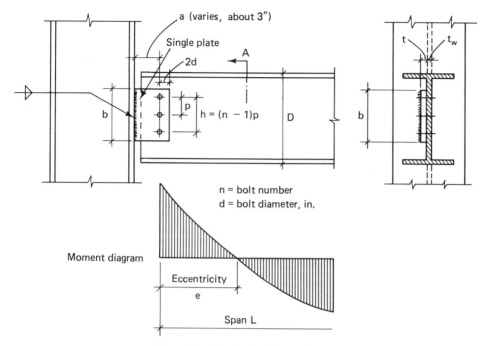

Figure 9.80 Single-plate Connection.

Table 9.6 d/t Ratios

Bolt size, in.	Beam Web or Plate Thickness, in.								
	1/4	5/16	3/8	7/16	1/2	9/16	5/8	11/16	3/4
3/4	3.0	2.4	2.0	1.71	1.5	1.33	1.20	1.09	1.00
7/8	3.5	2.8	2.33	2.0	1.75	1.56	1.40	1.27	1.17
1	4.0	3.2	2.67	2.29	2.0	1.78	1.60	1.45	1.33

Limits — A325's
Limits — A490's

Heavy line shows maximum web or plate thicknesses, t, for a given bolt diameter, d, for A325 and A490 bolts

knowing e/h and h, calculate the connection eccentricity, e, (Fig. 9.80). Compute the moment at the weldment as

$$M = V \times (e + a) \tag{9.80}$$

in which $V =$ beam shear force at inflection point
$e =$ eccentricity
$a =$ distance from the bolt line to the weldment

Check the plate nominal normal and shear stresses

$$f_b = \frac{M}{\frac{1}{4}tb^2} \qquad f_v = \frac{V}{bt} \tag{9.81}$$

Connection Design

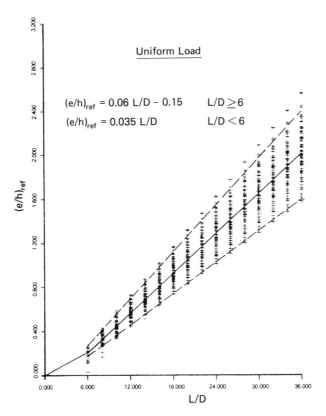

Figure 9.81 Design Curve with ±20% Bounds.

Finally, design the weldment based upon the resultant of the normal and shear stresses obtained from Eq. (9.81) as

$$f_r = (f_b^2 + f_v^2)^{0.5} \tag{9.82}$$

Example 9.4 A W24 × 68 beam of A36 steel, having a span $L = 24$ ft and uniformly loaded with $W = 102$ kips, is connected by a single plate as shown in Fig. 9.80.

Solution The beam web thickness is $t_w = \frac{7}{16}$ in.; therefore, the plate thickness will be $t = \frac{3}{8}$ in. Applying Eq. (9.72), the bolt diameter is

$$d = \sqrt{2 \times \tfrac{8}{3}} - 0.1 = 0.77 \text{ in.} \approx \tfrac{3}{4} \text{ in.}$$

Checking the d/t ratio, we have

$$d/t = \frac{3}{4} \times \frac{8}{3} = 2$$

The bolt capacities are

    In single shear (A325-N, *Manual*, p. 4-5) 9.3 kips

    In bearing (3.00-in. pitch, $t = \tfrac{3}{8}$ in., *Manual*, p. 4-6) 24.5 > 9.3.

Number of bolts in shear:

$$n = \frac{51}{9.3} = 5.5 \approx 6 \text{ bolts}$$

As the beam depth, D, equals 23.73 in., the span to depth ratio, L/D, is

$$\frac{24 \times 12}{23.73} = 12.1 > 6$$

Using Fig. 9.81 for $(e/h)_{\text{ref}}$ we calculate the value

$$(e/h)_{\text{ref}} = 0.06 \times 12.1 - 0.15 = 0.58$$

From Eq. (9.78), after substituting corresponding values, we find

$$e/h = 0.58\left(\frac{6}{5}\right) \times \left(\frac{100}{153}\right)^{0.4} = 0.59$$

With $p = 3$ in., $h = (6-1) \times 3 = 15$ in., $e = 0.59 \times 15 = 8.85$ in. For $a = 3$ in., $V = R = 51$ kips, from Eq. (9.80)

$$M = 51 \times (8.85 + 3.0) = 604.4 \approx 604 \text{ kips-in.}$$

and the stresses are

$$f_b = \frac{604}{\frac{1}{4}(\frac{3}{8})18^2} = 19.9 \text{ ksi}; \quad f_v = \frac{51}{(\frac{3}{8})18} = 7.56 \text{ ksi}$$

$$f_r = (19.9^2 + 7.56^2)^{\frac{1}{2}} = 21.3 \text{ ksi}, \quad \text{or} \quad 21.3 \times \tfrac{3}{8} = 7.99 \text{ kips/in.}$$

How many sixteenths of an inch are needed for weld size we shall get by dividing f_r by the capacity of a $\frac{1}{16}$-in. weld of electrode E-70XX

$$\frac{21.3 \times 3/8}{0.707 \times 1/16 \times 0.3 \times 70} - \frac{7.99}{0.93} = 8.6 \approx 9$$

Use $\frac{5}{16}$-in. fillets each side of the plate.

The connection sketch is shown in Fig. 9.82.

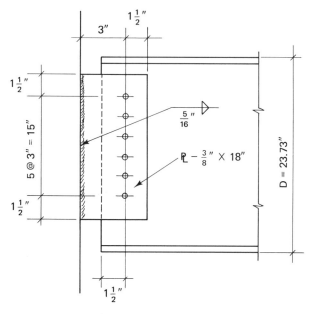

Figure 9.82 Example 9.4.

9.3.1.1 *Seated Beam Connections*

Figure 9.83 shows a typical seated connection, Seated connections should be used only when the beam is supported by a top angle. The outstanding legs of both angles are bent in the same way by the beam reaction. The exact position of this reaction, the location of the critical section in flexure, and the distribution of bending moment along this section are unknown. The angles are thus stressed in longitudinal and transversal bending, resulting in a combined torsional effect.

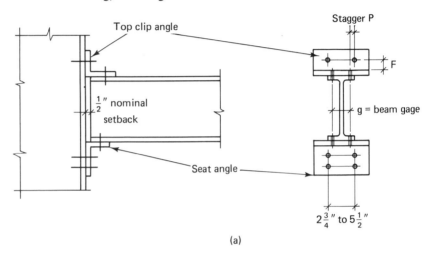

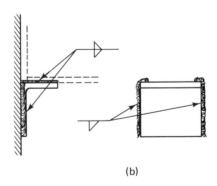

Figure 9.83 Seated Connection.

First, we shall consider bolted and next welded connections. In the design of the bolts connecting the vertical legs to the support, bending due to load eccentricity is neglected. In a welded connection it is accounted for. The assumptions made in current design practice will be discussed in the next example, Example 9.5.

Example 9.5 Design a seated connection of a W18 × 76 beam (web $t_w = \frac{1}{2}$ in. (13 mm)) of A36 material having a reaction of 32 kips (174 kN). Use ASTM A325-F bolts and AISC specification. The beam is connected to the flange of a W12 × 72 column.

Solution For a beam-web thickness of $\frac{7}{16}$ in. (*Manual*, p. 1-18) the bolt diameter is obtained from Eq. (9.72) as

$$d = \sqrt{2 \times \tfrac{7}{16}} - 0.1 = 0.84 \approx \tfrac{7}{8}\text{-in. dia. A325-F bolts}$$

The nominal beam setback is $\frac{1}{2}$ in. For possible mill underrun in beam length we shall assume a $\frac{3}{4}$-in. setback. Next, we assume that the beam reaction is uniformly distributed over the length necessary to develop the allowable stress in the beam at the toe of the fillet (length $b + k$ in Fig. 9.84). We also assume that the critical section for bending in the outstanding leg of the angle is at a distance of $k - t = \frac{9}{8} - \frac{3}{4} = \frac{3}{8}$ in. from the face of angle. The allowable bending stress in the angle is taken as

$$F_b = 0.75 F_y = 27 \text{ ksi} \tag{9.83}$$

(see *Manual*, Section 1.5.1.4.3, Solid Rectangular Sections Bent about Their Weaker Axis, p. 5-21). The allowable compressive bearing stress at the web toe of the fillet is also 27.0 ksi (*Manual*, section 1.10.10.1, p. 5-35). The required bearing length at the web toe of the fillet, b, is obtained from the equilibrium equation

$$F_b[t_w(b + k)] = R \tag{9.84}$$

Therefore

$$b = \frac{R}{F_b t_w} - k = \frac{32}{27(0.425)} - 1.375 = 1.41 \text{ in.} \tag{9.85}$$

The eccentricity, e, based on the required and not the actual value of b is

$$e = \frac{3}{4} + \frac{1.41}{2} - 1.125 = 0.33 \text{ in.}$$

For a length of angle of 6 in. and a leg thickness of $\frac{3}{4}$ in., the section modulus at the critical sections, S, equals

$$S = \frac{6 \times (3/4)^2}{6} = 0.56 \text{ in}^3.$$

The critical bending moment, M_{cr}, is

$$M_{cr} = Re = 32(0.33) = 10.6 \text{ kips-in.}$$

The bending stress at the critical section, f_b, is

$$f_b = \frac{M_{cr}}{S} = \frac{10.6}{0.56} = 18.8 < 27 \text{ ksi} \qquad \underline{\text{OK}}$$

The *Manual* on p. 4-40, gives in table V the allowable loads, R, in kips, for a given angle length and thickness. If we enter table V-A with a web thickness of $\frac{7}{16}$ in. we get $R = 33.7$ kips > 32 kips for an angle length of 6 in. and a leg thickness of $\frac{3}{4}$ in. This is the same value as calculated before. The bolt number in the vertical leg for a single shear capacity of 10.5 kip (*Manual*, p. 4-5) is

$$n = \frac{32}{105} = 3.05 \qquad \text{use 4 bolts, } \tfrac{7}{8}\text{-in. dia. A325-F}$$

Table V-C gives $R = 42.1$ kip > 32 kip for Type B (4 bolts in two gage lines). Therefore, the seat angle is L6 × 4 × $\frac{3}{4}$. The end distance for $\frac{7}{8}$ in. bolts from table 1.16.5.1 (*Manual*, p. 5-51) for a sheared edge is $1\frac{1}{2}$ in. or $1\frac{1}{4}$ in. for beam connection angles.

Connection Design 369

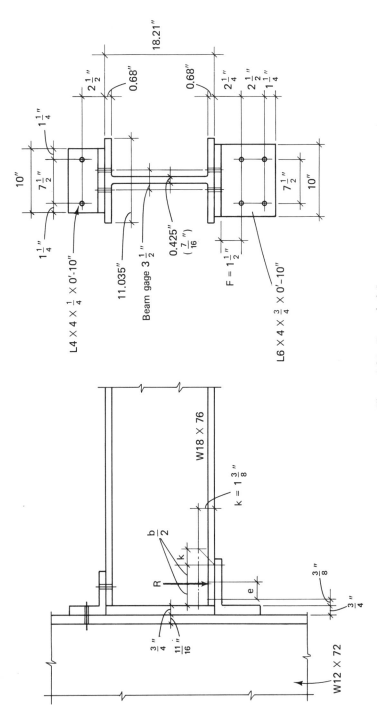

Figure 9.84 Example 9.5.

If the seat angle is welded (Fig. 9.83b) with welds at the ends of the vertical leg of the angles, the strength of this connection is affected by the thickness of the angle, the location of the resultant reaction of the beam, the length of the vertical leg of the angle, and the strength of the weld. Lyse and Schreiner (9.63) found that the strength of this connection varies roughly as the square of the thickness of the angle and inversely as the effective lever arm of the beam reaction, and is influenced by the length and the size of the weld on the vertical leg of the angle. Vertical shear has only a slight effect on the strength of the weld unless the angle is thick enough so that the bending deflection of the leg is reduced to a minimum. The tops of the welds are the most highly stressed and fractured first. The centers of rotation for the calculation of the resistance of the weld were not at the midheight of the weld. Increase in size of weld increases the strength of the connection, but not in proportion to the increase in weld size. The length of weld does not increase markedly the strength of the connection. It is probable that the weld needs to be only slightly longer than is necessary to prevent shear failure, figured simply as the total load divided by length of weld. The bending moments give rise to minor stresses in the lower portion of the welds. An increase in the radius of the fillet of the angle decreases the stresses at the fillet to a marked degree.

The seat angle connections with vertical fillet welds on the ends of the angles are designed first with respect to the strength of the angle itself and second with respect to the strength of the weld in combined shear and bending.

The next example will illustrate current procedure in this type of design, and will be compared with AISC table V1 (*Manual*, p. 4-43) design.

Example 9.6 A beam, W21 × 62 ($\frac{3}{8}$-in. web) has a reaction of 30 kips. Use seated beam connection to attach it to the flange of a W12 × 58 column ($\frac{5}{8}$-in. flange thickness). The beam flange is attached to the seat with bolts. Use 36-ksi material and E-70XX electrodes (Fig. 9.85).

Solution From Eq. (9.85) we obtain the bearing length, b,

$$b = \frac{30}{27(0.400)} - 1.375 = 1.40 \text{ in.}$$

The eccentricity of the reaction is (assuming the "k" value of the angle to be 1.25")

For critical angle section $e = \frac{3}{4} + \frac{1.40}{2} - 1.25 = 0.20$ in.

For vertical welds $e_1 = \frac{3}{4} + \frac{1.40}{2} = 1.45$ in.

Critical bending moment, M_{cr}, is

For angle $M_{cr} = Re = 30 \times (0.20) = 6.0$ kips-in.

For welds $M_{cr} = Re_1 = 30 \times 1.45 = 43.5$ kips-in.

For an angle thickness of $\frac{3}{4}$ in. and a length of 8 in., the stress in the outstanding angle leg in bending is

$$f_b = \frac{M_{cr}}{S} = \frac{6.0}{\frac{1}{6}(8)(\frac{3}{4})^2} = 8.0 \text{ ksi} < 27.0 \text{ ksi}$$

Connection Design

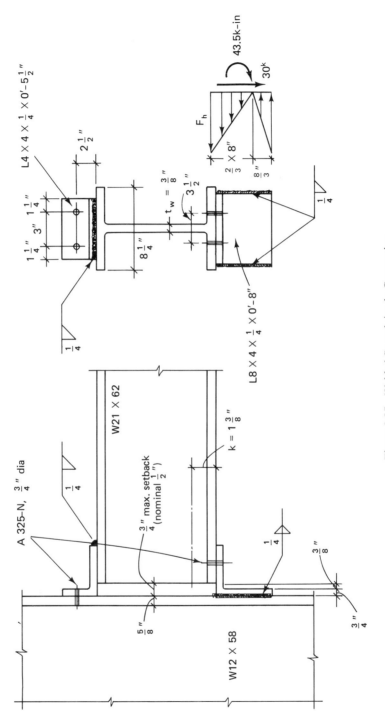

Figure 9.85 Welded Seated Angle Connection.

Checking the vertical weld size, the force per inch due to shear is

$$F_v = \frac{30}{2 \times 8} = 1.875 \text{ kips/in.}$$

If the neutral axis conservatively is assumed to be located at the lower one-third of the length of 8 in. (at 2.67 in. from bottom), the bending moment Re, is equal to

$$43.5 = 2(\tfrac{1}{2}F_h \times 5.33)(\tfrac{2}{3})8 = 28.43 F_h \tag{9.86}$$

in which F_h = unknown force at top of weld.

Solving Eq. (9.86) for F_h

$$F_h = 1.53 \text{ kips/in.}$$

The resultant force per unit length of weld (at the top) is

$$F_r = \sqrt{(1.53)^2 + (2.10)^2} = 2.42 \text{ kips/in.}$$

The number of sixteenths of inch of weld size required is $2.42/(0.93) = 2.6$. The minimum weld size depends on the column flange thickness, $\tfrac{5}{8}$ in. From table 1.71.2A (*Manual*, p. 5-52) the minimum size is $\tfrac{1}{4}$ in. The maximum size is $\tfrac{3}{4} - \tfrac{1}{16} = \tfrac{11}{16}$ in.

As the calculations show, a weld of $\tfrac{3}{16}$ in. will be adequate. We will choose the minimum size of $\tfrac{1}{4}$ in.

Instead of this calculation we could use table VI-A and C. (*Manual*, p. 4-43). From table VI-A for a beam-web $\tfrac{3}{8}$-in. thick we find an angle thickness of $\tfrac{3}{4}$ in. (capacity 31.2 kips) and an 8-in. angle. From table VI-C for a minimum weld size of $\tfrac{1}{4}$ in. we see that a 8×4 angle must be used (capacity 35.6 kips). The sizes of clip angles (top or side) are somewhat arbitrary. They only provide side support for the beam, as well as torsional restraint. In most cases angles of the order of $\tfrac{1}{4}$ to $\tfrac{3}{8}$ in. thick are adequate.

9.3.1.2 Stiffened Seated Beam Connections

Under heavy loads bending and twisting of the outstanding leg of an unstiffened seat would require thickness in excess of 1 in. If the outstanding leg or shelf is stiffened by two vertical angles (for bolted connections) or a T-section (in welded connections), as in Fig. 9.86a and b, then a seated connection may still be used in lieu of a web shear connection. The current practice in the design of such connections is illustrated in the next two examples, 9.7 and 9.8, following mainly suggestions in the AISC *Manual* (9.2).

In bolted connections the top of the vertical angles must be fitted to bear on the outstanding leg of the seat angle. The moment of eccentricity due to the position of the reaction, R, in calculating a bolt group is again neglected. It seems that a practical limit, after which this bending moment should be taken into account, is about 5 in. This is demonstrated in the AISC *Manual* given in table VII-A (p. 4-47), 5 in. as the Largest Stiffener Outstanding Leg. In welded connections, the eccentricity is always considered.

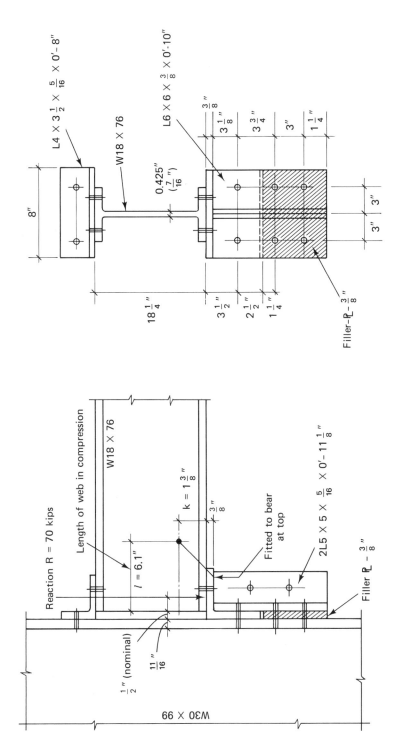

Figure 9.86a Bolted Stiffened Seat Beam Connection (Example 9.7).

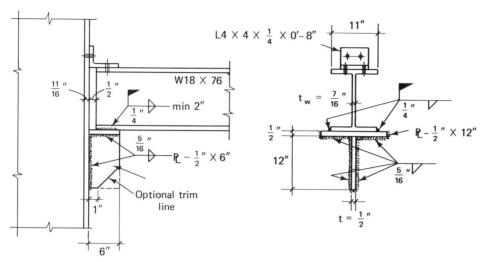

Figure 9.86b   Welded Stiffened Seat Beam Connection (Example 9.8).

Example 9.7   Design a stiffened seated beam connection of a W18 × 76 beam of A36 material to transmit a reaction of 70 kips (262 kN) to a W30 × 99 column as shown in Fig. 9.86a. Use ASTM A325-X bolts and AISC specification.

Solution   The required length of bearing, b, using Eq. (9.85), is

$$b = \frac{70}{27(0.425)} - 1.375 = 4.7 \text{ in.}$$

If the actual setback of the beam is $\frac{3}{4}$ in., then a seat angle of 6 in. will be adequate. The stiffener angle should be 5 in. Using a trial thickness of $\frac{5}{16}$ in. the bearing stress on the stiffeners is

$$f_p = \frac{70}{2 \times 5 \times \frac{5}{16}} = 22.4 < 0.90(36) = 32.4 \text{ ksi}$$

therefore, $\frac{5}{16}$ in. is adequate. The bolt diameter is

$$d = \sqrt{2 \times 3/8} - 0.1 = 0.77 \approx \tfrac{3}{4} \text{ in.}$$

The capacity of A325-X, $\frac{3}{4}$-in.-diameter bolts in single shear is 13.3 kips. In bearing for a $\frac{3}{8}$-in. thickness, even for a 2-in. pitch, the capacity is 17.7 kips > 13.3 kips. Therefore the number of bolts needed to transmit the reaction of 70 kips is

$$n = \frac{70}{13.3} = 5.3 \approx 6$$

To check the design, shown in Fig. 9.86a, use tables VII-A and VII-B (*Manual*, p. 4-47). For a reaction of 70 kips and thickness of stiffener outstanding legs of $\frac{5}{16}$ in., a maximum length in bearing of 4 in. is adequate, and 5 in. is provided. Six bolts (3 in one vertical row) can take 79.5 kips > 70 kips.   OK

Example 9.8   Design a welded stiffened seated beam connection with the data as in Example 9.7. Use E-70XX electrodes and AISC specification, Fig. 9.86b.

Connection Design   **375**

Solution On p. 4-48 of the AISC *Manual* the following suggestions on the connection's dimension are given. The minimum stiffener plate thickness, for A36 material and beams with unstiffened webs, should not be less than the supported beam-web thickness. The minimum stiffener plate thickness, t, should also be at least two times the required E-70XX weld size. In our case, minimum weld size for stiffener plate is dictated by the column flange thickness of $\frac{11}{16}$ in. From table 1.17.2A (*Manual*, p. 5-52) this size is $\frac{1}{4}$ in. Therefore, the stiffener plate cannot be thinner than $\frac{1}{2}$ in. As the web is $\frac{7}{16}$ in. $< \frac{1}{2}$ in., use for the stiffener $\frac{1}{2}$ in. The horizontal plate usually is of the same thickness, in this case $\frac{1}{2}$ in.

The actual setback is taken as $\frac{3}{4}$ in., Fig. 9.87a. The reaction is assumed to be uniformly distributed over the length necessary to develop the allowable compressive stress in the beam-web (27 ksi or 186 MPa). From Eq. (9.85), the bearing length is the same as in Example 9.7, 4.7 in. As $4.7 < (6 - \frac{3}{4})$ OK   The load eccentricity from vertical welds, e, is

$$e = 6 - \frac{4.7}{2} = 3.65 \text{ in.}$$

For a minimum weld size of $\frac{1}{4}$ in, we assume that one-half of its capacity will carry shear (the other half will carry the moment). The total length of welds required by shear is

$$\sum l = \frac{70}{\frac{1}{2}(0.707) \times \frac{1}{4} \times 0.3 \times 70} = 37.7 \text{ in.}$$

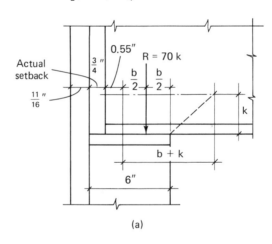

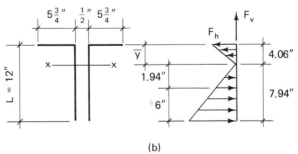

Figure 9.87 Detail for Bearing Length and Vertical Welds.

If the horizontal welds amount to $11\frac{1}{2}$ in., Fig. 9.87b, the vertical length, L, is

$$L = \frac{37.7 - 11.5}{2} = 13 \text{ in.}$$

We are not sure that exactly one-half of the capacity will be used for shear; therefore we can reduce this length to, say, 12 in.

Assuming elastic behavior, the weld centroid is at $\bar{y}$ from the top, i.e.

$$\bar{y} = \frac{2 \times 12 \times 6}{2(12 + 5.75)} = 4.06 \text{ in.}$$

The moment of inertia, I_x is

$$I_x = 2 \times 5.75 \times (4.06)^2 + 2\left[\frac{12^3}{12} + 12 \times (1.94)^2\right] = 568 \text{ in}^3.$$

The force in the top of the weld is

$$\text{Due to shear} \quad = F_v = \frac{70}{2 \times 17.75} = 1.97 \text{ kip/in.} \uparrow$$

$$\text{Due to moment} \quad - F_h = \frac{70 \times 3.65}{568} \times 4.06 = 1.83 \text{ kip/in.} \leftarrow$$

Using traditional vector analysis, the resultant force is

$$F_r = \sqrt{1.97^2 + 1.83^2} = 2.7 \text{ kip/in.} < 3.7: \quad \underline{\text{OK}}$$

The AISC *Manual* on pp. 4-52 and 4-53, table VIII gives for E-70XX electrodes the allowable loads for various weld sizes and seat widths.

9.3.1.3 End-Plate Shear Connections

A plate with a thickness between $\frac{1}{4}$ and $\frac{3}{8}$ in. and of a length, L, between 3 and $17\frac{1}{2}$ in. (Fig. 9.88) can often be used successfully instead of a double-angle framing connection. The plate is fillet-welded to the beam-web and bolted to the column. If no more than 6 bolts are used in one vertical row and the plate has the above thicknesses and lengths, the end rotation capacity and strength of such a connection closely

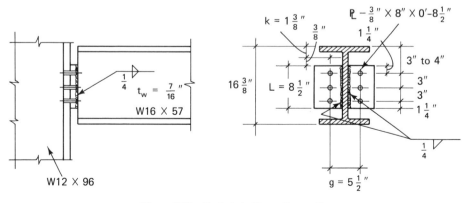

Figure 9.88 End-plate Shear Connection.

Connection Design

approximate the double-angle framing connection. Fabrication of the beam has to be under strict control to obtain an accurate beam length and square ends. The end plate must be parallel after fabrication and shims may be required on runs of beams to compensate for mill and shop tolerances.

The AISC *Manual* in table IX (p. 4-56) lists various capacities of connections. The next example will illustrate a design, based on this table.

Example 9.9 Design an end-plate shear connection for a W16 × 57 beam transmitting to a W12 × 96 column a reaction of 60 kips. The bolts are $\frac{7}{8}$-in. diameter ASTM-A325-F. Use A36 material, E-70XX electrodes, and AISC specification.

Solution One A325-F bolt of $\frac{7}{8}$-in. diameter has a shear (single) capacity of 10.5 kips, and 6 bolts are adequate (Fig. 9.88). For bearing capacity (3-in. spacing) table I-E (*Manual*, p. 4-6) gives 16.3 < 10.5 kips for a material thickness of $\frac{1}{4}$ in. We shall calculate the actual bearing thickness, the plate thickness, from the net shear requirement

$$t_{pl} = \frac{R}{2[L - n(d + \frac{1}{16})]F_v} \qquad (9.87)$$

in which t_{pl} = plate thickness; R = reaction in kips; L = plate length; n = number of bolts in one vertical line; d = bolt diameter; and F_v = allowable shear stress = 17.4 ksi = $0.3F_u$.

For our values, from Eq. (9.87) we obtain

$$t_{pl} = 0.30 \text{ in.} \quad \text{use } \tfrac{3}{8} \text{ in.}$$

The beam-web thickness is $\frac{7}{16}$ in. and the minimum weld size is $\frac{3}{16}$ in. (table 1.17.2A, *Manual*, p. 5-52). The required size (number of sixteenths) is obtained from the following equation:

$$N = \frac{R}{2L \times 0.707 \times \frac{1}{16} \times 0.3 \times 70} = \frac{R}{1.86L} \qquad (9.88)$$

In our case we have $N = 3.8 \approx 4$. Use $\frac{1}{4}$-in. welds. Looking at table IX (*Manual*, p. 4-56) we can easily check the design: For 3 bolts of $\frac{7}{8}$-in. diameter made of A325-F in one vertical line total capacity is 63.1 kips > 60: OK Minimum plate thickness is 0.32 < $\frac{3}{8}$ in. Plate length is $8\frac{1}{2}$ in., and $\frac{1}{4}$-in. weld capacity is 59.4 kips.

The minimum web thickness (AISC *Manual*, p. 4-56) is 0.51 in., which is in excess of $\frac{7}{16}$ in. Therefore the actual capacity is:

$$\frac{7/16}{0.51} \times 59.4 = 50.96 \text{ kips} < 60 \text{ kips} \qquad \text{N.G.}$$

We try next $L = 11\frac{1}{2}$ in. with 4 bolts. For one vertical line the total capacity is 84.2 kips > 60.0 kips: OK
Now

$$N = \frac{R}{1.86L} = 2.81 \quad \text{use } \tfrac{3}{16} \text{ in. welds.}$$

Since the required web thickness of 0.38 in (*Manual*, p. 4-56) is less than the actual web thickness of $\frac{7}{16}$ in. no further adjustment is needed. The weld capacity now is (*Manual* p. 4-56) 61.9 kips > 60.0 kips: OK

9.3.2 Rigid Moment Beam Connections

In Section 9.1.1, Types of Connections, it was pointed out that Type 1, rigid-frame construction, assumes that beam-to-column connections are so rigid that the original angles between intersecting members are practically unchanged. A steel frame with such connections can take horizontal load. This requires that columns must be designed to resist beam gravity moments and connections must be able to develop the full moment capacity of the girders (9.64). As columns are not efficient members when resisting moments and full moment connections are often expensive, especially if column stiffeners are required, this results in uneconomical frame structures. A better approach is to provide frame bracing with "wind connections."

Wind connections are designed to carry only the moments due to wind, without regard to the additional moments caused by gravity loading of the girders. This assumption implies that a connection is "intelligent" and "knows" which moment to carry and which not to carry. However this assumption can easily be proved by considering the moment-rotation characteristics of moment connections as shown in Fig. 9.89 acting through a full wind cycle (9.65). "Beam-line" plots show the end rotation of a beam, ϕ, as a function of the end moment. Since ϕ is directly proportional to M, the beam line is a straight line between point 1 (rigidly fixed beam, no rotation) and point 7 (simply supported beam, rotation ϕ_0).

Type 3 (semi-rigid) connections must be capable of furnishing, as a minimum, a reliable proportion of full end restraint. For bolted connections the use of type 2 connections with wind connections is a sound and economical approach and it would appear that the overall rigidity of the structure is not greatly reduced because of lighter connections. The satisfactory performance of some of the world's tallest buildings seems to confirm this view. For welded connections type 1 connections are quite adequate.

9.3.2.1 *Welded Connections*

For Type 1 connections, which must develop the frame moment, the following type of connection (Fig. 9.90) may be used. The flange plates, shop welded to the column and field-connected to the beam flanges, are transmitting the moment. The shear is assumed to be transferred from the beam to the column by a vertical plate, shop welded to the column and field bolted to the beam-web. Example 9.10 illustrates the design procedure.

> ***Example 9.10*** Design a moment connection for a W24 × 55 beam framed to a W14 × 82 column. The design moment 177 kips-ft and the end reaction of 43 kips are results of dead and live load only. All material is ASTM A36 steel with an allowable tension stress, F_t, of 22 ksi (for beams, $F_b = 24$ ksi). Use A325-N bolts and E-70XX electrodes.
>
> **Solution**
>
> BOLT DIAMETER. As the flange thickness of the beam is $t_f = 0.505$ in., the bolt diameter is
>
> $$d = \sqrt{2 \times 0.505} - 0.1 = 0.90 \text{ in.} \quad \text{use } \tfrac{7}{8}\text{-in. diameter bolts}$$

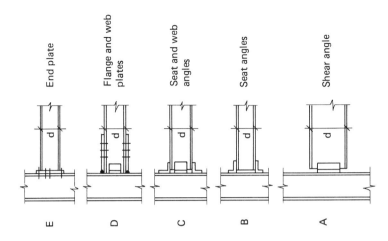
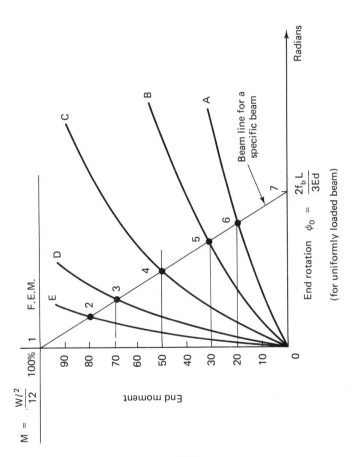

Figure 9.89 Rotation Characteristics of End-beam Connections.

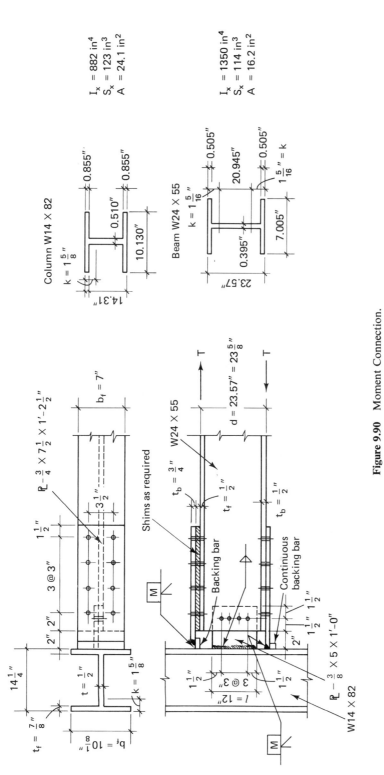

Figure 9.90 Moment Connection.

NET SECTION MODULUS OF THE BEAM. According to the AISC specification (*Manual*, p. 5-30) when the reduction of the area of either flange by field bolt holes exceeds 15% of the gross flange area, the excess shall be deducted from its moment of inertia. Therefore, because

$$A_f(\text{gross}) = 7.005 \times (0.505) = 3.54 \text{ in.}^2$$
$$A_f(\text{net}) = 3.54 - 2(0.875 + 0.125)(.505) = 2.53 \text{ in.}^2$$

The percentage loss: $\dfrac{(3.54 - 2.53)100}{3.54} = 28.5\% > 15\%$ (13.5% excess)

Thus the moment of inertia of the flanges must be reduced by 13.5%. The net moment of inertia is (see Fig. 9.90)

$$I_{(\text{net})} = 1350 - 2 \times 0.135 \times 3.54 \left(\frac{23.57 - 0.505}{2}\right)^2 = 1330 \text{ in.}^4$$

$$S_{(\text{net})} = \frac{1330}{23.57} \times 2 = 112.8 \text{ in.}^3$$

The stress in the beams

$$f_b = \frac{177 \times 12}{112.8} = 18.8 \text{ ksi} < 22 \text{ ksi:} \qquad \underline{\text{OK}}$$

HORIZONTAL FORCE AT BEAM FLANGE

$$T = (M \times 12)/d = \frac{177 \times 12}{23.57} = 90.1 \text{ kips}$$

FLANGE PLATES. The necessary net area of one plate, A_p, is

$$A_p = \frac{T}{F_t} = \frac{90.1}{22} = 4.10 \text{ in.}^2$$

Try a $\tfrac{3}{4}$-in. plate. The required width of the plate is

$$b = (4.10/0.75) + 2(\tfrac{7}{8} + \tfrac{1}{16}) = 7.34 \text{ in.} \qquad \text{use } 7\tfrac{1}{2} \text{ in.}$$

The net area of the plate $7\tfrac{1}{2} \times \tfrac{3}{4}$ ($A = 5.625$ in.^{2}) =

$$A_p = 5.625 - 2(\tfrac{7}{8} + \tfrac{1}{16}) \times \tfrac{3}{4} = 4.22 \text{ in.}^2 > 4.10 \text{ in.}^2: \qquad \underline{\text{OK}}$$

Check $0.85 \times 5.625 = 4.78 > 4.22$ in.^{2}: $\qquad \underline{\text{OK}}$

NET SECTION CONTROLS. Use: $\tfrac{3}{4} \times 7\tfrac{1}{2}$ in. flange plates top and bottom.

FLANGE CONNECTION. From the AISC *Manual*, pp. 4-5 and 4-6, the shear capacity of a $\tfrac{7}{8}$-in. dia. A325-N bolt in single shear is 12.6 kips. In bearing (on a $\tfrac{1}{2}$-in. flange thickness and 3-in. spacing) the capacity is 37.2 kips > 12.6 kips. The required number of bolts is

$$n = \frac{90.1}{12.6} = 7.2 \approx 8 \text{ bolts}$$

WEB CONNECTION. The number of $\tfrac{7}{8}$-in. dia. A325-N bolts required for shear (bearing capacity on $\tfrac{3}{8}$-in. web thickness and 3-in. spacing is 27.9 > 12.6)

$$n = \frac{43.0}{12.6} = 3.4 \approx 4 \text{ bolts}$$

From the AISC *Manual*, table I-F (p. 4-10) and $F_u = 58$ ksi, and $l_v = 1\tfrac{1}{2}$ in., the allowable load is

$$4 \times (0.395) \times 43.5 = 68.7 \text{ kips} > 43.0 \text{ kips:} \qquad \underline{\text{OK}}$$

Checking net shear in the web plate of length, l, 12 in.

$$l_{net} = 12 - 4(\tfrac{7}{8} + \tfrac{1}{16}) = 8.25 \text{ in.}$$

Allowable shear stress, $F_{vp} = 0.3 \times 58 = 17.4$ ksi and the required plate thickness is

$$t = 43/(17.4 \times 8.25) = 0.30 \text{ in.} \quad \text{use } \tfrac{3}{8} \text{ in.}$$

Bearing capacity of bolts on the plate (table I-F, *Manual*, p. 4-10)

For edge distance $l_v = l_h = 1\tfrac{1}{2}$ in.: $\quad 43.5 \times \tfrac{3}{8} \times 4 = 65.2 > 43.0$ kips

For 3-in. bolt spacing (table 1-E): $\quad 27.9 \times 4 = 111.6 > 43.0$ kips

Shear plate is connected to the column flange ($t_f = 0.855$ in. $> \tfrac{3}{4}$ in.) by two fillet welds. Minimum size is (table 1.17.2.A, *Manual*, p. 5-52) $\tfrac{5}{16}$ in. The shearing capacity of these welds is

$$2 \times 0.707 \times \tfrac{5}{16} \times 0.3 \times 70.0 \times 12 = 111.4 > 43 \text{ kips}: \quad \underline{\text{OK}}$$

To check the web shear in the column, within the boundaries of the rigid connection, we shall follow the suggestions given in the commentary on the AISC specification (*Manual*, p. 5-108), section 1.5.1.2, Shear. If the beam depth is d_b and the column depth is d_c, and if the moment M (the algebraic sum of clockwise and counterclockwise moments in kip-feet, applied on opposite sides of the connection boundary) is represented as a force couple in the beam flanges at an assumed distance of $0.95d_b$, with $0.4F_y$ as allowable shear stress, we have the following equilibrium equation with the unknown minimum web thickness, t_{min}:

$$0.4F_y t_{min} d_c = \frac{12M}{0.95d_b} \tag{9.89}$$

From there t_{min} is equal to

$$t_{min} = \frac{32M}{d_b d_c F_y} = \frac{32M}{A_{bc} F_y} \tag{9.90}$$

in which $A_{bc} = d_b d_c$ is the planar area of the connection web, (shaded in Fig. 9.91), expressed in square inches. In our case

$$A_{bc} = 14.31 \times 25.07 = 358.8 \text{ in.}^2$$

and

$$t_{min} = \frac{32(177)}{358.8 \times 36} = 0.44 \text{ in.} < 0.51 \text{ in.}: \quad \underline{\text{OK}}$$

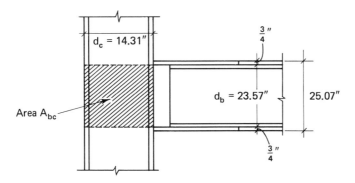

Figure 9.91 Web Shear Checking.

Connection Design

Column web stiffeners with cross-sectional area A_{st} are required opposite both the tension and compression flange connection of the beam whenever formula (1.15-1) (*Manual*, p. 5-47) yields a positive result. The horizontal force at the stiffeners is

$$\frac{M \times 12}{d_b + t_b} = \frac{177 \times 12}{23.57 + 0.75} = 87.3 \text{ kips}$$

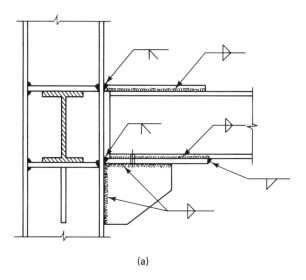

(a)

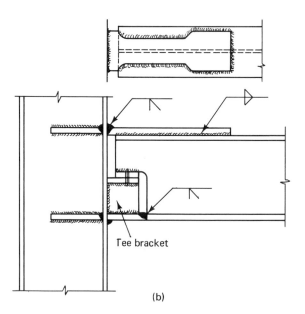

(b)

Figure 9.92 Rigid Frame Connections.

Therefore, from *Manual* Formula (1.15-1) we get

$$A_{st} = \frac{5/3(87.3) - 35(0.510)(0.75 + 5 \times 1.625)}{36} = -0.48 < 0$$

As this value is negative, no stiffeners are required.

Figures 9.92a and b show two more versions of Type 1 connections. Easy erection, with end clearance and flange plates butt welded to the column flange, is the main idea of Fig. 9.92a. The connection in Fig. 9.92b is similar to the previous one. The main problem with these connections is the danger of lamellar tearing in the column flange due to the shrinkage of groove welds in flange plates and column stiffeners. If the thickness of the column flanges is substantial (over 2 in.) the connection shown in Fig. 9.90 where the plates are prewelded to the column and site-bolted to the beam flanges is a better choice. In this case, the moment connection with end plate shown in Fig. 9.93 is also a good solution.

In spite of the fact that the end-plate moment connection represents a highly nonlinear, indeterminate, and complex situation, still it is of great practical significance (9.66). The AISC *Manual* (9.2) gives the design method based on the modified split-tee concept (pp. 4-111 to 4-119). Example 9.11 will demonstrate this procedure.

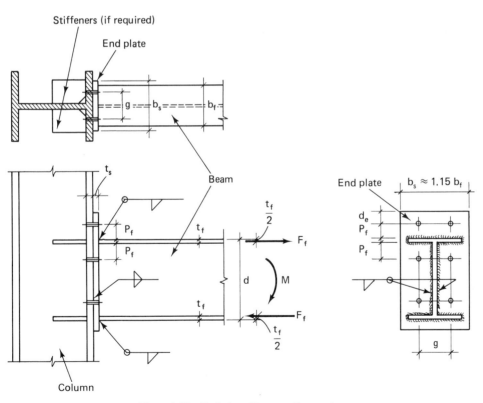

Figure 9.93 End-plate Moment Connection.

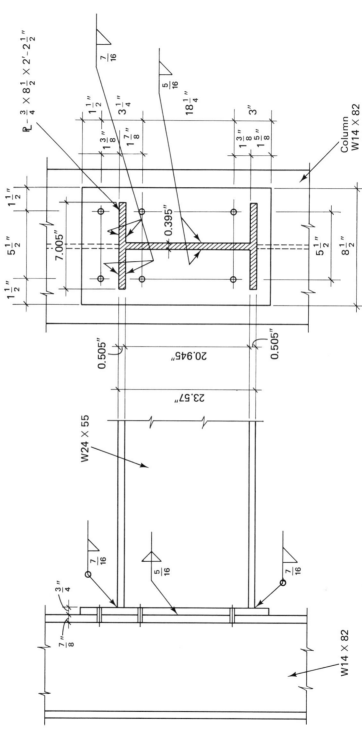

Figure 9.94 Example 9.11.

Example 9.11 For the beam and the column data, as well as for the negative moment and reaction data given in Example 9.10, design the beam-to-column connection using an end-plate connection and ASTM A325-N bolts. $F_y = 36$ ksi steel is used for all members. The end plate is shop welded to the beam with E-70XX electrodes (Fig. 9.94).

Solution
1. Find the nominal flange force F_f from

$$F_f = \frac{M}{d - t_f} = \frac{177 \times 12}{23.57 - 0.505} = 92.1 \text{ kips} \tag{9.91}$$

2. Find the required bolt area, A_b, for 2 bolts per row, from the equation

$$A_b = \frac{F_f}{2 \times 2 \times F_{bt}} = \frac{92.1}{2 \times 2 \times 44} = 0.524 \text{ in.}^2 \tag{9.92}$$

in which $F_{bt} = 44$ ksi = allowable tensile stress in bolt. A bolt of $\frac{7}{8}$-in. diameter has an area of 0.601 in. > 0.524 in. (*Manual*, table I-A, p. 4-3). Use $\frac{7}{8}$-in. dia. bolts.

3. Find the number of bolts required for reaction transfer. Single shear capacity of one A325-N bolt of $\frac{7}{8}$-in. diameter is 12.6 kips (*Manual*, table I-D, p. 4-5). The bearing capacity of one bolt on $\frac{3}{4}$-in. thickness (assumed plate thickness < column flange thickness) for 3-in. spacing is 55.7 kips $= 0.75 \times 74.3$ (table I-E, p. 4-6).

Therefore, the number is

$$n = \frac{43.0}{12.6} = 3.4 \quad \text{use 6 bolts}$$

4. Find the effective bolt distance P_e from the equation

$$P_e = P_f - \frac{d_b}{4} - w_t \tag{9.93}$$

in which P_f = bolt distance from the flange
d_b = bolt diameter
w_t = throat of the fillet weld between flange and plate.

Taking $P_f = \frac{7}{8} + \frac{1}{2} = 1\frac{3}{8}$ in. and $w_t = 0.707 \times \frac{5}{16} = 0.22$ in.

$$P_e = 1.375 - 0.875(0.25) - 0.220 = 0.936 \text{ in.}$$

5. Find the split-tee moment M_t from the equation

$$M_t = \left(\frac{F_f}{2}\right) \cdot \left(\frac{P_e}{2}\right) = 0.25 F_f P_e \tag{9.94}$$

$$M_t = 0.25(92.1)(0.936) = 21.56 \text{ kips-in.}$$

6. Find the moment modification factor α_m from the equation

$$\alpha_m = C_a C_b (A_f/A_w)^{0.32} (P_e/d_b)^{0.25} \tag{9.95}$$

in which coefficient C_a lumps all material interactions together, i.e.

$$C_a = 1.29 (F_y/F_{bu})^{0.4} (F_{bt}/F_p)^{0.5} \tag{9.96}$$

(F_y = yield stress of beam and plate material; F_{bt} = ultimate tensile stress of bolt; F_{bt} = allowable tensile stress in bolt; F_p = allowable bending stress in end plate) and

$$C_b = (b_f/b_s)^{0.5} \tag{9.97}$$

in which b_f = width of beam flange, b_s = width of end plate. In our case F_y = 36 ksi, F_{bu} = 93 ksi, F_{bt} = 44 ksi; F_p = 27 ksi. Substituting into Eq. (9.96)

$$C_a = 1.127 \approx 1.13$$

From Eq. (9.97) find C_b, the plate width correction

$$C_b = \left(\frac{7.005}{8.500}\right)^{0.5} = 0.908 \approx 0.91$$

As the ratio between the flange and web areas is

$$\frac{A_f}{A_w} = \frac{7.005 \times 0.505}{[23.57 - (2 \times 0.505)] \times 0.395} = 0.397$$

and the ratio between the effective span and bolt diameter

$$\frac{P_e}{d_b} = \frac{0.936}{0.875} = 1.070$$

the modification factor α_m from Eq. (9.95) is

$$\alpha_m = 1.13 \times 0.91(0.397)^{0.32}(1.070)^{0.25} = 0.778$$

7. Find the design effective moment M_e for the end plate from the equation

$$M_e = \alpha_m M_t \tag{9.98}$$

$$M_e = 0.778 \times 21.56 = 16.8 \text{ kips-in.}$$

8. Find the end-plate thickness. Required plate thickness t_s is (for allowable bending stress of 27 ksi = 0.75 × 36)

$$t_s = \left(\frac{6M_e}{b_s \times F_b}\right)^{\frac{1}{2}} = \left(\frac{6 \times 16.8}{8.0 \times 27}\right)^{\frac{1}{2}} = 0.68 \text{ in.} \qquad \text{use } \tfrac{3}{4} \text{ in.}$$

Check plate shear:

$$f_v = \frac{92.1}{2 \times 8\tfrac{1}{2} \times \tfrac{3}{4}} = 7.2 \text{ ksi} < 0.4F_y = 14.5 \text{ ksi}: \qquad \underline{\text{OK}}$$

9. Check the effective plate width b_e from the equation

$$b_e = b_f + 2w_s + t_s \tag{9.99}$$

in which b_f = width of beam flange; w_s = size of fillet weld (taken as zero for unreinforced groove welds); t_s = end-plate thickness.

In our example

$$b_e = 7.005 + 2 \times \tfrac{5}{16} + \tfrac{3}{4} = 8.38 \text{ in.} \approx 8\tfrac{1}{2} = b_s: \qquad \underline{\text{OK}}$$

10. Check fillet weld between top flange (tensile force F_f = 92.1 kips) and the end plate.

$$\text{Supplied length} = 2(7.005 + 0.505) - 0.395 = 14.625 \text{ in.}$$

$$\text{Required size (number of sixteenths of an inch)} =$$

$$\frac{92.1}{14.625 \times 0.707 \times \tfrac{1}{16} \times 21.0} = 6.8$$

Use 16-in. fillet welds.

11. Check the fillet weld between web and the end plate. Minimum weld size is $\tfrac{1}{4}$ in. (table 1.17.2A, *Manual*, p. 5-52)

The capacity of a $\tfrac{1}{4}$-in. weld is

$$0.707 \times \tfrac{1}{4} \times 0.3 \times 70 = 3.71 \text{ kips/in.}$$

The required length of weld on both sides of beam-web is

$$43.0/(2 \times 3.71) = 5.8 \text{ in.} < 20.945 \text{ in. (see Fig. 9.94)}$$

Required weld size to develop maximum bending stress (24 ksi) in web near flanges:

$$\frac{1.0 \times 0.395 \times 24}{2 \times 0.707 \times \frac{1}{16} \times 21.0} = \frac{0.395 \times 24}{2 \times 0.928} = 5.1$$

Use a $\frac{5}{16}$-in. weld continuous on both sides of beam-web. In Example 9.10 we found that column stiffeners are not required.

Instead of calculating the coeffient C_a and the ratio A_f/A_w of flange area and web area, the AISC *Manual* gives them in tables A and B (p. 4-113), respectively.

9.3.2.2 Bolted Connections

T-Stub flange-to-column connections: Describing in general the behavior of such connections, the T-stubs are assumed to transfer the end-beam-moment and the frame angles the beam-shear force. The bending moment is replaced by a force couple. On the compression side of a beam, a T-stub functions more or less as a bearing pad and may be treated as such, with major attention given to the possibility of the crippling of the column or T-stub web (9.39). The situation in the flange-to-column connections on the tension side of the beam is more complex. The main problems in the analysis of this part of the joint are those associated with the variation in tensile force in the bolts and the possible development of prying forces between the T-stub flange and the column. In Section 9.2.1.5, Concentric Tension on Bolt Groups, the prying phenomenon was discussed and, as was pointed out, it is important to predict whether or not prying forces occur at all and, if they do, to assess their magnitude, because prying forces add to the force on the bolt. Flange stiffness, rather than bending strength, the amount of pretension, and outer distance (Fig. 9.96) are the key to satisfactory performance of this type of connection. Here, in designing a T-stub connection, we shall follow the design method given in the AISC *Manual* on p. 4-89. This will be illustrated in Example 9.12.

Example 9.12 Design a T-stub connection of a W21 × 44 beam to a W36 × 230 column. The negative beam end moment is 100 kips-ft (136 kN-m) and the beam reaction is $V = 42$ kips (187 kN). All material is A36 and bolts are ASTM A325-F. Use AISC specification of 1978. The sketch of the connection is shown in Fig. 9.95.

Solution The tensile force in the top of the T-stub is

$$4T = \frac{M}{d} = \frac{100 \times 12}{20.66} = 58.1 \text{ kips}$$

Force in one bolt: $T = 58.1/4 = 14.5$ kips/bolt. From table I-A, *Manual*, p. 4-3, one $\frac{3}{4}$-in. A325 bolt may have a tensile force of $B = 19.4$ kips. Four bolts can take 77.6 kips $>$ 58.1 kips: OK

If the gage line on the column flange is 4 in., the flange length parallel to the stem is $2p = 2 + 4 + 2 = 8$ in., i.e., $p = 4$ in. On p. 4-88 of the *Manual* we

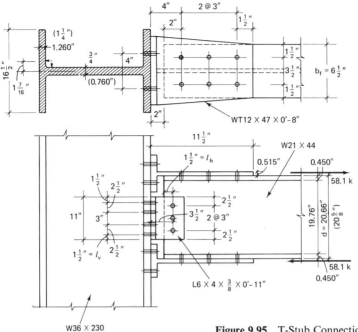

Figure 9.95 T-Stub Connection, Example 9.12.

enter the Preliminary Selection Table with

$$\frac{2T}{p} = \frac{2 \times 14.5}{4} = 7.25$$

and $b = 4.0/2$ minus one-half the stem thickness (see Fig. 9.96); assume $b = 1\frac{3}{4}$ in. The closest higher number to 7.25 is 7.88 and the preliminary thickness of the flange of the T-section is 7 in. A WT12 × 47 (*Manual*, p. 1-52) has a $\frac{7}{8}$-in. flange thickness. We shall try this section. The actual value of b now is

$$b = 2 - 0.258 = 1.742 \text{ in.} > 1\frac{1}{4}\text{-in. wrench clearance}$$

(AISC *Manual*, p. 4-132.)

$1.25b = 2.18$ in.

The outer span, a, is (Fig. 9.96)

$$a = (9.065 - 4.0)/2 = 2.532 < 1.25b \qquad \text{use } a = 2.18 \text{ in.}$$

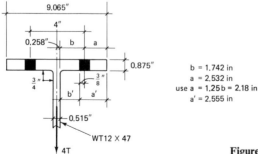

Figure 9.96 Detail of T-Stub Flange.

Further, we have
$$b' = b - (\tfrac{3}{4})/2 = 1.742 - 0.375 = 1.367 \text{ in.}$$
$$a' = a + \tfrac{3}{8} = 2.18 + 0.375 = 2.555 \text{ in.}$$
$$a' + b' = 3.922 \text{ in.}$$
$$d' = \tfrac{3}{4} + \tfrac{1}{16} = \tfrac{13}{16} = 0.8125 \text{ in.}$$

The ratio δ of net area of flange (at bolt line) and the gross area (at the face of the stem)
$$\delta = 1 - d'/p = 1 - (0.8125/4.0) = 0.797$$

The allowable bending moment caused in the T-stub flange by one bolt (kip-in.)
$$M = M_p/2 = pt_f^2 F_y/8 = [4.0(0.875)^2 \times 36]/8 = 13.78 \text{ kip-in.}$$

Then, the moment ratio α is
$$\alpha = (Tb'/M - 1)/\delta = [(14.5 \times 1.367/13.78) - 1]/0.797$$
$$= 1.05 > 1.0 \quad \text{use } \alpha = 1.0$$

Finally, the load per bolt including prying action (from Eq. (9.42) in Section 9.2.1.5)
$$B_c = T\left[1 + \frac{\delta\alpha}{(1+\delta\alpha)}(b'/a')\right] = 14.5\left[1 + \frac{0.797}{(1+0.797)}(1.367/2.555)\right]$$
$$= 1.238 \times 14.5 = 17.9 \text{ kips} < 19.4 \text{ kips:} \quad \underline{\text{OK}}$$

Required thickness of T-stub flange is
$$t_f = \left[\frac{8B_c a'b'}{pF_y[a' + \delta\alpha(a'+b')]}\right]^{\frac{1}{2}} = \left[\frac{8 \times 17.9 \times 2.555 \times 1.367}{4.0 \times 36[2.555 + 0.797 \times 3.922]}\right]^{\frac{1}{2}}$$
$$= 0.245 \text{ in.} < 0.875 \text{ in.:} \quad \underline{\text{OK}}$$

The prying force, Q, is
$$Q = B_c - T = 17.9 - 14.5 = 3.4 \text{ kips}$$

The connection of the bottom T-stub flange will be the same, i.e., with 4 A325-N bolts of $\tfrac{3}{4}$-in. diameter.

The stem of the T-stub has to take the force of the couple from the beam flange, i.e., 58.1 kips and transfer it through its flange to the column. From table I-D, *Manual*, p. 4-5, the single shear capacity of an A325-N bolt of $\tfrac{3}{4}$-in. dia. is 9.3 kips. From table I-E, *Manual*, p. 4-6, its bearing capacity (on $\tfrac{7}{16}$ in.) with 3-in. spacing is 28.5 > 9.3 kips. The number of bolts is
$$58.1/9.3 = 6.2 \quad \text{use 6 bolts (4\% overstress)}$$

Shear connection of the beam web has to transmit to the column the reaction $V = 42$ kips. The web thickness is 0.350 in. $\approx$ 3 in. The bearing capacity from table I-E is 24.5 kips > 2 × 9.3 kips. Therefore the number of double shear bolts in the web is
$$42.0/18.6 = 2.3 \quad \text{use 3 bolts}$$

The maximum length of framing angles is
$$20.66 - 9.065 + 0.515 = 12.11 \text{ in.}$$

A length of 11 in. will be used to allow staggering of the bolts in both angle legs. For a leg thickness of $\frac{3}{8}$ in. of the framing angles and a gage line (in 6-in. leg) at $3\frac{1}{2}$ in. the net distance F (see *Manual*, p. 4-132, A325 bolts stagger) is $3\frac{1}{2} - \frac{3}{8} = 3\frac{1}{8}$ in. and no staggering required. Anyway, for the ease of erection, the bolts in the outstanding angle legs will be staggered 1 in. The last is to check the edge distances $l_h = l_v = 1\frac{1}{2}$ in. From table I-F, Manual p. 4-10, for A36 material ($F_u = 58$ ksi) the allowable load is 43.5 kips $>$ 42.0 kips: OK

9.3.3 Column Bracket Connection

In Connection Performance, Section 9.2, both bolts and welds loaded in torsion and eccentric shear were discussed. Three approaches to the problem were given: the old elastic approach; its correction, the effective-arm approach; and the new ultimate strength method with the instantaneous center of rotation. For this last method, the tedious search for the instantaneous center of rotation can be avoided by using tables X through XVIII for bolts and tables XIX through XXVI for welds in the AISC *Manual*.

To illustrate the elastic and ultimate design methods some problems will be solved using both techniques, bolting and welding, and the results will be compared.

Example 9.13 Design a bolted connection of a bracket to a W36 × 245 column for the loading as shown in Fig. 9.97. All steel material is A36. Use ASTM A325-F bolts and AISC specification of 1978. Compare the elastic and ultimate designs.

Solution

ULTIMATE METHOD USING AISC TABLE XII (*Manual*, p. 4-64). For the bracket thickness $t = \frac{5}{8}$ in. $< t_f = 1\frac{3}{8}$ in. = column flange thickness, the bolt

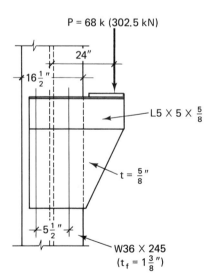

Figure 9.97 Example 9.13.

diameter according to Eq. (9.72) is

$$d \approx \sqrt{2 \times \tfrac{5}{8}} - 0.1 = 1.0 \text{ in.} \quad \text{use 1-in. diameter}$$

Single shear capacity is 13.7 kips. We have two vertical lines of bolts spaced at $5\tfrac{1}{2}$ in. with n bolts in each line. To find this number we use table XII (*Manual*, p. 4-64). The coefficient C, with which to enter in the table, is equal to

$$C = \frac{P}{r_v} = \frac{68}{13.7} = 4.96 \approx 4.94 \ldots \text{ for } n = 9.$$

The depth of the bracket plate is then

$$1\tfrac{1}{2} + 8 \times 3.0 + 1\tfrac{1}{2} = 27 \text{ in.}$$

It is good at this time to check the stresses in the bracket plate, namely, through the gage line closest to the force P (Fig. 9.98). The stress distribution in the brackets is fairly complex and the simple bending theory does not apply. But it is always better to have even a rough stress check than none. The net section modulus is given in the AISC *Manual*, p. 4-87, as 48 in.³. The net shearing area is

$$\tfrac{5}{8}[24 - 9 \times (1 + \tfrac{1}{8})] = 8.67 \text{ in.}^2$$

Bending moment at the vertical gage line closest to P is

$$68 \times (24 - 2\tfrac{3}{4}) = 1445 \text{ kip-in.}$$

Therefore, the approximate stresses in the bracket are:

$$\text{Due to bending } 1445/48 = 30.1 \text{ ksi} > 22 \text{ ksi} \quad \underline{\text{NG}}$$
$$\text{Due to shear } 68/8.67 \ \ = 7.8 \text{ ksi} < 14.4 \text{ ksi} \quad \underline{\text{OK}}$$

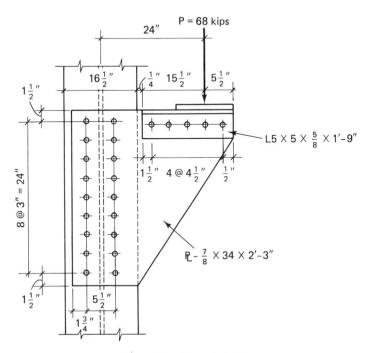

Figure 9.98 Example 9.13.

Because of bending, increase the bracket plate thickness to $\frac{7}{8}$ in. The section modulus is 67 in.³ and stress is 21.6 ksi < 22 ksi: **OK**

ELASTIC METHOD WITH BRACKET ROTATION ABOUT BOLT CENTER. The previous example can be a good guide how to start, namely how many bolts to take in each vertical gage line. We know that the elastic method approach is more conservative than the ultimate one. Therefore, let us assume that $n = 9 + 1 = 10$ bolts of 1-in. diameter, ASTM A325-F, Fig. 9.99. The centroid of the bolt group, due to the symmetry, is easy to locate. The polar moment of inertia I_p of bolts (the area of each one taken as a unit in terms of forces and not stresses) is equal to $I_x + I_y$, i.e.

$$I_p = \sum_{i=1}^{18}(x_i^2 + y_i^2) = 4 \times (5 \times 2.75^2 + 1.5^2 + 4.5^2 + 7.5^2 + 10.5^2 + 13.5^2) \stackrel{!}{=} I_p = 1636 \text{ in.}^2$$

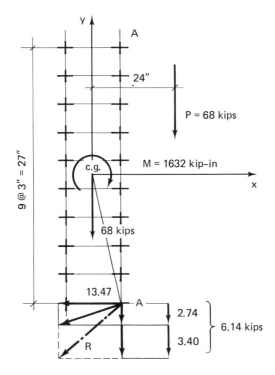

Figure 9.99 Elastic Approach.

The moment equals $M = 60 \times 24 = 1632$ kip-in.

1. *Vertical components of forces*

   a. From P: .... $68/20 = 3.40$ kips

   b. From M: $\dfrac{M}{I_p} \cdot x_{\max} = \dfrac{1632}{1636} \times 2.75 = 2.74$ kips

   Total vertical component = 6.14 kips ↓

2. *Horizontal component of forces*

Only the moment contributes to the horizontal component, i.e.,

$$\frac{M}{I_p} \times y_{max} = \frac{1632}{1636} \times 13.5 = 13.47 \text{ kips}$$

The resultant force is

$$R = \sqrt{13.47^2 + 6.14^2} = 14.8 \text{ kips} > 13.7: \quad \underline{\text{NG}}$$

Although we increased the number of bolts by 2, according to classical method, the connection is still not safe. Obviously, when the number of bolts in increased to 22 (from 18) the connection will be safe.

From this comparison we realize that the instantaneous center approach offers some savings. For instance, in this example the savings in number of bolts is 22%.

Example 9.14 Design a welded connection of the bracket shown in Fig. 9.97. All steel material is A36. Use for welding electrodes E-70XX and the AISC specification of 1978. Design by the instantaneous center method.

Solution To use table XXIII, Eccentric Loads on Weld Groups, *Manual*, p. 4-80, we first have to assume the vertical weld length, l, Fig. 9.100. Minimum weld size, governed by column flange thickness $t_f = 1\frac{3}{8}$ in., from table 1.17.2A (*Manual*, p. 5-52) is $\frac{5}{16}$ in. Maximum weld size, governed by bracket thickness of $\frac{7}{8}$ in. is $\frac{13}{16}$ in. (*Manual*, p. 5-53, section 1.17.3). We shall first try a weld of $\frac{5}{16}$ in. If horizontal parts of weld are about half of column flange width, $2 \times 8 = 16$ in., if $a \approx 0.9$ (large eccentricity) and $k \approx 0.3$ (a moderate value), from table XXIII we have a coefficient $C = 0.488$, where

$$C = \frac{P}{C_1 D l} \qquad (9.100)$$

in which $C_1 = 1.00$ for E-70XX electrodes; $D =$ number of sixteenths of an inch in fillet-weld size $= 5$; $P =$ permissible eccentric load in kips.

Solving Eq. (9.100) for vertical length of weld, l, we get

$$l = \frac{P}{CC_1 D} \qquad (9.101)$$

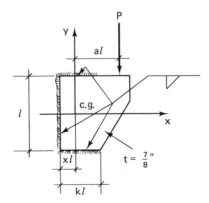

Figure 9.100 Example 9.14.

After substituting our values, we get $l = 28$ in. Checking the value of a (we assumed 0.9)

$$a = \frac{24}{28} = 0.86 \approx 0.9: \quad \underline{\text{OK}}$$

It looks as if the assumed values for a and k were acceptable. We have to round off obtained values and repeat the calculation. Therefore, for $l = 28$ in. and $k = 0.3$, the length of the horizontal welds is $0.3 \times 28 = 8.4 \approx 8\frac{1}{2}$ in. Then, the centroid is at $0.056 \times 28 = 1.57$ in. and al is equal to (Fig. 9.101)

$$al = 15\tfrac{3}{4} + 8\tfrac{1}{2} - 1.57 = 22.68 \text{ in.}$$

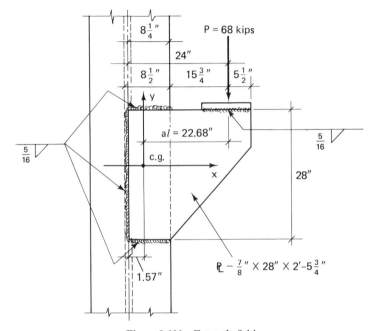

Figure 9.101 Example 9.14.

Therefore, $a = 22.68/28.0 = 0.81$. Now enter the same table XXIII by interpolation to obtain for C the following value

$$C = 0.488 + \frac{0.543 - 0.488}{0.10} \times 0.09 = 0.538$$

The permissible eccentric load, $P_{\max}$, from Eq. (9.100) is

$$P = CC_1 Dl = 0.538 \times 1.0 \times 5 \times 28.0$$
$$= 75.3 > 68.0 \text{ kips}: \quad \underline{\text{OK}}$$

Computer-aided design of bracket connections

The bracket-to-column flange connection design in both methods, elastic and ultimate, can be easily programmed on small programmable desk calculators.

In the following text, first the elastic design program for a Hewlett-Packard HP9820A calculator is described with the solution for Example 9.13(b). Next, a program in BASIC for a Hewlett-Packard HP9830A is given with the solution of Example 9.13(a).

The elastic bracket connection design was programmed for a desk calculator (in this case a Hewlett-Packard HP9820A). The solution is the same as previously given in Example 9.13(b). The program will determine the number of bolts in one vertical row of the bracket connection, as well as the maximum bolt stress and the approximate bending and shear stresses in the bracket plate.

The flowchart on page 398 gives the program operations and their sequences. The necessary input data are: bolt diameter, minimum thickness with respect to the bearing bolt capacity, vertical force and its eccentricity (force is positive if acting downward), horizontal force and its eccentricity (force is positive if acting from left to right), allowable shear and bearing bolt stresses, horizontal spacing gage between vertical bolt rows, vertical pitch, number of vertical rows (normally two or four), bracket plate thickness, bracket allowable bending and shear stresses. The complete program appears on page 399.

The instantaneous center (ultimate) design approach is programmed in BASIC for a HP9830A. This connection has two vertical lines of bolts $5\frac{1}{2}$-in. support with 3-in. spacing and up to 12 bolts in one line.

Input: The designer has to give the eccentricity, the force, the diameter of the bolt and its single shear capacity, the thickness t of the bracket plate, and the yield point stress of the steel used for the bracket.

Output: The results are: number of bolts in one vertical line, bending and shear stresses for bracket net section for the given plate thickness, and a message to increase the plate thickness if these stresses are larger than allowed. If the number of bolts in one vertical line is larger than 12, then an output message suggesting the increase of diameter d of the bolt and the single shear capacity will be received. Example 9.13(a) is solved by this program and Fig. 9.102 shows the results. The program listing is given on pages 400 and 401.

9.3.4 Beam and Girder Splices

Continuous beam splices

Any specific design of continuous beam splices will always depend upon the depth ratios of the stringers and floor beams or girders and their relative heights and positions.

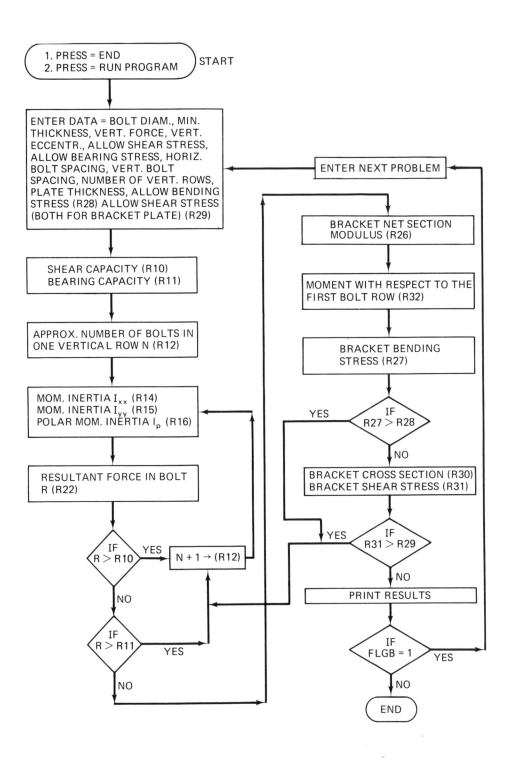

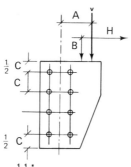

```
0:
TBL 4;TBL 5;FXD
3;CFG 13⊢
1:
ENT "BOLT DIAMET
ER?",R0,"THICKNE
SS MIN.?",R1,"VE
RT. FORCE?",R2⊢
2:
ENT "VERT. ECC.?
",R3,"HOR. FORCE
?",R4,"HOR. ECC.
?",R5⊢
3:
ENT "ALLOW SIN S
HEAR?",R6,"ALLOW
 BEARING?",R7,"H
OR BOLT SPACING"
,R8⊢
4:
ENT "VERT. SPACI
NG?",R9,"NUMBER
OF ROWS?",R13,"P
LATE THICKNESS?"
,R23⊢
5:
ENT "ALLOW BEND
STR.?",R28,"ALLO
W SHEAR STR?",R2
9⊢
6:
PRT "VERTICAL FO
RCE",R2,"VERTICA
L"," ECCENTRICIT
Y",R3⊢
7:
PRT "HORIZONAL F
ORCE",R4,"HORIZO
NAL"," ECCENTRIC
ITY",R5,"ALLOWAB
LE"⊢
8:
PRT "   SINGLE S
HEAR",R6,"   BEA
RING",R7,"MINIMU
M"," THICKNESS",
R1⊢
9:
π*R0↑2/4*R6→R10;
R1R0R7→R11⊢
10:
R2/2R10→R12;INT
(R12+1)→R12⊢
11:
R12*R9↑2*(R12↑2-
1)/12*R13→R14⊢
12:
R13*R8↑2*(R13↑2-
1)/12*R12→R15⊢
13:
R14+R15→R16⊢
14:
R2/R12R13→R17;
ABS R17→R17⊢
15:
R4/R12R13→R18;
ABS R18→R18⊢
16:
R2R3+R4*(R5+R12*
R9/2)→R19⊢
17:
R19*R8/2R16→R20;
ABS R20→R20⊢
18:
R19*(R12-1)*R9/2
R16→R21;ABS R21→
R21⊢
19:
√((R17+R20)↑2+(R
18+R21)↑2)→R22⊢
20:
IF R22>R10;R12+1
→R12;GTO 11⊢
21:
IF R22>R11;R12+1
→R12;GTO 11⊢
22:
R23*(R12*R9)↑2/6
→R24⊢
23:
R9↑2*R12*(R12↑2-
1)*R23*(R0+.125)
/6R12R9→R25⊢
24:
R24-R25→R26⊢
25:
R2*(R3-R8/2)+R4(
R5+R12*R9/2)→R32
⊢
26:
R32/R26→R27⊢
27:
IF R27>R28;R12+1
→R12;GTO 11⊢
28:
R12*R9*R23-R12*R
23*(R0+.125)→R30
⊢
29:
R2/R30→R31⊢
30:
IF R31>R29;R12+1
→R12;GTO 11⊢
31:
SPC 5⊢
32:
PRT "BOLT DIAMET
ER",R0⊢
33:
PRT "NUMBER OF B
OLTS","   PER RO
W",R12⊢
34:
PRT "NUMBER OF R
OWS",R13⊢
35:
PRT "VERTICAL SP
ACING",R9⊢
36:
PRT "HORIZONAL",
"   SPACING",R8⊢
37:
PRT "MAX. BOLT S
TRESS",R22⊢
38:
PRT "BENDING STR
ESS",R27,"SHEAR
STRESS",R31⊢
39:
PRT "PLATE DEPTH
",R12R9⊢
40:
PRT "PLATE THICK
NESS",R23⊢
41:
SPC 8⊢
42:
ENT "NEXT BRACKE
T?",R33;IF FLG 1
3=1;GTO 0⊢
43:
DSP "     FINISHE
D"⊢
44:
STP ⊢
45:
END ⊢
R236
```

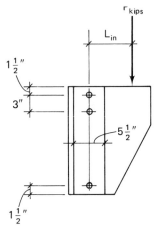

```
10 DIM A[36,12],B[14]
20 FIXED 2
30 PRINT TAB25"BOLTED BRACKET DESIGN"
40 PRINT TAB15"AISC SPECIFICATION NOV.1978, MANUAL 1980"
50 REM R1=SINGLE SHEAR CAPACITY,D=BOLT DIA., L=ECCENTRICITY,
60 REM N=NUMBER OF BOLTS IN 1 VERT.LINE,TOTAL BOLTS 2N, VERT.SPACING
70 REM 3 IN., HORIZONTAL SPACING 5.5 IN., A(L,N)=
80 REM COEFFICIENTS "C" FROM THE MANUAL 1980, C=P/R1,P.4-64
90 REM P=FORCE,INSTANTANEOUS CENTER METHOD
100 REM FORCE P IS IN KIPS, ECCENTR. IN INCHES, R1 CAPACITY IN KIPS
110 REM F1=BENDING STRESS,F2=SHEAR STRESS,F3=YIELD POINT STRESS
120 FOR L=3 TO 10
130 FOR N=1 TO 12
140 READ A[L,N]
150 NEXT N
160 NEXT L
170 DATA 0.94,2.32,3.92,5.8,782,9.9,12,14.1,16.2,18.3,20.4,22.4
180 DATA 0.8,1.99,3.39,5.1,698,9,11.1,13.2,15.3,17.4,19.5,21.6
190 DATA 0.7,1.74,2.96,4.51,6.24,8.15,10.2,12.2,14.4,16.5,18.6,20.8
200 DATA 0.62,1.54,2.62,4.03,5.6,7.39,9.3,11.3,13.4,15.5,17.7,19.8
210 DATA 0.55,1.38,2.36,3.63,5.07,6.72,853,10.5,13.5,14.6,16.7,18.8
220 DATA 0.5,1.25,2.14,3.3,4.61,6.15,7.84,9.67,11.6,13.6,15.7,17.8
230 DATA 0.46,1.14,1.96,3.01,4.22,5.66,7.23,8.97,10.8,12.8,14.8,16.9
240 DATA 0.42,1.04,1.8,2.78,3.89,5.23,6.7,8.34,10.1,12,13.9,15.9
250 FOR L=12 TO 24 STEP 4
260 FOR N=1 TO 12
270 READ A[L,N]
280 NEXT N
290 NEXT L
300 DATA 0.37,0.9,1.55,2.39,3.36,4.53,5.82,7.28,8.87,10.6,12.4,14.3
310 DATA 0.29,0.7,1.21,1.87,2.64,3.55,4.58,5.76,7.05,8.47,9.99,11.6
320 DATA 0.24,0.57,0.99,1.53,2.16,2.91,3.77,4.75,5.82,7.02,8.3,9.69
330 DATA 0.2,0.48,0.84,1.29,1.83,2.46,3.19,4.03,4.94,5.97,7.08,8.28
340 FOR L=30 TO 36 STEP 6
350 FOR L=30 TO 36 STEP 6
360 FOR N=1 TO 12
370 READ A[L,N]
380 NEXT N
390 NEXT L
400 DATA 0.16,0.39,0.68,1.04,1.48,2,2.59,3.27,4.02,4.86,5.77,6.77
410 DATA 0.14,0.33,0.57,0.88,1.24,1.68,2.18,2.57,3.39,4.1,4.87,5.72
420 DISP "L EQUALS";
430 INPUT L
440 DISP "FORCE P EQUALS";
450 INPUT P
460 DISP "SINGLE SHEAR R1 EQUA.";
470 INPUT R1
480 DISP "DIAMETER D EQUALS";
490 INPUT D
500 DISP "PLATE THICKNESS T EQUALS";
510 INPUT T
520 DISP "MATERIAL YIELD POINT STRESS F2 EQUALS";
530 INPUT F3
540 PRINT
```

```
550 FOR I=1 TO 14
560 READ B[I]
570 NEXT I
580 DATA 3,4,5,6,7,8,9,10,12,16,20,24,30,36
590 PRINT TAB23"POSSIBLE ECCENTRICITIES L="
600 FOR I=1 TO 14
610 PRINT B[I];
620 NEXT I
630 PRINT
640 PRINT
650 PRINT
660 PRINT
670 PRINT TAB33"PROBLEM"
680 PRINT
690 C=P/R1
700 PRINT "P="P;"L="L;"D="D;"R1="R1;"C REQ.="C
710 PRINT
720 PRINT
730 PRINT TAB33"SOLUTION"
740 PRINT
750 FOR N=1 TO 12
760 IF A[L,N] >= (0.971*C) THEN 830
770 NEXT N
780 N=N-1
790 PRINT "N="N;"C TABLE="A[L,N]
800 PRINT "MORE THAN 12 BOLTS NEEDED,INCREASE D AND R1"
810 PRINT "C REQUIRED="C
820 GOTO 460
830 M=P*(L-2.75)
840 S=(T*((3*N)↑2))/6-(9*N*((N↑2)-1)*T*(D+0.125))/(18*N)
850 F1=M/S
860 IF F1>0.618*F3 THEN 950
870 F2=P/(T*(3*N-N*(D+0.125)))
880 IF F2>(0.412*F3) THEN 930
890 PRINT "NUMBER OF BOLTS IN ONE VERT.LINE="N;"FOR R1="R1
900 PRINT "VERT.BOLT SPACING 3 IN.,HORIZ. 5.5 IN.,FORCE P="P;"ECC.L="L
910 PRINT "PLATE THICKNESS T="T;"STRESSES BENDING="F1;"SHEAR="F2
920 GOTO 950
930 PRINT "INCREASE PLATE THICKNESS T"
940 GOTO 460
950 END
```

```
                    BOLTED BRACKET DESIGN
            AISC SPECIFICATION NOV.1978; MANUAL 1980
                    POSSIBLE ECCENTRICITIES L=
 3.00     4.00     5.00     6.00     7.00     8.00     9.00    10.00
12.00    16.00    20.00    24.00    30.00    36.00


                            PROBLEM

P= 68.00    L= 24.00    D= 1.00    R1= 13.70    C REQ.= 4.96


                            SOLUTION

NUMBER OF BOLTS IN ONE VERT.LINE= 9.00    FOR R1= 13.70
VERT.BOLT SPACING 3 IN.,HORIZ. 5.5 IN.,FORCE P= 68.00    ECC.L= 24.00
PLATE THICKNESS T= 0.88    STRESSES BENDING= 21.59    SHEAR= 4.61
```

Figure 9.102

The easiest connection is obtained if the crossing beams are of the same height (Fig. 9.103). Two continuous beam splices are provided, and their cross section and connection to the beam flanges are designed for a force obtained by dividing the support moment by the depth, d, of the beams.

If the stringers have a lesser height than the floor beam, normally the top flanges

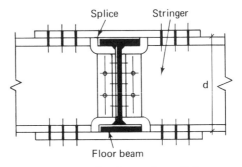

Figure 9.103 Continuous Beam Splice.

of stringers and floor beam are lying in one and the same plane to support the floor. In that case, the continuity splice is provided only at the top and either seat angles or chairs are employed to substitute for the bottom splice (Fig. 9.104). The web angles either could be made of two parts, covering separately the beam and its vertical extension, the chair (Fig. 9.104a), or the angles could go uninterrupted the whole depth. In the latter case, both flanges are cut to make place for the angle (Fig. 9.104b). First the support moment is divided by the stringer depth to obtain a force, T. Then the support moment is substituted by a couple, Td. The top splice cross section and the number of bolts, n, required to connect this splice to the stringer are calculated from this force, T. The chair is also connected with the same number of bolts, n, to the stringer flange. The number of bolts, m_w, in the web stringers will carry a portion, R_w, of the total vertical reaction, R. The difference, $(R - R_w) = R_c$, has to be transmitted through the chair to the floor beam. Therefore, all the connection bolts in the chair, m_c, are exposed to two eccentric forces: one vertical, R_c, acting at the mid-

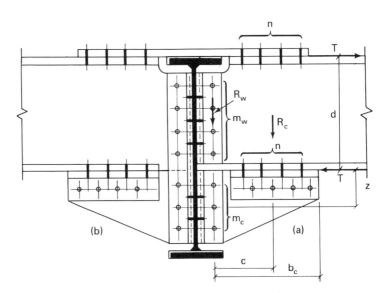

Figure 9.104 Splice and Chair Connection.

402 CONNECTIONS

point of the distance, c, and another horizontal one, T, acting with an eccentricity z equal to the distance between the centroid of bolts, m_c, and the top of the chair (Fig. 9.104a). Therefore, these bolts, m_c, have to transmit the vertical force, R_c, the horizontal force, T, and the moment, $M_c = R_c \times c = T \times z$. The chair actually represents a bracket connection, which was discussed previously in Section 9.3.3 and in Example 9.13. Therefore, this connection will not be discussed further.

Beam and girder splices

Rolled sections and plates have a limited length depending on their cross-sectional dimensions. If a beam or plate has to be longer than the normal available length, intermediate splices must be provided either in the workshop or at the site (in field). The camber of girders also requires splicing. Maximum lengths of obtainable shapes vary widely with producers, but a conservative range for all mills is from 60 to 75 ft. Some mills will accept orders for lengths up to 120 ft, but only for certain shapes and subject to special arrangements [AISC *Manual* (9.2), p, 1-8]. For workshop splices normally the designer can choose such locations where the bending moments are small. A girder web splice need not be accompanied by a flange splice. How much of the moment has to be covered by splicing is usually regulated by the corresponding specifications. The 1978 AISC specification (section 1.15.7, 3.35 p. 5-48) requires that the connections at ends of tension or compression members in trusses develop the force due to the design load, but not less than 50% of the effective strength of the member. Section 1.10.8 (AISC *Manual*, p. 5-35), on splices in beams and girders, requires that splices develop the strength required by the stresses, at the point of the splice.

The AASHTO specifications (9.41, section 1.7.15 and 16) on splices and connections state that splices, whether in tension, compression, bending or shear, shall be designed for not less than the average of the calculated stress at the point of splice and the strength of the member at the same point, but not less than 75% of the strength of the member. Where a section changes at a splice, the strength of the smaller section is to be used for the above splice requirements. The strength of the member must be determined by the gross section for compression members and by the net section for tension members and members primarily in bending.

Web plate connections must be covered symmetrically by splices on both sides. Splice plates for shear must extend to the full depth of the girder between the flanges. In a splice there must be not less than two rows of rivets or bolts on each side of the joint.

The AREA 1980 specifications (9.67) require that the connections have a strength not less than that of the member connected, based on the allowable unit stress in the member.

The British specification for steel girder bridges (1.36) stipulates similar requirements. Flange joints should preferably not be located at points of maximum stress. There must be enough rivets or bolts on each side of the splice to develop at least the load in the spliced plate. Splices in the webs of plate girders and rolled sections used as beams must be designed to resist the shearing forces and the moments in the web at the spliced section. Other foreign specifications have similar requirements.

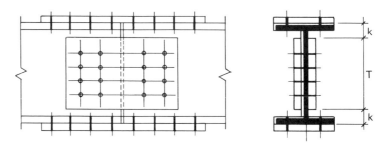

Figure 9.105 Rolled Beam Splice.

In bolted splices of rolled beams, the flanges are usually spliced (Fig. 9.105) only at the outside of the section, while the web is spliced symmetrically over the height, T (between roundings). A flange splice should be of the same or similar cross section as the flange, and the number of bolts is obtained as for any other group of bolts under axial shear.

Web splices have to transmit part of the moment, M, and the whole shear, V, exactly as in connections with eccentric shear. Part of the moment taken by the web is proportioned out of the total bending moment, M, acting at this cross section in the ratio of web to section moment of inertia; thus

$$M_w = M\left(\frac{I_w}{I_{\text{section}}}\right) \tag{9.102}$$

The whole shear, V, is taken by the web. Once the moment, M_w, and shear, V, are known the calculation of the bolts proceeds as shown before for eccentric shear.

In plate girders, even if they are welded, very often bolted field connections are made. It should be noted that the AASHTO specifications allow only turned bolts or ribbed bolts of low-carbon steel quality (A307), and not unfinished or black bolts as do the AISC specification. However, the latter do exclude the use of unfinished bolts in field connections (AISC *Manual*, section 1.15.12, p. 5-49). Depending on the height and slenderness of the structure, high-strength bolts must be used for column splices in tier structures 200 ft in height or more; but if the least horizontal dimension is less than 40% of the height, then this requirement extends to tier structures between 100 and 200 ft in height and, in structures over 125 ft in height, to connections of all beams and girders to columns and of any other beams and girders on which the bracing of the columns is dependent. In roof truss splices and connections of trusses to columns, column splices, column braces, knee braces and crane supports high strength bolts must also be used.

In flange connections the splicing of the flange is similar to that used for rolled beams. If the flanges are composed of two or more plates (at present mostly in bolted or riveted girders), the connection is extended so that in each section only one flange plate at a time is spliced. This could be done in a direct way (Fig. 9.106) or an indirect way (Fig. 9.107). Direct splicing is better but requires a much longer connection. Indirect splicing requires a shorter connection but the number of bolts, n, should be increased to n' to compensate for the bending of bolts sheared in two or more planes.

For the design of the web connection the procedure based on elastic behavior,

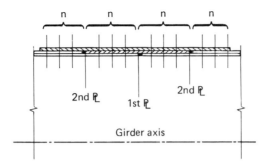

Figure 9.106

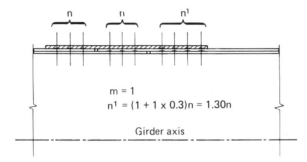

Figure 9.107 Indirect Flange Splice.

previously outlined, for the design of connections loaded in torsion and eccentric shear is modified. The entire bolt field on each side of the joint for plate girders has a much greater height (or depth) than width (Fig. 9.108). Therefore, in the calculation of the polar moment of inertia of the bolt group the term $\Sigma\, x_i^2$ is omitted, being small compared to $\Sigma\, y_i^2$, while also the forces, T_{iy}, in each bolt, which are proportional to small values of x_i are disregarded. In Fig. 9.109 only horizontal forces due to the action of a moment, M_w, are represented. Then each unknown force, H_i, is expressed in terms of vertical distance ratios

$$H_i = \frac{y_i}{y_{\max}} H_{\max} \tag{9.103}$$

The following equilibrium equation can now be written

$$M_w = \sum_{i=1}^{N} H_i y_i \tag{9.104}$$

When H_i from Eq. (9.103) is substituted into Eq. (9.104), the maximum horizontal force in the most stressed bolt is

$$H_{\max} = \frac{M_w y_{\max}}{\sum_{i=1}^{N} y_i^2} \tag{9.105}$$

Welded girders are normally spliced by full penetration butt welds. As they are

Connection Design **405**

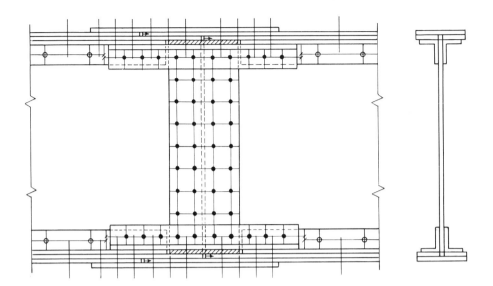

Figure 9.108 Field Splice of Plate Girder.

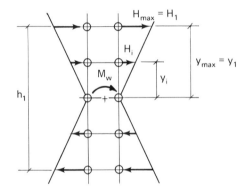

Figure 9.109 Horizontal Forces Due to Moment Action.

at least equivalent in strength to the base metal, no special calculations are required, except when checking for fatigue.

Next, for bolted splices, Example 9.15 is worked using the elastic approach, taking the rotation axis at the centroid of bolt group.

Example 9.15 Design a rolled beam splice for a W24 × 84 (61 cm × 1,226 kN/m) beam of A36 steel for a section with a bending moment of 322.5 kip-ft (437.4 kN-m) and a shear of 82 kips (364 kN). Use A325-N high-strength bolts with threads in shear planes.

Solution It is assumed that the total shear force has to be taken by the web splices and that the bending moment is distributed to the flange and the web in proportion to ratios of the respective moment of inertias of flange and web to the moment of inertia of the whole section.

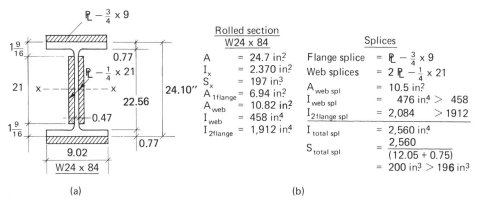

Figure 9.110 Data for Rolled Section and Splices.

DATA. Figure 9.110 gives all the data for a W24 × 84 (61 cm × 1,226 kN/m) beam [AISC *Manual* (9.2), p. 1-16]. Using Eq. (9.102) and substituting the appropriate values, $M_w = 322.5\ (458/2,370) = 62.3$ kip-ft (84.5 kN-m). The flanges carry the remainder of $322.5 - 62.3 = 260.2$ kip-ft (352.9 kN-m).

SPLICE CHOICE. As explained earlier, the splices will be chosen in accordance with Fig. 9.105. For flanges, the splices will have the same dimensions as the flange itself, rounded off to practical dimensions, which results in two plates $\frac{3}{4} \times 9$ in. (19 × 229 mm). As these plates are on top of the spliced flanges, their moment of inertia is larger than that of the flanges. The web is symmetrically covered by two plates extended to the flange fillets' roundings—that is, 21 in. (533 mm). First a thickness of one-half the web thickness is tried, or $\frac{1}{4}$ in. (6 mm). The AASHTO specifications (9.41) limit the minimum thickness for structural steel, except for rolled webs, to $\frac{5}{16}$ in. (8 mm); and the web thickness of rolled beams or channels must not be less than 0.23 in. (6 mm). The AISC specification (9.1) does not give any thickness limitations. In Europe [German DIN 4100 (9.52) for welded steel buildings] normally 6 mm (0.234 in.) is taken as a practical limit for structural elements. Therefore, let us select two PL-$\frac{1}{4} \times 21$ (6 × 533 mm). As the web splices carry both shear force and moment it is important to check both their cross-sectional area and moment of inertia. The data in Fig. 9.110(b) show that the dimensions are satisfactory.

BOLT DIAMETER. According to Fig. 9.74 bolt diameter is either $\frac{5}{8}$ in. (16 mm) or $\frac{3}{4}$ in. (19 mm) according to the smallest plate thickness. Because the flanges are about $\frac{3}{4}$ in. (19 mm), requiring 1 in. (26 mm) diameter, it is better to choose the next value larger than $\frac{5}{8}$ in. (16 mm)—that is, $\frac{3}{4}$ in. (19 mm). From the AISC *Manual* (pp. 4–5, 4–6) the following capacities are found for A325-N bolts

$$\text{In single shear:} \quad C_{SS} = 9.3 \text{ kips (41.4 kN)}$$
$$\text{In bearing (on web):} \quad C_p = 27.2 \text{ kips (121.0 kN)}$$
$$\text{In double shear:} \quad C_{DS} = 18.6 \text{ kips (82.7 kN)}$$

For the bolts in the web splices, it is noted that the governing bolt capacity is $C_{DS} = 18.6$ kips (82.7 kN) because the bolts are in double shear and bearing. Therefore, the maximum resultant force (from shear force and moment), T_{max}, cannot be more than C_{DS}. To get a first estimate of how many web bolts are

Connection Design

needed, the shear force is divided by C_{DS}

$$\frac{82}{18.6} = 4.4$$

This means the number must be much larger than 5 to be able also to accommodate the component due to the bending moment action.

Next we determine how many bolts can be put in one vertical row. A distance of 3 times the diameter between bolts equals $2\frac{1}{4}$ in. (57 mm), while the minimum edge distance is $1\frac{1}{4}$ in. (along a sheared edge—AISC *Manual*, p. 5–51). If 8 bolts in one vertical row are taken with 7 distances of $2\frac{1}{2}$ in. (64 mm) between them, larger than the AISC minimum of $2\frac{2}{3}$ times the nominal diameter, this will make a total of $17\frac{1}{2}$ in. (445 mm). If the remainder of splice depth equal to $3\frac{1}{2}$ in. (89 mm) is split (21 in. is the total height of the web splice), the edge distance is $1\frac{3}{4}$ in. $> 1\frac{1}{4}$ in. (45 mm $>$ 32 mm).

It seems that we can have 8 bolts in one vertical line, and we shall try two such lines with 16 bolts on each side of the splice. The reduced polar moment of inertia now is Eq. (9.105)

$$\sum_{i=1}^{16}(y_i)^2 = 4[(1.25)^2 + (3.75)^2 + (6.25)^2 + (8.75)^2] = 525.0$$

The maximum horizontal component force due to the moment is (from Eq. 9.105)

$$H_{max} = \frac{62.3 \times 12}{525} \times 8.75 = 12.46 \text{ kips (55.42 kN)}$$

The vertical component of force due to the shearing force is

$$V = \frac{82}{16} = 5.125 \text{ kips (22.8 kN)}$$

The maximum resultant force is

$$T_{max} = \sqrt{(12.46)^2 + (5.125)^2} = 13.5 \text{ kips (59.9 kN)}$$

As this is less than 18.6 kips, the double shear capacity of the web connection is OK.

BOLTS IN FLANGE SPLICES. If the moment taken by the flanges is divided by the distance between the centroids of the flange of $24.10 - 0.77 = 23.33$ in., the force of the equivalent couple is obtained as $260.2 \times 12/23.33 = 133.9$ kips (595.3 kN). To find out how may $\frac{3}{4}$-in.-diameter bolts are needed in single shear, this force is divided by the bolt capacity, $C_{SS} = 9.3$ kips (41.4 kN). The required number is then

$$\frac{133.9}{9.3} = 14.4 \qquad \text{(use 14 with 3\% overstress)}$$

or 7 pairs of bolts on both sides of the splices. As was pointed out previously, preferably not more than 6 bolts under axial shear should be used in one row. Here this situation cannot be avoided unless the bolts are forced to act in double shear. The flange thickness requires at least 1-in. (25-mm) diameter bolts, but the web splices reduced this size to a value of $\frac{3}{4}$ in. (19 mm). The bolts cannot be used to act in double shear unless flange splices are used on both faces of the flanges. This is more inconvenient because the bolting process will be much more difficult than in the case of

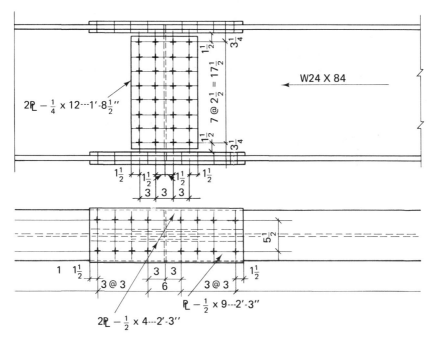

Figure 9.111 Beam-to-beam Splice.

single-flange splices. Therefore, in this situation of double shear only 8 bolts are required and 4 bolts in one row are acceptable (Fig. 9.111).

COMMENTS. The outlined procedure represents the standard method of design. Studies of the actual behavior of connections exposed to bending (9.68) show quite a different behavior. Figure 9.112 represents the results of tests of the relative rotation

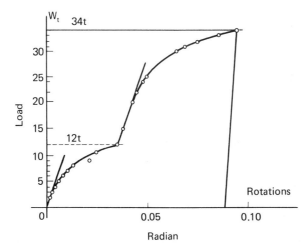

Figure 9.112 Rotations of the Joint.

of joint sections. The connected web parts came into contact under a load of 12-(117.68 kN). Figure 9.113 shows the deflections of the beam midpoint. The 12-t load marks the beginning of a new deformational mechanism. Bolts on both sides of the joint and web splices start to rotate around the common axis first positioned at the top of the web, but due to the plastic deformation of the web areas in contact it moves gradually downward. Strain gages on the web splice (Fig. 9.114) show this clearly.

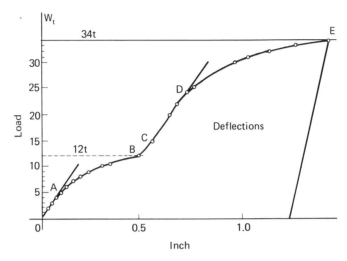

Figure 9.113 Midpoint Deflections.

Fatigue design of connections

Connections in buildings

The most recent AISC specification on fatigue (9.1, pp. 5-85 to 91), appendix B, requires the consideration of fatigue in the design of members and connections for buildings if they are subject to repeated variation of live load stress. The number of expected stress cycles, their range, and the type and location of the connection have to be taken into account.

From table B1 of the AISC specification, (9.1, p. 5-86) giving the anticipated number of cycles, the loading condition is established; the stress category is next determined from table B2 (*Manual*, pp. 5-87–89; see also illustrative examples, pp. 5-90–91). Then from table B3 (p. 5-89) the allowable range of stress

$$F_{sr} = f_{max} - f_{min}$$

is obtained and compared with the calculated value. The maximum stress is still controlled by the values given in sections 1.5 and 1.6 of the AISC specifications, and the only check against the danger of fatigue is the calculated stress range.

Example 9.16 The loads acting on the bracket considered in Example 9.14 (Fig. 9.101) might have the following values:

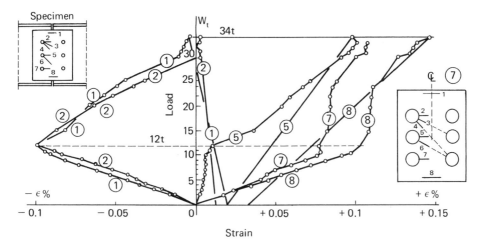

Figure 9.114 Strains in Web Splices.

$$V_{\max} = 68 \text{ kips } (302.5 \text{ kN}); \qquad V_{\min} = 38 \text{ kips } (169 \text{ kN})$$

The number of loading cycles is $N = 4 \times 10^5$.

Check the designed connection with respect to fatigue according to the AISC specification (9.1), appendix B.

Solution From AISC table B1, using $N = 4 \times 10^5$, this bracket has loading condition 2. From table B2 (p. 5-91: fillet-welded connections, third situation) and from figure B1 in illustrative example no. 21 the stress category is found to be F. From table B3, for loading condition 2 and category F, the allowable stress range is

$$F_{sr} = 12 \text{ ksi } (82.7 \text{ MPa})$$

When $V_{\max}$ is acting, the resultant force per unit length of weld is 4.30 kips/in. (753.05 kN/m) of weld. As the weld size was taken as $\tfrac{5}{16}$ in. (8 mm) with the throat area of

$$0.707(\tfrac{5}{16})(1.00) = 0.221 \text{ in.}^2 \; (1.42 \text{ cm}^2)$$

the maximum stress is

$$f_{\max} = \frac{4.30}{0.221} = 19.46 \text{ ksi } (134.27 \text{ MPa})$$

When $V_{\min}$ is acting, the force per linear inch of weld is 2.59 kips/in., or the stress is

$$f_{\min} = \frac{2.59}{0.221} = 11.72 \text{ ksi } (80.8 \text{ MPa})$$

Therefore the stress range is

$$f_{sr} = 19.46 - 11.72 = 7.74 \text{ ksi} < 12 \text{ ksi}: \qquad \underline{\text{OK}}$$

Connections in bridges

Section 1.7.3, Fatigue Stresses, of the 1973 AASHTO specifications was completely revised in Interim 8 (*Interim Specifications: Bridges*, Washington, DC, 1974, p. 86); see also Dr. J. W. Fisher, AISC, *Guide to 1974 AASHTO Fatigue Specifica-*

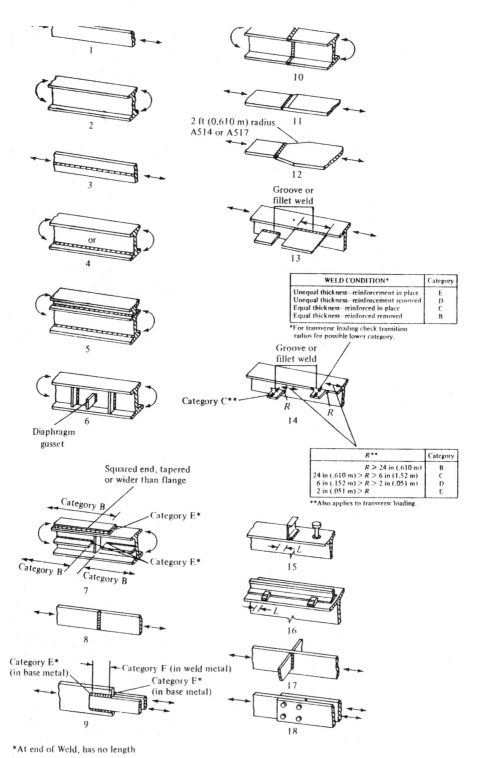

Figure 9.115 Illustrative Examples (From AASHTO, *Standard Specifications for Highway Bridges*, 12th ed., Washington, DC, 1977, p. 150).

tions, New York, 1974. The new specifications give the allowable range of stress for different categories according to the type and location of the material and the number of cycles. First the number of cycles of maximum stress to be used has to be selected. These are given in Interim 8 table 1.7.3(A), Stress Cycles. The allowable range of fatigue stress, F_{sr}, applies to dead loads in combination with live loads or wind loads or to wind loads only, and is given in the same Interim 8 table 1.7.3(B). (AASHTO table 1.7.2A1 (9.41, pp. 144–5) gives the information about the categories, also shown in Fig. 9.115. The use of these tables will be illustrated by the following example.

Example 9.17 In Fig. 9.116(a) and (b) the interior welded plate girder and the structural system of a highway bridge are shown. The loading is HS20-44 and HS lane loading (AASHTO specifications, pp. 18, 19) for case II (highways and streets). The material is high-strength, low-alloy structural steel with 50-ksi (345-MPa) minimum yield point to 4 in. thickness and $F_u = 70$ ksi (483 MPa), ASTM 588 (or AASHTO designation M222). In Fig. 9.116c the bending moment envelopes for maximum and minimum positive and negative moments are given. The field splicing is at 25 ft (7.6 m) and 40 ft (12.2 m) from the supports B and C, respectively. Check the design with respect to fatigue.

Solution According to table 1.7.2(B) Stress Cycles (AASHTO, p. 151) for truck loading the number of cycles is 5×10^5 and for lane loading is 10^5. Field splice no. 1 is falling into the range of stress reversals (the shaded area in Fig. 9.116c). The stress range, f_{sr} (i.e., maximum stress less minimum stress), at splice no. 1 is

$$f_{sr} = 8.85 - (-4.10) = 12.95 \text{ ksi (89.29 MPa)}.$$

Both flanges and web will be groove-welded with full-penetration groove welds. This type of connection classifies as weld metal or base metal at full-penetration groove-welded splices of welded sections having similar profiles when welds are ground flush and weld soundness established by nondestructive inspection (category B in AASHTO table 1.7.2A2 and sketch 10 in Fig. 9.115). From AASHTO table 1.7.2A1 the allowable stress range, F_{sr} is

$$27.5 \text{ ksi (189.61 MPa)} < 12.95 \text{ ksi (89.29 MPa)}: \quad \text{OK}$$

For two more detailed illustrative examples in bridge and building design the reader is referred to Chapter 12.

9.3.5 Truss Connections and Splices

Truss-type members are composed of angles or other rolled shapes joined with lacing, battens, or plates. They have to be connected either at both ends to a gusset plate by an end connection or spliced for length. In both cases, these connections are exposed to concentric shear. As was pointed out in Section 9.2.12, Bolt Groups in Concentric Shear, the end fasteners will take higher loads and therefore the maximum number of bolts in one row should not be more than 6. Tests by Chesson and Munse (9.60) confirm this fact. They state that there is considerable evidence that the shear strength of a joint is affected not only by the fastener strength and the ductility of the fasteners and connected material, but also by the length of the joint and the number of fasteners in line; however, the effect of joint length on ultimate strength appears small for less than five or six fasteners in line at normal spacings. Longer

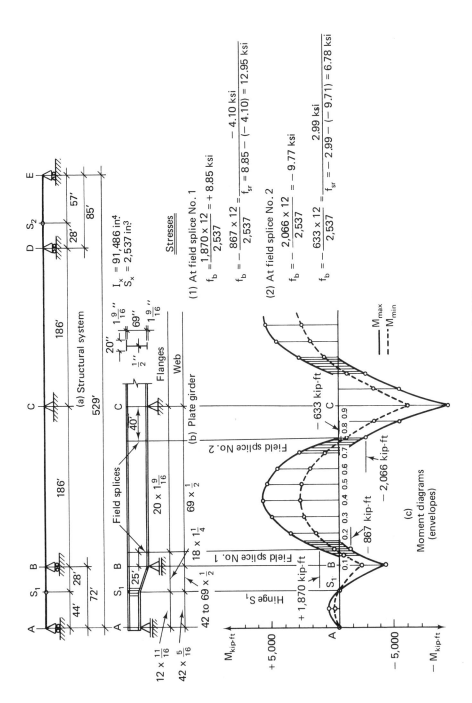

Figure 9.116 (Example 9.17).

joints which fail in shear fail progressively, beginning with the end fasteners. When the number of fasteners becomes 10 or more in line the average shear strength for the joint may be only three-fourths of the value predicted on the basis of single fastener shear strength. In view of these facts, it appears that wider and more compact joints rather than longer joints are preferable, provided the amount of indirectly connected materials is small.

A similar situation occurs when welding is used. The force distribution in welds in general is nonuniform and fillet welds parallel to the shear action have less strength than if they were oriented normal to it. When they are perpendicular to the shear action, the force distribution is affected by the shear lag phenomenon. Fortunately, at ultimate strength, the stress distribution across the plates, including flanges and webs, is almost uniform.

The above discussions are based on the assumption that the loading is static. If the loading is dynamic in nature, different behavior may occur.

Dynamic loading

Dynamic loading can take place with or without interruptions. Interrupted loading means that the loading is periodically repeated with certain rest periods in between. During the loading stage the connection elements are deformed, moved to one side; due to the elasticity of the bolt shank, during the unloading stage they are moved to the opposite side. If two successive loadings, P, are not larger than the friction force, F, then the connection is loaded in the elastic domain without permanent deformations. The force, P, is transmitted by friction from one element to another over the whole contact surface. If the force, P, is larger than the friction force, F (Fig. 9.117), then after the first unloading permanent deformation will occur. In this case unloading is elastic not only for the friction force, F, but for twice as large a force, due to the mechanical characteristics of friction. For movement in the opposite direction, protrusions of the contact surfaces are unloaded by the force, F, back to the neutral position and reloaded for another amount, F, while being bent in the opposite direction. This means that only after $P > 2F$ will the relationship between the force and deformation become curvilinear. The amount of permanent deformation depends

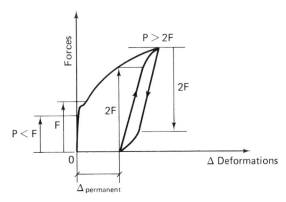

Figure 9.117 Force-deformation Diagram for Repeated (Dynamic) Loading.

upon the magnitude of the force when applied for the first time and the quality of the bolts.

If the dynamic loading is continuous, the load capacity of the connector is reduced because the plastic deformations have no time to develop. The fatigue design must be performed.

The actual design of end connections and splices will be illustrated by the following four examples using high-strength bolts and welding.

Example 9.18 Using the AISC specification (9.1) and ASTM A325-N bolts of the appropriate diameter, design the end connection of a truss diagonal of the cross section shown in Fig. 9.118. The tensile force acting on the diagonal is $P = 90$ kips (400 kN). Angles and gusset plate are of A36 material. The gusset plate is $\frac{1}{2}$ in. thick.

Solution From Fig. 9.118 it is recognized that bolts are sheared in two faying surfaces and that the bearing thickness is $\frac{1}{2}$ in.

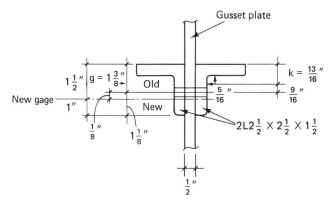

Figure 9.118 Example 9.18.

BOLT DIAMETER. From Fig. 9.74 and for a $\frac{1}{2}$-in. plate thickness, the bolt diameter is $\frac{3}{4}$ or $\frac{7}{8}$ in. From the AISC *Manual* (9.2, p. 4–135), the usual gage for a $2\frac{1}{2} \times 2\frac{1}{2}$ angle is $1\frac{3}{8}$ in. Half the width of a round washer according to the *Manual* (p. 4–132), is $\frac{3}{4}$ in. for a $\frac{3}{4}$-in. bolt diameter and $\frac{7}{8}$ in. for a $\frac{7}{8}$-in. bolt diameter. This washer can go only to a distance $k = \frac{13}{16}$ in. from the leg's root [*Manual*, p. 1–46 (Fig. 9.118)], or from the gage line for a distance of only $\frac{9}{16}$ in. $< \frac{3}{4}$ in., which even for a $\frac{3}{4}$-in. bolt diameter is too short. From table 1.16.5.1 of the *Manual*, p. 5–51, the minimum distance from a rolled edge of a $\frac{3}{4}$-in. bolt is 1 in. This means (Fig. 9.118) that the usual gage line could be moved an additional $\frac{1}{8}$ in. resulting in a g of $1\frac{1}{2}$ in. Therefore, the washer would go into the fillet $\frac{1}{16}$ in. (0.16 cm). This can be tolerated. A $\frac{7}{8}$-in. diameter bolt cannot be accommodated, as its washer would run into the fillet $\frac{3}{16}$ in. (0.48 cm), which cannot be tolerated for a fillet radius of $\frac{5}{16}$ in.

NUMBER OF BOLTS. The capacity of one A325-N bolt of $\frac{3}{4}$ in. diameter in double shear is 18.6 kips (82.7 kN) and in bearing (on $\frac{1}{2}$ in.) for 3d spacing 27.2 kips (121.0 kN) $>$ 18.6 kips. Therefore the shear capacity controls. The number of bolts, n, is

$$n = \frac{90}{18.6} = 4.8 \quad \text{use 5 bolts} < 6: \quad \underline{\text{OK}}$$

DETAILING OF THE CONNECTION. We already know the position of the new gage line (Fig. 9.118), i.e., at $1\frac{1}{2}$ in. Normal bolt spacing is 3d, or $2\frac{1}{4}$ in. This distance between centers of standard holes shall not be less than the following expression, AISC *Manual*, p. 5–50, Eq. (1.16–1)

$$2P/F_u t + d/2 \qquad (9.106)$$

in which P = force in one fastener = 90/5 = 18 kips; F_u = specified minimum tensile strength of the connected part = 58 ksi; t = thickness of the critical connected part = $\frac{1}{2}$ in. Substituting our values into Eq. (9.106) we have

$$2 \times 18/(58 \times \tfrac{1}{2}) + \tfrac{3}{8} = 1.62 \text{ in.} < 2.25 \text{ in.:} \quad \underline{\text{OK}}$$

The last dimension we have to find before detailing of the connection is the minimum-edge distance along the line of the transmitted force. This distance is given as Eq. (1.16-2) in the *Manual*, p. 5–50

$$2P/F_u t \qquad (9.107)$$

in which P, F_u, and t as given in Eq. (9.106). In our case,

$$2 \times (18/58) \times \tfrac{1}{2} = 1.24 \text{ in. } (3.2 \text{ cm}) < 2d = 1\tfrac{1}{2} \text{ in.}$$

The final detailing can now be done, see Fig. 9.119.

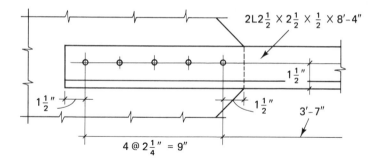

Figure 9.119 Detail of the End Connection, Example 9.18.

Example 9.19 Design the end-connection of Example 9.18 using welding with E-70XX electrodes.

Solution According to section 1.15.3, of the *Manual*, when a member is not subjected to repeated variation in stress (i.e., fatigue design), disposition of fillet welds to balance the forces about the neutral axis for end connections of single-angle, double-angles (our case), and similar type members is not required. Therefore, we have to find the size and length of the fillet weld and distribute it evenly (Fig. 9.120).

WELD SIZE. From table 1.17.2A (*Manual*, p. 5–52) we find the minimum size fillet weld. The plate and angle thickness is $\frac{1}{2}$ in. Therefore, the minimum size is $\frac{3}{16}$ in. From section 1.17.3 (*Manual*, p. 5–53) the maximum size of the fillet is

$$\tfrac{1}{2} - \tfrac{1}{16} = \tfrac{7}{16} \text{ in.}$$

Connection Design 417

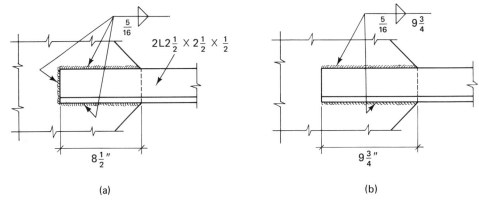

Figure 9.120 Example 9.19.

In Section 9.1.3.1, Welding Procedure for Lowest Cost, it was stated that a $\tfrac{5}{16}$-in. weld is the maximum size of fillet weld made in one pass in the horizontal position. The relative cost factor for fillet welds of $\tfrac{3}{16}$ in. is 0.85 and for $\tfrac{5}{16}$ in. it is 1.20 (9.69). Thus, the relative cost ratio is 1.20/0.85 = 1.412. The strength ratio, and therefore the length of the weld ratio, is 5/3 = 1.667 > 1.412. Hence a $\tfrac{5}{16}$-in. weld is more economical than a $\tfrac{3}{16}$-in. weld. We choose a $\tfrac{5}{16}$-in. weld.

WELD LENGTH. The shearing capacity of a $\tfrac{5}{16}$-in. fillet weld made with shielded metal-arc with E-7018 electrodes (which is very common in both shop and field) equals

$$0.707 \times 5/16 \times 0.3 \times 70 = 4.64 \text{ kips/in.}$$

The total weld length is

$$90/4.64 = 19.4 \text{ in. } (49.3 \text{ cm})$$

DETAILING THE CONNECTION. Figure 9.120 shows two possibilities: one consisting of two longitudinal welds and one transverse welding and the other of two longitudinal welds only. In the first case each of the parallel welds has a length of

$$\tfrac{1}{2}(19.4 - 2.5) = 8.45 \approx 8\tfrac{1}{2} \text{ in.}$$

Without a perpendicular weld, this length is

$$\tfrac{1}{2} \times 19.4 = 9.7 = 9\tfrac{3}{4} \text{ in.}$$

The first configuration is more compact than the second, and with the proper design of the gusset plate this weld perpendicular to the flow of stress in the plate is not harmful. When the transverse spacing of longitudinal fillet welds used in end connections is larger than 8 in. (20.3 cm), perpendicular welds must be used (*Manual*, p. 5–53).

Example 9.20 A vertical (tie) member of a large span roof truss has the cross section shown in Fig. 9.121. The tensile force in the vertical direction is 360.0 kips (1334.5 kN). Use AISC specification and ASTM A325-X bolts. The member is made of A36 steel.

Solution The tie has a built-up section and the whole sectional area has to be connected through outstanding legs of angles. The bolts in legs parallel to the

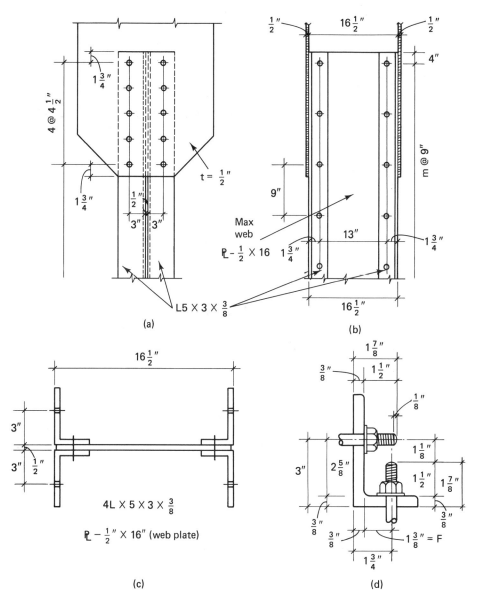

Figure 9.121 Example 9.20.

web plate cannot be spaced more than either 24 times their thickness ($\frac{3}{8}$ in.) = 9 in. or 12 in. (*Manual*, p. 5–56, section 1.18.3.1). In this case we can use 9 in.

BOLT DIAMETER. Figure 9.74 gives for $\frac{3}{8}$-in. thickness either $\frac{3}{4}$ in. or $\frac{7}{8}$ in. for the bolt diameter. As the force is sizeable a $\frac{7}{8}$-in. diameter will be adopted. We must have in mind what was said in Section 9.3.5 about the length of a joint and our accepted goal of a maximum of 6 bolts in one gage line.

BOLT NUMBER. Single shear capacity of ASTM A325-C $\frac{7}{8}$ in. bolt is (*Manual*, p. 4–5) 18.0 kips (80.1 kN), and bearing capacity on $\frac{3}{8}$-in. (angle) thick-

Connection Design 419

ness for a 3d spacing of 2.63 in. is 23.8 kips (105.9 kN) > 18.0 kips, use shear capacity. Number of bolts is then

$$n = \frac{360.00}{4 \times 18.0} = 5.0 \quad \text{use 5 bolts}$$

DETAILING OF THE CONNECTION. We have first to find how much stagger is required for $\frac{7}{8}$-in. bolts in both legs of one L5 × 3 × $\frac{3}{8}$. In Fig. 9.121d the situation is shown when both bolts (horizontal and vertical) are located in the same cross section without any stagger. The vertical clearance is $1\frac{1}{8}$ in. instead of the required $1\frac{3}{8}$ in. (see table, of Assembling Clearances, *Manual*, p. 4–132). The horizontal clearance, however, is worse, i.e., instead of having a positive clearance of $1\frac{3}{8}$ in. we have a negative clearance of $\frac{1}{8}$ in. (bolts' ends overlap). Therefore stagger is necessary. For a clear distance, F, of gage line of $1\frac{3}{8}$ in. and using a $\frac{7}{8}$-in. diameter bolt the stagger is $2\frac{3}{16}$ in. (table of Stagger for Impact Wrench Tightening, *Manual*, p. 4–132), use $2\frac{1}{4}$ in.

The minimum edge distance from table 1.16.5.1 (*Manual*, p. 5–51) is $1\frac{1}{2}$ in. for a sheared edge. The equation from the *Manual*, p. 5–50, (1.16–2), or Eq. (9.107) in this text (see also Example 9.18), is

$$2P/F_u t = 2 \times 360/(4 \times 5) \times 1/(58.0 \times 3/8) = 1.66 \text{ in.}$$

To have sufficient assembling and tightening thickness we shall start with an edge distance of $1\frac{3}{4}$ in. and proceed with four spacings of $2 \times 2\frac{1}{4} = 4\frac{1}{2}$ in. for the bolts, which transmit the load of 360 kips to the joint as shown in Fig. 9.121a. For stitch bolts we start with a distance

$$1\frac{3}{4} + (\tfrac{1}{2}) \times 4\frac{1}{2} = 4 \text{ in.}$$

and proceed with a maximum allowable distance of 9 in.

Example 9.21 Determine the plate thickness and specify the weld for the joint shown in Fig. 9.122 using AISC specification and A36 material. Welding is shielded metal-arc with E-70XX electrodes.

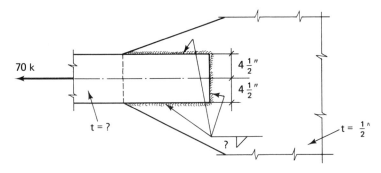

Figure 9.122 Example 9.21.

Solution

PLATE THICKNESS. Allowable stress in tension on the gross area is $0.6F_y = 22$ ksi (151.7 MPa) (*Manual*, p. 5–18). As the width of the plate is 9 in. (22.9 cm), and the given force is 70 kips (311.4 kN), the plate thickness t is

$$t = \frac{70}{9 \times 22} = 0.354 \text{ in.} \quad \text{use } \tfrac{3}{8} \text{ in.}$$

WELD SIZE. Again, we shall find the minimum and maximum fillet weld sizes and then decide on the proper size. The minimum size depends upon the thicker part, here $\frac{1}{2}$ in. From the *Manual*, p. 5–52, the minimum size is $\frac{3}{16}$ in. The maximum size is $\frac{5}{16}$ in. (*Manual* p. 5–53). On the reasoning given in Example 9.19 we choose $\frac{5}{16}$ in. The shear capacity is

$$0.707 \times 5/16 \times 0.3 \times 70 = 4.64 \text{ kips/in.}$$

WELD LENGTH. By dividing the magnitude of the load by the capacity per unit length, the total weld length is

$$70/4.64 = 15.1 \text{ in.}$$

The length of the fillet welds parallel to the force action (longitudinal welds) is

$$(15.1 - 9.0)/2 = 3.04 \text{ in. (7.73 cm)} \quad \text{use 3 in.}$$

Checking the minimum length of fillet welds (*Manual*, p. 5–53) we see that this length is at least equal to four times the nominal size, i.e.

$$4 \times \tfrac{5}{16} = 1\tfrac{1}{4} \text{ in.} < 3 \text{ in.:} \quad \underline{\text{OK}}$$

As the distance between longitudinal fillets is 9 in. > 8 in. perpendicular weld must be used to prevent excessive transverse bending in the connection (*Manual*, section 1.17.4, p. 5–53). Connection detailing is given in Fig. 9.123.

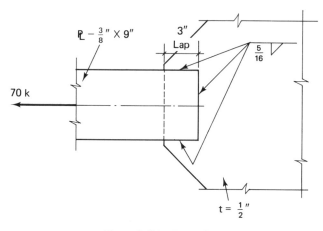

Figure 9.123 Example 9.21.

NOTATIONS

a, a'	= outer span in a T-hanger connection	A_f	= area of a flange
a	= leg size of a weld	A_p	= plate effective compression area around a high-strength bolt
A_b	= bolt effective area; required bolt area of two bolts per one row		
		A_{st}	= area of a stiffener
A_{bc}	= planar area of web in beam-to-column connection	A_t	= horizontal tear area in a coped beam

Connection Design

b, b'	= inner span in a T-hanger connection; plate width	f_t	= average tensile stress due to direct applied load to all bolts
b_c	= width of a chair in beam-to-beam joint	f_v	= calculated shear stress
b_e	= effective end-plate width	F	= force; friction force
b_f	= width of a beam flange	F_{bt}	= allowable tension stress in one bolt
b_s	= width of an end plate	F_{bu}	= ultimate bolt stress in tension
B	= allowable load on a bolt in a T-hanger connection	F_f	= force in a flange
B_c	= load per bolt, including prying force	F_p	= allowable bearing stress; in end-plate design allowable bending stress
C	= coefficient given in the tables for bracket design (*Manual*, pp. 4–62 to 4–83); reaction in a chair in beam-to-beam joint	F_{pcr}	= critical bearing stress
		F_r	= residual force
		F_{sr}	= allowable stress range (in fatigue design)
C_a, C_b	= coefficients in design of end-plate connection	F_u	= ultimate tensile strength
C_i	= initial compression load in plate	F_{us}	= plate shear strength
		F_{vc}	= AASHTO shear stress for combined shear and tension
C_m	= maximum loss of compression in a plate around bolt hole	F_{vu}	= pure shear failure stress
		F_y	= yield point stress
C_{SS}, C_{DS}	= bolt capacity in single shear; double shear	g	= gage line distance
		H_b	= resultant compression force below neutral axis
C_1	= adjustment for quality of electrode used in welding connection	H_i	= horizontal force component in a bolt
ΔC	= decrease in compression force	I, I_x, I_y	= moment of inertia
		I_p	= polar moment of inertia
d	= bolt diameter, beam depth	$I_{section}$	= moment of inertia of a rolled beam
d_b, d_c	= bolt diameter		
d_h, d'	= bolt-hole diameter	I_w	= moment of inertia of beam-web
d_i	= bolt distance from the neutral axis		
d_1, d_2	= radii of two cylindrical surfaces in contact	K_b, k_p	= spring constants of bolt and plate, respectively
D	= number of sixteenths of an inch in a weld; beam depth	l	= span, distance from a bolt line to a force; length of welds
e	= eccentricity; base for natural logarithms, 2.718 ...	l_a	= actual arm distance between a force and line of bolts or welds
f_{av}	= average stress		
f_b	= bending stress		
f_{cm}	= contact stress	l_e	= effective arm distance; distance along a line of force action
f_r	= residual stress		
f_{sr}	= calculated stress range (in fatigue design)	l_h, l_v	= minimum horizontal and

	vertical edge distances of bolt hole in beam web	R_{BS}	= resistance to block shear
L	= length of vertical welds	R_{ult}	= ultimate shear load of a single fastener
$\Delta L_{bi}, \Delta L_{pi}$	= initial change of bolt length and plate thickness, respectively	S	= section modulus; spacing of holes
		S_G	= uniform tension stress in a plate
$\Delta L'$	= increase in bolt length and plate thickness	t	= thickness of a plate or web
m	= number of active shear planes	t_b, t_f	= thickness of a beam flange
m_c, m_w	= number of bolts in a chair or a web	t_e	= throat of a fillet weld
		t_s	= end-plate thickness
M	= bending moment	T	= tensile force per bolt or per flange; resultant force in a bolt
M_c	= moment of eccentricity in a chair		
M_{cr}	= critical bending moment	T_b	= specified pretension load of a high-strength bolt
M_e	= effective moment in an end-plate joint		
		T_f	= final force in a bolt
M_p	= full plastic bending moment	T_i	= initial tensile force in a bolt; proof load
M_t	= split-tee moment		
M_u	= ultimate bending moment	ΔT	= increase of bolt tensile force
M_w	= part of bending moment taken by a web	V	= shear force; vertical force component in a bolt
n	= number of bolts in one vertical line	V_b	= shear force in a web below neutral axis
n'	= increased number of bolts in indirect splicing	w_s	= size of a fillet weld
		w_t	= size of the throat of a fillet weld
p	= length of flange parallel to stem in hanger type connection; pitch		
		x	= abscissa; distance to the neutral axis
P	= force; allowable eccentric load per bolt	z	= vertical eccentricity in a chair
		α	= moment ratio
P_e	= effective bolt distance	α_m	= moment coefficient in end-plate connection
P_f	= bolt's distance from a flange		
P_r	= beam reaction per one bolt	δ	= ratio of net area of flange and gross area in a T-stub joint
Q	= prying force		
Q_f	= final prying force	Δ, Δ_i	= joint deformation
r_i	= position vector of a bolt i	λ	= empirical coefficient
r_0	= distance between an instantaneous center	μ	= Poisson's ratio; empirical coefficients
R_c	= part of a beam reaction taken by a chair	θ_i	= angle between bolt or weld position vector and x-axis
R_i	= force in a bolt i		
R_{max}	= maximum reaction for a given number of bolts	θ_{ih}	= angle between resultant force R_i and weld axis
R_w	= part of a beam reaction taken by a web	ϕ_0	= end rotation of a simple beam

PROBLEMS

9.1. Design the double-splice butt connection shown in Fig. P-9.1 using the appropriate diameter of A325-N high-strength bolts (bearing type with threads included in shearing surfaces) for the maximum capacity of the spliced member of A36 steel. Use AISC specification (9.1). Detail this connection.

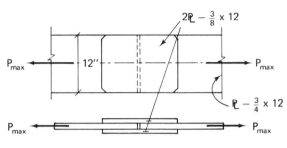

Figure P-9.1

9.2. Design the end connection of a diagonal composed of two angles L5 × 5 × $\frac{1}{2}$ in. of A36 steel, using $\frac{7}{8}$-in. A325-X high-strength bolts (with threads excluded from shearing surfaces) for the maximum allowable tension force in this diagonal according to AISC specification (9.1). Gusset plate is $\frac{3}{8}$-in. thick. Detail the connection.

9.3. A tension member is composed of one web plate $\frac{1}{2}$ in. × 12 in., two angles L4 × 4 × $\frac{1}{2}$ in. and two flange plates $\frac{1}{2}$ × 10 in. (see Fig. P-9.3). Design a field connection of that member for its maximum allowable capacity using $\frac{7}{8}$-in. A325-X bolts and AISC specification (9.1).

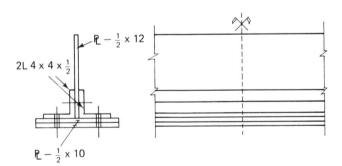

Figure P-9.3

9.4. Design a bracket connection to both flanges of a W12 × 40 column to transmit a vertical force of 50 kips acting at 15 in. from the column web axis. Use for brackets two plates-$\frac{1}{2}$ in. thick, $\frac{7}{8}$-in. A325-F bolts, and AISC specification (9.1) (see Fig. P-9.4).

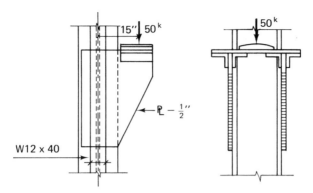

Figure P-9.4

9.5. Design a bracket connection to the web of a W12 × 40 column to transmit the loading as in Problem 9.4. Use for bracket $\frac{3}{4}$-in. plate, $\frac{7}{8}$-in. A325-X bolts, and AISC specifications (9.1), (see Fig. P-9.5).

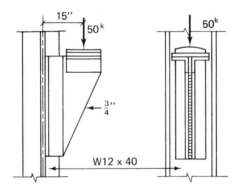

Figure P-9.5

9.6. Design a rigid moment connection of a W24 × 55 beam to a W14 × 68 column of A36 steel to transmit a reaction, $R = 60$ kips, and a moment, $M = 250$ kip-ft (see Fig. P-9.6).

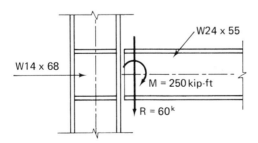

Figure P-9.6

9.7. Design a "knee" connection to a W14 × 84 column if the tension force in the knee member is 120 kips. Use 1-in. A325-X bolts, $\frac{7}{8}$-in. gusset plate, A36 steel, and AISC specification (9.1), and make connection symmetric with respect to the line of force action (see Fig. P-9.7).

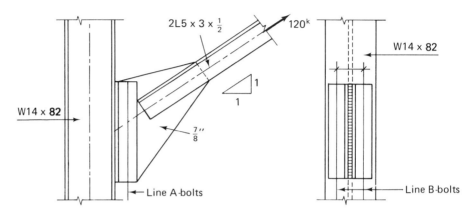

Figure P-9.7

9.8. Design a beam-to-beam connection for the full bending capacity of a spliced W24 × 55 beam. Use A36 steel, $\frac{7}{8}$-in. A325-F bolts, and AASHTO specifications (9.41). Number of stress cycles is 10^6 and live load moment is one-half the total moment (see Fig. P-9.8).

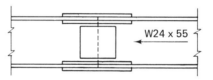

Figure P-9.8

9.9. Design the field connection of a continuous, welded plate girder for 50% of its capacity (the connection is located at the dead load inflection point) if girder dimensions are as those in Fig. P-9.9. Use AISC specification (9.1), 1-in. A325-X bolts, and double splicing of the flanges.

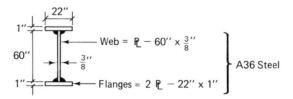

Figure P-9.9

9.10. Connect W21 × 50 stringers to a cross-beam girder if continuity of the stringers has to be assured. The stringer reaction force is 20 kips and the negative support moment (top flange in tension) is –157 kip-ft. Use chairs and continuity splice plate (fish plate) in A36 steel, $\frac{3}{4}$-in. A325-F bolts, and AISC specification (9.1) (see Fig. P-9.10).

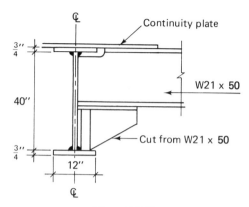

Figure P-9.10

9.11. Design an end connection of a truss member composed of two angles L5 × 3 × $\frac{1}{2}$ in. to a gusset plate $\frac{3}{4}$-in. thick for the full tensile capacity of the member. Use E-70 electrodes, manual shielded arc-welding process, and AISC specification (9.1). Balance fillet welds about the gravity axis.

9.12. A plate $\frac{1}{2}$ × 10 in. has to be connected to a plate $\frac{3}{4}$ × 15 in. to transmit a force of 105 kips. Steel is A36. Use E-70 electrodes, manual shielded arc process, and AISC specifications (9.1) (see Fig. P-9.12).

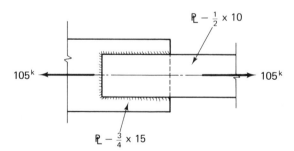

Figure P-9.12

9.13. Design the bracket connection of Problem 9.4, only now as a welded connection. Use E-70 electrodes, submerged arc process, and AISC specification (9.1).

9.14. Design the bracket connection of Problem 9.5 using E-70 electrodes, manual shielded arc process, and AISC specification (9.1).

9.15. Design the rigid beam-to-column connection of Problem 9.6 using E-70 electrodes, shielded arc process, and AISC specification (9.1).

9.16. Design the "knee" connection of Problem 9.7 using E-70 electrodes, shielded arc process, and AISC specification (9.1).

9.17. Design an end plate connection of a W21 × 44 beam to a W14 × 68 column for a shear force of 40 kips and bending moment of 50 kip-ft. Use E-70 electrodes, shielded arc process, and AISC specification (9.1).

9.18. Design a shear end connection of a W21 × 68 beam to a W14 × 120 column to transmit a reaction of 40 kips. Use E-60 electrodes, shielded arc process, and AISC specification (9.1).

9.19. Design an unstiffened beam seat connection for a simple W16 × 36 beam of a A36 steel with a reaction of 19 kips. Use E-60 electrodes, shielded arc process, and AISC specification (9.1).

9.20. Design a welded stiffened beam seat connection if the beam is a W24 × 62 and the column a W14 × 82. Beam reaction is 45 kips. Use E-70 electrodes, shielded arc process, and AISC specification (9.1) (see Fig. P-9.20).

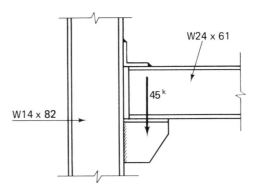

Figure P-9.20

9.21. Redesign the double splice butt connection of Problem 9.1 using A325-F high-strength bolts.

9.22. Design the framed beam connection of W18 × 50 beam to W12 × 72 column for a beam reaction of 65 kips, using $\frac{3}{4}$-in. diameter A325-N bolts in standard holes. Rolled sections are of A36 material. Check the design by using tables IIA (*Manual*, pp. 4–25) and IIC (pp. 4–27).

9.23. Redesign the framed beam connection of Problem 9.22 using E-70XX electrodes to weld frame angles to the beam web and A325-X bolts of $\frac{3}{4}$-in. diameter for outstanding legs.

9.24. Design the seated beam connection using bolts $\frac{7}{8}$-diameter A325-N. The beam is a W18 × 46, column flange is $\frac{3}{4}$ in. thick and they are both of A36 material. Reaction is 30 kips.

9.25. Redesign the seated beam connection of Problem 9.24 by welding it with E-70XX electrodes.

9.26. Design a stiffened seated beam connections of A36 steel to support a W30 × 108, also A36 material, with an end reaction of 92 kips. Use $\frac{7}{8}$-in. diameter ASTM A325-N bolts to attach the seat to a column (W14 × 109) web. A top angle is required.

9.27. Redesign Problem 9.26 by using welding instead of bolting. Use E-70XX electrodes, and traditional vector analysis. Check the design from the table VIII of the AISC *Manual*, pp. 4–52 and 4–53.

9.28. Design a bolted moment connection for a W24 × 84 beam framed to a W12 × 65 column. The design moment is 305 kip-ft and the beam reaction is 53.0 kips. All materials are A36 steel. Use $\frac{7}{8}$-in. diameter A325-N bolts.

9.29. Redesign Problem 9.28 by shop-welding an end plate to the beam and bolting it to the column flange. Use the table of Preliminary End-Plate Design (*Manual*, pp. 4–118 and 4–119).

REFERENCES

9.1. AISC, *Specification for the Design, Fabrication and Erection of Structural Steel for Buildings*, Nov. 1, 1978, New York, N.Y.

9.2. AISC, *Manual of Steel Construction*, 8th ed., Chicago, Illinois, 1980.

9.3. ASTM, *Book of Standards*, latest ed., published annually, Philadelphia, Pa.

9.4. STEWART, W. C., "History of the Use of High-Strength Bolts," *Trans. ASCE*, vol. 120, 1955, p. 1296.

9.5. WILSON, W. M., and F. P. THOMAS, *Bull. Eng. Exp. Sta.* (University of Illinois, Urbana), No. 302, May 31, 1938.

9.6. Research Council on Riveted and Bolted Structural Joints of the Engineering Foundation, *Specifications for Assembly of Structural Joints Using High Tensile Steel Bolts*, Jan. 1951, New York.

9.7. MUNSE, W. H., D. T. WRIGHT, and N. M. NEWMARK, "Laboratory Tests of Bolted Joints," *Trans. ASCE*, Vol. 120, 1955, p. 1299.

9.8. GALGOCZY, G., "Recent Experiments on the Application of High Strength Friction Grip Bolted Joints in Structural Steelwork, and Suggestions for their Design and Assembly," *Proc. 3rd Conf. on Dimensioning*, Hungarian Science Academy, Budapest, 1968, pp. 655–61.

9.9. BAKER, J. F., *The Steel Skeleton*, vol. 1, Cambridge University Press, Cambridge, 1954, p. 121.

9.10. Research Council on Riveted and Bolted Structural Joints of the Engineering Foundation, *Structural Joints Using ASTM A325 and A490 Bolts*, Sept. 1, 1966, New York.

9.11. UDIN, H., E. R. FUNK, and J. WULFF, *Welding for Engineers*, J. Wiley, New York, 1954, p. 1.

9.12. MORRIS, J. L., *Welding Processes and Procedures*, Prentice-Hall, Englewood Cliffs, NJ, 1954, p. 3.

9.13. Lincoln Electric Company, *Procedure Handbook of Arc Welding Design and Practice*, 11th ed., 1957 (reprinted 1967).

9.14. FLINTHAM, E., *Manual, Semi-Automatic, and Automatic Arc Welding*, British Oxygen Co., Hammersmith House, London, 1966, p. 343.

9.15. PATON, B. E., *Electroslag Welding*, 2nd ed., (trans. from Russian), AWS, New York, 1962.

9.16. AGIC, T., and J. A. HAMPTON, "Electroslag Welding with Consummable Guide on the Bank of America World Headquarters Building," *Welding Journal*, Vol. 47, No. 12, Dec., 1968, pp. 939–46.

9.17. Ibid. (ref. 9.14), p. 7.

9.18. AWS, *Specifications for Mild Steel Covered Arc Welding Electrodes* (AWS A5.1), latest ed., and *Specifications for Low Alloy Steel Covered Arc Welding Electrodes* (AWS A5.5), Miami, FL.

9.19. ELLIOTT, A. L., "Welding Structural Steel," *Contemporary Steel Design* (American Iron and Steel Institute), Vol. 1, No. 3, April 1964.

9.20. AWS, *Welding Handbook*, 7th ed., New York, 1976.

9.21. BLODGETT, O. W., *Design of Welded Structures*, Lincoln Arc Welding Foundation, Cleveland, Ohio, 1966, p. 1.1–3.

9.22. "Deutsche Industrie Normen," *Geschweisste Stahlbauten*, DIN 4100, Berlin, Dec., 1968.

9.23. STIELER, C., "Grundlagen des Schweissens," *Schweisstechnik im Stahlbau*, Vol. 1, Springer, Berlin, 1939, p. 119.

9.24. BLODGETT, O. W., *Design of Welded Structures* (see ref. 9.21), p. 7.2–4.

9.25. THORNTON, C. H., "Quality Control in Design and Supervision Can Eliminate Lamellar Tearing," *AISC Engineering Journal*, Vol. 10, No. 4, 1973.

9.26. AWS, *Structural Welding Code*, AWS D1.1–72, New York, 1972, Section 2, p. 2.

9.27. BIERETT, G., "Schrumpfung und Spannung," *Schweisstechnik im Stahlbau*, Vol. 1, Springer, Berlin, 1939, pp. 128–68.

9.28. STRUIK, J. H. A., A. O. OYELEDUN, and J. W. FISHER, "Bolt Tension Control with a Direct Tension Indicator," *Engineering Journal* AISC, Vol. 10, No. 1, first quarter, 1973, New York, pp. 1–5.

9.29. MCGUIRE, W., *Steel Structures*, Prentice-Hall, Englewood Cliffs, NJ, 1968, p. 803.

9.30. HERTZ, H., *Gesammelte Werke*, Vol. 1, Leipzig, 1895.

9.31. ROARK, R. J., *Formulas for Stress and Strain*, 4th ed., McGraw-Hill, New York, 1965, Table XIV, p. 320.

9.32. FISHER, J. W., and J. H. A. STRUIK, *Guide to Design Criteria for Bolted and Riveted Joints*, J. Wiley, New York, 1974.

9.33. HERTWIG, A., and H. PETERMANN, "Über die Verteilung einer Kraft auf die einzelnen Niete einer Nietverbindung," *Der Stahlbau*, Vol. 2, 1949, p. 289.

9.34. DEJONGE, R. A. E., "Riveted Joints: A Critical Review of the Literature Covering Their Development," *Res. Publ. ASME*, 1940. (Includes a bibliography and abstracts from the most important articles; see also *Trans. ASCE*, Vol. 99, 1934, p. 474).

9.35. CRAWFORD, S. F. and G. L. KULAK, "Eccentrically Loaded Bolted Connections," *Journal of the Structural Division*, ASCE, Vol. 97, No. ST3, March 1971, pp. 765–83.

9.36. KULAK, G. L., "Eccentrically Loaded Slip-Resistant Connections," *Engineering Journal*, AISC, Vol. 12, No. 2, 1975, pp. 52–5.

9.37. FISHER, J. W., "Behavior of Fasteners and Plates with Holes," *Journal of the Structural Division*, ASCE, Vol. 91, No. ST6, Dec. 1965, pp. 265–86.

9.38. AISC, *Specification for Structural Joints Using ASTM A325 and A490 Bolts*, April 26, 1978, Table 3.

9.39. DOUTY, R. T., and W. MCGUIRE, "High-Strength Bolted Connections," *Journal of the Structural Division*, ASCE, Vol. 91, No. ST2, April 1965, pp. 101–28.

9.40. HIGGINS, T. R., and W. H. MUNSE, "How Much Combined Stress Can a Rivet Take?," *Engineering News Record*, Dec. 4, 1952.

9.41. AASHTO, *Standard Specifications for Highway Bridges*, 12th ed., Washington, DC, 1977 (with 1978, 1979, 1980, and 1981 Interim Bridge Specifiations).

9.42. HENNIG, A., *Forsch. Ing. Wesen*, Vol. 4, 1933, p. 53.

9.43. ROŠ, M., and A. EICHINGER, Die Bruchgefahr fester Körper, bei wiederholter Beanspruchung-Ermüdung, Report No. 173, EMPA, Zurich, Sept. 1950.

9.44. LENZEN, K. H., "The Effect of Various Fasteners on the Fatigue Strength of a Structural Joint," *Proc. of the 49th Annual Convention of the American Railway Eng. Assoc.*, Chicago, March 14–16, 1950, pp. 1–28.

9.45. LEWITT, C. M., E. CHESSON, JR., and W. H. MUNSE, "Riveted and Bolted Joints: Fatigue of Bolted Structural Connections," *J. Struc. Div., Proc. ASCE*, Vol. 89, No. ST1, Feb. 1963, Part 1, p. 57.

9.46. MUNSE, W. H., "Research on Bolted Connections," *Trans. ASCE*, Vol. 137, 1956, p. 1265.

9.47. HANSEN, N. G., "Fatigue Tests of Joints of High-Strength Steel," *J. Struc. Div., Proc. ASCE*, Vol. 85, Paper No. 1972, March 1959, p. 750.

9.48. BLODGETT, O. W., "Detailing to Achieve Practical Welded Fabrication," *Engineering Journal*, AISC, Vol. 17, No. 4, 1980, pp. 106–19.

9.49. AWS, *Structural Welding Code*, Steel AWS D1.1–81, 5th ed., Miami, 1981.

9.50. L'institut Technique du Bâtiment et des Travaux Publics, *Règles de Calcul des Constructions en Acier* (Règles CM), ed. Eyrolles, Paris, 1974, p. 131.

9.51. KLÖPPEL, K., and PETRI, R., "Versuche zur Ermittlung der Tragfähigkeit von Kehlnähten," *Der Stahlbau*, Vol. 35, No. 1, 1966, pp. 9–25.

9.52. Deutsche Industrie Normen, *Vorschrifte für geschweisste Stahlhochbauten*, DIN 4100, Dec. 1956.

9.53. BUTLER, L. J. and G. L. KULAK, "Strength of Fillet Welds as a Function of Direction of Load," *Welding Journal, Welding Research Supplement*, Vol. 50, May 1971, pp. 231-S to 234-S.

9.54. BUTLER, L. J., SH. PAL, and G. L. KULAK, "Eccentrically Loaded Welded Connections," *Journal of the Structural Division*, Proc. ASCE, Vol. 98, No. ST5, May 1972, pp. 989–1005.

9.55. DAWE, J. L. and G. L. KULAK, "Welded Connections under Combined Shear and Moment," *Journal of the Structural Division*, Proc. ASCE, Paper No. 10457, Vol. 100, No. ST4, April 1974, pp. 727–41.

9.56. SCHREINER, N. G., "Behavior of Fillet Welds When Subjected to Bending Stresses," *Welding Journal*, Welding Research Council, Vol. 14, No. 9, Sept. 1935, pp. 1s–16s.

9.57. ARCHER, F. E., et al., "Strength of Fillet Welds," *UNICIV Report R6*, University of New South Wales, New South Wales, Australia, Nov., 1964.

9.58. FISCHER, J. W., *Guide to 1974 AASHTO Fatigue Specifications*, AISC, New York, May 1974.

9.59. LIPSON, S. L., "Single-Angle Welded-Bolted Connections," *Journal of the Structural Division*, ASCE, Vol. 103, No. ST3, Proc. Paper No. 12813, March 1977, pp. 559–71.

9.60. CHESSON, E., and W. H. MUNSE, Tensile Behavior of Large Riveted and Bolted Truss-Type Connections, private communications, University of Illinois, Urbana, Dec. 1961, p. 28.

9.61. Deutscher Stahlbau-Verband, *Stahlbau*, Vol. 2, 2nd ed., Cologne, 1964.

9.62. RICHARD, R. M., et al., "The Analysis and Design of Single Plate Framing Connections," *Engineering Journal*, AISC, Vol. 17, No. 2, 1980, pp. 38–52.

9.63. LYSE, I., and N. G. SCHREINER, "An Investigation of Welded Seat Angle Connections," *Journal American Welding Society*, Feb. 1935, pp. 1–15.

9.64. DISQUE, R. O., "Directional Moment Connections—A Proposed Design Method for Unbraced Steel Frames," *Engineering Journal*, AISC, Vol. 12, No. 1, 1975, pp. 14–18.

9.65. DISQUE, R. O., "Wind Connections with Simple Framing," *Engineering Journal*, AISC, Vol. 2, July 1964, pp. 101–3.

9.66. KRISHNAMURTHY, N., "A Fresh Look at Bolted End-Plate Behavior and Design," *Engineering Journal*, AISC, Vol. 15, No. 2, 1978, pp. 39–49.

9.67. AREA, *Manual for Railway Engineering*, Vol. II, Chapter 15, Steel Structures, Section 1.5.9, Connections and Splices, 1980, p. 15-1–20.

9.68. KUZMANOVIC, B. O., "Riveted Web Connections in Bending," *Publications, International Association for Bridge and Structural Engineering*, Vol. 20, 1960, pp. 151–78.

9.69. DONNELLY, J. A., "Determining the Cost of Welded Joints," *Engineering Journal*, AISC, Vol. 5, No. 4, 1968, pp. 146–67.

10

Built-Up Beams

10.1 PLATE GIRDERS

When the required bending capacity of a beam exceeds that of one of the heavier available rolled sections—say, a W36 × 194—then a plate girder has to be considered (Fig. 10.1). Figure 10.2 shows a plate girder bridge across the Sava River in Yugoslavia. It is the second largest span in the world, 856 ft (261 m) long.

In this text only single-walled welded girders will be discussed.

Plate girders are designed by the "moment-of-inertia" method, but stability and fatigue considerations very often are controlling design factors. Such stability considerations are concerned with either "buckling strength" or "load-carrying capacity." Most design codes and specifications use the first criterion, which does not consider the "postbuckling strength" of the web. The AISC specification (10.1) takes into account the "incomplete tension field action" and thus bases the design of a plate girder on its load-carrying capacity using a uniform factor of safety of 1.67. In the following discussion this method of design will be introduced. In Section 10.1.2 the buckling strength criterion will also be briefly discussed.

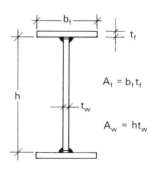

Figure 10.1 Welded Plate Girder.

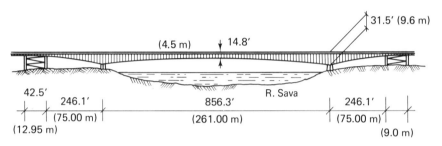

Figure 10.2 Plate Girder Bridge with Orthotropic Deck (Across the Sava River in Belgrade, Yugoslavia) with a Midspan of Length 856.3 ft (261 m).

Plate girders allow the designer to select the proper dimension for the flange and the web plates so as to achieve maximum economy. Their proportioning is discussed in detail in Section 10.1.3.

Although it is practically impossible to have a girder in pure shear without bending, for discussion purposes bending will be treated separately from shear strength for the sake of greater clarity.

10.1.1 Load-Carrying Capacity of Plate Girders

Bending strength

Web thickness

Experimental studies have shown that the web of a girder exposed to pure bending starts gradually at the early stage of bending to deflect laterally, and due to this deflection the web does not carry the full amount of stress predicted by the linear bending stress theory (Fig. 10.3). The stress in the compression flange therefore exceeds the value obtained by the beam-bending theory. This means that only a part of the compressed web portion is effective (about one-third of the compressed part, which is equivalent to $\frac{1}{6}A_w$, where A_w is the whole web area). Therefore, the bending strength

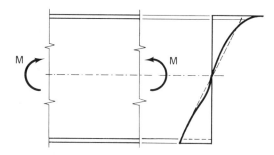

Figure 10.3 Increased Flange Compressive Stress.

of a girder depends mainly upon the compression flange, which may be treated as an isolated column.

There are three possible modes of buckling failure of a compression flange: torsional, lateral-torsional, and vertical (Fig. 10.4). The torsional buckling mode is mainly a local buckling phenomenon, and it is prevented by limiting the width-thickness ratio of the flange so as to make the critical compressive stress equal to or larger than the yield stress. The AISC (10.1, p. 20-22) states that this ratio must satisfy

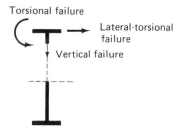

Figure 10.4 Three Modes of Collapse of Compression Flange.

$$\frac{65}{\sqrt{F_y}} < \frac{b_f}{2t_f} < \frac{95.0}{\sqrt{F_y}} \qquad (10.1)$$

in order to avoid a stress reduction (10.3, appendix C, Section C2, p. 5-94).

Lateral-torsional buckling of plate girders does not differ from that of beams, which was discussed in Section 7.1.4 and so will not be repeated here. The only mode so far not yet discussed is the vertical buckling of a compressed flange into the web (Fig. 10.4).

Due to the curvature of a plate girder in bending (Fig. 10.5) the web is exposed

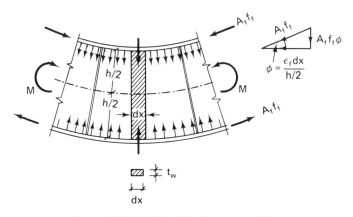

Figure 10.5 Web Buckling Due to Bending.

Plate Girders

to radial pressures exerted by both flanges. If the web plate is too thin and is very slender, as measured by the ratio h/t_w, it will buckle. To prevent this a maximum slenderness has to be specified.

If a strip of the web of width dx is considered (Fig. 10.5), then Eq. (5.68) in Chapter 5 gives for the critical, Euler stress the expression

$$F_{\text{crit}} = \frac{\pi^2 E}{12(1 - \mu^2)} \cdot \left(\frac{t_w}{h}\right)^2 \tag{10.2}$$

where t_w = web thickness, h = web depth, and μ = Poisson's ratio. The force exerted by the flanges on this column strip (Figs. 10.5 and 10.6a) is $A_f f_f \phi$, where f_f is the final stress in the flange (F_y for ultimate design). This stress has to be of sufficient duration first to overcome any existing residual stress, $+F_r$, and secondly to bring into yielding the contact surface between the top of the web and the bottom of the compression flange. If for simplicity a linear residual stress distribution is assumed (Fig. 10.6b), the linear contraction strain, ϵ_f, during this change of stress from $+F_r$ to $-F_y$ must be equal to (Fig. 10.6c)

$$\epsilon_f = \epsilon_r + \epsilon_y \tag{10.3}$$

or

$$\epsilon_f = \frac{F_r + F_y}{E} \tag{10.4}$$

Therefore, the angle ϕ is (Fig. 10.5)

$$\phi = \frac{2(F_r + F_y) \, dx}{hE} \tag{10.5}$$

For equilibrium, the applied force

$$A_f f_f \phi = \frac{2F_y(F_r + F_y) A_f \, dx}{hE} \tag{10.6}$$

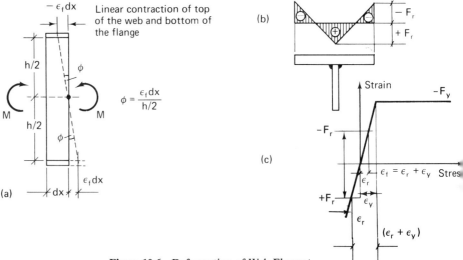

Figure 10.6 Deformation of Web Element.

should be equal to or smaller than the critical force, which is

$$F_{crit} t_w \, dx = \frac{\pi^2 E}{12(1-\mu^2)} \left(\frac{t_w}{h}\right)^2 \cdot t_w \, dx \tag{10.7}$$

Equating Eqs. (10.6) and (10.7) and solving for h/t_w yields

$$\frac{h}{t_w} \leq \sqrt{\frac{\pi^2 E^2}{24(1-\mu^2)} \cdot \frac{A_w}{A_f} \cdot \frac{1}{F_y(F_y + F_r)}} \tag{10.8}$$

When A7 steel was in use, the AISC specification assumed for F_r a value equal to $\frac{1}{2} F_y$ or $\frac{33}{2} = 16.5$ ksi. This value is still being used for all types of steel. For existing girders the minimum ratio of A_w/A_f (to be on the safe side) is around $\frac{1}{2}$. Using this value and 0.3 for Poisson's ratio, the final form of Eq. (10.8) is

$$\frac{h}{t_w} \leq \frac{13{,}785}{\sqrt{F_y(F_y + 16.5)}} \tag{10.9}$$

The AISC specification (10.1) rounds off the numerator to 14×10^3, and in section 1.10.2 of this specification the equation is given as

$$\frac{h}{t_w} \leq \frac{14{,}000}{\sqrt{F_y(F_y + 16.5)}} \tag{10.10}$$

The thickness, t_w, obtained from Eq. (10.10) is the minimum thickness. In practice other factors such as the required shear resistance, overall depth limitations, and possible stiffening of the web may increase this minimum thickness. This will be discussed in detail in Section 10.1.2.

Flange stress reduction

As mentioned above, side-deflection of the web transfers some compressive stress from the web into the compression flange. This will increase the flange stress beyond the stress calculated by the normal bending theory: $f = (M/I)y$. Thus to avoid lateral-torsional instability the maximum allowable stress obtained from pure torsional and warping torsional resistance may have to be reduced.

When buckling of the compression flange is prevented, experiments show, the relationship between the ultimate bending moments, M_u, and varying web slenderness ratios, h/t_w, for different ratios of the web to flange areas, A_w/A_f, is almost linear (10.2). A nondimensionalized diagram with ordinates M_u/M_y, where M_y is the moment required to initiate yielding at the centroid of the compression flange, and abscissas h/t_w (Fig. 10.7) shows various straight lines which pass through a point P with the abscissa β_0, which represents the critical web slenderness, h/t_w, for a web plate exposed to bending. The critical stress, F_{crit}, can thus be taken as

$$F_{crit} = \frac{k\pi^2 E}{12(1-\mu^2)} \left(\frac{t_w}{h}\right)^2 \tag{10.11}$$

where the buckling coefficient, k, depends upon the fixity of the web plate (boundary conditions). For full fixity, $k = 40$; for simply supported plate edges, $k = 24$. If an average value of 36 is taken and F_{crit} equated to F_y so as to avoid buckling before reaching yielding, the limiting (critical) web slenderness ratio, β_0, is obtained from

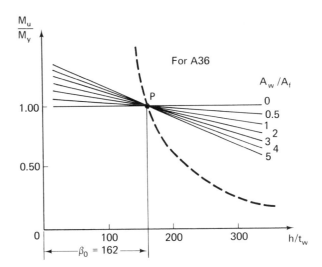

Figure 10.7 Ultimate Bending Moment, M_u, in Relation to Web Slenderness.

Eq. (10.11) as

$$\beta_0 = \left(\frac{h}{t_w}\right) = \frac{971.4}{\sqrt{F_y}} \qquad (10.12)$$

At working load level, F_y in this expression has to be divided by a factor of safety of 1.67 to yield F_b, and Eq. (10.12) now becomes

$$\beta_0 = \left(\frac{h}{t_w}\right) = \frac{752.4}{\sqrt{F_b}} \approx \frac{760}{\sqrt{F_b}} \qquad (10.13)$$

The equation representing the sloping lines shown in Fig. 10.7 is

$$M_u = M_y\left[1 - 0.0005\frac{A_w}{A_f}\left(\frac{h}{t_w} - \beta_0\right)\right] \qquad (10.14)$$

By substituting for β_0 its value from Eq. (10.13) and dividing both sides of this equation by $1.67S$, where 1.67 is the basic factor of safety and S the section modulus of the plate girder, the above equation is expressed in terms of allowable working stresses. Thus

$$F'_b = F_b\left[1 - 0.0005\frac{A_w}{A_f} \cdot \left(\frac{h}{t_w} - \frac{760}{\sqrt{F_b}}\right)\right] \qquad (10.15)$$

where F'_b is the reduced flange stress, F_b is the stress allowable in a beam segment under consideration between its lateral bracings, and A_w is the largest area of the web within the same unstiffened segment (10.1). Equation (10.15) is the same as given in section 1.10.6 of the AISC specification as equation (1.10-5).

Shear strength

If a web is stocky enough, it will not buckle under the action of pure shear (Fig. 10.8) before yielding occurs. The shear occurring in a web (V_b) is actually caused by bending or "beam action" and no postbuckling strength or "tension field action" is

438   BUILT-UP BEAMS

involved. Once the critical shear stress, $F_{v\text{-crit}}$ is reached as a result of the shear force, V_b, the web girders in the slender webs will buckle, postbuckling strength will be developed, and an additional shear force, V_t, can be carried (Fig. 10.9). Flanges and stiffeners of the plate girder will then be acting as chords and struts of an analogous Pratt truss. The ultimate shear force which a girder segment or panel can take thus

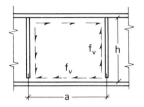

Figure 10.8 Web Plate Panel Under Pure Shear.

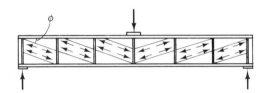

Figure 10.9 Tension Field Action (Postbuckling Behavior).

consists of a portion determined by "beam action," V_b, and a portion due to "tension field action," V_t, after shear buckling has occurred. Therefore, the ultimate shear-carrying capacity, V_u, is

$$V_u = V_b + V_t \tag{10.16}$$

The ultimate shear force capacity of a plate girder, V_p, is defined by

$$V_p = F_{ys} \cdot A_w \tag{10.17}$$

where F_{ys} is the shear yield-point stress. Using the Huber-Hencky-Mises yielding criterion, this stress is

$$F_{ys} = \frac{1}{\sqrt{3}} F_y \tag{10.18}$$

Beam action shear

Beam action shear, V_b, is

$$V_b = F_{v\text{-crit}} \cdot A_w \tag{10.19}$$

If A_w is found from Eq. (10.17) and its value substituted in Eq. (10.19), we obtain

$$V_b = V_p \cdot \frac{F_{v\text{-crit}}}{F_{ys}} = V_p \sqrt{3} \cdot \frac{F_{v\text{-crit}}}{F_y} \tag{10.20}$$

As for any other case of plate buckling the critical web-buckling shear stress, F_v, for elastic behavior is

$$F_{v\text{-crit}} = \frac{k\pi^2 E}{12(1-\mu^2)} \cdot \left(\frac{t_w}{h}\right)^2 \tag{10.21}$$

provided $F_{v\text{-crit}} \leq 0.8 F_{ys}$ and the proper buckling coefficient is used.

If the formula yields $F_{v\text{-crit}} > 0.8 F_{ys}$, then for the new critical stress the mean between the value obtained by Eq. (10.21) and the value of the limit of proportionality $(0.8 F_{ys})$ is taken—that is:

$$F_{v\text{-crit}} = \sqrt{0.8 F_{ys} \cdot \frac{k\pi^2 E}{12(1-\mu^2)} \cdot \left(\frac{t_w}{h}\right)^2} \tag{10.22}$$

Plate Girders

The corresponding buckling coefficient, k (for simply supported plates with an aspect ratio of a/h and subjected to shear) (Fig. 10.8), is

$$k = 4.00 + \frac{5.34}{(a/h)^2} \quad \text{(for } a/h \leq 1\text{)}$$
$$k = 5.34 + \frac{4.00}{(a/h)^2} \quad \text{(for } a/h > 1\text{)}$$

(10.23)

where a = length of the panel between vertical web stiffeners and h = web height.

Tension field action

The shear force, V_t, developed by tension field action is determined by considering the panel geometry and statics of the tension field. Isolating a part of the tension field symmetric about the vertical stiffener (section 1-1 in Fig. 10.10), we consider the forces (Fig. 10.10b). Along side B-C are acting the compression force in the stiffener, P_{stf}, and the tensile resultant force in the web, $f_t \cdot t_w a \sin \phi$, where f_t is the tensile force in the buckled web due to the tension field action. Along the vertical faces AB and CD two equal and opposite forces, T', are acting. They are equal because the same uniform tensile stress, f_t, is acting. Their vertical and horizontal components

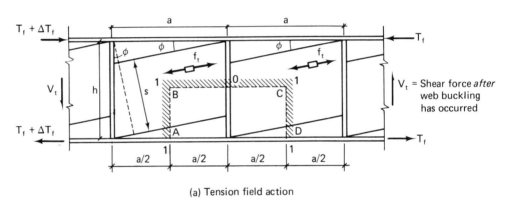

(a) Tension field action

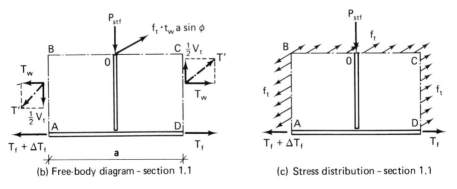

(b) Free-body diagram – section 1.1

(c) Stress distribution – section 1.1

Figure 10.10 Tension Field.

are $\tfrac{1}{2}V_t$ and T_w, respectively. On the tension flange (the bottom chord of the analogous Pratt truss) the tensile acting forces are T_f and $T_f + \Delta T_f$. It should be noted that no bending moments act along the vertical faces because shear strength is considered separately. Therefore, the forces in the top and bottom flanges are equal and opposite to one another (Fig. 10.10a). The width of the tension field in the buckled web, s, is

$$s = h\cos\phi - a\sin\phi \tag{10.24}$$

and the total tension force in the web, T, is $f_t \cdot st_w$. Therefore, the vertical component of that force, V_t, is

$$V_t = f_t st_w \cdot \sin\phi \tag{10.25}$$

Based on the principle of minimum potential energy, the angle ϕ of the tension field must be such as to give V_t its extreme value—that is

$$\frac{dV_t}{d\phi} = 0 \tag{10.26}$$

If Eq. (10.24) is substituted into Eq. (10.25) and differentiation is performed, after some manipulation the following expressions for 2ϕ and ϕ are obtained

$$\tan 2\phi = \frac{h}{a}$$

$$\tan\phi = -\frac{a}{h} + \sqrt{1 + \left(\frac{a}{h}\right)^2} \tag{10.27}$$

This yields the following for the angle ϕ

$$\phi = \tan^{-1}\left[\sqrt{1 + \left(\frac{a}{h}\right)^2} - \frac{a}{h}\right] \tag{10.28}$$

For limiting values of a/h—that is, for

$$0.5 < \frac{a}{h} < 3.0$$

the corresponding limiting values of ϕ are

$$9°13' < \phi < 31°43'$$

The change in the flange force, ΔT_f, is obtained by summing all horizontal forces of the free-body diagram (Fig. 10.10b)

$$\Delta T_f = f_t \cdot t_w \cdot a \cdot \sin\phi \cdot \cos\phi \tag{10.29}$$

Also, by summing moments about point 0 we obtain

$$V_t = \Delta T_f \left(\frac{h}{a}\right) \tag{10.30}$$

Substituting into Eq. (10.30) for ΔT_f the value from Eq. (10.29) and for $\tan 2\phi$ the expression in Eq. (10.27) and using the identity

$$\sin\phi \cdot \cos\phi = \frac{1}{2}\sin 2\phi = \frac{1}{2}\frac{\tan 2\phi}{\sqrt{1 + \tan^2 2\phi}} = \frac{1}{2} \cdot \frac{1}{\sqrt{1 + (a/h)^2}} \tag{10.31}$$

the final value of the vertical shear force, V_t, is

$$V_t = f_t \cdot A_w \cdot \frac{1}{2\sqrt{1 + (a/h)^2}} \tag{10.32}$$

This expression is modified using Eq. (10.17) for A_w, which yields

$$V_t = f_t \cdot V_p \cdot \frac{\sqrt{3}}{2F_y\sqrt{1+(a/h)^2}} \tag{10.33}$$

Finally, the ultimate shear capacity, V_u, is obtained when Eqs. (10.20) and (10.33) are substituted into Eq. (10.16)

$$V_u = V_p\left[\frac{F_{v\text{-crit}}\sqrt{3}}{F_y} + \frac{f_t\sqrt{3}}{2F_y\sqrt{1+(a/h)^2}}\right] \tag{10.34}$$

In the AISC specification (10.1, section 1.10.5.2) the value of $F_{v\text{-crit}}\sqrt{3}/F_y$, which is the ratio of the critical shear stress to the shear yielding stress, F_{ys}, is designated C_v. The tension stress, f_t, in the web cannot be equal to the yield-point stress, F_y, because the web material has already been exposed to critical shear. The equivalent tension stress corresponding to the critical shear stress (using the Huber-Hencky-Mises criterion) is $F_{v\text{-crit}}\sqrt{3}$. Therefore, the maximum value for f_t is

$$f_t = F_y - F_{v\text{-crit}}\sqrt{3} \tag{10.35}$$

or

$$\frac{f_t}{F_y} = 1 - C_v \tag{10.36}$$

When Eq. (10.36) is substituted into Eq. (10.34) we have

$$V_u = V_p\left[C_v + \frac{\sqrt{3}}{2}\frac{1-C_v}{\sqrt{1+(a/h)^2}}\right] \tag{10.37}$$

If instead of V_p in the above equation the value from Eq. (10.17) is substituted into Eq. (10.37) and both sides are divided by $1.67A_w$, the allowable shear stress, F_v, is equal to

$$F_v = \frac{F_y}{2.89}\left[C_v + \frac{1-C_v}{1.15\sqrt{1+(a/h)^2}}\right] \tag{10.38}$$

where $C_v \leq 1.00$. This is equation (1.10-2) in the AISC specification (10.1, section 1.10.5.2).

Of course, if no intermediate stiffeners are provided, the Pratt truss analogy is no longer valid because a tension field cannot be developed. Then in Eq. (10.38) only the first term can be used because the second term represents tension field action, which does not exist, and Eq. (10.38) is shortened to

$$F_v = \frac{F_y}{2.89} \cdot C_v \tag{10.39}$$

The values for the coefficient $C_v = F_{v\text{-crit}}/F_{ys}$ can be obtained from Eqs. (10.21) and (10.22) when both sides are divided by F_{ys} or $F_y/\sqrt{3}$. Bearing in mind the limits given in Eq. (10.21), the value for C_v for $C_v \leq 0.8$ becomes

$$C_v = \frac{\pi^2 E\sqrt{3}}{12(1-\mu^2)F_y}\frac{k}{(h/t_w)^2} = \frac{45,398k}{F_y(h/t_w)^2} \tag{10.40}$$

This above expression for any value of $0 < C_v < 0.8$ in the AISC specification, (10.1 p. 5-32) is rounded off to

$$C_v = \frac{45,000k}{F_y(h/t_w)^2} \tag{10.41}$$

When $C_v > 0.8$ its value is obtained from Eq. (10.22) as

$$C_v = \pi\sqrt{0.8\sqrt{3}\,\frac{E}{12(1-0.3^2)}} \cdot \frac{1}{h/t_w} \cdot \sqrt{\frac{k}{F_y}} = \frac{190.6}{(h/t_w)}\sqrt{\frac{k}{F_y}}$$

Again, this expression is slightly changed by rounding off 190.6 to 190, yielding

$$C_v = \frac{190}{h/t_w}\sqrt{\frac{k}{F_y}} \tag{10.42}$$

The buckling coefficient, k, has to be determined from Eq. (10.23)

Intermediate stiffeners

The force in the stiffener, P_{stf}, due to tension field action is obtained from the sum of vertical forces acting on the free body shown in Fig. 10.10b

$$P_{stf} = f_t \cdot t_w \cdot a \sin^2 \phi \tag{10.43}$$

If we substitute for f_t the value from Eq. (10.36) and use the identity

$$\sin^2 \phi = \frac{1 - \cos 2\phi}{2} = \frac{1}{2}\left(1 - \frac{1}{\sqrt{1 + \tan^2 2\phi}}\right) = \frac{1}{2}\left[1 - \frac{(a/h)}{\sqrt{1 + (a/h)^2}}\right] \tag{10.44}$$

then P_{stf} becomes

$$P_{stf} = \frac{(1 - C_v)}{2} F_y \cdot t_w \left[a - \frac{a^2/h}{\sqrt{1 + (a/h)^2}}\right] \tag{10.45}$$

If A_{stf} represents the cross-sectional area of the stiffener, then the ultimate force, P_{sy}, is equal to

$$P_{sy} = F_{y\text{-}stf} \cdot A_{stf} \tag{10.46}$$

where $F_{y\text{-}stf}$ is the yield-point stress of the material of the stiffener. When Eq. (10.46) is equated to Eq. (10.45) and solved for A_{stf} we have

$$A_{stf} = \left(\frac{1 - C_v}{2}\right) t_w \left[a - \frac{a^2/h}{\sqrt{1 + (a/h)^2}}\right] \frac{F_y}{F_{y\text{-}stf}} \tag{10.47}$$

If the right-hand side of this equation is multiplied and divided by the web depth, h, the expression is modified to

$$A_{stf} = \frac{1 - C_v}{2} \left[\frac{a}{h} - \frac{(a/h)^2}{\sqrt{1 + (a/h)^2}}\right] \frac{F_y}{F_{y\text{-}stf}} (ht_w) \tag{10.48}$$

If the stiffeners are arranged in pairs (Fig. 10.11a), then the force, P_{stf}, is equally divided among them and no eccentricity is occurring. If single stiffeners are used, due to the eccentricity of the force action with respect to the web the area of the stiffener,

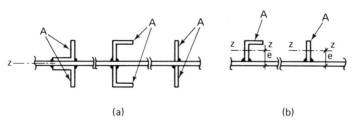

(a)　　　　　　　　(b)

Figure 10.11 Stiffener Arrangement.

A_{stf}, has to be increased. The AISC specification (10.1, section 1.10.5.4) therefore introduces in Eq. (10.48) a coefficient, D, for that purpose. Introducing Y for the ratio $F_y/F_{y\text{-stf}}$ using Eq. (10.48), the required area of the stiffener is now

$$A_{stf} = \frac{1-C_v}{2}\left[\frac{a}{h} - \frac{(a/h)^2}{\sqrt{1+(a/h)^2}}\right] Y \cdot D \cdot h t_w \qquad (10.49)$$

where D is 1.0 for a pair of stiffeners, 1.8 for single angle stiffeners, and 2.4 for single plate stiffeners. Equation (10.49) is the same as AISC equation (1.10-3) (10.1).

The spacing of intermediate stiffeners and their connection to the web is next discussed. The AISC specification (10.1, section 1.10.5.3) states that intermediate stiffeners are not required if the web slenderness, h/t_w, not only satisfies Eq. (10.10) but also is less than 260 and if the calculated shear stress, f_v, is less than the allowable shear stress, F_v, obtained from Eq. (10.39). If any of the above conditions is not satisfied, intermediate stiffeners are required. In that case the spacing of stiffeners, a, is guided by the value of the allowable shear stress, F_v, obtained from Eq. (10.38), but it cannot be more than either

$$\left(\frac{260}{h/t_w}\right)^2 h \quad \text{or} \quad 3h$$

The first distance of stiffeners from the end supports, which is called the "end panel length," has to be small enough to prevent tension field action to develop. This is done intentionally to avoid one-sided bending of the end stiffener, which otherwise would be exposed to bending around its minor axis and therefore could easily be deformed permanently. This end panel length is obtained from Eq. (10.39) by substituting for F_v the calculated maximum shear stress, f_v, and solving for a/h, as given in Eq. (10.23) for $a/h < 1.00$.

Instead of solving Eq. (10.38), tables 11 in appendix A of the AISC specification (10.1) can be used. Also, instead of solving Eq. (10.39) to determine if intermediate stiffeners are required and the end panel length, tables on p. 5-82 and 5-83 of the AISC *Manual* (10.3) can be used for A36 steel and 50 ksi yield steel.

The cross section of intermediate stiffeners obtained from Eq. (10.49) has to be checked by the condition (10.1, section 1.10.5.4) that the moment of inertia of a pair of intermediate stiffeners, or a single intermediate stiffener, with reference to an axis in the plane of the web must not be less than $(h/50)^4$.

To connect such stiffeners the practical maximum area, A_{stf}, and the corresponding force in the stiffener was first investigated for all practical combinations of the parameters h/t_w and a/h. It was found that the maximum ultimate stiffener force one could expect may be expressed approximately by

$$P_{stf} = 0.015h^2\sqrt{\frac{F_y^3}{E}} \qquad (10.50)$$

To ensure that this force is developed over each outer third of the stiffener—in other words, to connect the stiffener as soon as practical—the average ultimate shear flow per inch of total depth is

$$f_{vu} = \frac{3P_{stf}}{h} = 0.045h\sqrt{\frac{F_y^3}{E}} \qquad (10.51)$$

To change the level from ultimate to working stress, both sides of Eq. (10.51) are divided by the factor of safety (1.67). The expression for the allowable shear flow per unit depth, f_{vs}, is now

$$f_{vs} = \frac{0.045}{1.67\sqrt{E}} h\sqrt{F_y^3} = h\sqrt{\left(\frac{F_y}{339.1}\right)^3} \qquad (10.52)$$

The figure in the denominator is rounded off to 340, yielding the expression

$$f_{vs} = h\sqrt{\left(\frac{F_y}{340}\right)^3} \qquad (10.53)$$

which is given in the AISC specification (10.1) as equation (1.10-4). Section 1.10.5.4 of the specification, where equation (1.10-4) is given, also stipulates additional regulations about the stiffener length and connection details; the reader is referred to this particular section.

Bearing stiffeners

In Section 11.1, Beam Supports, bearing stiffeners are discussed together with web crippling. The reader is referred to this part of the text. In Fig. 10.12 a bearing end stiffener is shown with the characteristic cutouts around the fillet welds connecting the flange and the web. The requirement for interior bearing stiffeners and their design to support concentrated loads will be discussed in Example 10.1 in Section 10.1.3.

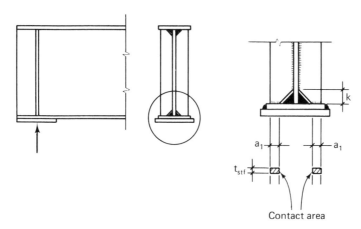

Figure 10.12 Bearing Stiffeners of Plate Girders.

Combined bending and shear

When a plate girder is designed with intermediate stiffeners and use is made of tension field action, part of the web material's tensile strength is used to develop such action (Eq. 10.35). Therefore, the tensile bending stress due to the bending moment acting in the web plane has to be further restricted to values below $0.6F_y$. Experimentally it was established that when the concurrent shear stress, f_v, is less than or equal to 60% of the allowable shear stress—that is, when

$$f_v \leq 0.6F_v \qquad (10.54)$$

—then the full allowable bending stress, F_b, can be used. Also, when the concurrent bending stress, f_b, is less than or equal to 75% of its allowable value—

$$f_b \leq 0.75 F_b \qquad (10.55)$$

—then the calculated shear stress, f_v is allowed to equal the full value of the allowable shear stress, F_v.

Beyond these limits a straight-line interaction formula (Fig. 10.13) [equation (1.10-7) in the AISC specification (10.1)] is used which states that bending tensile stress in the web must exceed neither $0.6 F_y$ nor

$$f_b \leq \left(0.825 - 0.375 \frac{f_v}{F_v}\right) F_y \qquad (10.56)$$

This relationship is represented in Fig. 10.13 by the dashed line CD introducing a factor of safety (1.67) to change from the ultimate to the service load level—that is, from the solid line AB to CD.

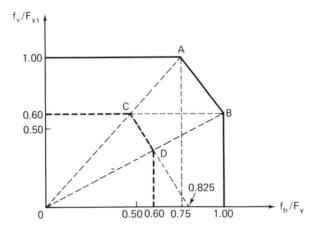

Figure 10.13 Combined Shear and Tension in Girder Web.

If the girder is fabricated of high-yield-strength quenched and tempered alloy steel plate suitable for welding, such as A514, use of tension field action is limited to regions where the concurrent bending stress is no more than $0.75 F_b$.

10.1.2 Buckling Strength of Plate Girders

In 1961 the maximum load-carrying capacity of a girder was introduced as a basis for design in buildings, as discussed in the previous section. For bridges AASHTO and some foreign specifications (discussed below) give design rules for plate girder still based on web buckling. The main objection to this practice is the necessity of using inconsistent values for safety factors in order to make the design agree better with the real behavior of the web plates.

AASHTO specifications (10.4): service load design method

The web plate thickness of plate girders without longitudinal stiffeners, t_w, is given by AASHTO specifications as a function of the web depth, h (in inches), and the calculated compressive bending stress in the flange, f_b (in lb/in.²). Thus

$$t_w = \frac{h}{23{,}000}\sqrt{f_b} \tag{10.57}$$

but not less than $h/170$. For stress expressed in ksi the above formula is

$$t_w = \frac{h}{727.32}\sqrt{f_b} \tag{10.58}$$

but not less than $h/170$.

Taking for the buckling coefficient, k, in Eq. (10.21) the lowest possible value (all four plate edges simply supported) of 23.9, and taking 29,000 for E, 0.3 for Poisson's ratio, and a factor of safety of only 1.2 (thereby recognizing the postbuckling web strength)—that is $F_{\text{crit}} = 1.2 F_b$—the following relation for the web slenderness, h/t_w, is obtained

$$\frac{h}{t_w} = \pi\sqrt{\frac{23.9 E}{12(0.91)1.2}} \cdot \sqrt{\frac{1}{F_b}} = 722.51\sqrt{\frac{1}{F_b}} \tag{10.59}$$

If Eq. (10.59) is compared with Eq. (10.58), it can be seen that the difference is less than 1%, which means that AASHTO on the one hand uses a very conservative buckling coefficient of 23.9, but on the other hand a very low factor of safety of 1.2 to bring the design into better agreement with test results. The usual factor of safety used by AASHTO (Chapter 1, Table 1.1 in this text) is 1.80.

Transverse intermediate stiffeners may be omitted if the web plate thickness is not less than

$$t_w = \frac{h\sqrt{f_v}}{7{,}500} \quad \text{or} \quad \frac{h}{t_w} = \frac{7{,}500}{\sqrt{f_v}} \tag{10.60}$$

where f_v = shear stress in lb/in.², but in no case to be less than $h/150$.

If longitudinal stiffeners are used, the thickness, t_w, can be $\frac{1}{2}$ the value obtained from Eq. (10.57). The position of a longitudinal stiffener is so chosen that its center line in the case of a plate, or the gage line if an angle is used as longitudinal stiffener, must be $h/5$ (in inches) from the inner surface or leg of the compression flange component (Fig. 10.14). Longitudinal stiffeners are usually placed only on one side of the

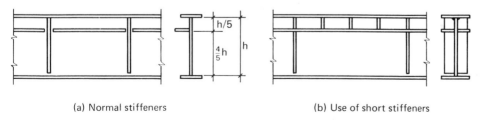

(a) Normal stiffeners (b) Use of short stiffeners

Figure 10.14 Longitudinal Stiffener.

web plate. They need not be continuous and may be cut at their intersections with the transverse stiffeners.

Longitudinal stiffeners should be proportioned so that

$$I = ht_w^3 \left[2.4 \left(\frac{a}{h} \right)^2 - 0.13 \right] \tag{10.61}$$

where I = minimum moment of inertia of the longitudinal stiffener about its edge in contact with the web plate. The thickness of the longitudinal stiffener, t_{stf}, must satisfy the following

$$t_{stf} \geq \frac{b'\sqrt{f_b}}{2,250} \tag{10.62}$$

where b' = width of stiffeners and f_b is in lb/in.². An illustrative example is given in Example 12.2 in the last chapter of this text.

Some foreign specifications

As the best representative of European specifications for plate girders, the German ones will be briefly discussed followed by some additional comments about British, Soviet, and French specifications.

German DIN 4114 (10.5)

For web plates a minimum factor of safety has to be provided. In considering web buckling, the web plate is assumed to be subdivided into rectangular panels of length a and height h (or h_1 and h_2 in the case of a longitudinal stiffener) and, as in the AASHTO specifications (10.4), to be simply supported along all four edges. The buckling coefficients, k, for such a plate are given for bending (three cases) and pure shear. In the case of bending the k-values are given as functions of a factor, ψ, which is the ratio of the edge bending stresses (Fig. 10.15), and of the panel aspect ratio, $\alpha = a/h$. In the case of pure shear the German DIN 4114 specifications give the same values for k as the AISC specification (10.1)—that is, Eq. (10.23).

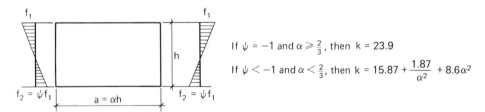

Figure 10.15 Buckling Coefficient, k.

Once a proper value for k is chosen, then the ideal buckling stresses, F_{1Bi} in bending and F_{vBi} in shear, are obtained as

$$F_{1Bi} = k \cdot F_e \quad \text{or} \quad F_{vBi} = k \cdot F_e \tag{10.63}$$

where F_e is a "coordinating" stress equal to the Euler buckling stress, as in Eq. (10.2).

In the case of combined bending and shear, both ideal buckling stresses must be calculated separately according to Eq. (10.63).

From these separate ideal buckling stresses, F_{1Bi} and F_{vBi}, and from the actual stresses, f_1 (maximum compressive stress in web) and f_v (shear stress in web), the ideal "comparative" stress, F_{CBi}, is calculated from

$$F_{CBi} = \frac{\sqrt{f_1^2 + 3f_v^2}}{\dfrac{1+\psi}{4} \cdot \dfrac{f_1}{F_{1Bi}} + \sqrt{\left(\dfrac{3-\psi}{4} \cdot \dfrac{f_1}{F_{1Bi}}\right)^2 + \left(\dfrac{f_v}{F_{vBi}}\right)^2}} \qquad (10.64)$$

For the special case in which $f_v = 0$, then simply $F_{CBi} = F_{1Bi}$, and when $f_1 = 0$, $F_{CBi} = F_{vBi} \cdot \sqrt{3}$.

In all cases in which F_{CBi} is beyond the proportionality limit of steel the ideal buckling stress must be reduced. Such a reduction is based on the following equation

$$F_{CB} = F_y \left(1 - \frac{0.046 F_y}{F_{CBi} - 0.57 F_y} \right) \qquad (10.65)$$

Finally, the factor of safety, v_B, against buckling is given by

$$v_B = \frac{F_{CB}}{\sqrt{f_1^2 + 3f_v^2}} \qquad (10.66)$$

If the ideal comparative buckling stress is $F_{CBi} < 53.36$ ksi (3,750 kg/cm²), this factor of safety in the case of gravity loads has to be $v_B \geq 1.35$ and in the case of gravity and wind loads $v_B \geq 1.25$.

For comparative stresses where $F_{CBi} > 53.36$ ksi (3,750 kg/cm²), the factor of safety can be less than the given values according to the following equation

$$v_B^1 \geq v_B \cdot \left[0.9 + 0.1 \cdot \left(\frac{53.36}{F_{CBi}} \right)^2 \right] \qquad (10.67)$$

The German specifications give detailed directions for subdividing the web plate which will be omitted here.

British BS449 (buildings) (10.6)

According to this specification, vertical stiffeners must be provided throughout the length of the girder at a distance not greater than $1\tfrac{1}{2}h$ apart when the thickness of the web is less than $h/85$ for structural steel or less than $h/75$ for high-strength steel. These stiffeners must be designed so that I satisfies the following

$$I \geq 1.5 \frac{h_1^3 \cdot t_w^2}{S^2} \qquad (10.68)$$

where $I =$ moment of inertia of the complete stiffeners about the center of the web, $S =$ maximum permitted clear distance between stiffeners for thickness t_w, and $t_w =$ minimum required thickness of web. Horizontal stiffeners may be used in addition to vertical stiffeners. Regulations about them are the same as for bridges, which are briefly discussed next.

British BS153 (bridges) (10.7)

According to this specification, vertical stiffeners must be provided in the same way as specified for buildings. The minimum web thickness for stiffened webs has to be $\frac{1}{180}$ the smaller clear panel dimension and $\frac{1}{270}$ the greater clear panel dimension. Horizontal stiffeners may be used in addition to vertical stiffeners.

Soviet SN and P, part 2, section 5 (10.8)

No proof of stability and no stiffening are required if

$$\frac{h}{t_w} \leq 110 \frac{\sqrt{2,100}}{F_b} \tag{10.69}$$

where F_b is in kg/cm² and when there is no locally applied stress. If only transverse stiffeners are used, the stability is established using a nondimensionalized comparative stress as some sort of interaction equation, but otherwise very similar to the German procedure. It is of interest to note the use of short stiffeners in combination with transverse and one longitudinal stiffener (Fig. 10.14b).

French Règles CM 66 1974 (10.9)

Local buckling of a compression flange is considered prevented if

$$\frac{b_f}{2t_f} < 15\sqrt{\frac{24}{F_y}} \tag{10.70}$$

where F_y is the yield stress in daN/mm² (decanewtons per square millimeter = kgf/mm²) and not in pascals (Pa). For mild steel $F_y = 24$ daN/mm² and the above ratio is quite close to 15.8, the value given in the AISC specification (10.1).

There is no need for web stiffeners if for all normal sections of the girder the calculated bending and shear stresses in daN/mm² satisfy the following expression

$$\left(\frac{f_b}{7}\right)^2 + f_v^2 < 0.015\left(\frac{1,000 t_w}{h}\right)^4 \tag{10.71}$$

If this condition is not satisfied, intermediate stiffeners are required.

10.1.3 Optimization of Plate Girder Dimensions

As was pointed out earlier, the height (or depth) of a plate girder is its most important dimension. Therefore, before the introduction of computer-aided design several approximate formulas were developed to help a designer determine this important dimension.

Less elaborate but of the same level of accuracy were several rules of thumb about the web height, h, the flange width, b_f, and their thicknesses, t_w and t_f, respectively.

As a result of the development of electronic computers and of optimization techniques the problem now is a matter of design rather than experience.

The following empirical rules may be used to get a first estimate of girder dimensions.

The height, h, for simply supported girders in buildings usually is

$$\frac{L}{8} > h > \frac{L}{12} \tag{10.72}$$

In bridges this height can be reduced to $L/15$. For continuous girders this value can be further reduced to $L/25$. Therefore, for continuous girders

$$\frac{L}{15} > h > \frac{L}{25} \tag{10.73}$$

If the girder is part of a frame, the height can be further reduced to

$$\frac{L}{30} > h > \frac{L}{40} \tag{10.73a}$$

An economical thickness for the web lies between the following limits

$$\frac{h}{120} > t_w > \frac{h}{260} \tag{10.74}$$

In general the web thickness is $t_w \approx h/110$.

The flange width, b_f, is

$$b_{\text{in.}} \approx \frac{h_{\text{in.}}}{4} + 2.5 \text{ in.} \tag{10.75}$$

or

$$\frac{h}{2.5} > b_f > \frac{h}{6} \tag{10.75a}$$

For lateral stability in terms of the span length the widths are

$$\frac{L}{60} < b_f < \frac{L}{25} \tag{10.75b}$$

The flange thickness, t_f, is normally estimated such that the ratio $b_f/2t_f$ prevents local buckling. It is suggested to choose t_f less than $1\frac{1}{2}$ in. If larger thicknesses must be used, a better steel quality (killed steel) has to be employed (see Flange Thickness in Section 2.5.3). Flange thicknesses greater than $1\frac{1}{2}$ in. (35 mm) are beyond normal practice and special provisions for the steel quality and fabrication procedures need to be made.

Web without stiffeners

Girders in buildings

The AISC specification (10.1) requires that two conditions be satisfied by the web plate to be safe without stiffeners

1. Slenderness of the web, h/t_w, must be less than or equal to 260.
2. Calculated shear stress, $f_v = V/A_w$, must be less than or equal to $F_v = (F_y/2.89)C_v$ (i.e., beam action shear without tension field action—Eq. 10.39).

To get the optimum dimensions—that is, a girder of minimum weight—following traditional procedures for the proportioning of plate girders several trial configurations have to be compared so as to determine the lowest weight. The above procedure does not guarantee that a global optimum will be reached. Computer-aided design enables us to change this situation. We shall now discuss a procedure developed by Holt and Heithecker (10.10). They considered symmetrical, laterally supported girders—that is, girders for which

$$F_b = 0.6F_y \quad \text{for} \quad \frac{L_b}{r_T} < \sqrt{\frac{(102 \times 10^3)C_b}{F_y}}$$

Due to the fact that true plate girders never consist of compact sections, $F_b = 0.6F_y$ is used rather than $F_b = 0.66F_y$ for laterally supported girders. The optimum values of the design variables (h = depth of web, t_w = thickness of web, and A_f = area of one flange) for a given moment, M, and shear, V, are determined by minimizing the area of the girder, A—

$$A = 2A_f + A_w \tag{10.76}$$

—subject to three constraints, two of which have already been mentioned and are required by AISC specification and a third constraint as a result of a limited stress being allowed in the compressive flange. This last condition yields either $F_b = 0.6F_y$ for a web slenderness of $h/t_w < \beta_0$ (Eq. 10.13) or F_b' from Eq. (10.15) in cases where $h/t_w > \beta_0$. The results of this proposed optimization are obtained in closed form, which can be directly applied to design problems and does not require the execution of a computer program. This simple result makes this method extremely valuable to any designer. The following step-by-step procedure has to be applied.

First: a moment parameter, X, is calculated from

$$X = \frac{(F_y)^{3/4} \cdot (M)}{(E)^{1/4} \cdot (V)^{3/2}} \tag{10.77}$$

where F_y = yield stress of steel in ksi,
M = maximum bending moment in kip-in.
E = Young's modulus in ksi,
V = maximum shear force in kips

Second: the optimum nondimensional depth, Y, is obtained from

$$Y = C_1 X^p \tag{10.78}$$

where C_1 and p are constants given in Table 10.1, parts (a) and (b).

Table 10.1 DATA FOR NONCOMPACT AND COMPACT SECTIONS

Case	X_{min}	C_1	p	C_2	C_3	q
(a) Noncompact sections—A36 steel						
1	0.58494	2.33976	0	1.42465	$\frac{5}{3}$	1
2	1.16988	2.16322	$\frac{1}{2}$	0.08182	0	$\frac{1}{2}$
3	2.44274	3.38096	0	0.98591	2.40833	1
4	3.25699	2.03827	$\frac{3}{7}$	2.86190	0	$\frac{4}{7}$
5	24.8816	8.08156	0	0.41246	7.69705	1
6	29.3494	2.30709	0.37097	2.49757	0	0.61270
7	192.459	16.23598	0	0.20090	24.01655	1
8	213.706	2.71532	$\frac{1}{3}$	1.87382	0	$\frac{2}{3}$
(b) Noncompact sections—A242 steel (F_y = 50 ksi)						
6	28.7123	2.28894	0.37545	2.53367	0	0.60845
7	279.440	18.74423	0	0.17394	29.44384	1
8	308.711	2.77342	$\frac{1}{3}$	1.82153	0	$\frac{2}{3}$
(c) Compact sections						
1	0.64343	2.33976	0	1.29513	$\frac{5}{3}$	1
2	1.28687	2.06255	$\frac{1}{2}$	2.93840	0	$\frac{1}{2}$
3	1.58238	2.59453	0	1.16796	1.84815	1
4	3.16475	1.76717	$\frac{1}{3}$	2.57216	0	$\frac{2}{3}$

Part (c) of Table 10.1 shows the corresponding values for compact sections. These apply to all steels and are theoretically exact, provided the section is compact. The value of X_{min} shown for each case is the smallest value of X to which that case is applicable. In practice this means that for any actual value of X we use X_{min} in the table such that $X_{min} \leq X$.

If $X < X_{min}$ for case 1, this means that the web is heavy enough to carry the shear and also will carry the bending moment without any flanges.

Third: the depth parameter, Y, expressed in terms of the unknown girder height, h, is calculated from

$$Y = \frac{(F_y)^{3/4} \cdot h}{(E)^{1/4} \cdot (V)^{1/2}} \tag{10.79}$$

Equating Eqs. (10.78) and (10.79) and solving for h, one obtains

$$h = \frac{C_1 X^p \cdot (E)^{1/4} \cdot (V)^{1/2}}{(F_y)^{3/4}} \tag{10.80}$$

Fourth: the optimum nondimensional cross-sectional area, Z, is computed from

$$Z = C_2 X^q + C_3 \tag{10.81}$$

where C_2, C_3, and q are constants (from Table 10.1).

Fifth: the area parameter, Z, expressed in terms of the unknown cross-sectional area, A, is computed from

$$Z = \frac{F_y \cdot A}{V} \tag{10.82}$$

Again, if Eq. (10.81) is equated with Eq. (10.82) and solved for A, one obtains

$$A = \frac{(C_2 X^q + C_3)V}{F_y} \tag{10.83}$$

Sixth: the web design is next performed. Once the web depth is known, whether optimum or not, the required web thickness follows immediately from two conditions of the AISC specification given at the start of this section.

The web thickness, U, in dimensionless form is calculated from the following expressions as a function of the depth parameter, Y

1. *For Noncompact Sections:*

   For $Y \leq 2.33976$: $\quad U = \dfrac{2.5}{Y}$ (10.84a)

   For $2.33976 \leq Y \leq 3.38096$: $\quad U = 1.06848$ (10.84b)

   For $3.38096 \leq Y \leq 0.60067(\alpha)^{3/2}$: $\quad U = 0.71190 \cdot (Y)^{1/3}$ (10.84c)

   For $0.60067(\alpha)^{3/2} \leq Y$: $\quad U = \dfrac{Y}{\alpha}$ (10.84d)

   where

   $$\alpha = 260 \cdot \sqrt{\frac{F_y}{E}} \tag{10.85a}$$

2. *For Compact Sections:* The same rules may be used, except that

$$U_{min} = 0.41182\, Y \tag{10.84e}$$

and

$$\alpha = \sqrt{\frac{F_y}{E}} \tag{10.85b}$$

The web thickness parameter, U, in terms of the unknown web thickness, t_w, is

$$U = \frac{(E)^{1/4} \cdot (F_y)^{1/4} \cdot t_w}{(V)^{1/2}} \tag{10.86}$$

Again, equating Eq. (10.86) with one of Eqs. (10.84) and solving for t_w, the web thickness will be obtained.

Seventh: the flange design is next performed. Once knowing h and t_w—that is, the web area, A_w—from Eq. (10.76), the required flange area, A_f, has to be

$$A_f = \tfrac{1}{2}(A - A_w) \tag{10.87}$$

The flange area, W, in dimensionless form is calculated from one of the following expressions:

1. *For Noncompact Sections:* If $Y/U \leq \sqrt{19.2/0.60} = 5.65685$, then

$$W = \frac{X}{0.60Y} - \frac{UY}{6} \tag{10.88a}$$

If $Y/U > 5.65685$, W is obtained from the following quadratic equation:

$$W^2 - \left[\frac{5}{3} \cdot \frac{X}{Y} - \frac{1}{6} \cdot UY + \frac{\gamma Y(Y - U\sqrt{32})}{2{,}000}\right]W - \frac{\gamma U(Y)^2(Y - U\sqrt{32})}{12{,}000} = 0 \tag{10.88b}$$

The positive root of this equation should be taken. Instead of solving Eq. (10.88b), the solution may be approximated by

$$W = \frac{5}{3} \cdot \frac{X}{Y} - \frac{1}{6} \cdot UY + \frac{\gamma Y(Y - U\sqrt{32})}{2{,}000} \tag{10.88c}$$

where γ is a material parameter calculated by

$$\gamma = \sqrt{\frac{E}{F_y}} \tag{10.89}$$

2. *For Compact Sections:* If $Y/U < 5.39360$, then

$$W = \frac{X}{0.66Y} - \frac{1}{6} \cdot UY \tag{10.88d}$$

If now the flange area parameter, W, in terms of A_f is computed from

$$W = \frac{F_y \cdot A_F}{V} \tag{10.90}$$

and next Eq. (10.90) is equated to the corresponding Eq. (10.88), the flange area, A_f, is obtained. Then, choosing the thickness of the flange, t_f, the necessary width, b_f, can be determined and the design checked for the slenderness ratio, $b_f/2t_f$, so as to prevent local buckling and for L_b/r_T to satisfy the condition that the girder be sufficiently laterally braced. This actually means that one determines the necessary spacing of the lateral bracings, L_b, so that the allowable stress is still $0.6F_y$—that is, so that

Eq. (7.18) is satisfied. Probably several trials are needed until a satisfactory flange ($b_f \times t_f$) with acceptable values of L_b is obtained.

The dimensions thus obtained have to be rounded off to actual available plate sizes. The following example will illustrate the procedure.

Example 10.1 A simply supported girder with loading, shear, and moment diagrams as shown in Fig. 10.16 has to be designed without stiffeners. (*Note*: At the supports bearing stiffeners must always be provided as well as at points of concentrated loads, points B and C in Fig. 10.16a.) The floor system of the building where this girder will be used provides lateral bracing at each 10 ft along the girder span. Use A36 steel and the AISC specification (10.1).

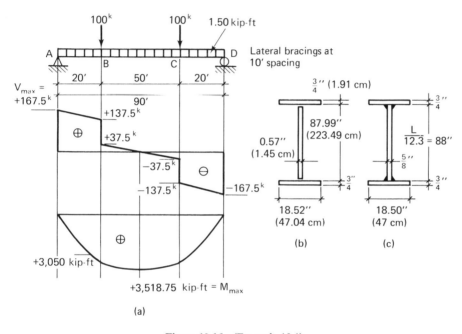

Figure 10.16 (Example 10.1).

Solution

Step 1: From Eq. (10.77):

$$X = \frac{(36)^{3/4} \cdot 3518.75 \cdot 12}{(29,000)^{1/4} \cdot (167.5)^{3/2}} = 1.12623(19.47814) = 21.93688$$

As $3.25699 < X < 24.8816$ (Table 10.1), case 4 applies with $C_1 = 2.03827$, $p = \frac{3}{7}$, $C_2 = 2.86190$, $C_3 = 0$, $q = \frac{4}{7}$.

Step 2: From Eq. (10.78):

$$Y = 2.03827(21.93688)^{3/7} = 7.65686$$

Step 3: From Eq. (10.79):

$$Y = 1.12623 \cdot \frac{h}{(167.5)^{1/2}} = 0.08702h$$

Plate Girders

Then
$$h = \frac{7.65686}{0.08702} = 87.99 \text{ in.}$$

Step 4: From Eq. (10.81) and Table 10.1:
$$Z = 2.86190 \cdot (21.93688)^{4/7} = 16.71245$$

Step 5: From Eq. (10.82):
$$Z = \frac{36}{167.5} \cdot A = 0.21492 A$$

Therefore
$$A = \frac{16.71245}{0.21492} = 77.76 \text{ in.}^2$$

Step 6: Assuming a noncompact girder section, and because $Y = 7.65686$ Eq. (10.84c) is applicable, and
$$\alpha = 260\sqrt{\frac{36}{29{,}000}} = 9.16064 \quad \text{and} \quad 0.60067(9.16064)^{3/2} = 16.65422 > Y$$

thus
$$U = 0.71190 \cdot (7.65686)^{1/3} = 1.40314$$

From Eq. (10.86):
$$U = \frac{(29{,}000)^{1/4} \cdot (36)^{1/4}}{(167.5)^{1/2}} \cdot t_w = 2.46983 \cdot t_w$$

which yields
$$t_w = \frac{1.40314}{2.46983} = 0.568 \text{ in.}; \quad \text{therefore:} \quad A_w = h \cdot t_w = 49.98 \text{ in.}^2$$

Check if $h/t_w < 260$:
$$h/t_w = \frac{87.99}{0.568} = 154.9 < 260$$

Check if $f_v \leq F_v$:
$$f_v = \frac{V}{A_w} = \frac{167.5}{49.98} = 3.35 \text{ ksi}$$

From Eq. (10.40) and $k = 5.34$ (taking $a/h \gg 4.00$)
$$C_v = \frac{45{,}000(5.34)}{36(154.9)^2} = 0.278 < 0.8$$

Therefore, according to Eq. (10.39), $F_v = (36/2.89)0.278 = 3.47 > 3.35$: <u>OK</u>

Step 7: In our case
$$A_f = \tfrac{1}{2}(77.76 - 49.98) = 13.89 \text{ in.}^2 \quad \text{(as a check)}$$

and
$$\frac{Y}{U} = \frac{7.65686}{1.40314} = 5.45695 < 5.65685$$

Therefore, from Eq. (10.88a):
$$W = \frac{21.93688}{0.60(7.65686)} - \tfrac{1}{6}(1.40314)(7.65686) = 2.98439$$

$$A_f = \frac{2.98439(167.5)}{36.0} = 13.89 \text{ in.}^2$$

If a thickness of $t_f = \tfrac{3}{4}$ in. is assumed, the flange width is $b_f = 18.52$ in.

All these theoretical dimensions (Fig. 10.16b) are now rounded off to practical values (Fig. 10.16c).

Check local buckling of flange (10.1, Table 6):

$$\frac{b_t}{2t_f} = \frac{18.5}{2 \cdot 0.75} = 12.3 < 15.8$$

Check if lateral bracing is sufficient for $F_b = 0.6F_y$:

$$A_f + \frac{1}{6}A_w = 13.88 + \frac{1}{6}(55.0) = 23.05 \text{ in.}^2$$

$$r_t = \sqrt{\frac{13.88(18.5)^2}{12(23.05)}} = 4.10 \text{ in.}; \quad \frac{L_b}{r_T} = \frac{10 \cdot 12}{4.10} = 29.2 < 53$$

(C_b is conservatively taken to be 1.00.)

Girders in bridges

The same method as in the preceding example could be applied to bridge girders with the constraints of the AASHTO specifications (discussed in Section 10.1.2). So far, no such analysis exists.

Web with stiffeners

In the previous section a method for the optimum design of girders without stiffeners was discussed. However, very often a global optimum is reached when transverse or longitudinal stiffeners are used. Therefore, it is important to have available a general method which will optimize a girder without or with stiffeners. Schilling (10.11) has developed such a general design procedure.

For an idealized cross section (Fig. 10.17) with the flange areas concentrated in lines of zero thickness at the top and bottom of the web, the elastic section modulus and the moment of inertia are functions of three geometric parameters: (1) the total cross-sectional area, A; (2) the ratio of the web area to the total area, A_w/A; and (3) the web slenderness, h/t_w.

Introducing the notations

$$a = \frac{A_w}{A} \quad \text{and} \quad b = \frac{h}{t_w} \quad (10.91)$$

the following expressions are obtained for the moment of inertia:

$$I = \frac{(3-2a)}{12}abA^2 \quad (10.92)$$

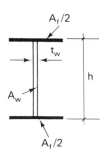

Figure 10.17 The Idealized Beam.

and for the elastic section modulus:

$$S = \frac{(3-2a)}{6}(ab)^{1/2}A^{3/2} \quad (10.93)$$

If Eqs. (10.92) and (10.93) are differentiated with respect to ratio a and the results equated to zero, two different values for the ratio a will be obtained as optimum values for the moment of inertia and the section modulus

For the moment of inertia: $a_{opt} = \dfrac{3}{4}$ and $I_{opt} = \dfrac{3dA^2}{32} = \dfrac{h_{opt}^4}{6b}$ (10.94)

For the section modulus: $a_{opt} = \dfrac{1}{2}$ and $S_{opt} = \dfrac{b^{1/2}A^{3/2}}{3\sqrt{2}} = \dfrac{2h_{opt}^3}{3b}$ (10.95)

Dividing Eqs. (10.92) and (10.93) by I_{opt} and S_{opt}, respectively, the following two non-dimensional relationships are obtained:

$$\dfrac{I}{I_{opt}} = \dfrac{8(3-2a)a}{9}$$ (10.96)

$$\dfrac{S}{S_{opt}} = \dfrac{3-2a}{2}\sqrt{2a}$$ (10.97)

Figure 10.18 shows the above ratios for different values of $a = A_w/A$ when the total area, A, and the web slenderness, h/t_w, are kept constant. It can be seen that A_w/A can vary significantly from its optimum value without greatly reducing I or S.

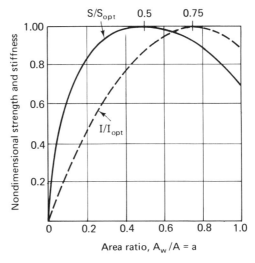

Figure 10.18 Variation of S/S_{opt} and I/I_{opt} with an a Variable and A and h/t_w = constant. [From (10.11).]

For instance, S is within 98% of its optimum value when A_w/A is varying between 0.39 and 0.62. At $a = 0.63$, S and I are both about 97.5% of their optimum values. For a given A and h/t_w each value of A_w/A corresponds to a particular beam depth, h. Consequently there is an optimum beam depth, defined by Eqs. (10.94) and (10.95). In general, h increases as A_w/A increases, and for a constant area, A, the optimum depths increase as h/t_w increases. For a constant h/t_w, the optimum elastic section modulus varies in proportion to $A^{3/2}$ and the optimum moment of inertia varies in proportion to A^2.

For a given M_{max}, the minimum cross-sectional area, A, required to achieve a desired strength or stiffness with a specified web slenderness can be calculated directly from Eqs. (10.94) and (10.95) by inserting the known values of $b = h/t_w$, $S_{opt} = M_{max}/F_b$, and I_{opt}. This last value, I_{opt}, has to be calculated by equating the expression for the maximum deflection, δ_{max} to the maximum allowable deflection. The flange and web area can thus be determined by using appropriate values of A_w/A—for example, 0.5 for the optimum section modulus. The web depth and thickness are defined by the known values of A_w and b as indicated in Eq. (10.91). Alternatively, h_{opt} for a desired S or I can be calculated from Eqs. (10.94) and (10.95).

For an idealized cross section, without considering the additional weight of stiffeners, the maximum bending strength for a given section area, A, is always obtained by using the maximum permissible web slenderness, b, and an optimum value of the area ratio, a. Therefore, Eqs. (10.94) and (10.95) cannot be used for optimization in a realistic situation where stiffeners have to be used and paid for. To achieve this it is sufficient that Eq. (10.95) be corrected to abolish the effect of zero thickness of the flange, and second, that an additional cross-sectional area, ΔA, be introduced as a penalty function for the web slenderness, b, to compensate for transverse stiffeners.

If Eq. (10.95) is solved for A and h_{opt}, the following expressions are obtained:

$$A = 2.621 \frac{S_{opt}^{2/3}}{b^{1/3}} \tag{10.98}$$

$$h_{opt} = 1.145(S_{opt} \cdot b)^{1/3} \tag{10.99}$$

The first equation is modified by a factor

$$\frac{1}{(1 - [b/2b_{max}])^{1/3}}$$

which will increase the required area for more slender webs to

$$A = 2.621 \left(\frac{S_{opt}^2}{b - (b^2/2b_{max})} \right)^{1/3} \tag{10.100}$$

The section area of stiffeners, ΔA, when their use is specified for the first time is about 6% of the total cross-sectional area, A. This means that for any given slenderness ratio, b, and a given actual shear force or shear stress and a specified stiffness requirement, the height h_1 can be determined for which stiffeners are used for the first time. For buildings, as discussed under "Web Without Stiffeners" in Section 10.1.3, according to the AISC specification (10.1, section 1.10.5.3) we do not have to use stiffeners when

$$b \leq 260 \quad \text{and} \quad f_v \leq \frac{F_y}{2.89} \cdot C_v \tag{10.101}$$

If f_v is replaced by $V/h_1 t_w$ and C_v by

$$\frac{45,000(5.34)}{36(h_1/t_w)^2}$$

in the above expression for f_v, the following relationship is obtained

$$\frac{V}{h_1 t_w} = \frac{45,000(5.34)}{2.89(h_1/t_w)^2} \tag{10.102}$$

which yields for h_1

$$h_1 = 83,150 \cdot \frac{t_w^3}{V} \tag{10.103}$$

For bridges [AASHTO specifications (10.4), section 1.7.71] transverse stiffeners have to be used when

$$b \geq 150 \quad \text{and} \quad \frac{h_1}{t_w} \geq \frac{7{,}500}{\sqrt{f_v}} \quad \text{(for } f_v \text{ in lb/in.}^2\text{)} = \frac{237}{\sqrt{f_v}} \quad \text{(for } f_v \text{ in ksi)} \tag{10.104}$$

Similarly, as for buildings, it can be shown that

$$h_1 = 56{,}170 \cdot \frac{t_w^3}{V} \tag{10.105}$$

where V is in kips For any height of $h > h_1$, the original 6% has to be increased by a factor of $(h/h_1)^{1/2}$. Therefore, the additional cross-sectional area, ΔA, is

$$\Delta A = 0.06 \left(\frac{h}{h_1}\right)^{1/2} A = 0.06 \left(\frac{b}{b_1}\right)^{1/2} A \tag{10.106}$$

where $h > h_1$ and h_1 is found from Eq. (10.103) or Eq. (10.105), and b_1 is the limiting slenderness for a given shear stress. The total area, A_{tot}, is therefore

$$A_{\text{tot}} = 2.621 \left(\frac{S_{\text{opt}}^2}{b - (b^2/2b_{\max})}\right)^{1/3} + 0.06 \left(\frac{b}{b_1}\right)^{1/2} A \tag{10.107}$$

Instead of finding the derivative of this equation, equating it to zero, and solving for an optimum, b, it is much simpler to try several web slenderness ratios, b, and find a practical optimum. In the above Eq. (10.107) the optimum section modulus, S_{opt}, is equal to $M_{\max}/F_b$, where F_b depends on the distance between lateral bracings.

The above procedure is demonstrated in Example 10.2. The data for this problem are the same as those used in Example 10.1 to allow a comparison of results.

Example 10.2 Given the data in Example 10.1, find the optimum girder proportions producing a minimum total area, A_{tot}. If required by the AISC specification (10.1), use stiffeners.

Solution As before it is assumed that lateral bracings at 10-ft distances will allow an F_b of 22 ksi.

The procedure now is to assume a slenderness, b, and to calculate h_{opt} from Eq. (10.99) and a corresponding web thickness, t_w, and the limiting height, h_1, from Eq. (10.103). Then, using Eqs. (10.100) and (10.106) obtain A and ΔA; summing these two values, the total area, A_{tot}, corresponding to the assumed slenderness is obtained.

The results of the calculations are shown in Table 10.2 and the results plotted in Fig. 10.19(a). Thus

$$S_{\text{opt}} = \frac{M_{\max}}{F_b} = \frac{3{,}518.75 \times 12}{22} = 1{,}919.32 \text{ in.}^3; \quad b_{\max} \text{ (for A36)} = 322$$

Table 10.2

b	h_{opt} (in.)	t_w (in.)	h_1 (in.)	A (in.²)	ΔA (in.²)	A_{tot} (in.²)
150	75.61	0.504	63.55	83.23	5.45	88.68
200	83.21	0.416	35.74	78.35	7.17	85.52
250	89.64	0.359	22.97	75.69	8.97	84.66
300	95.26	0.318	15.96	74.52	10.92	85.44

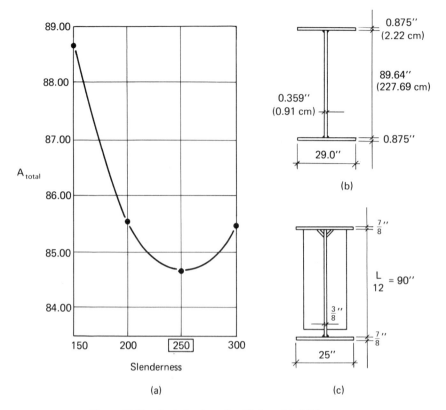

Figure 10.19 Optimization of a Girder with Stiffeners.

The optimum slenderness ratio was obtained as 250. The web area is $A_w = 89.64 \times 0.359 = 32.18$ in.² The area of girder section is $A = 75.69$ in.² This means that actually $a = 0.425$ is used and not $a = 0.5$; however, this is close enough. The area of one flange is $\frac{1}{2}A_f = \frac{1}{2}(75.69 - 32.18) = 21.76$ in.² If the dimensions for the web are rounded off, a plate $P_L - \frac{3}{8}$ in. × 90 in. is chosen with an actual area of $A_w = 33.75$ in.²

If the same flange thickness of $\frac{3}{4}$ in. (as in Example 10.1) is tried, the required width of the flange would be 29 in. The slenderness ratio is

$$\frac{b_f}{2t_f} = \frac{29}{2 \times 0.75} = 19.3 > 15.8: \quad \underline{\text{N.G.}}$$

If a thickness of $\frac{7}{8}$ in. is used, then $b_f = 25$ in. and

$$\frac{25}{2 \times 0.075} = 14.3 < 15.8: \quad \underline{\text{OK}}$$

The final results of the two methods—that is, without stiffeners and with stiffeners—are shown in Table 10.3.

The web dimensions are very similar, but due to a smaller web area in the case where stiffeners were used the flanges are stronger. It is preferable to avoid transverse stiffeners for ease of fabrication. Therefore, the first design is better than the second

Table 10.3

	Web	Flanges	Cross-sectional area, A (in.²)	Total area, A (in.²)	Remarks
Example 10.1	$\frac{5}{8} \times 88$	$\frac{3}{4} \times 18\frac{1}{2}$	82.75	82.75	Without stiffeners
Example 10.2	$\frac{3}{8} \times 90$	$\frac{7}{8} \times 25$	77.50	86.47	With stiffeners

because it uses a minimum number of stiffeners (at points of lateral bracing) and less steel.

Computer programs

Full-size computer programs

Appendix II shows a full listing of a computer program utilizing the previously discussed method by Holt and Heithecker (10.10) for optimization of girder proportions without stiffeners.

This program, written for a GE 635 in FORTRAN IV, level H language, was developed to solve the following problem.

Example 10.3 For the structural systems and loading shown in Fig. 10.20 find the optimum configuration of a plate girder without stiffeners designed according to AISC specification (10.1). The objective function to be optimized is the theoretical weight of the girder. Assume that the compression flange is sufficiently laterally supported so that $F_b = 22$ ksi for A36 steel.

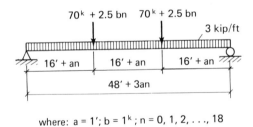

where: a = 1'; b = 1^k; n = 0, 1, 2, ..., 18

Figure 10.20 (Example 10.3).

Solution As input, Table 10.1(a) (Section 10.1.3) is read by the computer (from data cards) as well as data about the yield-point stress, F_y, and Young's modulus, E.

The computer next calculates M_{max} (MAXMUM), V_{max} (MAXSHR), and the moment parameter, X; Table 10.1 is now scanned to establish the proper case, and from the coefficients corresponding to this case the depth parameter, Y, is computed as well as the area parameter, Z. Solving Eqs. (10.80) and (10.83) the values of the depth (DEPTH) and the total area (AREAT) are obtained (card nos. 99 and 101 in the program). Next the web thickness parameter, U, and the flange parameter, W, are found and from these are rechecked the web thickness (WEBT), the web area (AREAW), the flange area (AREAFL), and the total

```
OPTIMAL VALUES FOR A GIRDER OF SPAN 48.00 FEET ARE AS FOLLOWS
THE MAXIMUM SHEAR WAS142.00 KIPS WHILE THE MAXIMUM MOMENT WAS 23808.00 INCH KIPS

THE MOMENT PARAMETER (X) IS          15.85        THE DEPTH PARAMETER (Y) IS           6.66
THE AREA PARAMETER (Z) IS            15.88        THE WEB THICKNESS PARAMETER (U) IS   1.34
THE FLANGE AREA PARAMETER (W) IS      2.48        THEORETICAL AREA                    54.74

DEPTH            =     70.4737  INCHES
WEB THICKNESS    =      0.4990  INCHES
WEB AREA         =     35.1677  SQUARE INCHES
FLANGE WIDTH     =     14.0947  INCHES
FLANGE THICKNESS =      0.5938  INCHES
FLANGE AREA      =      9.7789  SQUARE INCHES
TOTAL AREA       =     54.7255  SQUARE INCHES
SHEAR STRESS     =      4.0378  KSI
SECTION MODULUS  =   1102.2222  CUBIC INCHES
BENDING STRESS   =     21.5000  KSI
WEIGHT           =    186.0668  POUNDS PER FOOT
```

Figure 10.21 (Example 10.3).

area (AREATS). The width of the flange (BF) is taken as $\frac{1}{5}h$ (card no. 124) and its thickness (FLANGT) is next calculated. The section modulus (SX) is calculated (card no. 126) as well as the bending and shear stresses. The weight in lb/ft is obtained by multiplying the final total area (AREATS) by 3.4.

As output (Fig. 10.21) all the data for each span are given starting with $L = 48$ ft and ending with $L = 102$ ft.

In Appendix III the full listing of another program is shown. For the same loading and span conditions as in Example 10.3 the optimization of girder proportions is performed, but this time with stiffeners.

Example 10.4

Solution For input the result of the analysis in Example 10.3 is used; that is, data about the span (L), M_{max}, V_{max}, depth (h), total area, and section modulus are read in. Also read in are the data about the allowable maximum shear stress, aspect ratio and percentage of the web area in the stiffeners [actually table 11-36 from the AISC *Manual* (10.3), appendix A, p. 5-82] for four different slenderness ratios: 260, 280, 300, and 320. Next, the depth, h, obtained as an optimum for a girder without stiffeners is reduced by 12 in. Now for the assumed web slenderness and from the known section modulus the necessary flange areas are calculated. The shear stress is checked, and if $f_v > F_v$, the depth is increased by 1.0 in. and tried again until a height is found for which the shear stress is satisfactory. This height is used to start the iteration process, adding each time 1 in. to the previous depth and recalculating the necessary flange and stiffener areas and weight, transforming it to the total area ($A + \Delta A$) and thereby compensating for stiffeners. The computer next compares total areas and selects the minimum value. The iteration is made to always proceed an additional three steps beyond the optimum value to make sure that a global minimum is reached. The output shows the iterations indicating the optimum (Fig. 10.22).

It is of interest to compare the computer results of the optimization process for the same span—say, $L = 48$ ft for girders without and with stiffeners. The results are shown in Table 10.4. Of course the dimensions given in this table are purely theoretical and are used for comparison without any rounding off to practical dimensions. For girders without stiffeners the method given by Holt and Heithecker was used and for girders with stiffeners the method developed by the authors.

From the listings of computer programs used in Examples 10.3 and 10.4 it can be seen that only 7K of memory was needed. Therefore a programmable desk calculator can be used for the same purpose. Following the steps already discussed, the

Table 10.4

	Plate girder web	Plate girder flanges	Cross-sectional area	Total area	Weight (lb/ft)	Remarks	Method
Example 10.3	70.4737 × 0.4990	14.0947 × 0.6938	54.7255	54.7255	186.0668	Without stiffeners	From ref. (10.10)
Example 10.4	74.47 × 0.233	14.895 × 0.789	40.906	44.13	150.043	With stiffeners	By authors

THE DIMENSIONS FOR A CLASSICAL GIRDER WITH STIFFENERS OF SPAN 48.00 FEET AND WITH A SLENDERNESS RATIO 320.0 ARE

ITERATION	DEPTH (IN.)	WEB THICKNESS (IN.)	WEB AREA (SQ.IN.)	FLANGE WIDTH (IN.)	FLANGE THICKNESS (IN.)	FLANGE AREA (SQ.IN.)	TOTAL AREA (SQ.IN.)	SHEAR STRESS (KSI)	SECTION MODULUS (CUB.IN.)	BENDING STRESS (KSI)	WEIGHT (LBS/FT)
1	57.47	0.211	14.227	13.495	1.020	13.756	44.40	9.98	1102.22	21.60	150.946
2	58.43	0.214	14.652	13.695	0.983	13.462	44.40	9.69	1102.22	21.60	150.946
3	59.47	0.217	15.083	13.895	0.947	13.172	44.31	9.41	1102.22	21.60	150.656
4	70.47	0.220	15.526	14.095	0.913	12.886	44.24	9.15	1102.22	21.60	150.424
5	71.47	0.223	15.964	14.295	0.880	12.603	44.19	8.89	1102.22	21.60	150.248
6	72.47	0.226	16.414	14.495	0.849	12.329	44.15	8.65	1102.22	21.60	150.126
7	73.47	0.230	16.870	14.695	0.818	12.056	44.13	8.42	1102.22	21.60	150.059
OPTIMUM 8	74.47	0.233	17.332	14.895	0.789	11.787	44.13	8.19	1102.22	21.60	150.043
9	75.47	0.236	17.801	15.095	0.761	11.521	44.14	7.98	1102.22	21.60	150.079
10	76.47	0.239	18.276	15.295	0.734	11.259	44.17	7.77	1102.22	21.60	150.164
11	77.47	0.242	18.757	15.495	0.707	11.000	44.21	7.57	1102.22	21.60	150.298

Figure 10.22 (Example 10.4)

optimization method for girders without stiffeners (10.10) was programmed for both the HP9820A and the HP9830A; the program listings are given in Appendix IV. Input consisted of maximum bending moment in ft-kips, maximum shear force in kips, and type of steel used. Example 10.3 was also solved on an HP9820A and the results, quite identical to those obtained by a full size computer (Fig. 10.21), are shown in Fig. 10.23.

For girders with stiffeners, the authors' method (10.30) was programmed in

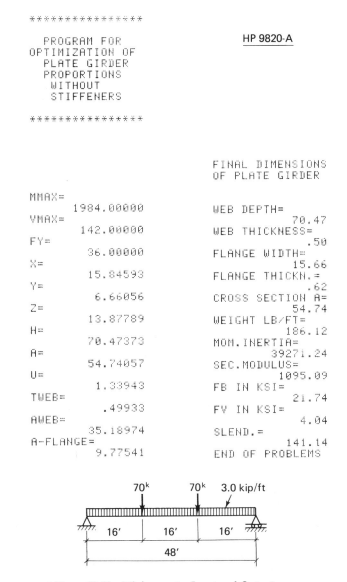

Figure 10.23 Minicomputer Input and Output.

BASIC for the HP9830A. As illustration of the method the following two examples are given.

Example 10.5 For easy comparison, as a first design example the same problem as discussed in AISC *Manual* (10.3, p. 2–90) (example 3) is used. The only change made in the problem is not restricting the section depth to 52 in. ($L/20$) and allowing the compression flange to be laterally unsupported to a length of l_u. A theoretical cross-sectional area $A = 55.094$ in.² was found, as compared to an AISC area of 62.375 in.², i.e., a decrease of 11.7%. If the theoretical dimensions are rounded off to the practical dimensions of a web of 72 in. $\times \frac{1}{2}$ in., two flange plates of 18 in. $\times \frac{5}{8}$ in., and an area of 58.50 in.², the saving is 6.2%.

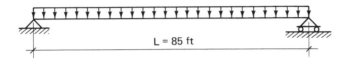

Girder of Example 10.5

```
CASE IS  5
                  SOLUTION FOR MAX.MOMENT  OF 2167.50      FT-KIP
MAX.SHEAR V= 102.00
STEEL IS FY= 36.00
                  GIRDER DEPTH H= 71.63
                  WEB THICKNESS T= 0.45
FLANGES BXT 17.91    X 0.64
DISCREPANCY D1= 0.11     %
GIRDER WEIGHT IS 187.32  LB/FT


            FINAL GIRDER DIMENSIONS
DEPTH H= 71.63    IN.WEB THICKNESS T= 0.45
FLANGE WIDTH BF= 17.91    THICKNESS TF= 0.64
UNBRACED LENGTH NOT LARGER THAN LU= 18.92    FT
FB= 22.25    FV= 3.17
```

Computer Output of Example 10.5

Example 10.6 As a second example, again the same problem as given in the AISC *Manual* on p. 2–92 (example 4) is solved. Design conditions are the same as given in Example 10.5 except that intermediate stiffeners are to be used. Theoretical dimensions yield an area without stiffeners of $A = 55.21$ in.² as compared to the AISC value of 59.38 in.², i.e., a reduction of 7.0%. Rounded-off dimen-

```
L= 90    M1= 3518.75    V= 167.5    FY= 36

FINAL OPTIMAL GIRDER DIMENSIONS AFTER NO. OF ITERATIONS B= 44.000

H= 89.740   TW= 0.360   BF= 25.640   TF= 0.860   A= 76.381   D= 9.034
TOT.AREA WITH STIF. 85.416   PERCENTAGE P= 11.828  %OF A
MOM.INERTIA= 108766.662     SEC.MOD.= 2376.447    RT= 6.636
ACT.STRESS FB= 17.753
UNBRACED LENGTH NOT LARGER THAN LU= 29.437    FT
FV= 5.190
```

Computer Output of Example 10.6

Plate Girders **467**

sions yield a web of 77 in. × $\frac{5}{16}$ in. and two flange plates of 22 in. × $\frac{3}{4}$ in. This yields an area of $A = 57.06$ in.^{2}, which represents a decrease of 3.9%. For the cross-sectional area of the stiffeners the value is 6.1 in.^{2}, or 11% of the total area A. The stiffener area is 4.5 in.^{2}

10.2 OPEN-WEB JOISTS

Open-web joists (Fig. 10.24) are suitable for the direct support of roof or floor decks in buildings. They have to be designed in accordance with standard specifications (10.12).

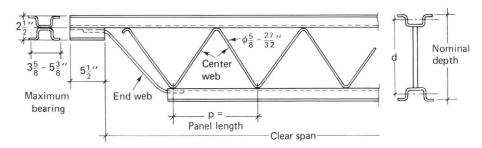

Figure 10.24 Open-Web Joist.

Joists are assumed to act as simply supported, uniformly loaded trusses supporting a floor or roof deck so constructed as to brace the top chord of the joists against lateral buckling. The bottom chord, which is always in tension, is designed as an axially loaded tension member. The top chord is designed only for axial compressive stress when the panel length clear of attachments does not exceed 24 in. Otherwise, it has to be considered as a continuous member subject to combined axial and bending stresses and must be designed so as to satisfy the following two interaction equations:

$$\text{At panel points:} \quad \frac{f_a}{0.515 F_y} + \frac{f_b}{F_b} \leq 1.00$$

$$\text{At midpanels:} \quad \frac{f_a}{F_a} + \frac{C_m f_b}{(1 - f_a/F'_e) F_b} \leq 1.00$$

(10.108)

in which

$$C_m = 1 - 0.3 \frac{f_a}{F'_e} \quad \text{for end panels}$$

$$C_m = 1 - 0.4 \frac{f_a}{F'_e} \quad \text{for interior panels}$$

$$F'_e = \frac{149 \times 10^3}{(l/r_x)^2}$$

where r_x is the radius of gyration about the axis of bending.

The web members are designed with respect to the vertical shear forces determined from full uniform loading, and they must not be less than $\frac{1}{4}$ the rated end reaction. The effect of eccentricity has to be taken into consideration. When bending due to eccentricity produces reversed curvature, C_m in Eq. (10.108) can be taken as 0.4.

For lateral stability, bridging is required as either horizontal, diagonal, or as sag rods approximately every 6 to 8 ft.

The deflection due to the design live load must not exceed the following

$$\text{Floors} = \frac{L}{360}$$

$$\text{Roofs} = \frac{L}{360} \quad \text{where a plaster ceiling is attached or suspended}$$

$$= \frac{L}{240} \quad \text{for all other cases}$$

Floors and roof decks may consist of cast-in-place or precast concrete or gypsum, formed steel, wood, or similar material. Cast-in-place slabs must not be less than 2 in. thick.

For typical open-web joists of the J-series ($F_b = 22$ ksi) and H-series ($F_b = 30$ ksi) the specification's "Standard Load Table" (10.12) gives the *total* safe uniformly distributed load in lb/ft. This table is valid for parallel chords installed to a maximum slope of $\frac{1}{2}$ in. per ft.

Chapter 12 includes an illustrative example of using typical open-web joists (Example 12.1). Many steel manufacturers have typical construction systems and so the designer does not need to "design" them, but instead merely has to check use according to the supplied details and loading capacities.

The lateral stability of open-web girders should be carefully studied. It was found (10.17) that the lateral stability is inversely proportional to the depth of the girder. For a given section of flange, by increasing the depth of the girder the lateral buckling load is reduced proportionately. Closer spacing of the web elements has very little influence on the lateral buckling strength of the chord members. The usual inclination of 45° for the web rods appears to be most efficient. Computer programs are available for the design of open-web joists.

Although most open-web joists are developed by manufacturers, there is still space for improvement of the design efficiency (10.18). Mathematical optimization has now evolved to a state such that it may be used for structural systems with many design parameters.

10.3 COMPOSITE BEAMS

10.3.1 Introduction

Composite construction consists of steel beams or girders supporting a reinforced concrete slab so interconnected that the beam and slab act together to resist bending. In the design of composite construction the specific properties of both materials—that is, of steel and concrete—must be taken into account, and according to this the corresponding specifications for both materials have to be applied. The basic design data are as follows:

Young's Modulus of Steel (structural members and reinforcing bars)

$$E_s = 29{,}000 \text{ ksi } (2{,}100 \text{ t/cm}^2;\ 200 \text{ MPa}).$$

Young's Modulus of Concrete, E_c, depends upon the quality of concrete, age, and the stress intensity and in general is between 2,000 and 5,700 ksi (140–400 t/cm²; 14–29 MPa). To determine E_c, the following formula is usually used

$$E_c = w^{1.5} 33(f'_c)^{1/2} \quad (\text{lb/in.}^2)$$

It is customary to use transformed areas by which concrete material is transformed to steel by dividing the width of the concrete by the modular ratio, $n = E_s/E_c$. The value of n is variable, lying between 5 and 14.5; the 1979 AASHTO *Interim Specifications: Bridges* (10.4), section 1.5.27, Flexure, state that it may be taken as the nearest whole number, but not less than 6.

Creep

Under long-term sustained loads concrete creeps and reaches equilibrium only after several years. The deformations of concrete may reach a value several times larger than the initial elastic deformations. As a measure of creep, the creep ratio, ψ = creep strain ÷ elastic strain, is used. Its value lies between 0.5 and 4.0 and depends mostly upon the water-cement ratio. At any time, t, after pouring concrete, the creep coefficient, ψ_t, is approximately

$$\psi_t = \psi_n(1 - e^{-t}) \tag{10.109}$$

where ψ_n is the expected creep coefficient at the time t_n. In Fig. 10.25a elastic and plastic strains of concrete due to creep are shown.

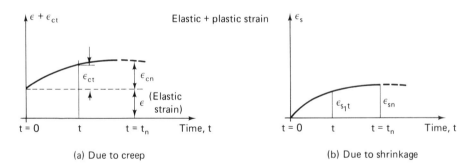

(a) Due to creep (b) Due to shrinkage

Figure 10.25 Plastic Strain.

Besides creep, the shrinkage of concrete affects the permanent stress distribution. These strains are shown in Fig. 10.25b, where ϵ_{sn} represents the final amount of shrinkage strain, ϵ_s.

Both these values, the total creep coefficient, ψ_n, and the final shrinkage strain, ϵ_{sn}, can be obtained from Table 10.5, which gives the German DIN 4227 specifications for plain concrete. The values are given as functions of environmental conditions to which the concrete will be exposed during the service life of the structure.

Table 10.5

Environmental conditions	Total creep coefficient, ψ_n	Final shrinkage strain, $\epsilon_{sn} \times 10^{-5}$
Under water	0.5K–1.0K	0
In very humid atmosphere—e.g., close to water surface	1.5K–2.0K	10
In open air	2.0K–2.0K	20
In dry air—e.g., in dry, interior spaces	2.5K–4.0K	30

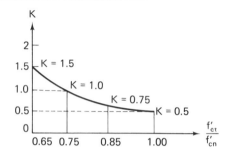

Figure 10.26 The Multiplicator, K.

The multiplication factor, K, can be taken from the graph in Fig. 10.26, where the abscissas represent the ratios of the concrete strength at time $t(f'_{ct})$ to the strength at infinite time (f'_{cn}).

The AISC specification (10.1) for buildings does not take into account stress redistribution with time due to creep and shrinkage. The AASHTO specifications (10.4) specify that the effect of creep be considered: the stresses and horizontal shears produced by dead loads acting on the composite section must be computed for n or for n taken 3 times the normal value, whichever gives the higher stresses and shears. The British Standard Code of Practice CP117 (10.13) states that in simply supported beams with shear connectors shrinkage and creep effects may be ignored. Where, in continuous beams, the stresses due to shrinkage modified by creep in concrete adversely affect the maximum resultant stress in the concrete or the shear connectors or the steel section, they must be calculated using a value of $\epsilon = 30 \times 10^{-5}$ (the same as the maximum value in Table 10.5), and a reduction of 0.5 for creep must be taken into account where applicable. The French specification Règles CM66 (10.9) recommends a value of $\epsilon = 40 \times 10^{-5}$.

10.3.2 Composite Steel-Concrete Construction

Composite construction is economical in buildings for longer spans (from 24 ft or 7.5 m up) and large live load intensities. The total height (the top of the slab to the bottom of the steel beam) should be at least $L/20$. The larger this "construction" height, the smaller the steel usage will be. For example, it was reported (10.14) that a composite steel and concrete structural system for parking garages cut costs as much as $300 per car space (size of garage was 590 cars) compared with other structural systems. The system consisted (Fig. 10.27) of long-span composite beams, steel columns, precast, prestressed concrete joists, and a special forming system for cast-in-place monolithic floor slabs. The 3.5-in. concrete slab, cast in reusable plywood panels, has wire-mesh reinforcement and spans 4 ft between joists. The slab is a one-way slab.

In the case of one-way slabs acting in a longitudinal direction, when joists intersect with the supporting girder (Fig. 10.28) the continuity of the slab is intentionally stopped by a joint. Also, in the direction of the slab spans, every 50–65 ft (15–20 m) joints should be provided to avoid building up of large tension stresses due to concrete shrinkage.

Figure 10.27 A Garage in Composite Construction (From Volume Indoor Parking, Inc., New York, N.Y., published in *Engineering News Record*, August 8, 1974, p. 42).

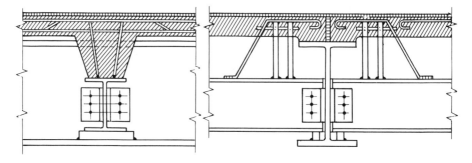

Figure 10.28 Joint in Concrete Slab.

For bridges, AASHTO specifications have allowed composite construction since 1944. The relative economy of composite rolled beams versus noncomposite ones for use in highway bridges consisting of simple beam structural system is shown in Fig. 10.29. Use of stay-in-place steel forms over and between stringers (e.g., Granco S-I-P Bridge Forms—Fig. 10.30) speed up the work by eliminating complicated form work which would otherwise be necessary (Fig. 10.31). There are several definite economic advantages resulting from composite steel-concrete action:

1. Savings in steel, ranging from 20 to 30%
2. Reduction in the depth of flexural members
3. Increased stiffness of a floor system
4. Increased overload capacity of a floor

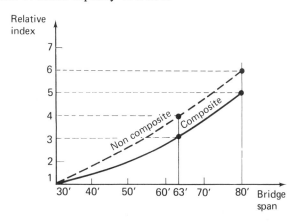

Figure 10.29 Relative Economies of Construction.

Figure 10.30 Stud Connectors (Three in One Row) and Stay-in-Place GRANCO S.I.P. Bridge Forms.

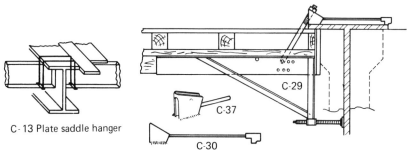

C-13 Plate saddle hanger

C-29 Bridge overhang bracket   C-30 Bridge overhang bracket hanger
C-37 Bridge overhang bracket half-hanger

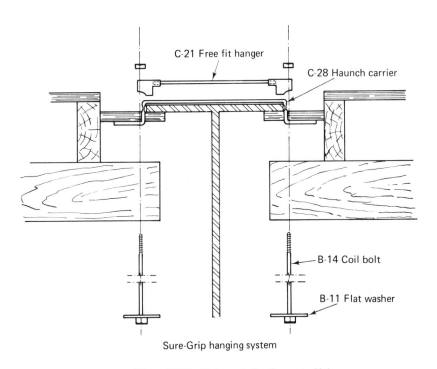

Sure-Grip hanging system

Figure 10.31 Falsework for Concrete Slab.

10.3.3 Shear Connectors

To transmit the horizontal shear forces acting in the contact plane between the steel beam and the concrete slab, special shear connectors are used. The most popular type of connectors are shown in Fig. 10.30. They are installed by the use of a welding gun. In that way, two to three studs can be welded each minute.

The AISC specification (10.1, section 1.11.4, Shear Connectors) requires that the entire horizontal shear be transmitted by welded connectors. This total shear has to be resisted between the point of maximum positive moment and points of inflection and is equal to the smaller of the two values obtained from the following two formulas:

$$V_h = \frac{0.85 f'_c \cdot A_c}{2}$$

$$V_h = \frac{A_s F_y}{2}$$

(10.110)

where f'_c = concrete strength, A_c = actual area of effective concrete flange, A_s = area of steel beam, and F_y = steel yield stress. The first formula uses the slab capacity and the second one the capacity of the steel. In both cases the same factor of safety of 2 is used.

The number of connectors resisting this force, V_h, on each side of the location of M_{max} must not be less than that determined by the relationship V_h/q, where q, the allowable shear load for one connector, is given in the AISC specification's table 1.11.4 (p. 5-39) for flat-soffit concrete slabs made with ASTM C33 aggregate. For flat-soffit concrete slabs made with rotary-kiln-produced aggregates and conforming to ASTM C330 with concrete unit weight not less than 90 lb/ft³, the allowable shear load for one connector is obtained by reducing the q-values given in AISC table 1.11.4 by multiplying them by factors less than 1.00 varying from 0.73 to 0.88 for $f'_c \leq 4.0$ ksi for an air-dry unit weight of from 90 to 120 lb/ft³; these factors are given in AISC table 1.11.4A, p. 5-39 (10.3).

The number of connectors required on each side of the point of M_{max} may be distributed uniformly. In the case of concentrated loads the AISC specification (10.1) gives a special number, N_2, of connectors which have to be between any concentrated load and the nearest point of zero moment. This number can be determined by using AISC equation (1.11-7) for this calculation.

Stud shear connectors must be at least 4 diameters in length and not greater than $\frac{7}{8}$ in. in diameter. They must have at least 1 in. of concrete cover in all directions. Unless located directly over the web, the diameter of studs must not be greater than $2.5 t_f$, where t_f = thickness of the flange to which they are welded. [British CP117, part 2 (10.13), limits this value to only $1.5 t_f$]. The minimum center-to-center spacing must not be less than 6 stud diameters along the longitudinal axis and 4 diameters transverse to the longitudinal axis of the supporting composite beam. The maximum distance is $8 t_s$, where t_s = concrete slab thickness.

The AASHTO specifications (10.4), Section 1.7.48(E), require that the shear connectors be designed for fatigue and checked for ultimate strength.

When the reinforced concrete slab is prefabricated, no shear connectors are used; instead, high-strength bolts are used to produce sufficient friction. Figure 10.32 shows a footbridge across the River Ems in Germany at Emsdetten, using sixteen high-strength bolts M24 per precast slab, equivalent to sixteen A325 bolts of $\frac{7}{8}$ in. diameter. The advantages are numerous: the possibility of producing cheaper reinforced concrete slabs of higher-quality concrete (e.g., with $f'_c = 8.5$ ksi); the reduction or elimination of shrinkage effects by storage before use; the reduction of creep effect; the avoidance of building up of stresses in the concrete in front connectors; the economy in formwork and falsework; and so on. For the friction coefficient according to Sattler's tests a value of 0.45 was taken. The assumption is that the slabs

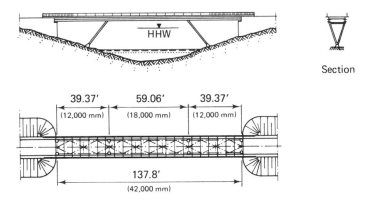
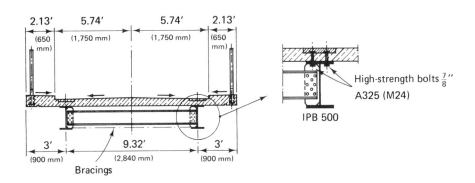
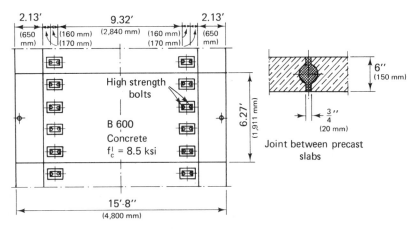

Figure 10.32 Composite Action without Connectors (Precast Slab Cassettes with High-Strength Bolts (From Deutscher Stahlbau-Verband, *Stahlbau*, vol. 2, 2nd ed., Cologne, 1964, p. 581).

will be in good contact with the steel rolled beams. In this example the effect of shrinkage was reduced by 50% after 6 weeks of slab storage before placement. It is important to produce mortar-filled joints of high strength to be able to take full advantage of the high strength of the concrete slabs.

The Yatsumichi railway bridge constructed in 1969 in Japan (10.19) was built with precast slabs bedded in epoxy mortar and connected to the steel girders by high-strength bolts. Static and fatigue tests of several model beams of similar construction showed that this type of beam is stronger under repeated loading than conventional composite beams.

Composite action can also be achieved through transverse prestressing. In this case the girder web extends into the slab to $\frac{2}{3}$ its depth, and the slab is compressed by transverse prestressing it into the web (Fig. 10.33). Transverse prestressing helps prevent premature failures of composite T-beams by longitudinal splitting of the concrete slab.

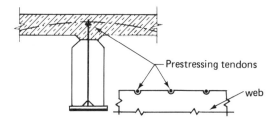

Figure 10.33 Composite Action Without Connectors and Without Top Steel Flange.

10.3.4 Design of Composite Beams

General considerations

Effective slab width

For slabs symmetrical about the beam, the effective width, b_{eff}, of a compressed concrete flange is the smallest dimension of the following

In buildings (AISC)	In bridges (AASHTO)
$\frac{L}{4}$	$\frac{L}{4}$
Center to center of beams	Center to center of beams
$16t_s + b_f$	$12t_s$

where $t_s =$ depth of concrete slab, $L =$ span of steel beam, and $b_f =$ top flange width of steel beam.

Foreign specifications are much more restrictive. For instance, the British CP117 specifications, part 2 (10.13), require the use of b_{eff} whenever the actual width, b, exceeds $L/20$—that is, in this case we must use

$$b_{\text{eff}} = \frac{b}{\sqrt{1 + 12(2b/L)^2}} \tag{10.111}$$

where b_{eff} may be not taken as less than $L/20$. For a span of 100 ft with beams at 9 ft center to center and a 9-in. slab, AASHTO specifications yield $b_{\text{eff}} = 108$ in. while the British standard allows only 91.6 in.

Composite section

American specifications are based on the elastic properties of the cross section (Fig. 10.34) and are usually calculated by the transformed area method. The basic material is the steel, and therefore the concrete area of the slab, $b_{\text{eff}} \cdot t$ has to be reduced by the modular ratio, n, such that

$$\frac{b_{\text{eff}}}{n} \cdot t_s = b_r \cdot t_s \tag{10.112}$$

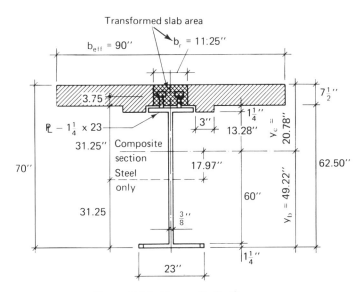

Figure 10.34 Composite Section.

In the following example in order to illustrate section property calculations, n was taken as 8, and therefore the reduced slab width is $b_r = 11.25$ in. The whole calculation is illustrated in table form (see Table 10.6). The first part is for the steel beam only; the second part is for the composite section with steel as the basic material. The dimensions are shown in Fig. 10.34.

The particular loading cases for which the composite section is applicable depend upon the method of construction. The structure may be shored or unshored. In the absence of shores under the steel beams during the pouring of concrete or the installation of precast panels, the stresses in steel beams are taken to be those due to the dead load of the steel and wet concrete; construction equipment and workers are

Table 10.6 CALCULATION OF SECTION PROPERTIES (FIG. 8.34)

Material		Name and dimensions	Area, A (in.²)	Distance to bottom of steel, y in	Statical moment about bottom of steel, Ay (in.³)	Distance of centroids to comp. axis, y_0 (in.)	Position moment of inertia Ay_0^2 (in.⁴)	Own moment of inertia, I_0 (in.⁴)	Moment of inertia I (in.⁴)
	Steel	P_L-1¼ × 23 Web P_L-⅜ × 60 P_L-1¼ × 23	28.75 22.50 28.75	61.875 31.25 0.625	1,778.906 703.125 17.969	} 17.97	25,834	6,750 53,935	86,519
		Steel section:	80.00	—	2,500.00	—	—	60,685	—
	Concrete	Slab 11.25 × 7.5	84.38	66.25	5,590.175	17.03	24,472	396	24,868
		Composite section	164.38	—	8,090.175	—	50,306	61,081	111,387

Moment of inertia

Centroid from bottom: Steel   Centroid composite section   Fiber distances from N.A.

$\bar{y}_s = \dfrac{2,500.00}{80.00} = 31.25$ in. $\bar{y}_c = \dfrac{8,090.175}{164.38} = 49.22$ in. $y_c = 20.78$ in. $y_b = 49.22$ in. $y = 13.28$ in.

Section modulus (in.³)

Steel section only   Composite section

Top   Bottom   Top concrete   Top steel   Bottom steel

$\dfrac{86.519}{31.25} = S_x$ $\dfrac{86.519}{31.25} S_s =$ $\dfrac{111,387}{20.78} = S_{tc}$ $\dfrac{111,387}{13.28} = S_t$ $\dfrac{111,387}{49.22} = S_b$

$= 2,769$ in.³ $= 2,769$ in.³ $= 5,360$ in.³ $= 8,388$ in.³ $= 2,263$ in.²

negligible. The shores when they are used, are usually left in place until the concrete reaches about $\frac{3}{4}$ its full strength (f'_c after 28 days). When the shores are removed, any past or additional load, either dead or live load, is taken by the composite section. The section properties of the steel beam alone do not come into consideration at all. More often, however, shores are not used and the steel beams act as the main supporting element for the necsssary formwork for the concrete slab. In that case the steel section alone is carrying its own weight and the additional dead weight of fresh concrete. Once the concrete is of sufficient strength to take stresses, the additional dead load: surfacing of the concrete, the floor in buildings or an asphalt layer in bridges, as well as railing on bridges, together with the live load are acting on the composite section.

In buildings (10.1), for construction without temporary shoring the stress in the steel may be computed from the total dead plus live load moment and the transformed section modulus, S_{trans}, provided that the numerical value of S_{trans} so used does not exceed

$$S_{trans} = \left(1.35 + 0.35 \frac{M_L}{M_D}\right) S_s \qquad (10.113)$$

which is equation (1.11-2) in the AISC specification. In this expression for the limiting value of S_{trans}, M_L is the moment caused by loads applied after the concrete has reached 75% of its required strength, M_D is the moment caused by loads applied before this time, and S_s is the section modulus of the steel beam referred to the flange where the stress is being computed. At sections subject to positive bending moment, the stress must be computed for the steel tension flange. At sections subject to negative bending moment, the stress must be computed for both flanges. These stresses must not exceed the appropriate value of stresses allowable in bending.

According to the AASHTO specifications for bridges (10.4) the maximum stresses in girders without shoring is the sum of the stresses produced by the dead loads acting on the steel girders alone and the stresses produced by the superimposed loads acting on the composite girders. For shored structures the same requirements as for buildings are valid.

In continuous spans the positive moment portion may be designed for composite action, as in simple spans. Shear connectors must be provided in the negative moment section, when the reinforcement steel embedded in the concrete is considered part of the composite section. If this reinforcement is not used in computing the section properties for negative moments, shear connectors need not be provided in these portions of the spans, but additional anchorage connectors must be placed in the region of the point of dead load contraflexure. The number of these additional connectors is computed by the formula

$$N_c = A_r^s \frac{f_r}{Z_r} \qquad (10.114)$$

where N_c = number of additional connectors for each beam at the point of contraflexure

A_r^s = total area of longitudinal slab reinforcement steel for each beam over interior support

f_r = range of stress due to live load plus impact (in lieu of more accurate computations, f_r may be taken as 10 ksi)

Z_r = allowable range of horizontal shear on an individual shear connector.

The additional connectors, N_c, must be placed adjacent to the point of dead load contraflexure and within a distance equal to $\frac{1}{3}b_{\text{eff}}$. Any field splices should be so located that they clear the connectors. These additional connectors are needed, in cases where the reinforcement in this region is continuous into the positive moment regions, in order to avoid overstressing and premature fatigue failures of the anchorage connectors well before the design life of the member has been reached. Thus additional anchorage connectors are required near the contraflexure points to develop the tension force in the longitudinal reinforcement (10.19).

Composite construction of concrete slabs on formed steel deck connected to steel beams or girders is designed in a similar way, but there are certain modifications. Readers should follow AISC specification, section 1.11.5 (10.1, p. 40–41).

Ultimate strength

Although, as mentioned before, elastic design methods for composite beams have been retained in American specifications, a number of important new findings based on ultimate strength have been introduced in elastic design since the 1960s (10.19). Therefore, the distance between the individual connectors could be uniform, because composite beams develop the full flexural capacity of the cross section as long as the sum of the ultimate strengths of the individual connectors between the points of zero and maximum moment is at least equal to the total horizontal shear.

Tests have shown that the ultimate strength of a composite beam is essentially independent of its history of loading. Accordingly, the ultimate strength of a beam built with temporary supports is identical to that of the same beam built without temporary supports. However, the presence or absence of temporary shoring during construction can have a pronounced effect on the magnitude of deflections.

Ultimate strength design procedures for composite beams have been used in bridge design since the early 1970s as a part of the load factor design criteria (AASHTO, 1971). In Great Britain a purely ultimate strength procedure for the design of composite beams in buildings has been adopted.

At the collapse stage, which is normally due to crushing of the concrete slab, a rectangular stress block diagram (Fig. 10.35) with a maximum concrete stress of $0.85 f'_c$ is assumed, along with one of F_y in the steel. It is assumed that for any position of the neutral axis within the concrete slab, the slab is cracked—that is, no tension is taken by the concrete. If the slab is "adequate," the neutral axis will stay within the slab (Fig. 10.35a). If it is "inadequate," then the neutral axis is either in the compression flange of the steel beam or somewhere below, passing through the web of the steel section (Fig. 10.35b).

In the first case, from equilibrium it follows that

$$C'_c = 0.85 f'_c A_c = 0.85 f'_c a b_{\text{eff}} \tag{10.115}$$

$$T' = F_y A_s \tag{10.116}$$

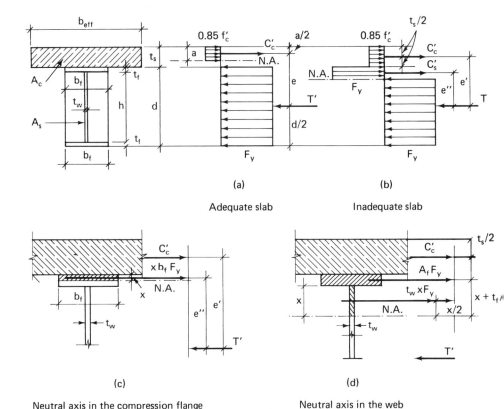

Figure 10.35 Ultimate Strength Procedure.

$$C'_c = T' \quad \text{and thus} \quad a = \frac{F_y A_s}{0.85 f'_c b_{\text{eff}}} \quad (10.117)$$

$$M'_u = T' \cdot e = T'\left(\frac{d}{2} + t_s - \frac{a}{2}\right) \quad (10.118)$$

In the second case, the force in the concrete slab is smaller than T'; that is, if

$$0.85 f'_c b_{cf} t_s < A_s F_y \quad (10.119)$$

then the neutral axis lies below the slab. The force in the concrete is known by its magnitude and its position. By trial and error the position of the neutral axis can now be established. First we try to determine whether it is passing through the compression flange at a distance x by establishing the following inequality

$$0.85 f'_c b_{\text{eff}} t_s + F_y A_f \lesseqgtr (A_s - A_f) F_y \quad (10.120)$$

If the compressive forces, represented by the left-hand side of Eq. (10.120), are larger than the right-hand side, then

$$0.85 f'_c b_{\text{eff}} t_s + F_y b_f x = (A_s - b_{\text{eff}} x) F_y \quad (10.121)$$

where x is the position of the neutral axis in the compression flange (Fig. 10.35c).

Solving for x we find

$$x = \frac{1}{2b_f F_y}(A_s F_y - 0.85 f'_c b_{eff} t_s) \tag{10.122}$$

If the left-hand side of Eq. (10.120) is smaller than the right-hand side, then the neutral axis is passing through the steel web (Fig. 10.35d), and

$$x = \frac{1}{2t_w}(A_s - 2A_f) - 0.425 \frac{f'_c}{F_y} \cdot \frac{t_s}{t_w} \cdot b_{eff} \tag{10.123}$$

Once x is obtained either from Eq. (10.122) or Eq. (10.123) the positions of forces can be found, and the ultimate moment capacity is (Fig. 10.35b)

$$M'_u = C'_c \cdot e' + C'_s \cdot e'' \tag{10.124}$$

The ratio M'_u/M_{max} (where M_{max} is the maximum sagging moment due to loads at the service level) represents the factor of safety. As was pointed out earlier, the AISC specification (10.1, section 1.11, Composite Construction) does not require ultimate capacity calculations, but the same specification when dealing with shear connectors uses the ultimate capacities of the concrete and the steel (Eq. 10.110) together with a factor of safety of 2. According to this, the above ratio of 2 can be considered satisfactory.

In the AASHTO specifications (10.4) under Load Factor Design (in section 1.7.61, Composite Beams and Girders) it is stated that composite beams must be proportioned so that the maximum strength of any section is not less than the sum of the computed moments at that section multiplied by the appropriate load factors. If the steel section satisfies some of the AASHTO compactness requirements [section 1.7.59(A)(1)(b): $h/t_w \leq 420/\sqrt{F_y}$; and (e): maximum shear force, $V \leq 0.55 F_y h t_w$], then the composite beam qualifies as compact and the calculation of positive values of M'_u as outlined before for regular beams is also applicable to composite beams. When the steel section does not satisfy the above compactness requirements, the maximum strength of the section is taken as the moment at first yielding.

If the section is exposed to a negative moment, it is assumed that the concrete does not carry any tensile stresses. In cases where the slab reinforcement is continuous over interior supports, the reinforcement may be considered to act compositely with the steel section. In AASHTO (10.4), article 1.7.48.D it is stated that the minimum longitudinal reinforcement including the longitudinal distribution reinforcement must equal or exceed 1% of the cross-sectional area of the concrete slab, $b_{eff} t_s$; $\frac{2}{3}$ of this required reinforcement is to be placed in the top layer of the slab within the effective width. When shear connectors are omitted from the negative moment region, the longitudinal reinforcement must be extended into the positive moment region beyond the anchorage connectors, at least 40 times the reinforcement diameter.

Local flange buckling and lateral buckling are important factors affecting the ultimate strength of composite beams subjected to negative moments (10.20). For beams with cover plates lateral buckling becomes more critical than local buckling. The ratio of the lateral buckling moment to the simple plastic moment decreases significantly with an increase in the span length and is only slightly affected by the amount of longitudinal slab reinforcement and size of the compression flange cover plate.

Deflections

It is known that the deflection substantially increases with time as a result of creep and shrinkage. In design, it is recommended that time-dependent deflection be taken at least equal to the instantaneous deflection.

Dead load deflections for unshored members are calculated with the elastic properties of the steel beam alone. Deflections due to any additional dead loads or to dead loads resulting from the removal of shoring, as well as long-term live loads, should be calculated using the composite section and a modular ratio of $3n$. For short-term live loads the modular ratio is n. In the case of continuous spans the structural members are nonprismatic. In making a first approximation this is ignored and an initial deflection calculation with prismatic members is normally performed. If it appears that the deflections may be critical for a particular design, then the calculation should be refined by considering the member to be nonprismatic. For composite sections the AASHTO specifications (10.4, section 1.7.4) require that the ratio of the overall depth of the girder (concrete slab plus steel girder) to the length of span preferably not be less than $\frac{1}{25}$, and that the ratio of the depth of the steel girder alone to the length of the span not be less than $\frac{1}{30}$. For continuous span-depth ratios the distance between the dead load points of contraflexure is taken instead of span length.

In composite beams shear deflection can reach almost $\frac{1}{3}$ the theoretical bending deflection, and therefore this effect cannot be ignored. The calculation is usually done using simple formulas available in texts on strength of materials.

In the analysis of a building frame where beams are of composite construction and columns are not, the stiffness of the composite member depends upon the bending moment diagram. If a positive (sagging) moment is acting over the full length of the composite member, the stiffness may be taken as I_c/L, where I_c is the moment of inertia of the composite section. If the positive moment is acting only over the midportion of the beam and negative moments are acting near the ends, the stiffness to be used is the stiffness of a nonprismatic member consisting of the steel beam in the negative moment regions and the composite section in the positive moment region. Most floor beams in buildings have a composite moment of inertia about 2.0 times larger than the moment of inertia of the steel section alone. This means that the stiffness of composite members is about 1.5 times the stiffness of steel beams in a frame analysis.

Composite floor systems

Composite buildings are hybrid structures formed by combining steel and concrete elements into a structural system. Usually the floors are of composite construction. Similar to the case of the majority of composite bridges built in this country, floor beams are designed as simple beams because greater economy is often achieved by simplifying fabrication and erection than is by reducing material costs. It is also a common practice in this country to omit shear connectors from a portion of, or throughout, the negative moment region of a continuous composite beam in order to avoid the significant drop in the fatigue strength of the tension flange as a result of the welds of the shear connectors.

In the following example the design of a typical floor panel using AISC specifications will be discussed.

Example 10.7 For the floor shown in Fig. 10.36(a) design a B2 floor beam using composite construction, A36 steel, and the AISC specifications (10.1). The live load is 125 lb/ft² and the reinforced concrete slab is 4 in. thick. Assume simply supported beam ends (Fig. 10.36c) and no shoring. Concrete is $f'_c = 3$ ksi.

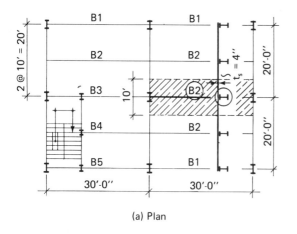
(a) Plan

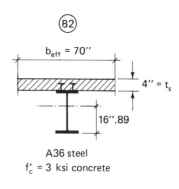
(b) Section of beam B2

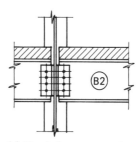
(c) Simple beam connection

Figure 10.36 Composite Floor.

Solution

LOAD ANALYSIS. As a dead load estimate for composite beams in buildings, 10 to 15% of the dead load supported by the steel beam (excluding its own weight) is a fair determination for the steel beam weight. A lower percentage is used for larger center-to-center beam distances. In our case, the dead load of the slab (Fig. 10.36) is

$$\tfrac{4}{12} \times 150 \times 10 = 500 \text{ lb/ft}$$

As a 10-ft center-to-center distance is large, we assume 10% of 500 lb/ft as the estimate for the steel beam weight—that is, 50 lb/ft. Thus

D.L.:	Slab	500 lb/ft	(7,297 N/m)
	Steel beam	50 lb/ft	(730 N/m)
		550 lb/ft $= W_D$	(8,027 N/m)
L.L.:	125 × 10	1,250 lb/ft $= W_L$	(18,242 N/m)
		1,800 lb/ft $= W_{tot}$	(26,269 N/m)

MOMENTS AND SHEARS. The moments are as follows:

D.L.: $M_{max} = M_D = 0.55 \dfrac{(30.0)^2}{8}$

$= 61.88$ kip-ft $= 742.5$ kip-in. (8,389.1 kN-cm)

L.L.: $M_{max} = M_L = 1.25 \dfrac{(30.0)^2}{8}$

$= 140.62$ kip-ft $= 1,687.5$ kip-in. (19,066.2 kN-cm)

$M_{tot} = 2,320$ kip-in. (27,455.3 kN-cm)

The shears are as follows:

D.L. + L.L.: $V_{max} = \tfrac{1}{2} \cdot (30.0)1.80 = 27.0$ kips (120.1 kN)

SELECTION OF SECTION. Assuming a compact section which is adequately braced, $F_b = 24$ ksi (165.5 MPa).

1. The required section modulus of the composite (transformed) section, $(S_{trans})_{rqd}$, is

$$(S_{trans})_{rqd} = \frac{M_{tot}}{F_b} = \frac{2,430}{24} = 101.25 \text{ in.}^3 \ (1,659.2 \text{ cm}^3)$$

From the AISC *Manual* (10.3, p. 2-109, Composite Design, no cover plate, 4-in. slab) W18 × 40 has $S_{trans} = 99.4$ in.³ or 1.83% less value than required; as this is less than 2% it is acceptable. Check steel weight: 40 lb/ft < 50 lb/ft: OK

2. The required section modulus of steel section, $(S_s)_{rqd}$, is

$$(S_s)_{rqd} = \frac{M_D}{F_b} = \frac{742.5}{24.0} = 30.9 \text{ in.}^3 \ (507 \text{ cm}^3)$$

The section W18 × 40 has $S_s = 68.4$ in.³ < 30.9 in.³

3. The maximum allowable value of the transformed section modulus, $(S_{trans})_{max}$ (from Eq. 10.113), is

$$(S_{trans})_{max} = \left(1.35 + 0.35 \frac{140.62}{61.88}\right) S_s = 2.145 S_s$$

If we substitute for S_s the value of $S_s = 68.4$ in.³ we have

$$(S_{trans})_{max} = 2.145(68.4) = 146.7 > 99.4: \quad \text{OK}$$

4. Check b_{eff}. In the AISC *Manual*'s composite design tables (10.3), $b_{eff} = 16 t_s + b_f = 70.0$ in. (177.8 cm).
Thus:

$$\tfrac{1}{4}(L) = \tfrac{1}{4}(30.0 \times 12) = 90 \text{ in.}$$

Center-to-center distance $= 10.0 \times 12 = 120$ in.

$$16 t_s + b_f = 64.0 + 6.0 = 70 \text{ in.}$$

which controls and represents b_{eff} from the *Manual* tables.

5. Check deflection, δ_{max}:

$$\delta_{max} = \frac{5}{384} W_L \cdot \frac{L^4}{EI_{trans}} = \frac{5}{384} \cdot (1.25) \frac{(30)^4 \times (12)^3}{29 \times 10^3 \times 1,680}$$

$$= 0.47 \text{ in.} < \frac{30 \times 12}{360} = 1 \text{ in.}$$

6. Check stresses:

In concrete: $f_c = \dfrac{1687.5}{335 \times 10} = 0.50$ ksi $< 0.45 \times 3 = 1.35$ ksi

In bottom of steel section: $f_b = \dfrac{2430}{99.4} = 24.4$ ksi $(1.86\% < 2\%$ overstress$)$

SHEAR CONNECTORS. The flange thickness of a W18 × 40 is 0.525 in. Therefore, the maximum stud diameter is $1.5 \times t_f = 0.786$ in. or $\tfrac{7}{8}$ in. (10.1). Select a diameter of $\tfrac{1}{2}$ in. From *Manual* table 1.11.4 (10.3, p. 5-39) the allowable horizontal shear load on one connector $\tfrac{1}{2}$ in. diameter × 2 in. headed stud is

$$q = 5.1 \text{ kips } (19.8 \text{ kN})$$

The entire horizontal shear, V_h, at the junction of the steel beam and the concrete slab, according to Eq. (10.110), is

$$V_h = \dfrac{0.85(3.0) \times 70 \times 4}{2} = 357 \text{ kips}$$

or

$$V_h = \dfrac{11.8 \times 36.0}{2} = 212.4 \text{ kips} < 357 \text{ kips}$$

Therefore, $V_h = 212.4$ kips (945 kN). If two connectors are used in one row, their capacity is $2 \times 5.1 = 10.2$ kips, and on each side of the point of maximum moment (on one-half the span) the total number of rows of connectors is

$$N = \dfrac{V_h}{2q} = \dfrac{212.4}{10.2} = 21$$

at a spacing of $s = (15.0 \times 12)/21 = 8.6$ in.: use $s = 8$ in. $> 6 \times \tfrac{1}{2} = 3$ in.

ULTIMATE STRENGTH. From Eq. (10.119):

$$0.85 \times 3 \times 70 \times 4 = 714 > 11.8 \times 36 = 424.8 \text{ kips}$$

Therefore, the slab is adequate and we have case 1.
According to Eq. (10.117)

$$a = \dfrac{36 \times 11.8}{0.85 \times 3 \times 70} = 2.38 \text{ in.} < 4.0 \text{ in.}$$

Finally, from Eq. (10.118)

$$M'_u = 36 \times 11.8 \left(\dfrac{17.90}{2} + 4.0 - \dfrac{2.38}{2}\right) = 4{,}995.6 \text{ kip-in.}$$

The factor of safety thus is

$$\text{F.S.} = \dfrac{4{,}995.6}{2430} = 2.06 \geq 2: \quad \underline{\text{OK}}$$

10.3.5 Concluding Remarks

In the introduction to this section (Section 10.3.1) it was pointed out that due to shrinkage and creep of concrete the maximum resultant stress in concrete, shear connectors, and the steel section may be adversely affected. The British specification CP117, part 2 (10.13), states that in simply supported beams with shear connectors this effect may be ignored, but not in continuous beams. The AISC specification (10.1) does not require that changes in stresses be checked. The AASHTO specifications

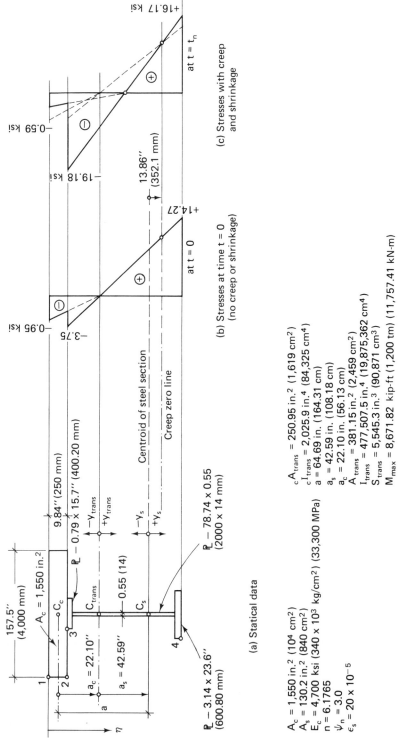

Figure 10.37 Changes in Stresses Due to Creep and Shrinkage (From Deutscher Stahlbau-Verband, *Stahlbau*, vol. 2, 2nd ed, Cologne, 1971, pp. 347–48).

(10.4) try to compensate for creep and shrinkage by requiring a threefold reduction in the concrete modulus—that is, by using $3n$ instead of n.

The German specifications—DIN 4239 (10.21) and DIN 1078 (10.22)—require that the stresses be determined with and without shrinkage and creep, as well as a detailed calculation of the redistribution forces.

As an illustration of how much the stresses are changed due to shrinkage and creep in a statically determinate system for the cross section shown in Fig. 10.37a, two stress diagrams, one for a stress state without shrinkage and creep (at the time $t = 0$) and one with shrinkage and creep, are shown in Fig. 10.37b and c, respectively (10.23).

10.4 LIGHT-GAGE STEEL MEMBERS

10.4.1 Design of Flexural Members

Cold-formed sections are made from flat sheets or strips (10.24), either by brake-forming or by roll-forming. Therefore, there are no residual cooling stresses, but there are effects of strain-hardening with nonuniform material properties. This strain-hardening is entirely concentrated in the "corners" of the section with very little in the flats (10.25). All that a designer needs are the average tensile yield-point values.

According to the shape of the cross section a compression flange can be stiffened or unstiffened. With unstiffened flanges the bending strength is determined by the sectional properties of the full (or reduced) section and the basic design stress, F_b, and by the allowable compression stress to avoid local buckling, F_c, or the allowable compression stress to avoid lateral buckling, F_b. For stiffened flanges local buckling considerations are omitted. Limiting values of width to thickness, w/t, still sufficient to allow $F_c = 0.60 F_y$, are given in American Iron and Steel Institute charts and tables (10.26, p.19). The allowable stresses for laterally unbraced beams of I-, L-, or Z-shapes are also given in these charts (10.26), which are indispensable for any practical design. Instead of describing the AISI 1980 specification (10.27) in great detail, an example of a beam with a stiffened flange section will be given to illustrate the design procedure. It will be assumed that the reader has access to the necessary charts and tables (10.26).

Example 10.8 For the "hat" section made of 16-gage steel in Fig. 10.38 with $F_y = 50$ ksi, check the design of a simply supported beam of 8.0 ft span with 20 lb/ft dead load and 300 lb/ft live load (10.28, example 15). The beam is laterally supported.

Solution

1. Maximum applied moment

$$M_{max} = w\frac{L^2}{8} = 0.32 \cdot \frac{8^2}{8} = 2.56 \text{ kip-ft (3.47 kN-m)}$$

Maximum compression stress

$$f_b = \frac{M_{max}}{S_x} = 2.56 \times \frac{12}{1.04} = 29.6 \text{ ksi} < 30 \text{ ksi}$$

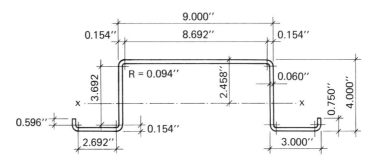

$I_x = 2.57$ in.^{4}; $S_{xc} = 1.04$ in.^{3}; $I_{yc} = 6.50$ in.^{4}
(at $F_b = 30$ ksi)

Figure 10.38 Stiffened Hat Section: Steel $F_y = 50$ ksi, Gage 16 ($t = 0.0598$ m.).

2. If the beam is unbraced, then according to section 3.3 of the AISI specification (10.27)

$$\frac{L^2 \cdot S_{xc}}{d \cdot I_{yc}} = \frac{8^2(12)^2(1.04)}{4 \cdot (6.50)} = 368.6 < 2{,}096.3$$

Therefore, no stress reduction from $0.6F_y$ is required.

$$\frac{0.36\pi^2 E C_b}{F_y} = \frac{0.36(\pi^2)29{,}500(1.0)}{50} = 2{,}096.3$$

For stress calculations the top flat width of 8.69 in. was reduced according to section 2.3.1.1 [10.26, chart 2.3.1.1(B)] to an effective width of $b = 43.0t = 2.58$ in.

3. For deflection calculations, section 2.3.1.1 [10.26, chart 2.3.1.1 (C)] gives

$$b = 54.5t = 3.27 \text{ in.}$$

Therefore, the moment of inertial, I_x, can be increased to a new value of 2.79 in.^{4}. The live load deflection is

$$\frac{5wL^4}{384EI_x} = \frac{5(0.3)(8)^4(12)^3}{384(29.5 \times 10^3)2.79} = 0.336 \text{ in.} \quad \text{or} \quad \frac{L}{286}$$

4. Shear stress in the web:

$$\text{End reaction} = 0.32 \times \frac{8}{2} = 1.28 \text{ kips (5.69 kN)}$$

$$h = d - 2t = 4.00 - 2(0.060) = 3.88 \text{ in. (9.8 cm)}$$

$$\text{Shear stress in web} = \frac{1.28}{2 \times 3.88 \times 0.060} = 2.75 \text{ ksi (18.96 MPa)}$$

According to section 3.4.1 (or chart 3) (10.26) for $h/t = 3.88/0.060 = 65$, $F_v = 16.5$ ksi > 2.75 ksi.

5. Regarding web crippling for end bearing, $N = 2$ in. [10.26, section 3.5 or charts 3.5(A)], maximum end reaction is

$$V_{max} = 1.51 \text{ kips} > 1.28 \text{ kips:} \quad \text{OK}$$

The section is satisfactory.

10.4.2 Steel Roof Decks

The steel roof deck is very much in use because it is an all-weather construction (given proper insulation), is light and economical, provides a good lateral diaphragm, and is fire-rated.

In the Steel Deck Institute's *Steel Roof Deck Design Manual* (10.29) the designer can find standard load tables for steel decks with $F_y = 33$ ksi, uniformly distributed load (with 10 lb/ft² of dead load inclusive) with maximum live load deflection not in excess of $L/240$. In these tables, for simple beams maximum moments are taken as $wL^2/8$ and for continuous beams $wL^2/10$.

Roofs having a slope of about 2% should have end laps of sheets of 2 in. minimum. Sheets are attached to supporting members by fusion-welding or screws. Standard-style deck sheets are available in widths of 18, 24, 30, or 36 in. For satisfactory attachment for all standard deck gages, welds or screws are required at all edge ribs and at those interior ribs necessary to produce a spacing of 18 in. between points of attachment. When any standard-style deck spans 7 ft or more, adjacent sheets should be fastened together at midspan. The attachment spacings are based on providing 30 lb/ft² gross uplift and lateral stability to the top flange of the supporting structural members.

Figure 10.39 shows a section through a typical roof deck with a 2-hr fire rating with roof insulation and with gypsum plaster on metal lath. The steel deck has a minimum depth of $1\frac{1}{2}$ in. on steel joists with a maximum of 5-ft-6-in. center-to-center.

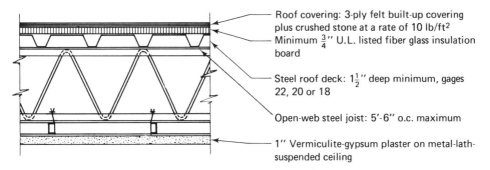

Figure 10.39 Insulated Steel Roof Deck of 2-hr Fire Rating with Gypsum Plaster on Metal Lath.

Fire protection is provided by a 1-in. vermiculite-gypsum plaster on metal-lath-suspended ceiling. Insulation is obtained by the use of $\frac{3}{4}$ in. thick fiber glass insulation board. Between the insulation and the steel deck a vapor barrier is usually used, although it is not required, in order to protect the deck components from condensation. When roofing is to be applied over the roof insulation, the means of attachment of the insulation (and vapor barrier if required) to the deck is just as important as the adhesion of the roofing membrane itself to the top surface of the insulation. To be considered a satisfactory base for insulation it is specified that the roof deck must not deflect more than $\frac{1}{8}$ in. at the center of the span when subjected to the maximum concentrated loads normally applied during construction of the roof. Deflections in excess of this amount will tear the insulation boards loose from the deck. This deflection criterion is independent of the span length and continuity and rather should be in terms of span length or, even better, curvature. On slopes in excess of 1 : 24 ($\frac{1}{2}$ in. per foot) only those types of insulation should be used which permit nailing for securing felts against slippage. There are five types of approved roof insulation boards: wood fiber (1 in. or more), perlite, glass fiber ($\frac{3}{4}$ in. or more), composite, and cellular glass ($1\frac{1}{2}$-in. minimum thickness).

10.4.3 Wall Studs

Wall panels or partitions constructed with light-gage steel studs faced on both sides with conventional wallboards (plywood or gypsum) are very much in use in buildings with floors made with light-gage steel members framed between supporting elements (Fig. 10.40). The rigidity of the wallboard material must be sufficient to restrain the studs against lateral buckling (in the plane of walls). The attachment of the boards, which is required on both sides, must be sufficient to provide support against in-plane buckling, and the distance a must not be [10.27, sections 5.1(b) and (c)] less than either

$$a_{max} = \frac{8EI_y k_w}{A^2 F_y^2} \quad \text{or} \quad \frac{Lr_y}{2r_x} \quad (10.125)$$

where

$$k_w = \frac{F_y^2 a A^2}{8EI_y} \quad \text{(modulus of elastic support)}$$

The lateral force (in kips) which each single attachment of the wall material must be capable of exerting on the stud to prevent lateral buckling of the stud is [10.27, section 5.1(d)]

$$P_{min} = \frac{k_w P_s L/240}{2\sqrt{EI_y k_w/a} - P_s} \quad (10.126)$$

where P_s = design load on stud in kips.

The following example will demonstrate the design procedure.

Example 10.9 (10.28) Calculate the allowable axial load per stud and the maximum spacing and required strength of wall sheathing attachments for a wall height of 15 ft using channel section studs with stiffened flanges $7 \times 2\frac{3}{4} \times 0.075$ (gage 14) of $F_y = 50$ ksi steel, as shown in Fig. 10.40.

Solution

1. From the AISI charts and tables (10.26, table 1, p. 65) we find: weight per foot = 3.50 lb/ft; area, $A = 1.00$ in.^{2}; full section maximum moment of inertia, $I_x = 7.66$ in.^{4}; minimum moment of inertia, $I_y = 1.00$ in.^{4}; radii of gyration, $r_x = 2.76$ in.; $r_y = 1.00$ in.

2. For shapes not subject to torsional-flexural buckling, the AISI specification (10.27, section 3.6.1.1) requires that a factor, Q, be computed. In our case, from the charts, $Q = 0.726$ for $F_c = 30$ ksi. Therefore, $(QF_y) = 0.726 \times 50 = 36.3$ ksi. When KL/r_x is less than $C_c/\sqrt{Q}$, the specification (10.27) gives the following formula for the average axial stress, P_s/A:

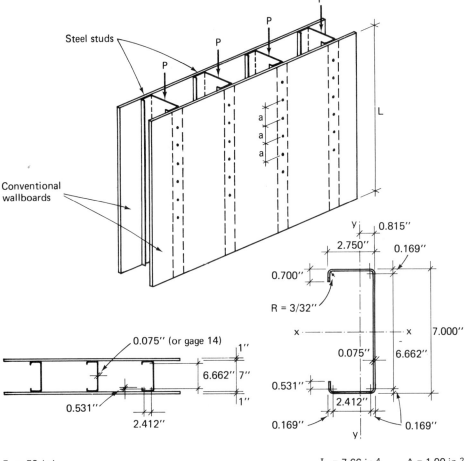

F_y = 50 ksi

7 x 2¾ x 0.075 Channel with stiffened flanges 3.50 lb/ft

I_x = 7.66 in.⁴   A = 1.00 in.²

I_y = 1.00 in.⁴

Figure 10.40 Studs with Conventional Wallboards.

$$F_{a1} = \frac{12}{23}QF_y - \frac{3(QF_y)^2}{23\pi^2 E}\left(\frac{KL}{r_x}\right)^2 \quad (10.127)$$

or

$$F_{a1} = 0.522(QF_y) - \left[\frac{(QF_y)LK/r_x}{1,494}\right]^2 \quad (10.128)$$

In our case:

$$C_c = \sqrt{\frac{2\pi^2 E}{F_y}} = \sqrt{\frac{2\pi^2(29.5)10^3}{50}} = 107.9 \quad \text{and} \quad C_c/\sqrt{Q} = 126.7$$

The slenderness of one stud about its major axis is

$$\frac{KL}{r_x} = \frac{1.0 \times 15.0 \times 12}{2.76} = 65.2 < 126.7$$

Light-Gage Steel Members   493

Therefore, according to Eq. (10.127):

$$F_{al} = 0.522(36.3) - \frac{36.3 \times 65.2^2}{1,494} = 16.4 \text{ ksi } (113.1 \text{ MPa})$$

The same value could be obtained using AISI charts [10.26, section 3.6.1.1(B), p. 31] for $KL/r = 65$ and $QF_y = 36$ ksi (248 MPa).

The allowable load on the stud is

$$P_s = F_{al}A = 16.4 \times 1.00 = 16.4 \text{ kips } (72.95 \text{ kN})$$

3. The modulus of elastic support, k_w, based on a test value is taken as 1 kip/in. Then, from Eq. (10.125), either

$$a_{max} = \frac{8(29.5 \times 10^3)1.0 \times 1.0}{(1.0)^2(50)^2} = 94.4 \text{ in. } (240 \text{ cm})$$

or

$$a_{max} = \frac{15 \times 12 \times 1.0}{2 \times 2.76} = 32.6 \text{ in. } (83 \text{ cm})$$

Therefore, use $a = 32$ in.

4. Check the minimum modulus of elastic support [10.27, section 5.1.(c)]

$$k_w = \frac{(50)^2 \times 32 \times (1.0)^2}{8(29.5 \times 10^3)1.0} = 0.34 < 1.00: \quad \underline{\text{OK}}$$

5. From Eq. (10.126) the force on each attachment is

$$P_{min} = \frac{1.0 \times 16.4 \times \frac{180}{240}}{2\sqrt{29,500 \times 1.00 \times 1.00/32} - 16.4} = 0.28 \text{ kip } (1.25 \text{ kN})$$

One A307 bolt of $\frac{1}{4}$ in. diameter in single shear can carry 0.49 kip; in bearing for a material with $F_y = 50$ ksi, 1.96 kips. Therefore, bolts of $\frac{1}{4}$ in. diameter spaced 32 in. apart will satisfy.

6. Check spacing of connectors in compression elements:

$$S = 200\frac{t}{\sqrt{f}} = 200 \times \frac{0.075}{\sqrt{16.4}} = 4.05 \text{ in } (10.3 \text{ cm})$$

This requirement (10.27, section 4.4) does not apply to the cover sheets, which act only as sheathing material and are not considered load-carrying elements.

NOTATIONS

a	= spacing of vertical stiffeners; ratio of web and total girder areas	A_r^s	= total area of longitudinal slab reinforcement steel over interior support
A	= total girder cross-section area	ΔA	= equivalent section area of stiffeners
A_c	= area of concrete flange		
A_f	= area of the flange	b	= web slenderness ratio
A_s	= area of steel beam	b_{eff}	= effective flange width
A_{stf}	= area of stiffener	b_f	= flange width
A_{tot}	= cross-section area of girder and stiffeners	b_r	= reduced slab width
		b'	= stiffener's width
A_w	= web area	C_c	= limiting slenderness ratio

C_m	= coefficient for combined stresses	F_y	= yield point stress
		F_{ys}	= yield point stress in shear
C_v	= ratio of shear buckling stress and shear yielding stress	F_{ystf}	= yield point stress of stiffener
		h	= web clear depth
C_1, C_2, C_3	= constants	h_1	= web depth for which vertical stiffeners are used for the first time
C'_c	= ultimate compressive force in concrete slab		
C'_s	= ultimate compressive force in part of steel section	I	= girder moment of inertia; minimum moment of inertia of longitudinal stiffener
e	= eccentricity; base of natural logarithms, 2.71828 ...		
		I_x	= moment of inertia about maximum axis
E	= Young's modulus		
E_c	= Young's modulus of concrete	k	= buckling coefficient
E_s	= Young's modulus of steel	K	= multiplication factor; effective buckling length factor for prismatic columns
f_b	= bending stress		
f'_c	= specified compression strength of concrete		
		L	= span
		L_b	= unbraced length of a beam
f'_{cn}	= stress in concrete at infinite time	M	= moment
		M_{max}	= maximum bending moment
f'_{ct}	= stress in concrete at time t	M_y	= first yielding moment
f_f	= final stress in flange	M_u	= ultimate bending moment
f_r	= range of stress due to live load plus impact	M'_u	= ultimate bending moment of a composite beam
f_t	= tensile stress		
f_v	= shear stress	n	= modular ratio, integer
f_{vs}	= allowable shear flow per unit length of web	N_c	= number of additional connectors
f_1	= maximum compressible stress in web	p	= constant
		P_{sy}	= ultimate force on a stiffener
F_{all}	= allowable compressive stress	P_{stf}	= force in a stiffener
F_b	= allowable bending stress	q	= allowable shear load per one connector; constant in girder optimization
F'_b	= reduced allowable bending stress		
F_{CB}	= reduced compressive buckling stress	r_T	= radius of gyration of flange and part of web
F_{CBi}	= comparative ideal buckling stress	s	= width of tension field
F_{crit}	= Euler critical stress	S	= maximum permitted clear distance of stiffeners; section modulus
F_e	= coordinating stress		
F_{1Bi}	= ideal buckling stress in bending		
		S_{opt}	= optimum section modulus
F_r	= residual stress	S_{trans}	= section modulus of transformed area section
F_v	= allowable shear stress		
F_{vcrit}	= critical shear buckling stress	S_b	= section modulus at bottom of steel section
F_{vBi}	= ideal buckling stress in shear		

Notations 495

S_s	= steel beam section modulus in composite construction	U	= nondimensional web thickness
S_t	= section modulus at top	V_b	= beam action shear
S_{tc}	= section modulus at the top of concrete slab	V_h	= horizontal shear in composite construction
t	= time period	V_p	= ultimate shear force capacity
t_f	= flange thickness	V_t	= tension field action shear
t_n	= time at the end of creep or shrinkage	V_u	= ultimate shear carrying capacity
t_{stf}	= thickness of stiffener	V_{max}	= maximum shear
t_s	= thickness of concrete slab	w	= volume weight of concrete, uniform dead load intensity
t_w	= web thickness		
T'	= ultimate tensile force in a steel beam in composite construction	W	= nondimensional flange area parameter
		X	= nondimensional web parameter
T	= total tension of tension field action	Y	= nondimensional depth parameter
T_f	= tensile flange force due to tension field action (web buckled)	Z	= nondimensional cross-sectional area parameter
T_w	= horizontal component in the web of tension field	Z_r	allowable range of shear acting on an individual connector

PROBLEMS

10.1. Design a welded plate girder using A36 steel and AISC specification (10.1) for the loading shown in Fig. P-10.1. Lateral bracings are located at 10-ft intervals. Try one cross section without and another with transverse stiffeners. Compare these two designs.

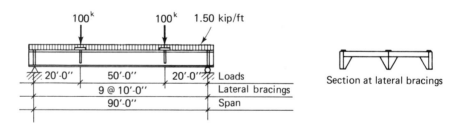

Figure P-10.1

10.2. Design a welded plate girder to support a 120-kip uniformly distributed load and a 125-kip concentrated load at midspan (see Fig. P-10.2). The girder is simply supported with a span of 50 ft–0 in. and has sufficient lateral support for its compressive flange. Use A36 steel, E70 electrodes, and AISC specification (10.1).

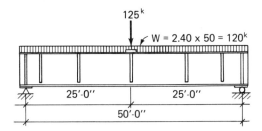

Figure P-10.2

10.3. A plate girder of A36 steel used in a building has cross section and loads as shown in Fig. P-10.3. Calculate the maximum bending and shear stresses and then decide if intermediate transverse stiffeners are needed. If the answer is yes, determine their location. Use AISC specification (10.1). Lateral support is provided by the floor slab construction every 5 ft.

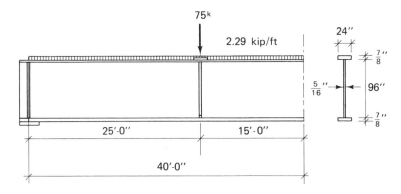

Figure P-10.3

10.4. A plate girder is loaded as shown in Fig. P-10.4. If the compression flange is continuously supported by a concrete slab, design this girder using the optimization technique given by Holt and Heithecker (without stiffeners) (Section 10.1.3) using AISC specification (10.1) and A36 steel.

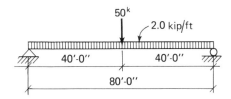

Figure P-10.4

10.5. Design an open-web built-up beam to be used as a purlin at 2 ft–0 in. center to center on a slope of 1 : 24 (see Fig. P-10.5). The roofing has a weight of 36 lb/ft²; the snow load is 25 lb/ft². The simple span length of the purlin is 30 ft–0 in. Use rods for diagonals and a pair of angles for both chords.

Problems   **497**

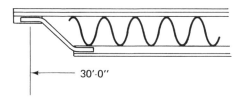

Figure P-10.5

10.6. For the cold-formed beam section with unstiffened flanges shown in Fig. P-10.6 and made of grade-50 steel, calculate the section modulus, S_x, about the major axis.

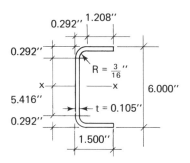

Figure P-10.6

10.7. If a cold-formed section as in Problem 10.6 is used as a beam with an unbraced length of 6 ft, calculate the allowable resisting moment, M_x.

10.8. Redesign the girder in problem 10.1 using high-strength, low-alloy ASTM A441 steel.

10.9. Redesign the girder in problem 10.2 using ASTM A588 weathering steel.

10.10. Redesign the girder in problem 10.2 using ASTM A514 steel.

10.11. Same as Problem 10.1, with lateral bracings at 15 ft apart.

10.12. Same as Problem 10.2, with lateral bracings at 10 ft apart.

10.13. Same as Problem 10.2, with lateral bracings 25 ft apart.

10.14. Same as Problem 10.2, but unbraced length is 50 ft.

10.15. Design a welded plate girder to support a uniform load of 3 kips/ft and two concentrated loads of 60 kips located 17 ft from each end. The simply supported girder has a span of 48 ft. The compression flange of the girder will be laterally supported only at points of concentrated load. Use A36 steel.

10.16. Same as Problem 10.15, with lateral bracings 12 ft apart.

10.17. Same as Problem 10.15, but use A588 steel.

10.18. Using $F_y = 36$ ksi, design the section of a welded plate girder with no intermediate stiffeners to support a uniform load of 2.5 kips/ft on a span of 85 ft. The girder is framed between columns and its compression flange is laterally supported for its entire length.

10.19. Same as Problem 10.18, with lateral bracings at 15 ft.

10.20. Same as Problem 10.18, but use steel with $F_y = 50$ ksi.

REFERENCES

10.1. AISC, *Specification for the Design, Fabrication, and Erection of Structural Steel for Buildings*, New York, 1978.

10.2. BASLER, K., "Strength of Plate Girders in Bending," *J. Struc. Div., Proc. ASCE*, Paper no. 2913, vol. 87, no. ST6 Aug. 1961 pp. 153–61.

10.3. AISC, *Manual of Steel Construction*, 8th ed., Chicago, 1980.

10.4. AASHTO, *Standard Specifications for Highway Bridges*, 12th ed., Washington, DC, 1977 (with 1978, 1979, 1980, and 1981 Interim Bridge Specifications).

10.5. Deutsche Industrie Normen, *Berechnungsgrundlagen für Stabilitätsfälle im Stahlbau: Knickung, Kippung, Beulung*, DIN 4114, Berlin, 1952.

10.6. British Standards Institution, *Specification for the Use of Structural Steel in Buildings*, new ed., BS 449, London, 1965.

10.7. British Standards Institution, *Specification for Steel Girder Bridges*, new ed., London, 1966, part 3B and 4.

10.8. Stroitelnye normy i pravila, Part II, Section V, Chapter 3: "Stalnye Konstruktsii, Normy Proektirovaniia," Governmental Structural Committee of the Council of Ministers of USSR, Moscow, 1963; 2nd ed. 1966.

10.9. L'Institut Technique du Batiment et des Travaux Publics and Le Centre Technique Industriel de la Construction Métallique, *Règles de Calcul des Constructions en Acier*, 5th ed., Règles CM66–1974, ed. Eyrolles, Paris, 1974.

10.10. HOLT, E. D., and G. L. HEITHECKER, "Minimum Weight Proportions for Steel Girders," *J. Struc. Div., Proc. ASCE*, Paper no. 6838, vol. 95, no. ST10, Oct. 1969, pp. 2205–16.

10.11. SCHILLING, C. G., "Optimum Properties for I-Shaped Beams," *J. Struc. Div., Proc. ASCE*, Paper no. 11007, vol. 100, no. ST12, Dec. 1974, pp. 2385–401.

10.12. Steel Joist Institute, *Standard Specifications, Load Tables and Weight Tables for Steel Joists and Joist Girders*, Richmond, VA, 1981.

10.13. British Standards Institution, *Composite Construction in Structural Steel and Concrete*, British Standard Code of Practice, CP117, part 2, "Beams for Bridges."

10.14. Volume Indoor Parking, "Composite Structural Design Cuts Building Costs," *Engineering News Record*, Aug. 8, 1974, p. 42.

10.15. "Bridge No. 27 Across Maas River at Herstal [Belgium]," *Acier-Stahl-Steel*, vol. 28, 1963, p. 198.

10.16. TOPRAC, A. A., and D. G. EYRE, "Composite Beams with a Hybrid Tee Steel Section," *J. Struc. Div., Proc. ASCE*, Paper no. 5518, vol. 93, no. ST5, Oct. 1967, pp. 309–22.

10.17. VARGHESE, P. C., and C. G. CHETTIAR. "Lateral Stability of Plane Open-Web Structures," *J. Struc. Div., Proc. ASCE*, Paper no. 10924, vol. 100, no. ST11, Nov. 1974, pp. 2223–33.

10.18. DOUTY, R. T., and J. O. CROOKER, "Optimization of Long-Span Cold-Formed Truss Purlins," *J. Struc. Div., Proc. ASCE*, Paper no. 10952, vol. 100, no. ST11, Nov. 1974, pp. 2275–88.

10.19. Task Committee on Composite Construction, "Composite Steel-Concrete Construction," *J. Struc. Div., Proc. ASCE,* Paper no. 10561, vol. 100, no. ST5, May 1974, pp. 1085–139.

10.20. HAMADA S., and J. LONGWORTH, "Buckling of Composite Beams in Negative Bending," *J. Struc. Div., Proc. ASCE,* Paper no. 10917, vol. 100, no. ST11, Nov. 1974, pp. 2205–22.

10.21. Deutsche Industrie Normen, *Verbundträger-Hochbau,* Richtlinien für die Berechnung und Ausbildung, DIN 4239, Berlin, 1956.

10.22. Deutsche Industrie Normen, *Verbundträger-Strassenbrücken,* Richtlinien für die Berechnung und Ausbildung, DIN 1078.

10.23. Deutscher Stahlbau-Verband, *Stahlbau: Ein Handbuch für Studium und Praxis,* 2nd ed., Cologne, 1971, p. 348.

10.24. DE WOLF, J. T., et al., "Local and Overall Buckling of Cold-Formed Members," *J. Struc. Div., Proc. ASCE,* Paper no. 10875, vol. 100, no. ST10, Oct. 1974, pp. 2017–36.

10.25. WINTER, G., "Commentary on the 1968 Edition of the Specifications for the Design of Cold-Formed Steel Structural Members," in *Cold-Formed Steel Design Manual—Part V,* American Iron and Steel Institute, New York, 1970.

10.26. American Iron and Steel Institute, charts and tables for use with the 1968 edition of the specifications for the design of cold-formed steel structural members, in *Cold-Formed Steel Design Manual—Part IV,* New York, 1977.

10.27. American Iron and Steel Institute, "Specifications for the Design of Cold-Formed Steel Structural Members," New York, 1980.

10.28. American Iron and Steel Institute, illustrative examples, in *Cold-Formed Steel Design Manual—Part III,* New York, 1977.

10.29. Steel Deck Institute, *Steel Roof Deck Design Manual,* no. 18, Westchester, Ill., 1972.

10.30. KUZMANOVIĆ, B. O., and N. WILLEMS, "Mini-Computers in Structural Optimum Design," pp. 82–4. *Civil Engineering,* ASCE, May 1982.

11

Beam and Column Supports

11.1 BEAM SUPPORTS

Beams can be supported in several ways, and supports can be classified according to their location and the type of connection involved. Distinction can thus be made among beam supports as being

1. Beam-to-column connections (Fig. 11.1)
2. Beam-to-beam connections (Fig. 11.2)
3. Beam-to-foundation supports (Fig. 11.3)

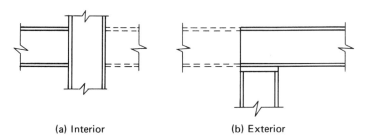

(a) Interior   (b) Exterior

Figure 11.1 Beam-to-column Connections.

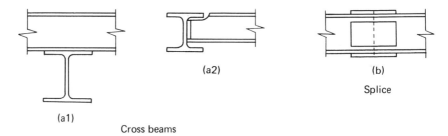

(a1) (a2) (b) Splice

Cross beams

Figure 11.2 Beam-to-beam Connections.

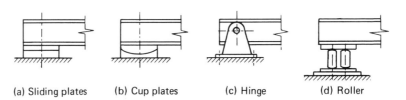

(a) Sliding plates (b) Cup plates (c) Hinge (d) Roller

Figure 11.3 Beam-to-foundation Supports.

In Chapter 9 most of the beam-to-column and beam-to-beam connections, including splices, were discussed in detail. In this chapter the beam-to-beam connection shown in Fig. 11.2a1 will be discussed in addition to the beam-to-foundation supports shown in Figs. 11.3a,b, c, and d. Beam end supports can be classified, according to the reaction components that will occur, as

1. Tangential (simple) supports
2. Hinged supports
3. Roller supports
4. Fixed (built-in) supports

These types can be represented diagramatically as shown in Fig. 11.4. Another means of classification, based on the possibility of horizontal motion, divides supports into two categories

1. Fixed supports (no horizontal motion)
2. Expansion supports (horizontal motion possible)

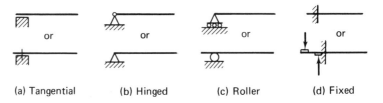

(a) Tangential (b) Hinged (c) Roller (d) Fixed

Figure 11.4 Diagrammatic Representation of Beam End Supports.

11.1.1 Tangential (Simple) Supports

Tangential supports (Fig. 11.5), which are used for relatively short spans (less than 30 ft) and moderate loads, do not allow beam end rotation. If no special provision is made, horizontal motion of the beam relative to the supports requires overcoming considerable friction. Sometimes bronze or other low-friction sliding surfaces are provided. A much better type of bearing that allows rotation is shown in Fig. 11.6. By providing slotted holes the bearing allows rotation as well as sliding. The AASHTO specifications (11.2, section 1.7.32) specify that spans of 50 ft and greater must be provided with rollers, rockers, or sliding plates for expansion purposes as well as with a type of bearing employing a hinge, curved bearing plates or pin arrangement for deflection purposes.

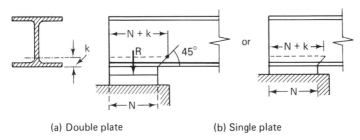

Figure 11.5 Tangential (Simple) Support.

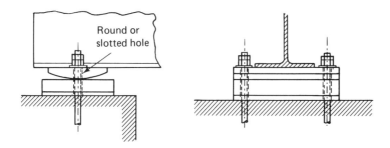

Figure 11.6 Fixed and Expansion Bearings.

All beam supports have to meet certain basic requirements:

1. There must be no web crippling.
2. The allowable foundation bearing stress must not be exceeded.
3. The support must fulfill its bearing objectives.

Web crippling

As shown in Fig. 11.5(a) the dimension $(N + k)$ determines the bearing load intensity transmitted to the web. If R is the transmitted reaction, we have [AISC specifications (11.1), equation 1.10.9]

$$\frac{R}{t(N + k)} \leq 0.75 F_y \tag{11.1}$$

where R = concentrated reaction
t = web thickness
N = length of bearing plate
k = distance from outer face of flange to web toe of fillet

If this equation is not satisfied, bearing stiffeners have to be provided.

Bearing stiffeners

According to the AISC specifications (11.1) bearing stiffeners must be provided in pairs at unframed ends on the webs of plate girders and where required at points of concentrated loads. Such stiffeners must have a close bearing against the flange transmitting the load and must extend approximately to the edge of the flange. They are designed as columns. The column is assumed to consist of the pair of stiffeners and a centrally located strip of web with a width 25 times its thickness at interior stiffeners and 12 times its thickness when the stiffeners are located at the end of the web; the effective length is taken as $\frac{3}{4}$ the length of the stiffener. The AASHTO specifications (11.2, section 1.7.42 and 1.7.43) also require bearing stiffeners at supports, preferably in pairs. To calculate the radius of gyration a centrally located strip of web not exceeding 18 times the web thickness can be taken into account. The AASHTO specifications do not specify the effective length. The thickness of a bearing stiffener plate, t_s, must satisfy the following:

$$t_s \geq \frac{b'}{12}\sqrt{\frac{F_y \text{ [psi]}}{33,000}} \quad \left(\frac{b'}{12}\sqrt{\frac{F_y \text{ [MPa]}}{227.54}}\right) \tag{11.2}$$

where F_y is the yield stress in psi or MPa as indicated, b' is the width of the plate, and the allowable compressive stress is as given in Section 1.7.1A of the AASHTO specifications. Both the AISC and AASHTO specifications limit the bearing pressure on the stiffeners; AASHTO allows $0.80F_y$ while AISC uses $0.90F_y$.

Bearing stress on foundations

The allowable bearing stress depends on the type of foundation. The AISC specifications (11.1, section 1.5.5) give the following values:

On sandstone and limestone: $F_p = 0.40$ ksi (2.76 MPa)

On brick in cement mortar: $F_p = 0.25$ ksi (1.72 MPa)

On the full area of a concrete support: $F_p = 0.35 f'_c$

On less than the full area of a concrete support: $F_p = 0.35 f'_c \sqrt{A_2/A_1} \leq 0.7 f'_c$

where f'_c is the specified compression strength of the concrete and A_1 is the bearing area in square inches and A_2 is the full cross-sectional area of concrete support in square inches. The AASHTO specifications (11.2) allow the following unit bearing stresses:

Granite: $F_p = 800$ lb/in.^{2} (5.52 MPa)

Sandstone and limestone: $F_p = 400$ lb/in.^{2} (2.76 MPa)

These stresses are on the basis of at least a 3-in. (7.6-cm) projection of the bridge seat beyond the edge of shoe or plate; otherwise a 25% reduction has to be applied.

Concrete:

1. Allowable Stress Design: $F_p = 0.30 f'_c$, on loaded area
2. Load Factor Design: $F_p = 0.85 \phi f'_c$ on loaded area

For both designs the following adjustments to F_p apply:

1. When the supporting surface is wider on all sides than the loaded area, the allowable bearing stress on the loaded area may be increased by $\sqrt{A_2/A_1}$, but not more than 2.
2. When the supporting surface is sloped or stepped, A_2 may be taken as the area of the lower base of the largest frustrum of a right pyramid or cone wholly contained within the support, having for its upper base the loaded area and having side slopes of 1 vertical to 2 horizontal.
3. When the loaded area is subjected to high edge stresses due to deflection or eccentric loading, the allowable bearing stress on the loaded area shall be multiplied by a factor of 0.75. The requirement of (1) and (2) shall also apply.

Bearing objectives

The main objective of a bearing is to transfer the load from the beam to a support in such a manner that the type of reaction components are those assumed by the designer.

In the case of a tangential support if the support is a fixed support, both a horizontal and vertical reaction will occur provided beam end rotation is negligible. If a tangential support is of the expansion type, horizontal reactions equal to μR will occur if R is the reaction and μ the friction coefficient between the bearing plates (Figs. 11.5 and 11.6).

11.1.2 Hinged Supports

Two types of hinged supports (fixed and expansion) are shown in Fig. 11.7a and b.

11.1.3 Roller Supports

The expansion hinge type of support is in large bridges replaced by a roller type of support. Types of these are shown in Fig. 11.8a and b. Provision has to be made for preventing lateral movement, skewing, and creeping. Both AASHTO and AISC specify the allowable bearing stress. The AASHTO specifications (11.2, Section 1.7.42) give a maximum allowable bearing, p, in pounds per linear inch, which for diameters up to 25 in. (63.5 cm) is

$$p = \frac{F_y - 13,000}{20,000} \cdot 600d \tag{11.3}$$

and which for diameters from 25 to 125 in. (63.5 to 317.5 cm) is

$$p = \frac{F_y - 13,000}{20,000} \cdot 3,000\sqrt{d} \tag{11.4}$$

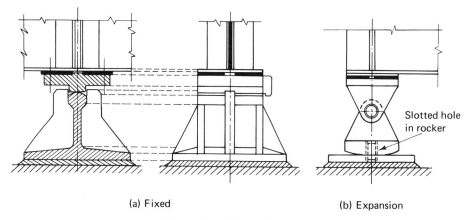

(a) Fixed (b) Expansion

Figure 11.7 Hinged Supports.

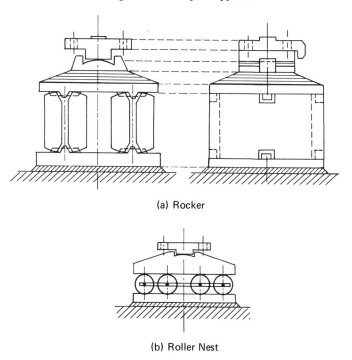

(a) Rocker

(b) Roller Nest

Figure 11.8 Roller Supports.

where d is the diameter of the roller or rocker in inches and F_y is in lb/in.². The AISC specifications (11.1, Section 1.5.1.5) give the allowable stress in kips per linear inch as

$$F_p = \frac{F_y - 13}{20} \cdot 0.66d \tag{11.5}$$

where F_y now is in ksi and d in inches.

To reduce friction in bearings materials other than steel are often used. Lubrite bearings, for instance, are patented types made of bronze, sintered metal, or other material provided with inserts of a special graphite lubrication compound. Another material is Teflon. Elastomeric bearings made of neoprene or rubber or similar compounds are also used. Bearings are either plain bearings consisting of elastomer only or laminated bearings consisting of layers of elastomer restrained at their interfaces by bonded laminates such as steel. These bearings are specified in the AASHTO specifications (11.2, Sections 12 and 25). Several types of bearings employing elastomer and/or Teflon are shown in Fig. 11.9.

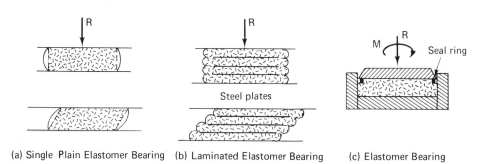

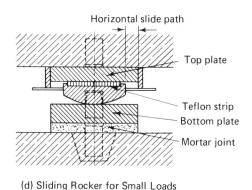

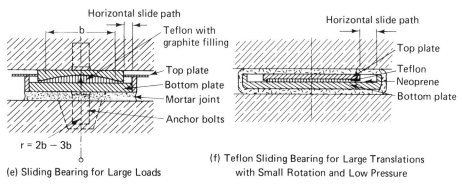

Figure 11.9 Bearings Employing Elastomer and Teflon.

Beam Supports 507

Example 11.1 For the supports shown in Figs. 11.10, 11.11, and 11.12 design (1) an expansion rocker for a load of 308 kips (1,370 kN); (2) a bolster for a load of 442 kips (1,966 kN) and a horizontal load of 8.15 kips (36.2 kN); and (3) a pinned bolster for 94 kips (42 kN) and a horizontal load of 7.5 kips (33.4 kN). Use AASHTO specifications (11.2) and A588 steel.

Solution

EXPANSION ROCKER (FIG 11.10). The axial load is $R = 308$ kips (1,370 kN). The bearing stress on the foundation, when centered, is

$$F_p = \frac{308}{18 \times 27} = 0.634 \text{ ksi (4.4 MPa)}$$

The length of the rocker is $27 - (2 \times 1.75) = 23.5$ in. (60 cm). Assume that $d = 2 \times 9 = 18$ in. (45.7 cm). The bearing pressure is

$$p = \frac{308}{23.5} = 13.1 \text{ kips/in. (2.29 kN/m)}$$

The allowable bearing stress is

$$p = \frac{F_y - 13,000}{20,000} \times 600d = \frac{70 - 13}{20} \times 600(18)$$

$$= 30.78 \text{ kips/in. (5.4 kN/m):} \quad \underline{\text{OK}}$$

BOLSTER (FIG. 11.11). The axial load is $R = 442$ kips (1,966 kN); $H = 8.15$ kips (36.2 kN). The minimum area of the sole plate allowing 1 ksi (6.9 MPa) on concrete is 442 in.² (2,851.6 cm²). The girder flange is $20 \times 1\frac{3}{4}$ in. (50.8 × 4.5 cm). Assume the length of the sole plate to be 27 in. (68.6 cm). The width has to be $\frac{442}{27} = 16.37$ in. (41.6 cm). Take

$$18 \text{ in.} \times 27 \text{ in.} \rightarrow F_p = \frac{442}{18 \times 27} = 0.909 \text{ ksi (6.3 MPa)}$$

Check for edge pressure due to horizontal loads (Fig. 11.11c):

$$e = \frac{8.15}{442} \times 11.75 = 0.2167 \text{ in. (0.55 cm)}$$

$$F_{toe} = \frac{R}{L}\left(1 + \frac{6e}{L}\right) \times \frac{1}{27} = \frac{442}{18}\left(1 + \frac{6 \times 0.2167}{18}\right) \times \frac{1}{27}$$

$$= 0.975 \text{ ksi} < 1 \text{ ksi:} \quad \underline{\text{OK}}$$

PINNED BOLSTER (FIG. 11.12). The axial load is $R = 94$ kips (418 kN); $H = 7.5$ kips (33.4 kN). The flange width is 12 in. (30.5 cm). Use a 10 × 11 in. sole plate (25.4 × 27.9 cm).

Assume a pin radius of 1 in. (2.5 cm). The bearing stress on the pin is 20 ksi (138 MPa). The minimum bearing area is $\frac{94}{20} = 4.7$ in.²—say, 5 in.² (32.2 cm²); $t = 5/(2 \times 1) = 2.5$ in.—say, two each 2.0 in. and four each 1.0 in. Check for edge pressure due to horizontal load (Fig. 10.12c):

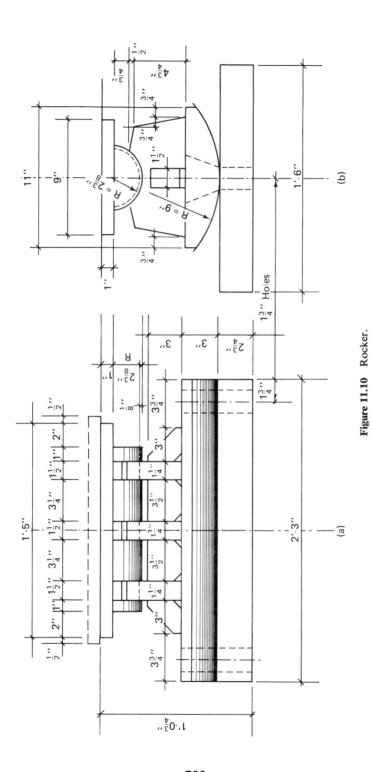

Figure 11.10 Rocker.

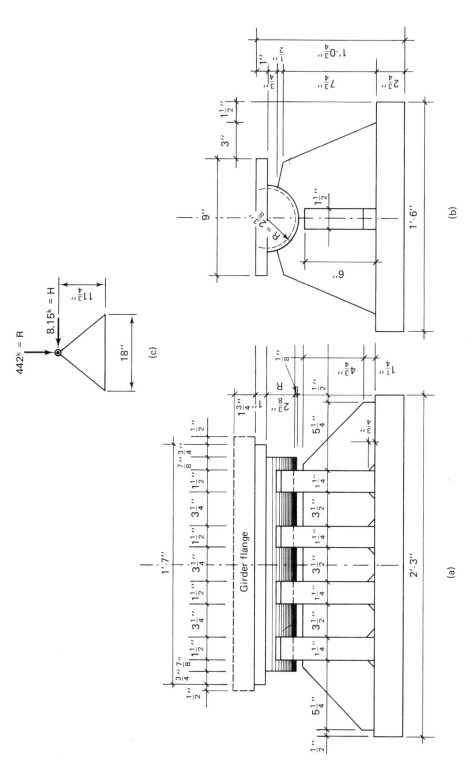

Figure 11.11 Bolster.

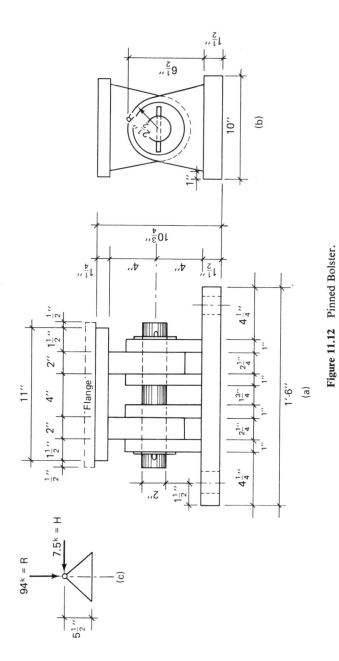

Figure 11.12 Pinned Bolster.

$$e = \frac{7.5 \times 5.5}{94} = 0.439 \text{ in. } (1.1 \text{ cm})$$

$$F_{\text{toe}} = \frac{R}{L}\left(1 + \frac{6e}{L}\right) \times \frac{1}{18} = \frac{94}{10}\left(1 + \frac{6 \times 0.439}{10}\right)\frac{1}{18}$$

$$= 0.66 \text{ ksi} < 1 \text{ ksi:} \quad \underline{\text{OK}}$$

Check base plate:

$$M = \frac{94}{18} \times \frac{4.25^2}{2} = 47.16 \text{ kip-in.}$$

$$S_{\text{rqd}} = \frac{47.16}{25} = 1.887 \text{ in.}^3 = \frac{bt^2}{6}$$

$$b = 10 \text{ in. } (25.4 \text{ cm}); \qquad t^2 = \frac{6 \times 1.887}{10} = 1.13 \rightarrow t = 1.06 \text{ in.}$$

Use $1\frac{1}{2}$ in. (3.8 cm).

Example 11.2 For the load of 442 kips (1,966 kN) design the bearing stiffeners in A588 steel for the plate girder shown in Fig. 11.13. Use AASHTO specifications (11.2).

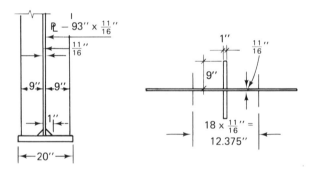

Figure 11.13 Bearing Stiffener.

Solution For the 9-in. plate width shown

$$t_{\min} = \frac{9}{12}\sqrt{\frac{50{,}000}{33{,}000}} = 0.923 \text{ in., } \quad \text{use 1 in. (2.5 cm)}$$

$$F_p = \frac{442}{(9-1)(1) \times 2} = 27.6 \text{ ksi} < 0.8 \times 50 \text{ ksi:} \quad \underline{\text{OK}}$$

$$A = 2 \times 9 \times 1 + 12.375 \times 0.6875 = 26.51 \text{ in.}^2 \ (171 \text{ cm}^2)$$

$$I = \tfrac{1}{12} \times 1 \times 18.6875^3 = 543.2 \text{ in.}^4 \ (22{,}610 \text{ cm}^4)$$

$$r = \frac{543.2}{26.51} = 20.5 \text{ in. } (52.1 \text{ cm})$$

$$\left(\frac{l}{r}\right)^2 = \left(\frac{93}{20.5}\right)^2 = 421$$

$$F_a = \frac{F_y}{2.12}\left(1 - \frac{421 F_y}{4\pi^2 E}\right) = \frac{50}{2.12}\left(1 - \frac{421 \times 50}{4\pi^2 \times 29{,}000}\right)$$

$$= 23.15 \text{ ksi } (159.6 \text{ MPa})$$

The allowable capacity is

$R = 23.15 \times 26.51 = 614$ kips (2,731 kN) > 442 kips (1,966 kN): OK

11.1.4 Beam-to-Beam Supports

When two beams cross over one another (Fig. 11.2a) the bearing stress between the beams and web crippling need to be checked. Intermediate bearing stiffeners may be needed [AISC (11.1), section 1.10.5.1; AASHTO (11.2), section 1.7.42 and 43].

11.2 COLUMN SUPPORTS

Column supports are either of the pinned type or the flat base plate type. Although the pinned base is essentially the same as for a beam, the main difference is the design of the base plate.

11.2.1 Pinned Column Base

The design of the pin is standard procedure. To demonstrate the design of the base plate the procedure recommended by the AISC specifications is illustrated in the following example.

Example 11.3 For the column base shown in Fig. 11.14 design the base plate.

Solution The eccentricity e is $4/\sqrt{3} = 2.31$ in. (5.9 cm). Also

$V = 200 \sin 60° = 173.2$ kips (770 kN); $\quad H = 100$ kips (445 kN)

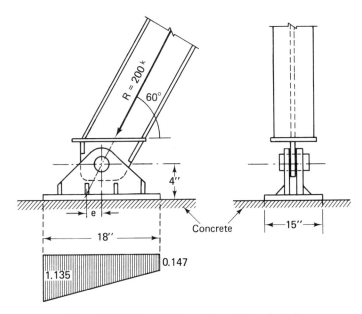

Figure 11.14 Pinned Column Base (Example 11.3).

$$f = \frac{173.2}{18 \times 15} \pm \frac{173.2 \times 2.31}{15 \times 18^2/6} = 0.641 \pm 0.494 \text{ ksi}$$

$$= +1.135 \quad \text{or} \quad +0.147 \text{ ksi} < 0.25 f_c' = 1.2 \text{ ksi}$$

$$(= +7.8 \quad \text{or} \quad +1.0 \text{ MPa} < 0.25 f_c' = 8.3 \text{ MPa})$$

11.2.2 Flat Column Base Plates

The design of a column base plate depends to a large extent on the ratio of moment to axial load. Accordingly, it is customary to distinguish among the following classes:

Class I: Compression over the entire area (Fig. 11.15a):
$e/H < \frac{1}{6}$

Class II: Tension over an area of one-third or less (Fig. 11.15b):
$\frac{1}{6} < e/H < \frac{1}{3}$

Class III: Tension over an area of more than one-third (Fig. 11.15c):
$e/H > \frac{1}{3}$

These three cases will be discussed next.

Class I base plates (Fig. 11.15a)

Working stress design

Using normal combined bending theory the stress in the concrete is

$$f_c = \frac{R}{A} \pm \frac{6Re}{BH^2} = \frac{R}{BH}\left(1 \pm \frac{6e}{H}\right) \tag{11.6}$$

which should not exceed the specified values (AISC: $0.35 f_c'$; AASHTO: $0.3 f_c'$ ksi). In this case anchor bolts have no computable stress.

Ultimate load design

Although it is difficult to determine the actual stress distribution at ultimate load, the one shown in Fig. 11.15a (iii) is a conservative estimate provided e remains small. Depending upon the loading at ultimate load the value of e may increase, in which case the approach discussed for class III base plates will have to be used.

Example 11.4 Consider a column with an axial load of $R = 140$ kips (623 kN) and a moment of 300 kip-in. (34 kN-m) resting on a base plate with an assumed H-value of 18 in. (45.7 cm); $f_c' = 3$ ksi (20.7 MPa). Use AISC specifications (11.1).

Solution

WORKING STRESS DESIGN. The eccentricity is $e = \frac{300}{140} = 2.143$ in.; $H/6 = 3$ in.

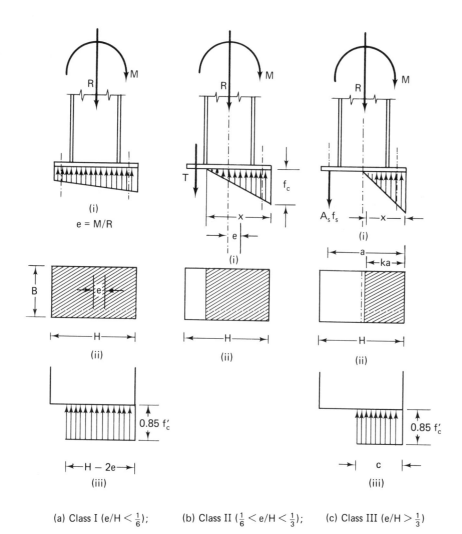

Figure 11.15 Column Base Plates.

Therefore, this is a class I base plate. Calculating B using Eq. (11.6) yields

$$0.35(3) = \frac{140}{18B}\left(1 + \frac{6 \times 2.143}{18}\right) = \frac{13.33}{B}$$

$$B = 12.7 \text{ in.}$$

Use $B = 15$ in. (38.1 cm).

CHECK FOR ULTIMATE STRENGTH:

$$e = 2.143 \text{ in. } (5.4 \text{ cm}); \quad H - 2e = 18 - 4.286 = 13.714 \text{ in.}$$

$$R = 13.714 \times 15 \times 0.85 \times 3 = 524 \text{ kips } (2{,}331 \text{ kN})$$

If we assume that M is proportional to R throughout the loading range, then $M_{\text{ult}} = \frac{524}{140} \times 300 = 1{,}123$ kip-in. (126 kN-m). Assuming that $\frac{1}{4}$ of all design loads are dead loads, then

$$R_{ult} = 1.3 \times \frac{R}{4} + 1.7 \times \frac{3R}{4} = 1.6R = 224 \text{ kips (996 kN)}$$

524 kips (2,331 kN) ≫ 224 kips (996 kN):   <u>OK</u>

This implies also that the moment capacity is adequate.

Class II base plates (Fig. 11.15b)

Working stress design

These base plates are usually calculated in the working design method using a concrete beam analogy. According to Fig. 11.15(b)

$$x = 3\left(\frac{H}{2} - e\right) \qquad (11.7)$$

Now

$$R = \tfrac{1}{2}Bxf_c$$

which yields

$$B = \frac{2R}{xf_c} \qquad (11.8)$$

In this method the tension in the anchor bolts is neglected.

Example 11.5 Consider the same column as in Example 11.4 but with a moment of 560 kip-in. (63.3 kN-m).

Solution Again assume H is 18 in. (45.7 cm). According to Eq. (11.7)

$$x = 3(9 - \tfrac{560}{140}) = 15 \text{ in. (38.1 cm)}$$

and applying Eq. (11.8)

$$B = \frac{2 \times 140}{15 \times 1.05} = 17.78 \text{ in. (45.1 cm)}$$

Use $B = 18$ in. (45.7 cm).

In class II cases a nominal size is chosen for the anchor bolts.

Ultimate load design

It is customary for class II base plates to use the technique developed for class III base plates, discussed in the next section.

Class III base plates (Fig. 11.15c)

Working load design

To determine x using a concrete beam analogy,

$$R = \tfrac{1}{2}f_c xB - A_s f_s \qquad (11.9)$$

Taking moment equilibrium about the right edge,

$$M - \frac{RH}{2} = A_s f_s a - \frac{f_c xB}{2} \cdot \frac{x}{3} \qquad (11.10)$$

Assuming a linear strain distribution and setting $x = ka$,

$$\frac{\epsilon_s}{\epsilon_c} = \frac{a-x}{x} = \frac{1-k}{k}$$

We now set

$$\epsilon_s = \frac{f_s}{E_s}; \quad \epsilon_c = \frac{f_c}{E_c}; \quad \text{and} \quad \frac{E_s}{E_c} = n$$

which yields

$$\frac{f_s}{f_c} = n\left(\frac{1-k}{k}\right) \tag{11.11}$$

In actual design a trial-and-error procedure is usually followed to satisfy the above equations.

Ultimate load design

Equilibrium of the vertical forces yields

$$R_u = 0.85 f'_c c B - A_s f_y \tag{11.12}$$

For moment equilibrium

$$M_u - R_u\left(\frac{H-c}{2}\right) = A_s f_y\left(a - \frac{c}{2}\right) \tag{11.13}$$

Example 11.6 A 14WF 142 is resting on a W14 × 142 steel base plate; $f'_c = 3{,}000$ lb/in.² (20.7 MPa). Use A36 bolts.

Solution

$$R = 100 \text{ kips (445 kN)} \quad \text{and} \quad M = 2{,}400 \text{ kip-in. (271.2 kN-m)}$$

Assume $H = 30$ in. (76.2 cm) (Fig. 11.16).
  WORKING STRESS DESIGN. Assume $B = 30$ in. (76.2 cm). Also
$A_s = 2 \times 3.14 = 6.28$ in.² (40.5 cm²)

$$a = 15 + \frac{14.75}{2} + 2.50 = 24.87 \text{ in. (63.2 cm)}$$

$$e = \frac{2{,}400}{100} = 24 \text{ in. (61 cm)} > \frac{H}{3} = 10 \text{ in. (25.4 cm)}$$

$$n = 9$$

$$f_s = 0.33 F_u = 19.1 \text{ ksi (132 MPa)}$$

$$f_c = 0.7 f'_c = 2.10 \text{ ksi (14.6 MPa)}$$

Next the following steps are taken:

1. Obtain a trial value for k from Eq. (11.11)

$$\frac{1-k}{k} = \frac{f_s}{nf_c} = \frac{19.1}{9 \times 2.10} = 1.01$$

$$k = 0.497$$

2. Calculate the new value of f_c from Eq. (11.9):

$$R = \tfrac{1}{2} f_c x B - A_s f_s$$

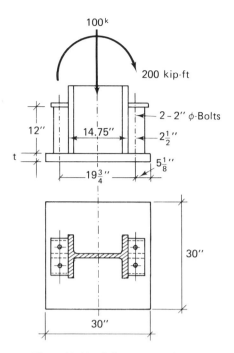

Figure 11.16 Column Base Plate.

$$100 = \frac{0.497 \times 30 \times 24.87 f_c}{2} - 6.28 \times 19.1$$

$$f_c = 1.19 \text{ ksi (8.2 MPa)}$$

3. Calculate the new value of f_s from Eq. (11.10):

$$2{,}400 - 100 \times 15 = 6.28 f_s \times 24.87 - \frac{1.19(0.497)^2 \times 30 \times 24.87^2}{6}$$

$$f_s = 11.6 \text{ ksi (79.9 MPa)}$$

4. Calculate the new value of k

$$\frac{1}{k} - 1 = \frac{11.6}{9 \times 1.19}$$

$$k = 0.48$$

5. Repeat steps 1–4. First assume

$$k = \frac{0.48 + 0.497}{2} = 0.489$$

$f_c = 0.95$ ksi (6.6 MPa); $f_s = 10.3$ ksi (71.0 MPa); $k = 0.45$
this now yields

$f_c = 0.98$ ksi (6.8 MPa); $f_s = 9.71$ ksi (67.0 MPa); $k = 0.476$

Assume $k = 0.465$. Then

$f_c = 0.93$ ksi $< 0.35 \times 3\sqrt{\dfrac{1}{0.465}} = 1.54$ ksi; $f_s = 9.75$ ksi; $k = 0.463$: <u>OK</u>

The bolt force is $2 \times 3.14 \times 9.75 = 61.2$ kips (272 kN). To check the base plate itself we assume that the critical section occurs at the bolt line

$$x = ka = 0.463 \times 24.87 = 11.5 \text{ in. (29.2 cm)}$$

And the stress at the bolt line is

$$f_c = \frac{6.38 \times 0.93}{11.5} = 0.52 \text{ ksi (3.58 MPa)}; \quad \text{assume} \quad t = 2 \text{ in. (5 cm)}$$

$$M = 30\left[\frac{0.52}{2} \times \frac{5.12^2}{3} + \frac{0.93}{2} \times \frac{2 \times (5.12)^2}{3}\right]$$

$$= 312 \text{ kip-in. (35.2 kN-m)}$$

$$f_s = \frac{312}{30 \times 2^2/6} = 15.6 \text{ ksi (108 MPa)} < 27 \text{ ksi (186.2 MPa)}: \quad \underline{\text{OK}}$$

ULTIMATE LOAD DESIGN. According to Fig. 11.15 and Eq. (11.12):

$$R_u = 0.85 \times 3 \times 30c - 6.28 \times 36 = 76.5c - 226$$

$$c = 2.96 + \frac{R_u}{76.5} \tag{a}$$

From Eq. (11.13):

$$\left(\frac{R_u}{100}\right)(2,400) - R_u\left(15 - \frac{c}{2}\right) = 226\left(24.87 - \frac{c}{2}\right)$$

$$9R_u + \frac{R_u c}{2} = 5,630 - 113c$$

Substitute c from Eq. (a):

$$9R_u + 1.48R_u + \frac{R_u^2}{153} = 5,630 - 335 - 1.48R_u$$

$$R_u^2 + 1,830R_u - 808,000 = 0$$

$$R_u = 367 \text{ kips (1,632 kN)}$$

Assuming proportional increase in the moment:

$$M_u = 24 \times 367 = 8,800 \text{ kip-in. (993.5 kN-m)}$$

Again assuming that the dead load represents $\frac{1}{4}$ the total loads, the overall load factor is

$$\frac{1.3}{4} + \frac{1.7(3)}{4} = 1.6$$

$$\text{L.F.} = \frac{367}{100} = 3.67 > 1.6: \quad \underline{\text{OK}}$$

$$c = 2.96 + \frac{367}{76.5} = 7.75 \text{ in. (20.2 cm)}$$

Next we check the moment in the base at the bolt line:

$$M = \frac{0.85 \times 3 \times (5.12)^2}{2} = 33.5 \text{ kip-in./in. (149 kN-m/m)}$$

$$M_p = \tfrac{1}{4} \times 36 \times 2^2$$

$$= 36 \text{ kip-in./in. (160 kN-m/m)}$$

$$> 33.5 \text{ kip-in./in. (149 kN-m/m)}: \quad \underline{\text{OK}}$$

NOTATIONS

b'	= width of stiffener plate	A, A_1, A_2	= area
c	= depth of stress block	A_s	= area of steel
d	= diameter of roller or rocker	B	= width of base plate
e	= load eccentricity	E_c	= modulus of elasticity for concrete
f_c	= concrete stress		
f'_c	= concrete cylinder strength	E_s	= modulus of elasticity for steel
f_s	= steel stress	F_p	= bearing stress
f_y	= yield stress of steel	F_y	= yield stress
k	= distance from outer face of flange to web toe of fillet; factor	H	= length of base plate, horizontal force
		I	= moment of inertia
l	= length of column	L.F.	= load factor
n	= E_s/E_c	M_p	= plastic moment capacity
p	= allowable bearing per linear inch	N	= length of bearing
		R	= axial load
r	= radius of gyration	S	= section modulus
t_s	= stiffener thickness	σ	= stress
t	= web thickness	ϵ	= strain
x	= depth of stress block		

PROBLEMS

11.1. Design a bearing plate for a beam W27 × 84 to transmit a reaction of 50 kips to a reinforced concrete wall of $f'_c = 3$-ksi strength (see Fig. P-11.1). Check web crippling of the beam. Use A36 steel and AISC specification (11.1).

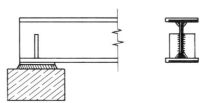

Figure P-11.1

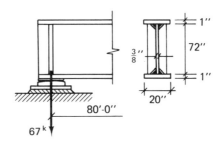

Figure P-11.2

520   BEAM AND COLUMN SUPPORTS

11.2. A girder of 80-ft span is simply supported and the maximum vertical reaction is 67 kips (see Fig. P-11.2). Design tangential support for that girder on a masonry wall. Use AISC specification (11.1).

11.3. For a large-span roof truss design the hinged support to carry a vertical force of 70 kips and a transverse horizontal force of 10 kips. The supporting wall is a concrete wall of 12 in. thickness. The bottom chord of the truss is as shown in Fig. P-11.3. Use AISC specification (11.1).

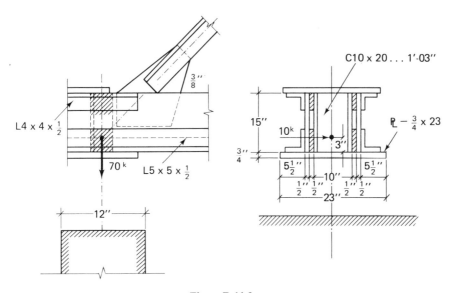

Figure P-11.3

11.4. For the truss of Problem 11.3 design a rocker support. The lower chord section is the same as in Fig. P-11.3. Use AISC specification (11.1).

11.5. A beam is supported by a plate girder as shown in Fig. P-11.5. Design this support if the beam reaction is 50 kips. Use AISC specification (11.1).

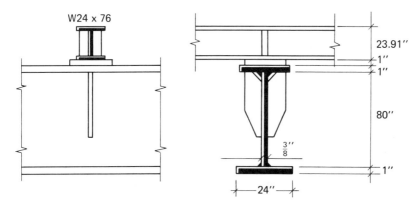

Figure P-11.5

11.6. For a highway bridge of 180 ft simple span design the hinged support at the abutment for a vertical reaction of 83.6 kips and a horizontal one of 25.0 kips. Dimensions of the welded plate girder are given in Fig. P-11.6. Use AASHTO specifications (11.2).

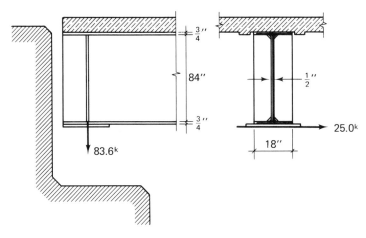

Figure P-11.6

11.7. For the bridge in Problem 11.6 design a roller support capable of allowing a horizontal displacement of $\Delta L = 2$ in. in both directions from the equilibrium position. Use AASHTO specifications (11.2).

11.8. Design a pinned column base plate of A36 steel resting on concrete ($f'_c = 4$ ksi) with reactions $V = 150$ kips and $H = 50$ kips. Use AISC specification.

11.9. Design column base plates for a W14 × 132 (A36) to carry an axial load of 120 kips with eccentricities of (a) 3 in. (b) 6 in., and (c) 9 in. Use AISC specification. Design by working stress method and check for ultimate strength (L.F. = 1.6) $f'_c = 3.6$ ksi.

11.10. Repeat Problem 11.1 for a W24 × 76 and a reaction of 60 kips.

11.11. Repeat Problem 11.2 for maximum vertical reaction of 100 kips.

11.12. Repeat Problem 11.3 for a vertical reaction component of 100 kips and a transverse horizontal reaction component of 25 kips.

11.13. Design a rocker support for the truss of Problem 11.12.

11.14. Repeat Problem 11.5 if the beam reaction is 80 kips (also design stiffeners as needed).

11.15. Repeat Problem 11.6 if the vertical reaction is 120 kips and the horizontal reaction equals 30 kips. Use AASHTO specifications.

11.16. For the support of Problem 11.15 design a roller support capable of allowing a horizontal displacement of $\Delta L = 1$ in. in both directions from the equilibrium position.

11.17. Design a pinned column base plate of A36 steel resting on concrete ($f'_c = 5$ ksi) with reactions $V = 200$ kips and $H = 75$ kips. Use AISC specifications.

11.18. Design column base plates for a W14 × 68 (A36) to carry an axial load of 200 kips with eccentricities of (a) 2 in., (b) 5 in., and (c) 9 in. Use AISC specification. Design by working stress method and check for ultimate strength $(L.F. = 1.6)$ $f'_c = 4$ ksi.

REFERENCES

11.1. AISC, *Manual of Steel Construction*, 8th ed. Chicago, 1980.

11.1. AASHTO, *Standard Specifications for Highway Bridges*, Washington, DC, 1977 (1978, 1979, 1980, and 1981 Interim Bridge Specifications).

11.3. ANDRÄ, W., and F. LEOHNARDT, "Neue Entwicklungen für Lager von Bauwerken Gummi- und Gummitopflager," *Die Bautechnik*, vol. 39, pp. 37–50, Feb. 1962.

12

Summary: Design of Simple Structures

12.1 INTRODUCTION

In Chapter 1 a brief outline of the design process was given. In the later chapters most design examples were limited to individual members. In this final chapter two practical examples are discussed to demonstrate how to apply the design of various individual elements so as to arrive at an integrated design of an assembled structure as a whole.

As any structure is designed to carry some external forces in the form of applied loads, local codes or ordinances which prescribe loads must be followed in assessing loading conditions. In all cases the code requirements are considered to be the minimum legal requirements.

As already discussed in Chapter 1, loads are either dead loads or live loads (service loads). Vertical loads produced by the weight of the structure itself and any component permanently attached to it constitute the dead loads. Throughout this text the difficulty of assessing dead loads has been stressed and several methods discussed on how to estimate these loads. The live load is usually also an approximation of the true service load. In buildings uniformly distributed loads of various intensities, also called occupancy loads, are mostly considered. Many codes include a separate requirement for concentrated loads (workers during erection or repair or some horizontal loads) to be placed on the system for certain occupancies where loads of this type are probable. Most structural engineering handbooks (for example, references

12.1, 12.2) give dead loads and various live loads (occupancy, wind, snow, seismic, crane, impact, vibration, and temperature loads), and the reader is referred to those.

For bridges the loads are prescribed by the corresponding specifications. As pointed out in Chapter 1, the designer must use the latest edition of any given specification, which are periodically revised (for example, references 12.3, 12.4).

In this chapter, two design examples will be given: design of a building and of a highway bridge.

12.2 SINGLE-STORY BUILDINGS

In Section 1.1.1.2 dealing with steel buildings and their parts, it was pointed out that all steel building structures fall into one of the following four different types: wall bearing (Figs. 1.3a and b), beam-column framing (Figs. 1.4a, b, c, d, e, and 12.2), shell-type (Figs. 1.7 and 12.1), or suspension system (Figs. 1.8 and 12.3). For the first type only the roof is constructed of steel. Shells and suspension systems are too advanced for a basic design text, and therefore are not discussed here. For that reason, the building in the first example is of the beam-column framing type, actually a single-story building.

12.2.1 Structural Concepts of Building Systems

Before starting to calculate loads and their "actions" or stresses and deflections the designer must select the material to be used and have a clear concept of the way in which the transfer of vertical and horizontal loads from the point of their application to the ground will take place. The soundness of this structural concept actually determines the success or the failure of a design, with respect to not only its safety, but also its economy. In any actual design situation the feasibility of various systems and the extent to which they reduce or increase the cost of other elements of the structure must be established first. The intelligent selection of the best concept can only be made after the criteria for a particular problem have been thoroughly investigated and defined.

In any framing system a deck element is supported by regularly spaced beams or joists, which in turn are supported by other beams or directly by columns or walls (Fig. 12.2). Vertical loads thus carried by these elements are finally transmitted to columns and to their foundations. Horizontal loads, wind forces, or any other dynamic forces are usually transferred by rigid floor or roof diaphragms directly to frames, provided that the diaphragm capabilities of the deck and the adequacy of the assembly details make this possible.

12.2.2 Design Example 12.1

Design a furniture warehouse of rectangular shape covering an area of 50 × 150 ft (15.24 × 45.72 m) with a possible extension in length in both directions. (See Fig. 12.4.) The main frame consists of a steel two-hinged portal frame 30 ft (9.14 m)

Figure 12.1 Steel Shell Roof (From *Modern Steel Construction*, vol. VIII, no. 3, 1968, pp. 10–11).

on center which supports a built-up roof of 2-hr fire rating; use open-web steel joists. For walls insulated, cold-formed steel or aluminum panels are to be used. The floor in the building consists of a concrete slab 8 in. thick directly resting on the soil. There are no elevated cranes, and a clearance to the ceiling of 20 ft–0 in. (6.1 m) should be preserved. Foundation conditions are favorable, and at a depth of 5 ft (1.5 m) below the floor level, for square-shaped isolated footings of about 50 ft^2, a load of 4 kips/ft^2

Figure 12.2 Steel Deck on Joists and Girders (From AISI, *Contemporary Structures*, p. 19).

(20.0 t/m² or 1.65 kN/m²) is safe. The building is located in the central midwest part of the United States.

SOLUTION

Performance requirements

Warehouses, like any other type of commercial building, must be designed and constructed in such a way as to ensure maximum utilization of floor space in accordance with the physical characteristics peculiar to the commodities it will be expected

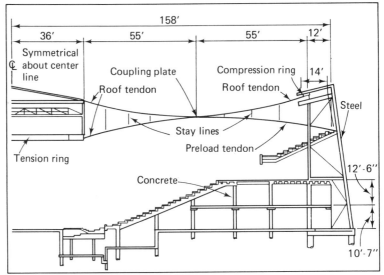

(a) Tendons connected at midpoint of span with coupling plates

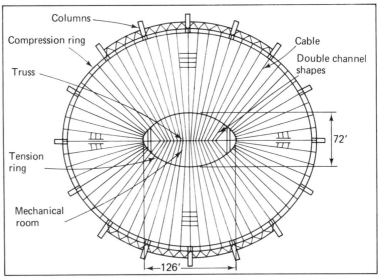

(b) Cables closer together at ends of tension ring

Figure 12.3 Elliptical, Cable-hung Roof of Activities Center—Structural Engineer, Lloyd W. Abbott, Tulsa, Okla. (From *ENR*, Jan. 13, 1972, p. 16).

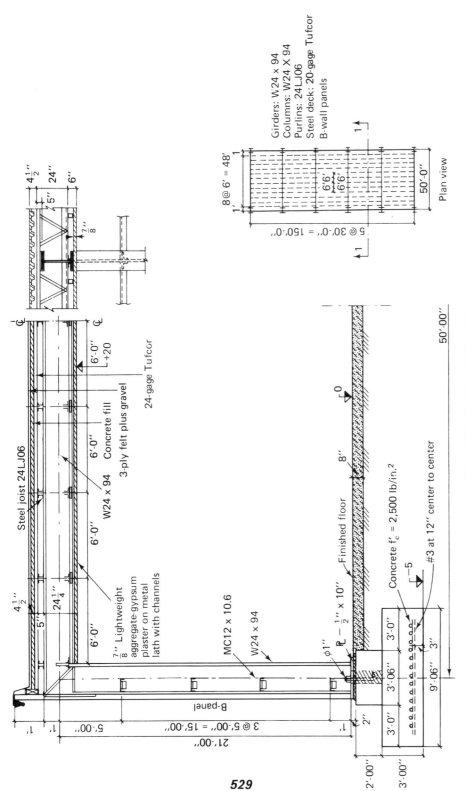

Figure 12.4 Section and Plan View.

to house. Furniture requires a single large area free of obstructions. Therefore, no interior columns are to be used. Thus a structural system consisting of a two-hinged frame is a satisfactory structural system in this case. To reduce the number of these frames to a minimum, a 30-ft spacing was chosen which will require only six equal frames. Given the expressed intention of a possible future extension of the building length, frames will also be used at the building ends. For a 30-ft (9.14-m) span, as purlins open-web steel joists are chosen, spanning between frames as simply supported beams spaced at 6 ft on center maximum.

A very important consideration is to provide sufficient glass area for adequate light intensity inside the building. This study will be omitted in this example.

The structural system will now consist of steel deck panels acting as continuous beams over more than three spans which will transmit dead and live load to open-web steel joists. These in turn will transfer loads across 30-ft simple beam spans together with the additional load due to their own weight plus that of the suspended ceiling to the main two-hinged frames. Frames will carry these loads plus their own weight as well as any horizontal loads acting on tributary areas of 30-ft width and transmit those to the foundations and through them to the soil.

Local conditions

Roofing

As already stated the steel deck forms the main element of the roof. Granco (12.5), Strongform (12.6), Smith (12.7), Inryco (12.8), and many other systems can be applied. Our local conditions require a 2-hr fire rating. To protect the open-web joists and the steel deck a suspended ceiling should be used. To obtain the necessary slope for roof drainage a light weight insulating concrete poured on a high-tensile (say, Granco's Tufcor) steel deck will be used. Because of the protection afforded by the suspended ceiling no wire mesh will be used. For a slope of $\frac{1}{8}$ in. in 12 in. (or 1%) the concrete thickness varies by a total amount of 3 in., resulting in an average total thickness of insulating concrete of $4\frac{1}{2}$ in. and yielding an average thickness of $3\frac{13}{16}$ in. (Fig. 12.5). The dead load on the steel deck is

Three-ply felt and gravel:	5.5 lb/ft²
Lightweight concrete fill:	6.3 lb/ft²
Tufcor, gage 20:	2.1 lb/ft²
	13.9 lb/ft² (66.6 kN/m²) or 83.4 lb/ft (1,217.1 kN/m) of joist

Joists

Armco steel joists of 24LJ-series will be used—that is, 24 in. of nominal depth and made of A36 steel of 19 lb/ft weight. Table 12.1 gives for a clear span of 33 ft as a safe total load 443 lb/ft (controlled by deflection). The dead load on the joists is

Roofing:	6 × 13.9 =	83.4 lb/ft
Joist:		19.0 lb/ft
Suspended ceiling:	6 × 10 =	60.0 lb/ft
		162.4 lb/ft—say, 165 lb/ft (2,410 kN/m)

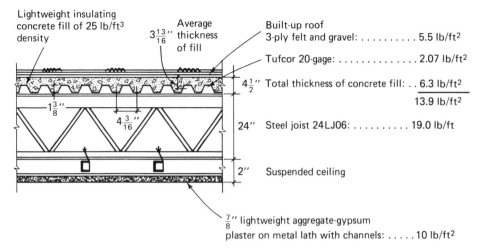

Figure 12.5 Built-up Roof with Suspended Ceiling for 2-hour Fire Rating.

Frame girder

The approximate depth of the girder of the main frame is estimated (as given in Section 10.1.3) to be

$$\frac{L}{35} = \frac{600}{35} = 17.14 \text{ in.} \approx 18 \text{ in. (45.7 cm)}$$

Because of the large distance between the frames (30 ft) an increase of 33% is taken; that is, take 24 in.

This would indicate that a wide-flange section of the 24 series could be applied. From the AISC *Manual's* tables of properties (12.14) the weight of a W24 lies between 55 and 162 lb/ft, with an average of about 100 lb/ft. For a 2-hr fire rating a sprayed mineral fiber insulation of $1\frac{1}{4}$-in. thickness is required only for the frame columns because the girder is protected by the suspended ceiling.

Frame columns

The total approximate vertical reaction of the frame leg is estimated as follows

From snow + live load:	$45 \times 30.0 \times 25.0 \times 10^{-3} =$	33.75 kips
From roofing:	$13.9 \times 30.0 \times 25.0 \times 10^{-3} =$	10.42 kips
From joists:	$19 \times 30.0 \times 6 \times 10^{-3} =$	3.42 kips
From suspended ceiling:	$10.0 \times 30.0 \times 25.0 \times 10^{-3} =$	7.50 kips
From girder:	$100.0 \times 25.0 \times 10^{-3} =$	2.50 kips
From column:	$100.0 \times 20.0 \times 10^{-3} =$	2.00 kips
	Total:	59.59 kips (265.1 kN)

This relatively small value would indicate that bending and not combined bending and compression, (i.e., beam-column action) will be the controlling factor. With girts for the wall panels at 5-ft intervals no lateral stability problems are expected. Therefore, a 24W rolled section of the same section as the girder is assumed.

Single-Story Buildings

Table 12.1 STANDARD LOAD TABLE/LONGSPAN STEEL JOISTS, LJ-SERIES (Based on allowable stress of 22,000 psi)

Joist Designation	Approx. Wt. in Lbs. per Linear Ft.	Depth in Inches	SAFE LOAD in Lbs. Between	CLEAR OPENING OR NET SPAN IN FEET																			
				22–24	25	26	27	28	29	30	31	32	33	34	35	36	37	38	39	40			
20LJ03	14	20	9900		388/368	369/350	351/331	334/304	318/280	304/258	290/239	276/220	260/200	245/184	232/169	219/156	208/143	197/133	187/123	178/114			
20LJ04	16	20	11900		464/440	439/417	416/390	395/357	375/328	356/300	337/272	317/248	298/227	281/208	266/190	251/175	238/161	226/150	215/138	204/128			
20LJ05	17	20	13000		510/484	485/460	461/430	439/395	419/364	399/336	381/310	364/287	346/265	326/242	308/222	291/204	276/189	262/174	249/162	237/150			
20LJ06	19	20	15400		602/571	579/550	559/521	530/474	498/427	466/386	437/351	410/320	386/292	364/267	344/246	326/226	308/209	293/192	278/178	265/165			
20LJ07	20	20	16500		643/610	619/588	597/553	576/515	557/481	521/435	489/396	459/360	432/329	408/302	385/277	365/255	345/235	328/217	311/201	296/186			
20LJ08	22	20	17900		699/664	673/639	648/600	626/560	605/520	585/486	567/442	533/402	502/367	473/337	447/309	423/285	401/262	380/242	361/225	344/209			
20LJ09	24	20	18600		726/689	699/664	674/624	650/578	628/540	608/505	588/473	570/437	554/399	522/366	493/336	467/309	442/285	420/264	399/244	379/227			
20LJ10	27	20	20100		783/743	753/715	726/669	701/623	677/581	655/544	634/510	615/477	597/448	580/411	548/377	519/346	491/320	466/296	443/274	421/254			
20LJ11	29	20	20700		807/766	776/737	748/689	722/642	698/599	675/561	654/523	634/492	615/463	597/436	580/410	565/383	550/354	522/326	496/303	472/281			
20LJ12	32	20	22100		861/817	829/787	799/736	771/685	745/637	721/596	698/558	677/525	656/491	637/463	620/438	603/414	587/383	572/355	543/328	517/304			
20LJ13	36	20	23100		901/855	867/823	836/770	807/714	779/665	754/623	730/584	708/546	687/514	667/485	648/457	631/433	614/408	598/387	583/368	569/348			
				28–32	33	34	35	36	37	38	39	40	41	42	43	44	45	46	47	48			
24LJ04	16	24	11000		329/307	315/285	302/266	290/248	279/232	268/217	258/204	248/190	236/176	225/164	215/152	206/143	197/133	188/125	181/117	173/110			
24LJ05	17	24	12100		361/335	347/313	334/292	322/274	310/257	297/240	284/223	271/208	259/194	248/181	237/169	227/158	218/149	209/140	201/131	194/124			
24LJ06	19	24	14900		443/411	424/382	407/356	390/332	375/308	356/285	339/264	322/245	307/228	293/211	279/197	267/184	255/172	244/161	234/152	225/142			
24LJ07	20	24	16400		488/450	468/419	449/391	431/365	414/341	398/320	379/297	360/275	343/255	327/238	312/222	298/207	285/194	273/181	262/170	251/160			
24LJ08	22	24	19300		574/527	549/490	525/455	503/421	482/388	460/359	438/333	418/308	399/286	380/267	363/249	347/232	322/218	318/204	305/191	292/180			
24LJ09	24	24	20500		609/559	592/528	575/498	560/460	537/424	513/393	487/363	464/337	442/313	421/292	402/272	384/254	368/238	352/223	337/209	324/196			
24LJ10	27	24	21700		646/593	627/559	610/528	593/497	577/471	563/439	540/406	514/378	490/351	467/326	446/304	426/285	408/266	391/249	374/234	359/220			
24LJ11	29	24	22800		680/624	660/586	641/553	624/523	607/495	592/471	577/445	563/418	542/388	521/361	499/337	477/315	457/294	437/276	419/259	402/243			
24LJ12	32	24	24400		727/664	706/626	686/591	667/559	650/528	633/501	617/476	602/453	587/421	574/392	548/366	524/342	501/320	480/300	460/281	441/264			
24LJ13	36	24	25200		648/683	727/645	706/606	687/573	669/543	651/515	635/490	619/465	604/443	590/421	577/402	564/384	552/366	528/342	506/322	486/302			
24LJ14	38	24	26200		780/712	757/669	736/632	716/598	697/566	679/537	662/510	646/484	630/460	615/439	601/419	588/401	575/383	563/359	539/337	517/317			

SOURCE: Steel Joist Institute and AISC Standard Specifications and Tables, pp. 31–2 Arlington, Va., 1975. Reprinted by permission.

Wall panels

A Smith metal wall of the B-panel system (Fig. 12.6) is chosen. From Table 12.2, 0.040-in. aluminum is chosen. As the maximum length of the exterior panels is 40 ft > 23 ft, this is satisfactory. For insulation 1-in.-thick glass fiber board is selected. The completely assembled wall system including side joints has a U-factor not exceeding 0.15. The vinyl sealing gaskets will be factory-applied to the interlocking side joints of the interior panel.

Table 12.2 LOAD SPAN TABLE (Shadowall exterior)

Insulated B-panel			Maximum allowable span						
Interior	Exterior No. Spans		0.032" Aluminum	0.040 Aluminum	0.050" Aluminum	24 Gage Steel	22 Gage Steel	20 Gage Steel	18 Gage Steel
0.032" Aluminum	1		6'-4"	6'-5"	6'-5"	6'-6"	6'-7"	6'-8"	6'-11"
	2		8'-0"	8'-0"	8'-0"	8'-0"	8'-0"	8'-0"	8'-0"
0.040" Aluminum	1		6'-9"	6'-10"	6'-10"	6'-11"	7'-1"	7'-1"	7'-3"
	2		8'-0"	8'-0"	8'-0"	8'-0"	8'-0"	8'-0"	8'-0"
0.050" Aluminum	1		7'-3"	7'-4"	7'-4"	7'-5"	7'-6"	7'-7"	7'-9"
	2		8'-0"	8'-0"	8'-0"	8'-0"	8'-0"	8'-0"	8'-0"
24-gage steel	1		7'-2"	7'-2"	7'-3"	7'-4"	7'-5"	7'-5"	7'-7"
	2		8'-0"	8'-0"	8'-0"	8'-0"	8'-0"	8'-0"	8'-0"
22-gage steel	1		7'-11"	7'-11"	7'-11"	8'-0"	8'-0"	8'-0"	8'-0"
	2		8'-0"	8'-0"	8'-0"	8'-0"	8'-0"	8'-0"	8'-0"
20-gage steel	1		8'-0"	8'-0"	8'-0"	8'-0"	8'-0"	8'-0"	8'-0"
	2		8'-0"	8'-0"	8'-0"	8'-0"	8'-0"	8'-0"	8'-0"
18-gage steel	1		8'-0"	8'-0"	8'-0"	8'-0"	8'-0"	8'-0"	8'-0"
	2		8'-0"	8'-0"	8'-0"	8'-0"	8'-0"	8'-0"	8'-0"

Wind Load = 20 lb/ft²   Deflection Limitation = $\frac{1}{180}$

NOTE: Values are for Shadowall exterior profile.
SOURCE: Elwin G. Smith Division, Cyclops Corporation, *Metal Wall and Roof Systems*, Catalog 75A17, pp. 10–11, Pittsburgh, Pa., 1975.

Design

Roofing

Table 12.3 [Granco catalog (12.5), table 4, p. 31] shows maximum allowable loads if continuity of the steel deck over three spans is assumed. In our case with 25-ft-long deck panels this is always true. In addition to dead load there are also snow loads and a minimum live load on flat roofs of 20 lb/ft² (Uniform Building Code, International Conference of Building Officials, 1979). The snow load for the central midwest United States on roof slopes less than 1 in 4 is 25 lb/ft². Therefore, the total load on the steel deck is

Dead load: 13.9 lb/ft² (from calculation of dead load on the steel deck)
Snow load: 25 lb/ft²
Live load: 20 lb/ft²
58.9 ≈ 60 lb/ft² (287 kN/m²)

Single-Story Buildings

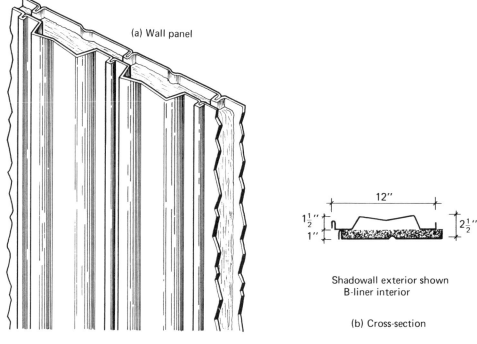

Figure 12.6 B-panel System (From Elwin G. Smith Division, *Metal Wall and Roof Systems*, Catalog 75A17, p. 10, Pittsburgh, Pa., 1975).

From Table 12.3 a Tufcor of 26 gage can be chosen, as 65 lb/ft² > 60 lb/ft² on 6-ft–0-in. spans (center to center). We shall use galvanized 24-gage Tufcor of 1.41 lb/ft² weight.

Steel joist

Loads per linear foot of joist are
Dead load: 165 lb/ft (from calculation of dead load on the joists)
Snow load: $25 \times 6 =$ 150 lb/ft
Live load: $20 \times 6 =$ 120 lb/ft
                     435 lb/ft < 443 lb/ft (6465.1 kN/m) (from Table 12.1)

Also

Maximum moment: $M_{max} = 0.44 \times \frac{(30)^2}{8} = 49.5$ kip-ft (67.1 kN-m)

Maximum stress: $f_b = \frac{49.5 \times 12}{465} \times 12.00 = 15.33 \, \text{ksi} \, (105.7 \, \text{MPa}) < 22 \, \text{ksi} \, (151.7 \, \text{MPa})$

Maximum deflection: $\delta_{max} = (1.15) \frac{5(0.44)(30)^4(12)^3}{384(29,000)(465)} = 0.68$ in. $< \frac{30 \times 12}{360} = 1.00$ in. (2.54 cm)

Use 24LJ06 joists, 19 lb/ft (277.3 kN/m)

Table 12.3 TUFCOR STEEL DECKS

MAXIMUM ALLOWABLE LOADS (lb/ft²)—THREE-SPAN CONDITION

Product	Spans (Center to center)															
	3′–0″	3′–6″	4′–0″	4′–6″	5′–0″	5′–6″	6′–0″	6′–6″	7′–0″	7′–6″	8′–0″	8′–6″	9′–0″	9′–6″	10′–0″	10′–6″
Standard Corruform	99	73	56	44												
26-gage Tufcor		192	147	116	94	78	65	56								
24-gage Tufcor				165	134	110	93	79	68	59	52	46				
22-gage Tufcor					170	141	118	101	87	76	67	59	53	47	43	38
20-gage Tufcor							141	121	104	91	80	70	63	56	51	46

NOTES: 1. All values based on Tufcor and Corruform supporting total load in flexure without benefit of any composite action between concrete fill and Tufcor deck.
2. Values given are total allowable loads in lb/ft².
3. Values for stress are based on limiting stress of 30 ksi, and are derived by $M = wl^2/10$.

PHYSICAL PROPERTIES

Description	Unit	Standard Corruform	26-gage Tufcor	24-gage Tufcor	22-gage Tufcor	20-gage Tufcor
Nominal corrugation pattern (pitch × depth)	(in.)	$2\frac{1}{2} \times \frac{9}{16}$	$4\frac{5}{8} \times 1\frac{5}{16}$	$4\frac{9}{16} \times 1\frac{5}{16}$	$4\frac{9}{16} \times 1\frac{5}{16}$	$4\frac{9}{16} \times 1\frac{5}{16}$
Nominal cover width	(in.)	30	28	32	32	32
Weight (based on galvanized coating)	lb/ft.²	0.85	1.11	1.41	1.74	2.07
Moment of inertia (per ft width)	in.⁴	0.010	0.067	0.091	0.114	0.137
Section modulus (per ft width)	in.³	0.034	0.094	0.134	0.170	0.204

SOURCE: Granco Steel Products, *Floor/Roof Construction*, Catalog 99–10, no. 5: *Metal Decking*, St. Louis, 1972, tables 3 and 4, p. 31.

Main frame

The frame will be analyzed for two loading conditions

1. Full gravity load: $D + L + S$, where D = dead load, L = live load, S = snow load.

2. Three-quarters gravity plus wind load: $\frac{3}{4}(D + L + \frac{1}{2}S + W)$.

LOAD CONDITION 1: $[D + L + S]$ (Fig. 12.7). Each joist will cause a reaction of
$$P = 0.44 \times 30.0 = 13.2 \text{ kips (58.7 kN)}$$
This load is converted to a uniformly distributed load of
$$w = \frac{13.2}{6} = 2.2 \text{ kips/ft (32.1 kN/m)} \quad \text{of girder} \quad \text{(Fig. 12.7b)}$$
With the weight of the girder itself assumed to be 0.1 kip/ft, the total load on the frame is
$$w = 2.3 \text{ kips/ft (33.6 kN/m)}$$
The reactions and bending moments are given in Fig. 12.7b. Assuming an adequately braced compact section of A36 steel with $F_b = 24$ ksi, the minimum required section modulus is $S_x = 187.2$ in.³. A preliminary selection for the column and the girder is now a W24 × 84.

LOAD CONDITION 2: $[\frac{3}{4}(D + L + \frac{1}{2}S + W)]$ (Fig. 12.8). The basic wind load (horizontal pressure on vertical surfaces) is (Uniform Building Code, International Conference of Building Officials, 1979 edition) 30 lb/ft². The shape factors for the windward and leeward walls are, for $h/w = \frac{21}{150} = 0.14$, 0.7 and −0.4. Therefore, the wind load is

On windward wall: $\quad \frac{3}{4} \times 0.7 \times 30 = 15.75$ lb/ft²
On leeward wall: $\quad -\frac{3}{4} \times 0.4 \times 30 = -9.00$ lb/ft²

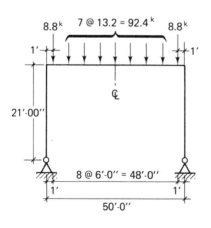

(a) Actual loading

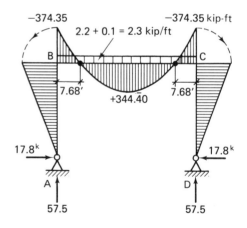

(b) Approximated loading (D + L + S)

Figure 12.7 Loading Condition 1.

The intensity of the wind load is

On windward wall: $15.75 \times 30.00 \times 10^{-3} = 0.473$ kip/ft (6.9 kN/m) of column

On leeward wall: $-9.00 \times 30.00 \times 10^{-3} = -0.270$ kip/ft (3.94 kN/m) of column

The intensity of the vertical load is

$$\tfrac{3}{4}[2.3 - \tfrac{1}{2}(25 \times 30.0 \times 10^{-3})] = 1.44 \text{ kips/ft (21.0 kN/m)} \quad \text{of girder}$$

The bending moments and reactions are shown in Fig. 12.8a for $\tfrac{3}{4}$ the wind load only and for load condition 2 in Fig. 12.8b. Comparing bending moment diagrams it is seen that loading case 1 controls the design.

Final design of frame

Girder

Although the girder is also a beam-column, due to the fact that the axial force in our case is small, only 17.8 kips, producing a small axial compressive stress of $f_a = 17.8/24.7 = 0.72$ ksi, the girder can be considered in bending only.

The trial section was a W24 × 84 with $A = 24.7$ in.^{2}, $S_x = 197$ in.^{3}, $r_x = 10.7$ in., $r_y = 2.06$, and $b_f = 9.015$ in. The lateral bracings are at 6-ft intervals. This section is compact and adequately braced because

$$12.7 b_f = 12.7 \times 9.015 \times \tfrac{1}{12} = 9.54 \text{ ft} > 6.0 \text{ ft}$$

and

$$\frac{555.6}{3.46 \times 12} = 13.38 \text{ ft} > 6.0 \text{ ft}$$

Therefore, $\tfrac{9}{10}$ reduction of the negative moment can be utilized. But because

$$\tfrac{9}{10}(374.35) = 336.92 \text{ kip-ft} < 344.40 \text{ kip/ft}$$

the reduction will be not 0.9, but only as much as needed to equalize the absolute values of the extreme negative and positive moments—that is, to a value 359.38 kip-ft (487.32 kN/m).

Now

$$(S_x)_{rqd} = \frac{M_{max}}{F_b} = \frac{359.38 \times 12}{24} = 179.7 \text{ in.}^3 \rightarrow \underline{\text{W24} \times 84}$$

Column

The frame is unbraced against sidesway. To get the effective length factor, K_x, from the chart on p. 5-125 in the commentary of the AISC specification (10.3) coefficients G_A and G_B have to be computed. As A is a hinge, $G_A = 10$, and having the same section in column and girder,

$$G_B = \frac{1/21}{1/50} = 2.38$$

from the AISC chart, K_x is obtained as $K_x = 2.18$. Buckling in the bending plane (frame plane) is controlling because

$$\frac{r_x}{r_y} \times L_y = \frac{10.7}{2.06} \times 5.0 = 25.97 \text{ ft} < 2.18 \times 21.0 = 45.8 \text{ ft}$$

Single-Story Buildings

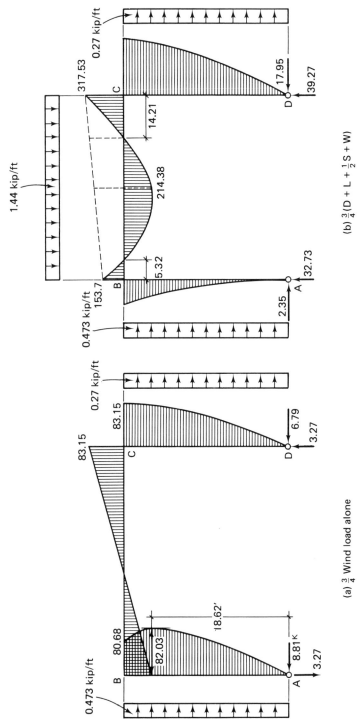

Figure 12.8 Loading Condition 2.

Therefore, for
$$\frac{K_x L_x}{r_x} = \frac{2.18 \times 21 \times 12}{10.7} = 51.3$$
from table 3-36 of the AISC *Manual* the allowable compression stress is $F_a = 18.23$ ksi. The actual bending stress is
$$f_{bx} = \frac{374.35 \times 12}{197} = 22.8 \text{ ksi } (157.2 \text{ MPa}) < 24.0 \text{ ksi } (165.5 \text{ MPa})$$
The compression stress is $f_a = 57.5/24.7 = 2.33$ ksi. As $f_a/F_a = 2.33/18.23 = 0.13 < 0.15$, AISC *Manual* equation (1.6-2) (eq. 8.16) is used instead of equation (1.6-1a) (eq. 8.14) or equation (1.6-1b) (eq. 8.15 in this text) and
$$0.13 + \frac{22.8}{24.0} = 1.08 > 1.00: \quad \underline{\text{N.G.}}$$
If the section is increased to a W24 × 94 with $A = 27.7$ in.^{2}, $S = 222$ in.^{3}, $r_x = 9.87$ in., and $r_y = 1.98$ in., then
$$\frac{K_x L_x}{r_x} = \frac{2.18 \times 21.0 \times 12}{9.87} \times 55.7; \quad F_a = 17.84 \text{ ksi } (123 \text{ MPa})$$
$$f_a = \frac{57.5}{27.7} = 2.08 \text{ ksi } (14.3 \text{ MPa}); \quad f_{bx} = \frac{374.35 \times 12}{221.0} = 20.33 \text{ ksi } (140.2 \text{ MPa})$$
and $f_a/F_a = 2.08/17.84 = 0.12 < 0.15$.

Checking formula (1.6-2) (eq. 10.16 in this text):
$$0.12 + \frac{20.33}{24} = 0.96 < 1.00: \quad \underline{\text{OK}}$$

Use for frame W24 × 94

The bottom chord of all joists are to be connected to the bottom flange on both sides of the girder web (see Fig. 12.4) using short angles.

Corner connection (Fig. 12.9)

The shear transfer in a square knee (Fig. 12.9b), under the assumption that only the web takes shear and the flanges only bending, is
$$T = \frac{M}{0.95 d_g} = V_{ab} = F_v t_w d_c$$
The required t_w is therefore
$$t_x = \frac{M}{0.95 d_g d_c F_v} = \frac{374.35 \times 12}{0.95(24.29)(24.29)0.40(36)}$$
$$= 0.556 \text{ in. } (1.41 \text{ cm}) > 0.516 \text{ in. } (1.31 \text{ cm})$$
and a diagonal stiffener must be used. Its cross section is obtained from the following equation:
$$V_{ab} + T_{stf} \cos 45° = T$$
$$A_{stf} = \frac{1}{\cos 45°} \left[\frac{M}{0.95 d_g (0.60) F_y} - \frac{0.4 F_y}{0.6 F_y} t_w d_c \right] = 2.83 \text{ in.}^2 (18.26 \text{ cm}^2)$$

Use for stiffener PL–$\frac{3}{4}$ in. × 4 in.

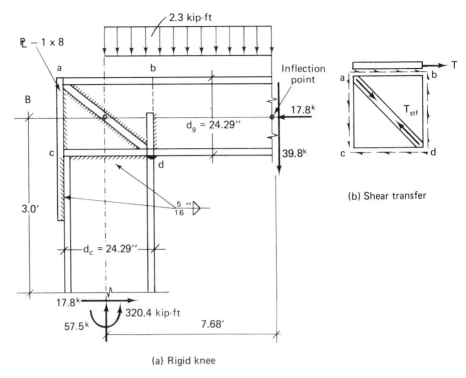

Figure 12.9 Corner Connection.

At point d it is proposed to groove-weld the flanges and fillet-weld ($a = \frac{5}{16}$ in.) the web of the column. Splice PL–1 in. × 8 in. is fillet-welded ($a = \frac{5}{16}$ in.) and tapered toward point a.

Foundations

Hinged column base (Fig. 12.10)

Although the column base was considered to be pinned (see Chapter 11, Section 11.2), only in cases when the reactions become very large (for frame spans larger than 150 ft for normal loads) must the column base have a hinge; otherwise it is executed as a flat-column class I base plate. To ensure that it will not offer any substantial moment restraint, the anchor bolts are placed in the column axis. The anchor bolts are always used irrespective of the type of column end, because they make the erection of columns much easier. The anchors are hooked to a small angle, $\lfloor 1 \times 1 \times \frac{1}{4}$, already in the foundation (Fig. 12.10) during the erection, and tightened when the column is plumb (regulated by steel wedges). Next the empty channels and a cushion of about 2 in. thickness are concreted.

For the plate dimensions shown in Fig. 12.10 the bearing pressure is

$$f_p = \frac{57.5 + 0.094 \times 20.0}{27 \times 10} = \frac{59.38}{270} = 0.22 \text{ ksi (1.52 MPa)}$$
$$< 0.35 f'_c = 1.05 \text{ ksi (7.24 MPa)}$$

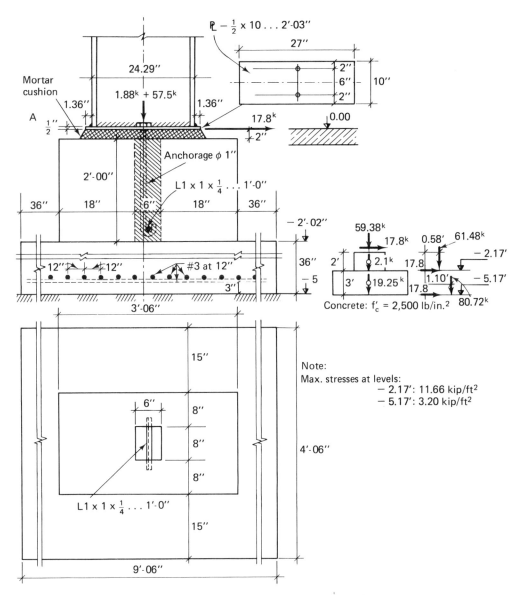

Figure 12.10 Hinged Column End and Foundation.

The plate thickness can be obtained by considering the cantilevered edge (1.36 in.) or the plate between the column flanges as a plate supported along three edges and free along its fourth edge. Considering a 1-in.-wide strip of the cantilevered edge, the thickness is

$$t = 1.36\sqrt{\frac{3 \times 0.22}{22}} = 0.24 \text{ in. } (0.61 \text{ cm})$$

which is too small to produce any uniform stress distribution.

Single-Story Buildings 541

A three-edge plate analysis will yield similar impractical small values. Therefore, we shall use a $\frac{1}{2}$-in.-thick (1.27-cm) plate.

Concrete footing

Concrete for footing, $f'_c = 2,500$ psi (17.2 MPa), at the -2.17-ft (-0.66 m) level (in concrete) stress is

$$f = \frac{61.48}{10.50} \pm \frac{17.8(2.0)}{6.125} = 5.85 \pm 5.81 = 11.66 \text{ kips/ft}^2 = 0.08 \text{ ksi} (.56 \text{ MPa})$$

$$\text{or} \quad +0.04 \text{ kip/ft}^2 = 0.0003 \text{ ksi} (0.002 \text{ MPa})$$

and at the -5.17-ft (-0.66 m) level (on soil) is

$$f = \frac{80.72}{42.75} \pm \frac{17.8(5.0)}{67.69} = 1.89 \pm 1.31$$

$$= 3.20 \text{ kips/ft}^2 (0.15 \text{ MPa}) < 4.0 \text{ kips/ft}^2 (0.19 \text{ MPa}): \quad \underline{\text{OK}}$$

$$\text{or} \quad 0.58 \text{ kip/ft}^2 (0.028 \text{ MPa})$$

The concrete footing is reinforced in both directions by #3 bars at 12 in. center to center. There is no tension anywhere in the foundation.

12.3 BRIDGES

As a second design example we shall use the design of a highway bridge. As an introduction we shall first very briefly discuss typical girder bridge structures.

12.3.1 Girder Bridges

In Section 1.1.1.3 on steel bridges and their parts (Figs. 1.10 and 1.15), it was pointed out that girder type bridges, of simple or box section, and trusses are the two main types of bridge structures. For spans of small and moderate length (250–300 ft) (75–90 m) simple girders are the main carrying bridge members. For larger spans (over 300 ft [90 m]), when there is sufficient clear height available, trusses will take over from girders. Box girders are used even for moderate spans if esthetics or other local conditions require it, as well as for really large spans (over 800 ft [245 m]).

Simple girders are now usually welded plate girders with a reinforced concrete slab on top, known as deck bridges (Fig. 1.15). The number of parallel girders depends upon the roadway width. Such a typical highway girder bridge is discussed in the next example.

12.3.2 Design Example 12.2

Design the steel superstructure for a two-span continuous highway bridge for HS20-44 load and two traffic lanes with a roadway of 26 ft (7.92 m) and with safety curbs. Each span equals 120 ft (36.6 m). There is no composite action between the 7-in. (17.8 cm²) slab of $f'_c = 3,000$ lb/in.² (20.7 MPa) and the girders. Assume a future concrete wearing surface of $1\frac{1}{2}$ in. (3.8 cm) on top of the slab. Use A36 steel (M183) E-70 electrodes, and the 1977 AASHTO specifications with interim specifications (12.3). The bridge is on a major highway with the following stress cycles:

For truck loading: 5×10^5

For lane loading: 10^5

(This corresponds to case II in the AASHTO specifications, 1977, p. 151.) There are no sidewalks for pedestrians. There is also no danger of earthquakes. The depth of the girders should be kept to around 5 ft.

SOLUTION

In Fig. 12.11a the cross section of the bridge superstructure is shown. From *Standard Plans for Highway Bridges*, vol. 2 (12.10), a type with several girders (here four) was chosen. Girder B (Fig. 12.11a) will be designed and all four girders fabricated in the same way. This is very often cheaper than separate fabrication of exterior and interior girders. Service load design method (allowable stress design) will be used.

Load analysis

Dead load

To get a first approximation of the dead load of the girder itself, it is often useful to develop a cross section based on known proportions, starting from an assumed web depth (as discussed in Section 10.13), and to assess the weight in lb/ft of the section obtained in this way. The AASHTO specifications 1977 (section 1.7.4, "Depth Ratios") require for beams or girders that the girder depth to span length ratio preferably be not less than $\frac{1}{25}$. For continuous spans the span is the distance between dead load points of contraflexure. If a ratio of $\frac{1}{24}$ is assumed in conformity with the 5-ft limit for the real span length, then we have

$$D = \frac{L}{24} = \frac{120 \times 12}{24} = 60 \text{ in. } (152.4 \text{ cm})$$

where D = web depth. Because the depth is pretty much restricted by $L/24$, a ratio of $D/b_f = 3$ is chosen and a flange width of $b_f = 20$ in. (50.8 cm) ($L/72$) is obtained. Taking from AASHTO, 1977 (section 1.7.43(B), "Flanges") $b_f/t_f = 20$, 1-in. (2.54 cm) plate thickness is obtained. For web thickness calculation [AASHTO, 1977 section 1.7.43(C)], we select $D/t_w = 160$, as 170 is the maximum. Thus the web thickness is $\frac{60}{160} = \frac{3}{8}$ in. (0.95 cm) $< \frac{5}{16}$ in. (0.79 cm), which according to AASHTO 1977 (section 1.7.7) is the minimum thickness of structural steel. The cross-sectional area of such a girder (Fig. 12.11c) is 62.5 in.^{2} (403 cm^2) (or 212.5 lb/ft [3.1 MN/m]). This section represents the average section, with the understanding that near the abutments and the midpier the section will change to become eventually lighter and heavier, respectively. If the weight of stiffeners, splices, and bracings is taken to be approximately 17%, the first trial weight of one girder is

$$1.17 \times 212.5 = 248.6 \approx 250 \text{ lb/ft} = 0.25 \text{ kip/ft } (3.65 \text{ kN/m})$$

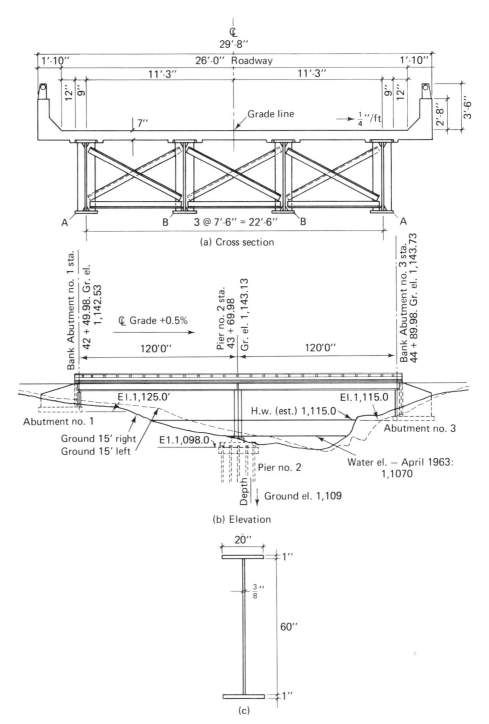

Figure 12.11

Other dead loads are

$$\text{Slab:} \quad 7.5 \times \frac{7}{12} \times 0.15 = 0.656 \text{ kip/ft}$$

$$\text{Future wearing surface:} \quad 7.5 \times \frac{1.5}{12} \times 0.15 = 0.141 \text{ kip/ft}$$

$$\text{Haunch around girder:} = 0.030 \text{ kip/ft}$$

$$\text{Barrier curb:} \quad \tfrac{1}{2} \times 0.40 = 0.200 \text{ kip/ft}$$

$$w_D = 1.277 \text{ kips/ft} \approx \underline{\underline{1.28 \text{ kips/ft}(18.7 \text{ kN/m})}}$$

Live load

The AASHTO 1977 specifications (section 1.3.1, Distribution of Wheel Loads to Longitudinal Beams) require that due to the stiffness of the deck and partial grid work action caused by the girders, live load bending moments for each interior girder must be found by multiplying the wheel (not axle) load by the fraction $S/5.5$ for concrete on steel girder, where S is the distance between the girders. Thus

$$\frac{7.5}{5.5} = 1.364; \quad \text{for two wheels:} \quad \frac{7.5}{2 \times 5.5} = 0.682 \text{ lanes}$$

Truck-trailer loading HS20-44 or lane loading HS20-44 will be used. The impact factor for HS20-44 loading (AASHTO, 1977 section 1.2.12, Impact) is

$$I = \frac{50}{120 + 125} = 0.204 < 0.300$$

Structural analysis (moments and shears)

Preliminary analysis

For a constant moment of inertia for girders, the easiest way of calculating the maximum bending moments and shear forces is to use the AISC tables (12.11). A variable moment of inertia (stiffer section at interior support) will increase moments to some extent, and it is advisable to use some existing programs like STRESS or STRUDL; but for input at least the ratio of moments of inertia has to be known. To assess properly such "relative" moments of inertia it is best to obtain bending moments from the AISC handbook (12.11) in order to compute the ratio of the bending moments. If this is done for a preliminary design using AISC table 2.0 (12.11), the value obtained for maximum positive moment is

D.L.: $\quad 0.0703 \times 1.28(120)^2 = 1,295.8 \approx 1,300$ kip-ft
L.L. + I: $\quad 1,530.3 \times 1.204 \times 0.682 = 1,256.6 \approx 1,260$ kip-ft
$\qquad\qquad\qquad$ Total: $\qquad\qquad \approx 2,560$ kip-ft (3,470 kN-m)

and that for maximum negative moment (at the pier) is

D.L.: $\quad -0.1250 \times 1.28(120)^2 = -2,304.0 \approx -2,300$ kip-ft
L.L. + I: $\quad -1,567.7 \times 1.204 \times 0.682 = -1,287.3 \approx -1,300$ kip-ft
$\qquad\qquad\qquad$ Total: $\qquad\qquad \approx -3,600$ kip-ft (4,880 kN-m)

The ratio K_2 is thus

$$K_2 = \frac{3,600}{2,560} = 1.41$$

Final analysis

A computer program developed by Howard, Needles, Tammen, and Bergendoff in their Kansas City office (12.12) was used, and part of the output is shown in Fig. 12.12—that is, the envelopes for shear forces and bending moments.

It can be seen that the final moments are increased as compared to the preliminary extreme values as a result of variable moments of inertia.

The computer output also gives the areas of the flange plates for a depth of 60 in. according to the programmed restraints for allowable stresses—that is, to 1965 AASHTO with 1967 modification, and 1966 AWS (computer run was in May 1969).

Table 12.4 shows these results, including the calculated allowable stresses.

Table 12.4 FLANGE AREAS AND STRESSES

```
HNTB-KC  126           REF. NO.   1 - 4       BY: AM 19 05 69    JOB NO.      0  0  0
FILE NO.                                                         SEC. NO.
RUN    05/20/69        SPECIAL PROBLEM "TWO SPAN BRIDG DESIGN    SHEET NO.   16


NON-COMPOSITE PLATE AREA REQUIRED A36  F=20.0 KSI   MAT. CODE= 6   CYCLE CODE= 2
1965 AASHO W/1967 MOD.)
                                         POINT                            LD.MOD.FAC.
     SPAN   0.1    0.2    0.3    0.4    0.5    0.6    0.7    0.8    0.9    1.0    A    B

       1    7.4   15.0   19.1   20.0   17.9   14.5   10.0   10.0   21.4   37.8  1.00 1.00


ALLOWABLE DESIGN STRESSES           A36   F=20.0 KSI   MAT. CODE= 6   CYCLE CODE= 2
(1965 AASHO W/1967 MOD.)
                                         POINT                            LD.MOD.FAC.
     SPAN   0.1    0.2    0.3    0.4    0.5    0.6    0.7    0.8    0.9    1.0    A    B

       1   20.00  20.00  20.00  20.00  20.00  18.30  13.04  18.41  20.00  20.00  1.00 1.00

  **  170 57   20.000        11.000       0.785       0.000        1.000       0.000 INPUT
```

Girder design

From the literature it is known (12.13) that under present fabrication practices it is seldom worthwhile to place more than two butt splices in the flange of normal plate girders for simple beam spans up to 120 ft. Beyond this span length in continuous beams, as in our case, it may be economical to provide a total of four splices in a flange—that is, to use three different flange thicknesses in both spans. An approximate guide is to assume that about 800 lb of flange material must be saved for each additional butt splice placed in the flange. Studies have also shown that the minimum volume of flange steel is obtained in a two-splice flange when the cross-sectional area of the thinner flange plate is about one-half that of the thicker plate. Therefore, we shall change flange thickness in each span only once, from 1 in. to 2 in. A 1-in. thickness will be used from 0 to 93 ft (283 m) and 2-in. thickness from 93 to 120 ft (36.6 m) (Fig. 12.13). These dimensions will next be checked according to the newest AASHTO specifications (especially for fatigue). Table 12.5 gives the extreme bending and shear stresses which are needed for checking with allowable stresses and for transverse stiffener distribution. It is known (12.13) that the weight of web material

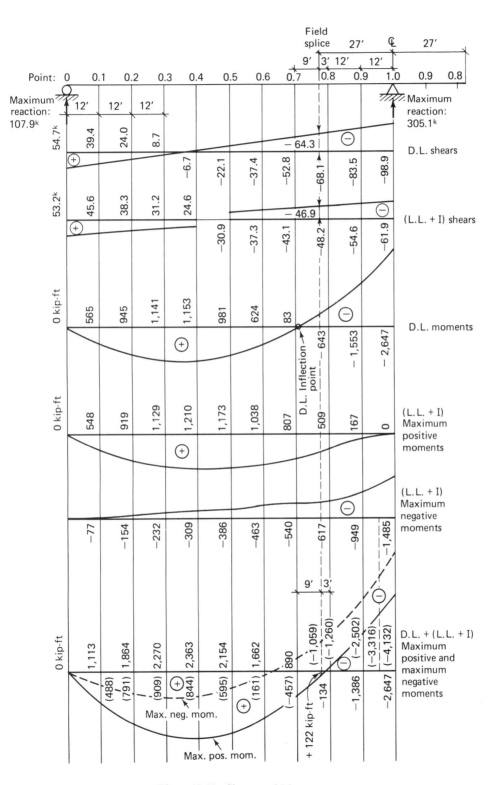

Figure 12.12 Shears and Moments.

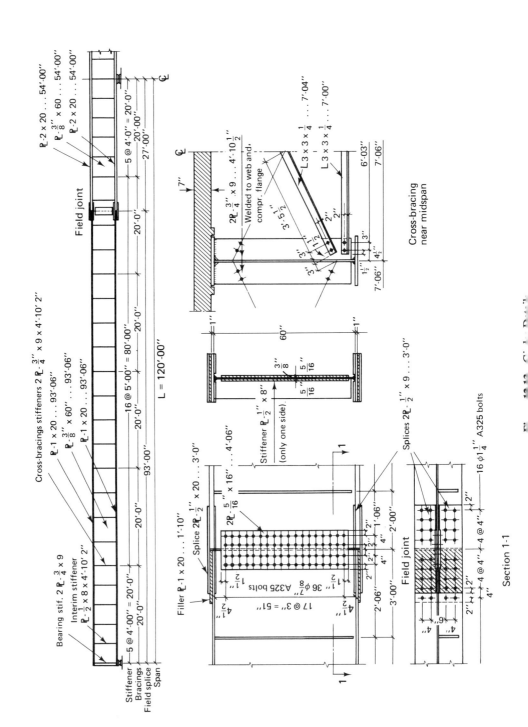

Table 12.5

Section	Point	Bending moments		Shear forces (kips)	Bending stresses		Bending stress range, f_{sr} (ksi)	Shear stresses, f_v (ksi)	Stiffener spacing (maximum space-60") (in.)
		Maximum positive (kips/ft)	Maximum negative (kips/ft)		Maximum positive (ksi)	Minimum negative (ksi)			
1	2	3	4	5	6	7	8	9	10
$I_x = 43{,}960$ in.^{4}	0.1	1,113	488	85	9.42	4.13	5.29	3.78	65
	0.2	1,864	791	62.3	15.77	6.69	9.08	2.77	77
$S_x = 1{,}418$ in.^{3}	0.3	2,270	909	39.9	19.21	7.69	11.52	1.77	93
	0.4	2,363	844	17.9	20.00	7.14	12.86	0.80	105
	0.5	2,154	595	−53	18.23	5.03	13.20	2.36	85
	0.6	1,662	161	−74.7	14.06	1.36	12.70	3.32	72
	0.7	890	−457	−95.9	7.53	−3.87	11.40	4.26	62
$I_x = 83{,}630$ in.^{4}	0.8	−134	−1,260	−116.3	−0.62	−5.79	5.17	5.17	57
	0.9	−1,386	−2,502	−138.1	−6.37	−11.49	5.12	6.14	53
$S_x = 2{,}613$ in.^{3}	1.0	−2,647	−4,132	−160.8	−12.16	−18.98	6.80	7.15	48

saved by using a horizontal stiffener must be at least 5 times the weight of that stiffener before any meaningful economy is achieved. For normal girder designs this means that girder depths of about 80 in. are required before horizontal stiffeners really start to become economical. For that reason, if needed, only vertical stiffeners will be used. As $D/t_w = 160 > 7{,}500/\sqrt{7{,}150} = 89$ (Eq. 10.60), stiffeners are required. The value for $f_v = 7{,}150$ lb/in.^{2} is obtained from Table 12.5, column 9, as the maximum shear stress.

Stiffener spacing

Stiffener spacing in Table 12.5 was obtained from AASHTO (12.3, 1977, fig. 1.7.43D1, p. 179). At the abutment, $f_v = 4.80$ ksi (33.1 MPa), and the spacing is 58 in.—say, 4 ft–06 in. (1.47 m). Next to the pier the distance is 4 ft, while all others are at 5 ft. Therefore, splicing of girders will be at 93 ft symmetrically from both sides of the abutments. The abscissa is $\frac{93}{120}$ = Section Point 0.78. The stress range at that point by interpolation is approximately

$$\frac{11.40 - 5.17}{10} \times 8 = 4.98 \text{ ksi (34.34 MPa)}$$

According to AASHTO, 1977 Table 1.7.2A1 it is seen that this stress range is safe for any category from A to F for any number of cycles. Full fatigue consideration will be discussed later. The chosen stiffener spacing is shown in Fig. 12.13.

Lateral bracing will be placed (AASHTO, 1977 section 1.7.17, Diaphragms, Cross Frames, and Lateral Bracing) each 20 ft. The distance from interior support to the dead load inflection point is 35 ft > 20 ft (AASHTO, 1977 footnote 2 to table 1.7.1A). Therefore, the unbraced length in the zone of negative moments is $l = 20$ ft. (6.1 m). The ratio l/b_f is $(20 \times 12)/20 = 12 < 36$ (AASHTO 1977 table 1.7.1A), and the allowable compression stress in the bottom flange in the braced length of 20 ft on both sides of the interior support is

$$F_b = 20.00 - 7.5(12)^2 = 18.44 \text{ ksi (127.14 MPa)}$$

From Fig. 12.12 the maximum negative moment at $\frac{1}{4}$ the unbraced length is $-3{,}316$ kip-ft and the stress is

$$\frac{3{,}316 \times 12}{2{,}613} = 15.23 \text{ ksi (105 MPa)} < 18.4 \text{ ksi (126.9 MPa)} \quad \underline{\text{OK}}$$

Check of the stress ranges with respect to fatigue

OUTSIDE THE FIELD JOINT. From Fig. 9.115 it can be seen that illustrative examples nos. 4 and 6 are applicable while the stress categories are B and C. According to the AASHTO specification, Table 1.7.2A1, the allowable range of stress for 5×10^5 cycles and category B is 27.5 ksi and for C it is 19 ksi. From Table 12.5, column 8, it is found that the maximum stress range was

$$F_{sr} = 13.20 \text{ ksi (91 MPa)} < 19 \text{ ksi (131 MPa)} \quad \underline{\text{OK}}$$

Actually, the design is satisfactory for a number of cycles up to 2×10^6. At any workshop butt-welded flange or web (which is always at middistance between stiffeners) category B is acceptable.

AT THE FIELD JOINT. This case is illustrative example no. 18 in Fig. 9.115 and stress category B, which gives

$$F_{sr} = 27.5 \text{ ksi } (189.6 \text{ MPa}) > 4.98 \text{ ksi } (34.3 \text{ MPa}): \quad \underline{\text{OK}}$$

Field splicing

The location of the field joint was chosen with respect to practical considerations for transportation of the girders from the workshop to the erection site and structural considerations of stress magnitude and stress range at the joint. With two symmetrically positioned field joints, the total girder length of 241 ft is divided into three pieces of 93.5 ft + 54 ft + 93.5 ft. According to Fig. 12.12 (last diagram) the maximum moments are

$$\text{Positive:} \quad +122 \text{ kip-ft } (165.4 \text{ kN-m})$$
$$\text{Negative:} \quad -1{,}059 \text{ kip-ft } (-1{,}435.8 \text{ kN-m})$$

The AASHTO 1977 specifications (section 1.7.15, Splices and section 1.7.16, Connections) require splicing for not less than 75% of the strength of the member (in our case of the weaker section). In terms of moments this is equivalent to

$$0.75(2{,}363) = 1{,}772 \text{ kip-ft } (2{,}403 \text{ kN-m}) > 1{,}059 \text{ kip-ft } (1{,}435.8 \text{ kN-m})$$

Therefore, the location is satisfactory with respect to structural strength considerations (fatigue was already checked).

High-strength ASTM A325 bolts (AASHTO M-164) will be used in a bearing type of connection with threads excluded from the shear planes of the faying surfaces between the connected parts. According to the AASHTO 1977 specifications (section 1.7.41C) the allowable stress in bearing is $F_p = 40$ ksi (275.8 MPa) and in shear 20 ksi (137.9 MPa). The last value should be reduced by 20% for a splice length more than 2 ft (0.61 m) from the joint covered by that splice.

FLANGE SPLICING. The moment that needs to be transferred is 1,772 kip-ft. This will produce a force in the flanges (if the web contribution is disregarded) of

$$\frac{1{,}772 \times 12}{61} = 348.6 \text{ kips } (1{,}550.6 \text{ kN})$$

Following the procedure outlined in Example 9.1 of Chapter 9, for the diameter of the bolts $1\frac{1}{4}$ in. is indicated for 1-in.-thick flanges ($\frac{3}{4}$-in.-thick for web). Splicing of flanges will be achieved by using one plate on top, PL–20 × $\frac{1}{2}$, and two PL–9 × $\frac{1}{2}$ underneath the flange (see Fig. 12.13). This will cause the fasteners to act in double shear. The capacities of one bolt are

$$\text{In double shear:} \quad 20(2)\frac{(1.25)^2}{4}\pi = 49.1 \text{ kips } (218.4 \text{ kN})$$

$$\text{In bearing:} \quad 40(1.25)1.0 = 50.0 \text{ kips } (222.4 \text{ kN})$$

$$\text{Use } \underline{49.1 \text{ kips } (218.4 \text{ kN})}$$

The number of $1\frac{1}{4}$-in.ϕ bolts is

$$n = \frac{348.6}{49.1} = 7.1$$

That is, two rows with 4 bolts in each. This number has to be increased (AASHTO, 1977, section 1.7.15, Splices), by two extra transverse lines of fasteners for one filler of 1-in. thickness used to compensate for the thicker flange of 2 in. The filler will be extended for one row beyond the end of the splices. Therefore, on each side of the flange joint there will be four rows with a total of 16 fasteners of $1\frac{1}{4}$-in. diameter. The bolt layout is obvious. There are four gage lines, as indicated in Fig. 12.13, with a pitch of 4 in. (10 cm).

WEB SPLICING. Shear force at the web joint (Fig. 12.12) is 111.2 kips (494.6 kN). Using Eq. (9.102), part of the moment to be transferred by the web splices is

$$M_w = 1{,}772 \times \frac{6{,}750}{37{,}210} = 321.4 \text{ kips/ft (4.69 MN/m)}$$

where $I_w = 6{,}750$ in.^{4} (web moment of inertia). The web joint will be spliced symmetrically using two splices of $\frac{5}{16}$ in. thickness. Their moment of inertia is 8,201 in.^{4} > 6,750 in.^{4} (web). The shear stress in the splices is

$$f_v = \frac{111.2}{2 \times 54 \times \frac{5}{16}} = 3.29 \text{ ksi (22.7 MPa)}$$

The bending stress is

$$f_b = \frac{321.4 \times 12}{8{,}201} \times 27 = 12.70 \text{ ksi (87.6 MPa)}$$

The principal stress is

$$f_1 = \frac{3.29 + 12.70}{2} + \sqrt{\left(\frac{12.70 - 3.29}{2}\right)^2 + (3.29)^2}$$

$$= 13.74 \text{ ksi (94.7 MPa)} < 20 \text{ ksi (137.9 MPa):} \quad \underline{\text{OK}}$$

If bolts of $\frac{7}{8}$-in. diameter are used at a vertical spacing of 3 in., starting with the first row $4\frac{1}{2}$ in. from both ends of the web, there will be 18 bolts in one row with a maximum distance h_1 between the end bolts of $60 - 2(4.5) = 51$ in. From Eq. (9.106) the maximum horizontal force in a bolt due to the moment is

$$H_{\max} = \frac{M_w}{\sum y_i^2} \cdot y_{\max} = \frac{321.4 \times 12}{8{,}722}(25.5) = 11.28 \text{ kips (50.18 kN)}$$

Due to web shear force there is acting in each bolt a vertical force of

$$V = \frac{111.2}{2 \times 18} = 3.09 \text{ kips (13.75 kN)}$$

The maximum resultant force in one bolt (see also Example 9.15) is

$$T_{\max} = \sqrt{(3.09)^2 + (11.28)^2} = 11.70 \text{ kips (52.04 kN)} < 13.12 \text{ kips:} \quad \underline{\text{OK}}$$

As bolts are in double shear, the capacities of one bolt are

In double shear: $\quad 20 \times \frac{(0.875)^2 \pi}{4} \times (2) = 24.05$ kips (107 kN)

In bearing: $\quad 40 \times 0.875 \times \frac{3}{8} = 13.12$ kips (58.4 kN)

Bearing stiffeners

According to AASHTO, 1977 (section 1.7.43(F), Bearing Stiffeners (1) Welded Girders) the bearing column consists of two stiffener plates and an $18t_w$-long strip of the web. The allowable bearing stress is $0.80F_y = 29$ ksi. The minimum thickness of the stiffener, if $b_{stf} = 9$ in. (to prevent local buckling), is

$$(t_{stf})_{min} = \frac{b_{stf}}{12}\sqrt{\frac{36}{33}} = \frac{9.00}{11.40} = 0.78 \text{ in.} \approx \tfrac{3}{4} \text{ in. (2 cm)}$$

The area of the column is $2 \times 9 \times \tfrac{3}{4} + 18 \times \tfrac{3}{8} = 20.25$ in.² The moment of inertia of the column for a pair of stiffeners with respect to the center line axis of the web is

$$\tfrac{1}{12}[(18.375)^3 - (0.375)^3]\tfrac{3}{4} = 388 \text{ in.}^4$$

The radius of gyration now is

$$r_{stf} = \sqrt{\frac{388}{20.25}} = 4.38 \text{ in. (11.1 cm)}$$

while the column slenderness is $60/4.38 = 13.7$. The allowable stress (AASHTO (12.3) Table 1.7.1A) is

$$F_a = 16.980 - 0.00053(13.7)^2 = 16.88 \text{ ksi (116.38 MPa)}$$

The stress in the bearing column now is

At the abutment: $f_a = \dfrac{107.9}{20.25} = 5.3$ ksi (36.5 MPa)

At the pier: $f_a = \dfrac{305.1}{20.25} = 15.1$ ksi (104.1 MPa) < 16.9 ksi (116.5 MPa) OK

Bearing pressure on the stiffener end (1-in. clip) is

$$f_p = \frac{305.1}{2 \times 8.0 \times \tfrac{3}{4}} = 25.4 \text{ ksi (175.1 MPa)} < 29 \text{ ksi (200 MPa):} \quad \text{OK}$$

Two $\tfrac{5}{16}$-in. fillet welds will transmit to the web the whole reaction as AASHTO requires over a stiffener length ($F_v = 0.27 \times 70 = 18.9$, AASHTO, 1977 section 1.7.41(B)) of

$$l = \frac{305.1}{2 \times 0.707 \times \tfrac{5}{16} \times 18.9} = 36.5 \text{ in.} < 60 - 2 = 58 \text{ in. (147.3 cm)}$$

Transverse intermediate stiffeners

The minimum moment of inertia of one stiffener (AASHTO, 1977 section 1.7.43(D)) is

$$I = \frac{60 \times (0.375)^3}{10.92}\left[25\left(\frac{60}{60}\right)^2 - 20\right] = 1.45 \text{ in.}^4 \text{ (60.35 cm}^4\text{)}$$

If for a stiffener a single plate of $b_{stf} = 8$ in. is used, then the required minimum thickness of the plate is

$$\tfrac{8}{16} = \tfrac{1}{2} \text{ in. (1.2 cm)}$$

Therefore, use PL-$\tfrac{1}{2}$ in. × 8 in. for intermediate stiffeners.

Cross-frames (Fig. 12.14)

The wind load (AASHTO, 1977 section 1.2.14) is 50 lb/ft² (240 kN/m²) and the wind force at the center pier cross-frame is

$$W = 0.050 \times 120.0 \times 9.0 = 54 \text{ kips (240.2 kN)}$$

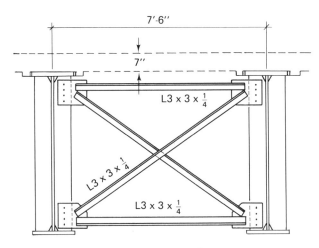

Figure 12.14 Cross-frames.

For one girder this is $\frac{54}{4} = 13.5$ kips; AASHTO, 1977 (section 1.7.5) requires a slenderness of secondary members less than 140, and the allowable compression stress for that maximum slenderness is

$$F_a = 16 - 0.3(140)^2 \times 10^{-3} = 10.12 \text{ ksi (69.8 MPa)}$$

The required cross section is

$$A = \frac{13.5}{10.12} = 1.344 \text{ in.}^2 \text{ (8.67 cm}^2\text{)}$$

The required radius of gyration for a horizontal element is

$$r_{min} = \frac{6 \times 12}{140} = 0.514 \text{ in. (1.3 cm)}$$

and for a diagonal (length $= \sqrt{6^2 + 4^2} = 7.25$ ft) is

$$r_{min} = \frac{7.25 \times 12}{140} = 0.622 \text{ in. (1.6 cm)}$$

Use $\lfloor$ 3 × 3 × ¼ with $A = 1.69$ in.² > 1.334 in.² and $r = 0.69 > 0.622$ in. For intermediate cross-frames use the same.

Deflections and camber

Maximum deflection due to dead load of 1.28 kips/ft occurs at 0.4L and equals 1.65 in. or L/873 [obtained from the same computer program (12.12)]. Deflection due to live load was less than that amount. Table 12.6 gives these calculated deflections, $\Delta_{D.L.}$, at all tenth points. According to AASHTO (12.3, article 1.7.14(C), Camber)

Table 12.6 DEFLECTIONS DUE TO DEAD LOAD

Point	0	0.1	0.2	0.3	0.4	0.5	0.6	0.7	0.75	0.8	0.9	1.0
Deflections, $\Delta_{D.L.}$ (in.)	0.000	0.639	1.170	1.520	1.650	1.558	1.277	1.878	0.666	0.463	0.138	0.000

to compensate for possible loss of camber of heat-curved girders the actual amount of camber in inches, Δ, at any section is

$$\Delta = \Delta_{D.L.}\left[1 + \frac{(0.02)120^2(12)^2(36)}{(29{,}000) \times (31) \times 1.650}\right] = 2.00\, \Delta_{D.L.}$$

In practice this means that for actual camber the values from Table 12.6 should be doubled.

Roller supports (at abutments) (Fig. 12.15a)

For expansion rollers use material A-668, class F (AASHTO, 1977 table on p. 137). Using Eq. (11.3) and a roller diameter of $d = 8$ in. (20.3 cm) the allowable linear bearing pressure, p, is obtained as

$$p = \frac{50 - 13}{20}0.6(8) = 8.88 \text{ kips/in. } (1555 \text{ kN/m})$$

The stress in the roller for a roller length of 22 in. is

$$p = \frac{107.9}{22 - 2 \times 1\frac{1}{4}} = 5.53 \text{ kips/in. } (968.5 \text{ kN/m}) < 8.88 \text{ kips/in. } (1555 \text{ kN/m})$$

For the base plate try PL–$1\frac{1}{2} \times 8 \ldots 30$ in. The bearing pressure when centric position (AASHTO, 1977 section 1.5.26(3)) is

$$f_p = \frac{107.9}{8 \times 30} = 0.45 \text{ ksi } (3.1 \text{ MPa}) < 0.3(3.0) = 0.9 \text{ ksi } (6.2 \text{ MPa})$$

The bending moment in the plate (per inch of length of plate) is

$$M\, C_L = 0.61 \times 4.0 \times 2.0 = 4.88 \text{ kip-in. } (0.55 \text{ kN-m})$$

The section modulus of the plate is $S = \frac{1}{6}(1.5)^2 1.0 = 0.375$ in.^{3}. The bending stress in the plate is

$$f_b = \frac{4.88}{0.375} = 13.0 \text{ ksi } (89.6 \text{ MPa}) < 20 \text{ ksi } (137.9 \text{ MPa}): \quad \underline{\text{OK}}$$

Fixed bearing (at midpier) (Fig. 12.15b)

Because the reaction is still not too large, a simple fixed rocker bearing support of A36 steel will be designed. The diameter of the rocker is 50 in.

From Eq. (11.4) the allowable bearing per linear inch, p, is

$$p = \frac{36 - 13}{20}(3\sqrt{50}) = 24.4 \text{ kips/in. } (4{,}273 \text{ kN/m})$$

With two $1\frac{1}{4}$-in. ϕ-pintles the net bearing length is $22 - (2 \times 1.25) = 19.5$ in. (49.5 cm). Thus

$$f_p = \frac{305.1}{19.5} = 15.6 \text{ kips/in.} < 24.4 \text{ kips/in. } (4{,}273 \text{ kN/m}): \quad \underline{\text{OK}}$$

(a) Expansion bearing details (on abutments)

Figure 12.15 Bearings.

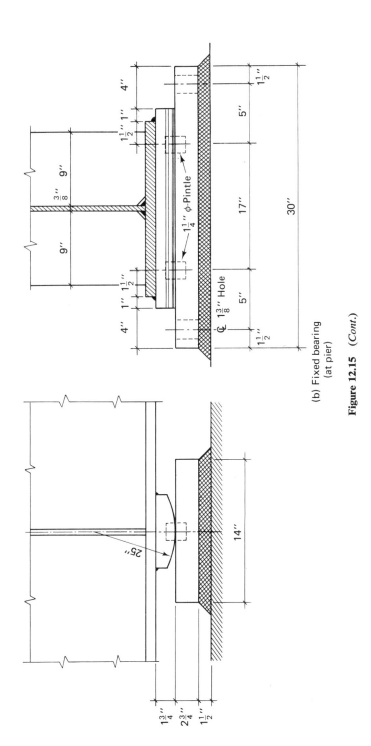

(b) Fixed bearing (at pier)

Figure 12.15 (*Cont.*)

For the base plate try PL-$2\frac{3}{4} \times 14 \ldots 30$. The bearing pressure is

$$\frac{305.1}{14 \times 30} = 0.73 \text{ ksi (5.03 MPa)} \approx 0.70 \text{ ksi (4.82 MPa):} \qquad \underline{\text{OK}}$$

The bending moment for a 1-in. plate width is

$$M\,C_L = 1.00 \times \tfrac{14}{2} \times \tfrac{14}{4} = 24.5 \text{ kip-in. (2.77 kN-m)}$$

The section modulus of the plate (for 1-in. width) is

$$S = \frac{1 \times (2.75)^2}{6} = 1.26 \text{ in.}^3$$

The bending stress in the plate is

$$\frac{24.5}{1.26} = 19.4 \text{ ksi (133.7 MPa)} < 20.0 \text{ ksi (137.9 MPa):} \qquad \underline{\text{OK}}$$

REFERENCES

12.1. GAYLORD, E. H., JR., and C. N. GAYLORD, *Structural Engineering Handbook*, 2nd ed., McGraw-Hill, New York, 1979.

12.2. MERRITT, F. S., ed., *Structural Steel Designer Handbook*, McGraw-Hill, New York, 1972.

12.3. AASHTO, *Standard Specifications for Highway Bridges*, 12th ed., Washington, DC, 1977 (with 1978, 1979, 1980, and 1981 Interim Bridge Specifications).

12.4. AREA, *Manual for Railway Engineering*, Chicago, 1980.

12.5. Granco Steel Products, *Floor/Roof Construction*, Catalog 99–10, no. 5: *Metal Decking*, St. Louis, 1972.

12.6. Bowman Building Products, *New Steel Decks for Floor and Roofs*, no. 5: *Metal Decking*, Pittsburgh, 1970.

12.7. E. G. Smith Division, Cyclops Corporation, *Metal Wall and Roof Systems*, Catalog 75A17, Pittsburgh, 1975, pp. 10–1.

12.8. Inland-Ryerson Construction Products Company, *Inryco Floor/Ceiling Systems*, Chicago, 1971.

12.9. Armco Steel Corporation, *Armco Steel Joists*, no. 5: *Steel Joists*, Kansas City, 1969.

12.10. U.S. Department of Transportation, Federal Highway Administration, Bureau of Public Roads, *Standard Plans for Highway Bridges*, vol. 2: *Structural Steel Superstructure*, Washington, DC, 1968.

12.11. AISC, *Moments, Shears, and Reactions for Continuous Highway Bridges*, New York, 1966.

12.12. HOWARD, NEEDLES, TAMMEN, and BERGENDOFF, *Continuous Beam Analysis*, Program no. 126, Kansas City, 1969.

12.13. Bethlehem Steel Corporation, *Economics of Simple Span Highway Bridges*, Bethlehem, PA, 1967; and *Bridge Design Aids*, Guidelines for Economical Bridge Girder Design, 1968.

12.14. AISC, *Manual of Steel Construction*, 8th ed., Chicago, 1980.

Appendices

APPENDIX I-1 FORTRAN IV PROGRAM FOR THE DESIGN OF SIMPLY SUPPORTED BEAMS, AISC SPECIFICATION

PAGE I-I.1

25033 01 11-03-71   DESIGN OF A36 STEEL SIMPLY SUPPORTED BEAMS WITHOUT OR WITH
                     3 CONCENTRATED FORCES AND UNIF. DISTRIBUTED LOAD IN

```
 1     C      DESIGN OF A36 STEEL SIMPLY SUPPORTED BEAMS WITHOUT OR WITH
 2     C      UP TO 3 CONCENTRATED FORCES AND UNIF. DISTRIBUTED LOAD IN
 3     C      SIMPLE BENDING, WORKING STRESS DESIGN METHOD.
 4     C      IF CHECK = 0 NONCOMPACT SECTION, IF CHECK = 1 COMPACT SECTION,
 5            REAL MAXMOM, LCOMP, L, LB, LU
 6            INTEGER CHECK, WF, WEIGHT
 7            DIMENSION WF(50), WEIGHT(50), CHECK(50), S(50), AWEB(50),
 8           1 LCOMP(50), LU(50)
 9     C      INPUT DATA FOR ROLLED WF SECTIONS
10            READ(5,100) (WF(I), WEIGHT(I), CHECK(I), S(I), AWEB(I),                11
11           1 LCOMP(I), LU(I), I=1,50)
12     100    FORMAT(3I8, 4F8.2)                                                      6
13            WRITE(6,101)                                                            6
14     101    FORMAT(6X,2HWF,2X,6HWEIGHT,3X,5HCHECK,7X,1HS,4X,                        8
15           1 4HAWEB,3X,5HLCOMP,6X,2HLU//)
16            WRITE(6,100) (WF(I), WEIGHT(I), CHECK(I), S(I), AWEB(I),                8
17           1 LCOMP(I), LU(I), I = 1,50)
18     C      INPUT DATA FOR THE PROBLEM
19     99     READ(5,102) L, LB, X1, X2, X3, W, P1, P2, P3                           13
20     102    FORMAT(9F8.2)                                                          16
21            IF(L.EQ.0.0) STOP                                                      16
22            WRITE(6,103) L, LB, X1, X2, X3, W, P1, P2, P3                          19
23     103    FORMAT(1H1, 2HL=F8.2, 2HFT/ 3HLB=F8.2, 2HFT/ 3HX1=F8.2, 2HFT/          22
24           1 3HX2=F8.2, 2HFT/3HX3=F8.2, 2HFT/ 2HW=F8.2, 10HKIP PER FT/
25           2 3HP1=F8.2, 3HKIP/ 3HP2=F8.2, 3HKIP/ 3HP3=F8.2, 3HKIP)
26     C      CALCULATE THE REACTION RA (AT THE LEFT HAND SIDE)
27            RA= W*L/2.0 +P1*(L-X1)/L + P2*(L-X2)/L + P3*(L-X3)/L                   22
28     C      MAXMOM LOCATION X0 AND MAXMOM
29            IF (X1+ X2 + X3.EQ.0.0) GO TO 8                                        23
30     1      X0= RA/W                                                               26
31            IF (X0.LE.X1) GO TO 9                                                  27
32            IF (RA - W*X1 - P1) 2,2,3                                              30
33     2      X0 =X1                                                                 31
34            MAXMOM = RA*X1 - W*X1**2/2.0                                           32
35            GO TO 12                                                               33
36     3      X0= (RA - P1)/W                                                        34
37            IF (X2 + X3.EQ.0.0) GO TO 10                                           35
38            IF (X0.LE.X2) GO TO 10                                                 38
39            IF (RA - P1 - P2 - X2*W) 4,4,5                                         41
40     4      X0= X2                                                                 42
41            MAXMOM= RA*X2 - P1*(X2 - X1) - W*X2**2/2.0                             43
42            GO TO 12                                                               44
43     5      X0= (RA - P1 - P2)/W                                                   45
44            IF (X3.EQ.0.0) GO TO 11                                                46
45            IF (X0.LE.X3) GO TO 11                                                 49
46            IF (RA - P1 - P2 - P3 - X3*W) 6,6,7                                    52
47     6      X0= X3                                                                 53
48            MAXMOM= RA*X3 - P1*(X3-X1) - P2*(X3-X2) - W*X3**2/2.0                  54
49            GO TO 12                                                               55
50     7      X0= (RA - P1 - P2 - P3)/W                                              56
```

Appendices **561**

APPENDIX I-1 (Cont.)

PAGE I-1.2

25033 01 11-03-71   19,254   OF A36 STEEL SIMPLY SUPPORTED BEAMS WITHOUT OR WITH
                              3 CONCENTRATED FORCES AND UNIF; DISTRIBUTED LOAD IN

```
51                 MAXMOM= RA*X0 - P1*(X0-X1) - P2*(X0-X2) - P3*(X0-X3)                57
52               1 - W*X0**2/2.0
53                 GO TO 12                                                            58
54         8       X0= L/2.0                                                           59
55                 MAXMOM= W*L**2/8.0                                                  60
56                 GO TO 12                                                            61
57         9       MAXMOM= RA*X0 - W*X0**2/2.0                                         62
58                 GO TO 12                                                            63
59        10       MAXMOM= RA*X0 - P1*(X0-X1) - W*X0**2/2.0                            64
60                 GO TO 12                                                            65
61        11       MAXMOM= RA*X0 - P1*(X0-X1) - P2*(X0-X2) - W*X0**2/2.0               66
62        12       WRITE(6,104) RA, X0, MAXMOM                                         67
63       104       FORMAT(1H1,3HRA=F8.2,3HKIP/3HX0=F8.2,2HFT/7HMAXMOM=F8.2,             70
64               1 6HKIP-FT)
65         C       ROLLED SHAPE DESIGN PROCEDURE
66         C       SUPPOSE FIRST A COMPACT,ADEQUATELY BRACED SECTION
67         C
68                 SREQRD =(MAXMOM * 12.0)/24.0                                        70
69                 DO 13 I=1,50                                                        71
70                 IF (S(I) .GE. SREQRD) GO TO 14                                      72
71        13       CONTINUE                                                            75
72                 I= 50                                                               77
73                 GO TO 19                                                            78
74        14       WRITE(6,105) I, WF(I), WEIGHT(I), CHECK(I), S(I), LCOMP(I), LU(I)  79
75       105       FORMAT(1H0,4X,2HI=I2,5X,I2,1X,2HWF,5X,7HWEIGHT=I8/                 82
76               1 6HCHECK=I2,5X,2HS=F8.2,5X,6HLCOMP=F8.2,5X,3HLU=F8.2//)
77                 IF (CHECK(I) .EQ. 0 .OR. LCOMP(I) .LT. LB) GO TO 15                 82
78         C       SOLUTION IS COMPACT AND ADEQUATELY BRACED SECTION
79                 WRITE(6,106) WF(I), WEIGHT(I), S(I)                                 85
80       106       FORMAT(1H0,24HFINAL WIDEFLANGE SECTIONI2,5X,                        88
81               1 7HWEIGHT=I8,5X,2HS=F8.2)
82                 GO TO 18                                                            88
83        15       SREQRD = (MAXMOM * 12.0)/22.0                                       89
84                 J= I + 1                                                            90
85                 DO 16 I=J,50                                                        91
86                 IF (SREQRD .LE. S(I) .AND. LU(I) .GE. LB) GO TO 17                  92
87        16       CONTINUE                                                            95
88                 I=50                                                                97
89                 GO TO 19                                                            98
90        17       WRITE(6,106) WF(I), WEIGHT(I), S(I)                                 99
91                 THE ACTUAL CALCULATED STRESSES
92        18       FB = (MAXMOM * 12.0)/S(I)                                          102
93                 SHEAR= RA/AWEB(I)                                                  103
94                 WRITE(6,107) FB, SHEAR                                             104
95       107       FORMAT(1H0,15H3ENDING STRESS=F8.2,2X,3HKSI/                        107
96               1 21HAVERAGE SHEAR STRESS=F8.2,2X,3HKSI)
97                 GO TO 99                                                           107
98        19       WRITE(6,108)                                                       108
99       108       FORMAT(1H0,50HSELECTION IS NOT POSSIBLE,DESIGN AS A PLATE GIRDER)  110
100                GO TO 99                                                           110
```

PAGE I-1.3

25033 01 11-03-71   19,254   OF A36 STEEL SIMPLY SUPPORTED BEAMS WITHOUT OR WITH
                              3 CONCENTRATED FORCES AND UNIF; DISTRIBUTED LOAD IN

```
101                END                                                                111
```

23677 WORDS OF MEMORY USED BY THIS COMPILATION

APPENDIX I-2 MINICOMPUTER PROGRAM FOR THE DESIGN OF SIMPLY SUPPORTED BEAMS, FOR HP9820A:

I-2.1. Moments, Shears, and Modifiers, C_b

```
0:
FXD 5;GTO 59⊢
1:
ENT "LENGTH",R1,
"DIST. LOAD",R2,
"UNBRACED SPANS"
,A⊢
2:
1→Y;PRT "LENGTH"
,R1,"DIST. LOAD"
,R2,"UNBRACED SP
ANS",A;SPC 2⊢
3:
ENT "POINT LOADS
",R(3+Y),"LOCATI
ON",R(6+Y)⊢
4:
PRT "POINT LOAD"
,R(3+Y),"LOCATIO
N",R(6+Y);SPC 2;
IF 3>Y;Y+1→Y;
JMP -1⊢
5:
R2R1/2+R4(R1-R7)
/R1+R5(R1-R8)/R1
+R6(R1-R9)/R1→R1
0⊢
6:
IF R4+R5+R6=0;R1
/2→R14;GTO 19⊢
7:
R7-0.001→X;7→Y⊢
8:
IF X-R7≤0;0→R11;
JMP 2⊢
9:
1→R11⊢
10:
IF X-R8≤0;0→R12;
JMP 2⊢
11:
1→R12⊢
12:
IF X-R9≤0;0→R13;
JMP 2⊢
13:
1→R13⊢
14:
(R10-R4R11-R5R12
-R6R13)/R2→R14⊢
15:
IF R14>X;R(Y+1)→Y
)-0.001→X;GTO 8⊢
16:
0→R(Y+4);R(Y-1)+
0.005→X⊢

17:
R10-R4R11-R5R12-
R6R13-R2X→R20⊢
IF R20>0;X+0.005
→X;GTO 17⊢
18:
X-0.005→R14⊢
19:
PRT "LOCATION OF
";PRT "    MAX.
MOMENT",R14;SPC
2;R1/A→R15;0→R16
⊢
20:
IF R15=R14;SFG 1
⊢
21:
IF R15-R7≤0;0→R1
1;JMP 2⊢
22:
R15-R7→R11⊢
23:
IF R15-R8≤0;0→R1
2;JMP 2⊢
24:
R15-R8→R12⊢
25:
IF R15-R9≤0;0→R1
3;JMP 2⊢
26:
R15-R9→R13⊢
27:
R10R15-R2*R15↑2/
2-R4R11-R5R12-R6
R13→R17⊢
28:
IF R15≤R14;GTO 4
0⊢
29:
IF FLG 1=1;GTO 4
0⊢
30:
IF R15>R1;SFG 1⊢
31:
IF R14-R7≤0;0→R1
1;JMP 2⊢
32:
R14-R7→R11⊢
33:
IF R14-R8≤0;0→R1
2;JMP 2⊢
34:
R14-R8→R12⊢
35:
IF R14-R9≤0;0→R1
3;JMP 2⊢

36:
R14-R9→R13⊢
37:
R10R14-R2*R14↑2/
2-R4R11-R5R12-R6
R13→R19⊢
38:
1→R18;SFG 3⊢
39:
PRT "MAX. MOMENT
",R19,"C",R18;
SPC 1;SFG 1⊢
40:
IF R16>R17;GTO 4
5⊢
41:
1.75-1.05(R16/R1
7)+0.3*(R16/R17)
↑2→R18⊢
42:
IF R18>2.3;2.3→R
18⊢
43:
IF FLG 3=0;PRT "
MAX. MOMENT",R17
,"C",R18;SPC 1⊢
44:
GTO 48⊢
45:
1.75-1.05(R17/R1
6)+0.3*(R17/R16)
↑2→R18⊢
46:
IF R18>2.3;2.3→R
18⊢
47:
IF FLG 3=0;PRT "
MAX. MOMENT",R16
,"C",R18;SPC 1⊢
48:
R17→R16;CFG 3⊢
49:
R15+R1/A→R15;IF
R15=R14;SFG 1⊢
50:
IF R15>R1;GTO 52
⊢
51:
GTO 21⊢
52:
R2R1/2+R4(R1-R7)
/R1+R5(R1-R8)/R1
+R6(R1-R9)/R1→R3
0⊢
53:
R1R2+R4+R5+R6→R3
1⊢
```

APPENDIX I-2.1 (Cont.)

```
54:
R31-R30→R32⊢
55:
IF R30>R32;PRT "
MAX. SHEAR",R30;
JMP 2⊢
56:
PRT "MAX. SHEAR"
,R32⊢
57:
SPC 8⊢
58:
CFG 1;CFG 2;CFG
3;STP ⊢
59:
PRT "  THIS PROG
RAM"," DETERMINE
S THE","  C VALU
ES AND"⊢
60:
PRT " MAXIMUM MO
MENT","  FOR EAC
H UN-","  BRACED
 LENGTH"⊢
61:
PRT " AS WELL AS
 THE","MAXIMUM S
HEAR IN","  THE
SECTION."⊢
62:
SPC 4;GTO 1⊢
63:
END ⊢
R199
```

I-2.2 Selection of WF-Section

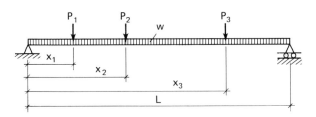

```
0:
ENT "MAX. MOMENT
",R151,"MAX. SHE
AR",R152;0→X;
FXD 2;0→Y⊢
1:
ENT "C",R153,"LE
NGTH",R154,"UNBR
ACED LENGTH",R15
5;GTO 45⊢
2:
PRT "MAX. SHEAR"
,R152;SPC 1;PRT
"LENGTH",R154;
SPC 1;PRT "UNBRA
CED LENGTH",R155
⊢
3:
SPC 2;PRT "MAX.
MOMENT",R151;
SPC 1;PRT "C",R1
53;SPC 2⊢
4:
INT (RX/10↑4)/10
→R170⊢
5:
(RX-R170*10↑5)/1
00→R171⊢
6:
INT (R(X+50)/100
0)/100→R172⊢
7:
R(X+50)-R172*10↑
5→R173⊢
8:
INT (R(X+100)/10
↑7)/100→R174⊢
9:
R(X+100)-R174*10
↑9→A⊢
10:
INT (A/10↑5)→R17
5⊢
11:
A-R175*10↑5→A⊢
12:
INT (A/10↑4)→R17
6⊢
13:
(A-R176*10↑4)/10
0→R177⊢
14:
IF R176=0;GTO 23
⊢
15:
IF R155>R172;
GTO 23⊢
16:
R151*12/R170→A;
IF A>24;GTO 43⊢
17:
IF A>Y;A→Y⊢
18:
R152/R171→B;IF B
>14.5;GTO 43⊢
19:
GTO 47⊢
20:
PRT "DEPTH",R175
;SPC 1;PRT "WEIG
HT",R173;SPC 1⊢
21:
PRT "BENDING STR
ESS",Y;SPC 1;
PRT "SHEAR STRES
S",B⊢
22:
SPC 8;DSP "    F
INISHED";STP ⊢
23:
IF R154/R174≤53*
R153;22→R165;
GTO 38⊢
24:
IF R154/R174≤119
*R153;JMP 2⊢
25:
170000*R153/(R15
4/R174)↑2→R162;
JMP 2⊢
26:
24-(R154/R174)↑2
/1181*R153→R161⊢
27:
12000*R153/R154R
177→R163⊢
28:
IF R154/R174≤119
*R153;JMP 6⊢
29:
IF R162>R163;
JMP 3⊢
30:
IF R163>22;22→R1
65;JMP 9⊢
31:
R163→R165;JMP 8⊢
32:
IF R162>22;22→R1
65;JMP 7⊢
33:
R162→R165;JMP 6⊢
34:
IF R161>R163;
JMP 3⊢
35:
IF R163>22;22→R1
65;JMP 4⊢
36:
R163→R165;JMP 3⊢
37:
IF R161>22;22→R1
65;JMP 2⊢
38:
R161→R165⊢
39:
R151*12/R170→A;
IF A>R165;GTO 43
⊢
40:
IF A>Y;Y→A⊢
41:
R152/R171→B;IF B
>14.5;GTO 43⊢
42:
GTO 47⊢
43:
X+1→X;IF X=50;
GTO 49⊢
44:
GTO 4⊢
45:
PRT "  THIS PROG
RAM"," DETERMINE
S THE","BEST SEC
TION FOR"⊢
46:
PRT "THE GIVEN L
OADS.";0→R169;
SPC 8;GTO 2⊢
47:
R169+R155→R169;
IF R169+1>R154;
GTO 20⊢
48:
ENT "MAX. MOMENT
",R151,"C",R153;
GTO 3⊢
49:
PRT "  SELECTION
NOT","POSSIBLE,D
ESIGN","AS PLATE
 GIRDER.";SPC 8⊢
50:
DSP "   NO SOLUTI
ON";STP ⊢
51:
END ⊢
R245
```

I-2.3 Storage of Data

```
0:
ENT "NUMBER OF B
EAMS?",X⊢
1:
X3→Y⊢
2:
ENT "SECTION MOD
ULUS",A,"AREA OF
 WEB",B⊢
3:
ATN↑ 4+B→RY⊢
4:
ENT "COMPACT LEN
GTH",A,"WEIGHT",
B⊢
5:
ATN↑ 3+B→R(Y+50)
⊢
6:
ENT "RADIUS OF G
YR.",A,"DEPTH",B
⊢
7:
ENT "COMPACT (YE
S=1)",C,"D/AREA
FLANGE",Z⊢
8:
ATN↑ 7+BTN↑ 5+C
TN↑ 4+Z→R(Y+100)
⊢
9:
IF Y+2<X;Y+1→Y;
GTO 2⊢
10:
STP ⊢
11:
END ⊢
R397
```

APPENDIX I-3 PROGRAM FOR DESIGN OF SIMPLY SUPPORTED BEAMS FOR HP9830A

I-3.1 Structural Analysis

```
10 COM B[9],C:10],M:10],R,R5,M1,M2,M3,M4,G
20 PRINT "DESIGN OF SIMPLE BEAMS OF F36 W.FLANGESTEEL SECTIONS"
30 PRINT
40 PRINT
50 PRINT
60 PRINT TAB10"PART I:STRUCTURAL ANALYSIS"
70 WRITE (15,80)
80 FORMAT 80"*"
90 PRINT
100 REM THIS PROGRAM CAN HANDLE UP TO 3 POINT LOADS AND UNIF.DISTRIBUTED LOAD"
110 REM UNBRACED LENGTHS MUST BE EQUAL;IF NO POINT LOAD ABSCISSA IS THEN X1=L
120 REM ALL LENGTHS IN FT, FORCES IN KIPS AND UNIF.LOAD B(M,9)=K/FT
130 REM ONLY ONE PROBLEM AT THE TIME CAN BE HANDLED
140 REM PROBLEM DATA:L,LB,X1,X2,X3,P1,P2,P3,W;ARE STORED AS ARRAY B(M,N)
150 REM PROGRAM WILL FIND ABS.MAX.MOMENT,SHEAR,MOMENTS AT BRACING POINTS AND
160 REM POINTS OF FORCE APPLICATION AND MODIFIER COEF.C(T)=CB FOR EACH LB
170 PRINT
180 PRINT
190 PRINT
200 PRINT
210 PRINT
220 STANDARD
230 PRINT TAB20"PROBLEM"
240 FOR N=1 TO 9
250 DISP "ROW"N;
260 INPUT B[N]
270 NEXT N
280 PRINT "L="B[1];"LB="B[2];"X1="B[3];"X2="B[4];"X3="B[5];
290 PRINT "P1="B[6];"P2="B[7];"P3="B[8];"W="B[9];
300 PRINT
310 PRINT
320 PRINT
330 PRINT TAB16"SOLUTION OF THE PROBLEM"
340 PRINT
350 REM LEFT REACTION R=R1+R2+R3+R4;RIGHT R5=SUM OF V - R
360 R1=B[9]*B[1]/2
370 IF B[3]=B[1] THEN 430
380 R2=B[6]*((B[1]-B[3])/B[1])
390 R3=B[7]*((B[1]-B[4])/B[1])
400 R4=B[8]*((B[1]-B[5])/B[1])
410 R=R1+R2+R3+R4
420 GOTO 450
430 R=R1
440 REM RIGHT HAND SIDE REACTION IS R5
450 R5=B[9]*B[1]+B[6]+B[7]+B[8]-R
460 PRINT "R-LEFT="R"KIP","R-RIGHT="R5"KIP";
470 PRINT
480 PRINT
490 PRINT "LOCATION OF ABS.MAX.MOMENT X0 AND ITS MAGNITUDE M1"
500 IF*B[3]<B[1] THEN 540
510 X0=B[1]/2
520 M1=B[9]*B[1]↑2/8
530 GOTO 780
540 X0=R/B[9]
550 IF X0>B[3] THEN 580
560 M1=R*X0-B[9]*X0↑2/2
570 GOTO 780
580 IF R-B[9]*B[3]-B[6]>0 THEN 620
590 X0=B[3]
```

APPENDIX I-3.1   (Cont.)

```
 600 M1=R*B[3]-B[9]*B[3]↑2/2
 610 GOTO 780
 620 X0=(R-B[6])/B[9]
 630 IF B[7]=0 THEN 650
 640 IF X0>B[4] THEN 670
 650 M1=R*X0-B[6]*(X0-B[3])-B[9]*X0↑2/2
 660 GOTO 780
 670 IF R-B[9]*B[4]-B[6]-B[7]>0 THEN 710
 680 X0=B[4]
 690 M1=R*X0-B[9]*X0↑2/2-B[6]*(X0-B[3])
 700 GOTO 780
 710 X0=(R-B[6]-B[7])/B[9]
 720 IF B[8]=0 THEN 740
 730 IF X0>B[5] THEN 760
 740 M1=R*X0-B[6]*(X0-B[3])-B[7]*(X0-B[4])-B[9]*X0↑2/2
 750 GOTO 780
 760 X0=(R-B[6]-B[7]-B[8])/B[9]
 770 M1=R5*(B[1]-X0)-B[9]*(B[1]-X0)↑2/2
 780 PRINT TAB27"X0="X0,TAB48,"M1="M1
 790 PRINT
 800 G=B[1]/B[2]
 810 PRINT "NUMBER OF UNBRACED SECTIONS IS"G
 820 PRINT
 830 FIXED 2
 840 Q=X0/B[2]
 850 P=ABS(Q-INT(Q))
 860 STANDARD
 870 IF G>1 THEN 920
 880 C[1]=1
 890 PRINT "UNBRACED LENGTH=SPAN L","MODIFIER CB=1.0";
 900 PRINT
 910 GOTO 1440
 920 FOR T=1 TO G
 930 X=B[2]*T
 940 IF X0<X THEN 970
 950 IF B[6]=0 THEN 1180
 960 GOTO 1040
 970 IF X0<B[2]*(T-1) THEN 950
 980 IF P<0.05 THEN 950
 990 M[T]=M1
1000 C[T]=1
1010 PRINT "MAX.MOMENT IS INBETWEEN THE BRACING POINTS M1="M1
1020 PRINT
1030 GOTO 1380
1040 IF B[3]>X THEN 1190
1050 M[T]=R*X-B[9]*X↑2/2-B[6]*(X-B[3])
1060 M2=R*B[3]-B[9]*B[3]↑2/2
1070 IF B[7]=0 THEN 1210
1080 IF B[4]>X THEN 1240
1090 M[T]=M[T]-B[7]*(X-B[4])
1100 M3=R*B[4]-B[9]*B[4]*2/2-B[6]↑(B[4]-B[3])
1110 IF B[8]=0 THEN 1230
1120 IF B[5]>X THEN 1240
1130 M[T]=M[T]-B[8]*(X-B[5])
1140 M4=R*B[5]-B[9]*B[5]↑2/2-B[6]*(B[5]-B[3])
1150 M5=B[7]*(B[5]-B[4])
1160 M4=M4-M5
1170 GOTO 1240
1180 M2=M3=M4=0
1190 M[T]=R*X-B[9]*X↑2/2
1200 GOTO 1240
1210 M3=M4=0
1220 GOTO 1240
1230 M4=0
1240 IF T=1 THEN 1290
```

APPENDIX I-3.1 (*Cont.*)

```
1250 IF T=G THEN 1290
1260 IF ABS(M[T-1])>ABS(M[T]) THEN 1310
1270 K=M[T-1]/M[T]
1280 GOTO 1320
1290 C[T]=1.75
1300 GOTO 1360
1310 K=M[T]/M[T-1]
1320 C[T]=1.75-1.05*K+0.3*K↑2
1330 IF C[T]>2.3 THEN 1350
1340 GOTO 1360
1350 C[T]=2,3
1360 PRINT "BRACING POINT X=",X,"MOMENT M(X)=",M[T];
1370 PRINT
1380 PRINT "SECTION NO.=",T"BETWEEN THE BRACINGS","MODIFIER CB=",C[T];
1390 PRINT
1400 NEXT T
1410 PRINT "X1=",B[3];"M(P1)=",M2;"X2=",B[4];"M(P2)=",M3;"X3=",B[5];"M(P3)=",M4;
1420 PRINT
1430 PRINT
1440 LOAD 3
1450 END
```

I-3.2 Beam Design

```
10 COM B[9],C[10],M[10],R,R5,M1,M2,M3,M4,G
20 DIM A[40,8]
30 PRINT
40 PRINT
50 PRINT
60 PRINT TAB10"PART II: STRUCTURAL DESIGN"
70 WRITE (15,80)
80 FORMAT 80"*
90 PRINT
100 REM THIS PROGRAM WILL SELECT THE MOST ECONOMICAL WF A36 STEEL SECTION
110 REM FOR MOMENTS AND SHEARS FOUND IN PART I: M(T);C(T),R, AND R5
120 REM SECTION PROPERTIES ARE STORED AS ARRAY A(I,J), AND THEY ARE:
130 REM NO,WEIGHT,SEC.MOD.,WEB AREA,LC,LU,RT,D/AF
140 REM DATA ARRAY A(44,8) IS STORED IN THE SAME FILE 3
150 REM PROGRAM PART I IS STORED IN FILE 2, PART II IN FILE 3
160 PRINT
170 PRINT "SOLUTION OF THE DESIGN PROBLEM PART II"
180 FOR I=1 TO 40
190 FOR J=1 TO 8
200 READ A[I,J]
210 NEXT J
220 NEXT I
230 PRINT
240 PRINT
250 DATA 12,14,14.9,2.382,3.5,4.2,0.95,13.3
260 DATA 12,16,17.1,2.638,4.1,4.3,0.96,11.3
270 DATA 12,19,21.3,2.858,4.2,5.3,1,8.67
280 DATA 12,22,25.4,3.201,4.3,6.4,1.02,7.19
290 DATA 14,22,29,3.16,5.3,5.6,1.25,8.2
300 DATA 14,26,35.1,3.547,5.3,7,1.28,6.59
310 DATA 16,26,38.4,3.922,5.6,6,1.36,8.27
320 DATA 14,30,42,3.737,7.1,8.7,1.75,5.37
330 DATA 16,31,47.2,4.367,5.8,7.1,1.39,6.53
340 DATA 14,34,48.6,3.984,7.1,10.2,1.76,4.56
350 DATA 18,35,57.6,5.31,6.3,6.7,1.49,6.94
360 DATA 18,40,68.4,5.639,6.3,8.2,1.52,5.67
370 DATA 21,44,81.6,7.231,6.6,7,1.57,7.06
380 DATA 18,50,88.9,6.386,7.9,11,1.94,4.21
390 DATA 21,50,94.5,7.915,6.9,7.8,1.6,5.96
400 DATA 24,55,114.9,9.31,7,7.5,1.68,6.66
410 DATA 24,62,131,10.208,7.4,8.1,1.71,5.72
420 DATA 21,68,140.9,9.086,8.7,12.4,2.12,3.73
430 DATA 24,68,154,9.848,9.5,10.2,2.26,4.52
440 DATA 24,76,176,10.525,9.5,11.8,2.29,3.91
450 DATA 24,84,196,11.327,9.5,13.3,2.31,3.47
460 DATA 27,84,213,12.287,10.5,11,2.49,4.19
470 DATA 24,94,222,12,52,9.6,15.1,2.33,3.06
480 DATA 27,94,243,13.191,10.5,12.8,2.53,3.62
490 DATA 30,99,269,15.418,10.9,11.4,2.57,4.23
500 DATA 30,108,299,16.257,11.1,12.3,2.61,3.75
510 DATA 30,116,329,16.956,11.1,13.8,2.64,3.36
520 DATA 33,118,359,18.073,12,12.6,2.84,3.87
530 DATA 33,130,406,19.192,12.1,13.8,2.88,3.36
540 DATA 36,135,439,21.33,12.3,13,2.53,3.77
550 DATA 33,141,448,20.147,12.2,15.4,2.92,3.01
560 DATA 36,150,504,22.406,12.6,14.6,2.99,3.18
570 DATA 36,160,542,23.407,12.7,15.7,3.02,2.94
580 DATA 36,170,580,24.596,12.7,17,3.04,2.73
590 DATA 36,182,623,26.339,12.7,18.2,3.05,2.55
600 DATA 36,194,664,27.915,12.8,19.4,3.07,2.39
610 DATA 33,201,684,24.081,16.6,24.9,4.12,1.86
620 DATA 36,210,719,30.452,12.9,20.9,3.09,2.21
630 DATA 33,221,757,26.296,16.7,27.6,4.15,1.68
640 DATA 36,230,837,27.284,17.4,26.8,4.3,1.73
650 S1=M1/2
```

APPENDIX I-3.2 (Cont.)

```
660  IF S1>837 THEN 1330
670  FOR I=1 TO 40
680  S2=A[I,3]
690  IF S2<S1 THEN 1240
700  D=B[2]*12/A[I,7]
710  E=B[2]*A[I,8]*12
720  FOR T=1 TO G
730  IF C[T]=1 THEN 760
740  F1=M[T]*12/S2
750  GOTO 780
760  M[T]=M1
770  GOTO 740
780  H=53*SQR(C[T])
790  L=119*SQR(C[T])
800  IF A[I,5]<B[2] THEN 830
810  F[T]=24
820  GOTO 960
830  IF A[I,6]<B[2] THEN 850
840  GOTO 1090
850  IF D <= H THEN 1090
860  IF D >= L THEN 890
870  F[T]=24-D↑2/(1181*C[T])
880  GOTO 900
890  F[T]=170000*C[T]/D*2
900  IF F[T] >= 22 THEN 1090
910  S[T]=12000*C[T]/E
920  IF S[T] >= 22 THEN 1090
930  IF F[T] >= S[T] THEN 950
940  F[T]=S[T]
950  IF F1>1.03*F[T] THEN 1240
960  IF R<R5 THEN 1070
970  F2=R/A[I,4]
980  IF F2>14.5 THEN 1240
990  W[T]=A[I,2]
1000 PRINT "UNBRACED BEAM SEGMENT NO."T;"AL.BEND.FB="F[T];
1010 PRINT
1020 PRINT "WF"A[I,1];"X"A[I,2];"ITERATION'S CYCLE"I;
1030 PRINT
1040 PRINT "NOM.BEND.STRESS IS"F1;"NCM.SHEAR STRESS IS"F2
1050 PRINT
1060 GOTO 1140
1070 F2=R5/A[I,4]
1080 GOTO 1080
1090 F[T]=22
1100 S1=M[T]*6/11
1110 IF S1>1110 THEN 1330
1120 IF S2<S1 THEN 1240
1130 GOTO 960
1140 NEXT T
1150 FOR T=1 TO G
1160 IF T=G THEN 1180
1170 IF W[T] <= W[T+1] THEN 1200
1180 W=W[T]
1190 GOTO 1210
1200 W=W[T+1]
1210 NEXT T
1220 A=A[I,1]
1230 GOTO 1260
1240 NEXT I
1250 GOTO 1330
1260 PRINT
1270 PRINT
1280 PRINT "FINAL SELECTED WF SECTION IS"
1290 PRINT TAB30,A;"X"W;
1300 PRINT
```

APPENDIX I-3.2 (*Cont.*)

```
1310 PRINT
1320 GOTO 1360
1330 PRINT "NO WF SOLUTION FROM THE SUPPLIED ROLLED BEAMS"
1340 GOTO 1360
1350 REM IF A NEXT PROBLEM FOLLOWS INPUT Z=1
1360 DISP "NEXT PROBLEM? Z=";
1370 INPUT Z
1380 IF Z=1 THEN 1410
1390 REWIND
1400 END
1410 LOAD 2
```

APPENDIX II   FORTRAN IV PROGRAM FOR OPTIMIZATION OF PLATE GIRDER WITHOUT STIFFENERS (EXAMPLE 10.3)

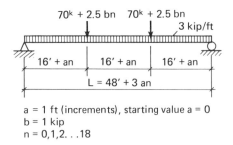

$a = 1$ ft (increments), starting value $a = 0$
$b = 1$ kip
$n = 0, 1, 2 \ldots 18$

PAGE II-1.1

75650 01   04-30-71   20,792   PROGRAM BY RUSSELL L. SILL AND JAMES T. BANKS

OPTIMIZATION OF CLASSICAL GIRDER WITHOUT STIFFENERS

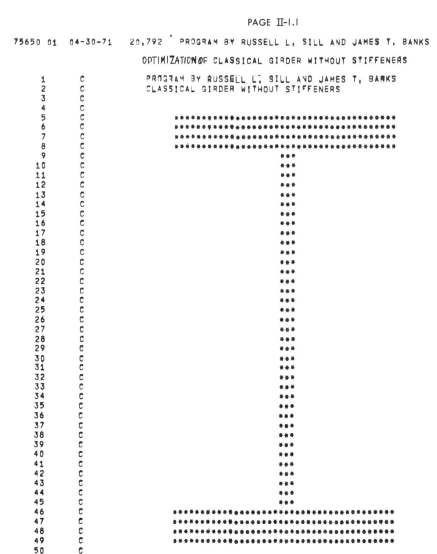

APPENDIX II (Cont.)

PAGE II-1.2

```
75650 01   04-30-71   20,792    PROGRAM BY RUSSELL L, SILL AND JAMES T, BANKS
                                        CLASSICAL GIRDER WITHOUT STIFFENERS

 51      C      THIS PROGRAM UTILIZES AN OPTIMIZATION PROCESS DEVELOPED BY EDWARD
 52      C      C, HOLT AND GARY L, WEITHECKER,  IT MAY BE FOUND IN THE ASCE
 53      C      JOURNAL OF THE STRUCTURAL DIVISION, OCTOBER,1969 ON PAGES 2205 TO
 54      C      2217
 55     10      FORMAT(1H ,6F10,5)
 56     30      FORMAT(1H ,2F10,1)
 57     40      FORMAT(1H1,52HNO VALUES COULD BE DETERMINED FOR THE GIRDER OF SPAN
 58             1,F6,2,5H FEET)
 59     50      FORMAT(1H1,35HOPTIMAL VALUES FOR A GIRDER OF SPAN,F7,2,20H FEET AR
 60             1E AS FOLLOWS//22H THE MAXIMUM SHEAR WAS,F6,2,35H KIPS   WHILE THE M
 61             2AXIMUM MOMENT WAS,F9,2,10H INCH KIPS//)
 62     60      FORMAT(1H0,27HTHE MOMENT PARAMETER (X) IS,9X,F6,2,10X,26HTHE DEPTH
 63             1 PARAMETER (Y) IS,10X,F6,2//26H THE AREA PARAMETER (Z) IS,11X,F6,2
 64             2,10X,34HTHE WEB THICKNESS PARAMETER (U) IS,2X,F6,2//33H THE FLANGE
 65             3 AREA PARAMETER (W) IS,4X,F6,2,10X,16HTHEORETICAL AREA,21X,F6,2//)
 66     70      FORMAT(1H0,5HDEPTH,13X,1H=,1X,F12,4,2X,6HINCHES//14H WEB THICKNESS
 67             1,5X,1H=,1X,F12,4,2X,6HINCHES//9H WEB AREA,10X,1H=,1X,F12,4,2X,13HS
 68             2QUARE INCHES//13H FLANGE WIDTH,6X,1H=,1X,F12,4,2X,6HINCHES//17H FL
 69             3ANGE THICKNESS,2X,1H=,1X,F12,4,2X,6HINCHES//12H FLANGE AREA,7X,1H=
 70             4,1X,F12,4,2X,13HSQUARE INCHES//11H TOTAL AREA,8X,1H=,1X,F12,4,2X,1
 71             53HSQUARE INCHES//13H SHEAR STRESS,6X,1H=,1X,F12,4,2X,3HKSI//16H SE
 72             6CTION MODULUS,3X,1H=,1X,F12,4,2X,12HCUBIC INCHES//15H BENDING STRE
 73             7SS,4X,1H=,1X,F12,4,2X,3HKSI//7H WEIGHT,12X,1H=,1X,F12,4,2X,15HPOUN
 74             8DS PER FOOT)
 75             DIMENSION R(8),C1(8),P(8),C2(8),C3(8),Q(8)
 76             REAL MAXMOM,MAXSHR
 77             DO 260    J = 1,8
 78             READ(5,10) R(J),C1(J),P(J),C2(J),C3(J),Q(J)                            2
 79    260      CONTINUE                                                               5
 80             READ(5,30) FY,E                                                        7
 81             B1 = FY**0.75/E**0.25                                                 10
 82             B2 = (E**0.25)*(FY**0.25)                                             11
 83             GAMMA = (E/FY)**0.50                                                  12
 84             ALPHA = 260.0/GAMMA                                                   13
 85             DO 1220 I = 1,19                                                      14
 86             AI = I - 1                                                            15
 87             SPAN = 48.0 + 3.0*AI                                                  16
 88             MAXSHR = 142.0 + 7.0*AI                                               17
 89             MAXMOM = 12.0*(1984.0 + 218.0*AI + 5.875*AI**2)                       18
 90             B3 = MAXMOM/MAXSHR**1.5                                               19
 91             X = B1*B3                                                             20
 92             II = 1                                                                21
 93    520      IF(X.LE.R(II)) GO TO 560                                              22
 94             II = II + 1                                                           25
 95             IF(II.EQ.9) GO TO 1210                                                26
 96             GO TO 520                                                             29
 97    560      IF(II.EQ.1)  GO TO 1210                                               30
 98             Y = C1(II-1)*X**P(II-1)                                               33
 99             DEPTH = (Y/B1)*MAXSHR**0.5                                            34
100             Z = C2(II-1)*X**Q(II-1) + C3(II-1)                                    35
```

APPENDIX II (Cont.)

PAGE II-1.3

```
75650 01   04-30-71   20,792   PROGRAM BY RUSSELL L. SILL AND JAMES T. BANKS
                               CLASSICAL GIRDER WITHOUT STIFFENERS
101          AREAT = Z*MAXSHR/FY                                              36
102          IF(Y.LT.2.33976) GO TO 760                                       37
103          IF(Y.LT.3.38096.AND.Y.GT.2.33976) GO TO 780                      40
104          ALPHA1 = ALPHA**1.5                                              43
105          IF(Y.LT.ALPHA1.AND.Y.GT.3.38096)  GO TO 800                      44
106          IF(Y.GT.ALPHA1)   U = Y/ALPHA                                    47
107          GO TO 810                                                        50
108   760    U = 2.5/Y                                                        51
109          GO TO 810                                                        52
110   780    U = 1.06848                                                      53
111          GO TO 810                                                        54
112   800    U = 0.7119*Y**0.333                                              55
113   810    WEBT = (MAXSHR**0.5)*U/B2                                        56
114          AREAW = DEPTH*WEBT                                               57
115          B4 = Y/U                                                         58
116          IF(B4.GT.5.66) GO TO 900                                         59
117          W = (X/(0.6*Y))-(U*Y/6.0)                                        62
118          GO TO 930                                                        63
119   900    W = ((5.0*X/(3.0*Y)-U*Y/6.0 + GAMMA*Y*(Y - 5.6569*U)/2000.0) + ((5 64
120          1.0*X/(3.0*Y)-U*Y/8.0 + GAMMA*Y*(Y - 5.6569*U)/2000.0)**2 + 4.0*GAM
121          2MA*U*Y**2*(Y-5.6569*U)/12000.0)**0.5)/2.0
122   930    AREAFL = MAXSHR*W/FY                                             65
123          AREATS = AREAW + 2.0*AREAFL                                      66
124          BF = DEPTH/5.0                                                   67
125          FLANGT = AREAFL/BF                                               68
126          SX = AREAFL*DEPTH + DEPTH**2*WEBT/6.0                            69
127          FV = MAXSHR/AREAW                                                70
128          FB = MAXMOM/SX                                                   71
129          WEIGHT = AREATS*3.4                                              72
130          WRITE(6,50) SPAN,MAXSHR,MAXMOM                                   73
131          WRITE(6,60) X,Y,Z,U,W,AREAT                                      76
132          WRITE(6,70) DEPTH,WEBT,AREAW,BF,FLANGT,AREAFL,AREATS,FV,SX,FB,WEIG 79
133          1HT
134          GO TO 1220                                                       82
135   1210 WRITE(6,40) SPAN                                                   83
136   1220 CONTINUE                                                           86
137          STOP                                                             88
138          END                                                              89
23707 WORDS OF MEMORY USED BY THIS COMPILATION
```

APPENDIX III OPTIMIZATION OF PLATE GIRDER WITH STIFFENERS (EXAMPLE 10.4)

PAGE III-1.1

75651 01 04-30-71 20,869 PROGRAM BY RUSSELL L. SILL AND JIM BANKS

OPTIMIZATION OF CLASSICAL GIRDER WITH STIFFENERS

```
C     PROGRAM BY RUSSELL L. SILL AND JIM BANKS
C     CLASSICAL GIRDER WITH STIFFENERS
```

APPENDIX III (Cont.)

PAGE III-1.2

```
75651 01   04-30-71   20.869   PROGRAM BY RUSSELL L. SILLI AND JIM BANKS
                                CLASSICAL GIRDER WITH STIFFENERS

51    C
52    C    FORMAT AND DIMENSION STATEMENTS
53    C
54    50  FORMAT(1H1,51HOPTIMAL VALUES FOR A GIRDER WITH STIFFENERS OF SPAN,
55        1F7.2,20H FEET ARE AS FOLLOWS//22H THE MAXIMUM SHEAR WAS,F6.1,35H K
56        2IPS  WHILE THE MAXIMUM MOMENT WAS,F8.2,10H INCH KIPS//)
57    70  FORMAT(1H0,5HDEPTH,15X,1H=,1X,F12.4,2X,6HINCHES//14H WEB THICKNESS
58        1,5X,1H=,1X,F12.4,2X,6HINCHES//9H WEB AREA,10X,1H=,1X,F12.4,2X,13HS
59        2QUARE INCHES//13H FLANGE WIDTH,6X,1H=,2X,F12.4,2X,6HINCHES//17H FL
60        3ANGE THICKNESS,2X,1H=,1X,F12.4,2X,6HINCHES//12H FLANGE AREA,7X,1H=
61        4,1X,F12.4,2X,13HSQUARE INCHES//11H TOTAL AREA,8X,1H=,1X,F12.4,2X,1
62        53HSQUARE INCHES//13H SHEAR STRESS,6X,1H=,1X,F12.4,2X,3HKSI//16H SE
63        6CTION MODULUS,3X,1H=,1X,F12.4,2X,12HCUBIC INCHES//15H BENDING STRE
64        7SS,4X,1H=,1X,F12.4,2X,3HKSI//7H WEIGHT,12X,1H=,1X,F12.4,2X,15HPOUN
65        8DS PER FOOT)
66    80  FORMAT(3F10.3)
67    90  FORMAT(6F12.4)
68    100 FORMAT(1H1,61HTHE DIMENSIONS FOR A CLASSICAL GIRDER WITH STIFFENER
69        1S OF SPAN,F7.2,34H FEET AND WITH A SLENDERNESS RATIO,F6.1,4H ARE//
70        2,126H ITERATION   DEPTH      WEB          WEB       FLANGE    FLANGE
71        3   FLANGE    TOTAL     SHEAR    SECTION    BENDING    WEIGHT/
72        4,115H                        THICKNESS    AREA       WIDTH    THICKNES
73        5S       AREA      AREA      STRESS   MODULUS    STRESS/127H
74        6         (IN.)     (SQ.IN.)   (IN.)     (IN.)
75        7 (SQ.IN.)   (KSI)    (CUB.IN.)    (KSI)    (LBS/FT)/)
76    110 FORMAT(1H0,4X,I2,6X,F6.2,5X,F5.3,6X,F6.3,4X,F6.3,5X,F5.3,6X,F6.3,5
77        1X,F6.2,5X,F4.2,5X,F7.2,5X,F5.2,6X,F7.3)
78    120 FORMAT(1H0,8X,1H*//11H ***********,12H OPTIMUM ***,1X,F6.2,5X,F5.3,6
79        1X,F6.3,4X,F6.3,5X,F5.3,6X,F6.3,5X,F6.2,5X,F4.2,5X,F7.2,5X,F5.2,5X,
80        2F7.3,//11H ***********//9X,1H*)
81        DIMENSION SHRSTR(4),ASPRAT(4),PCWAFS(4),SLENDR(4),DEPTH(4),WEBT(4)
82        1,AREAW(4),FLANW(4),FLANT(4),AREAF(4),AREAT(4),FV(4),FB(4),WGT(4)
83        REAL MAXSHR,MAXMOM,NUMSTF
84
85    C    READ IN THE VALUES FOR MAXIMUM SHEAR STRESS, ASPECT RATIO, AND
86    C    THE PERCENTAGE OF THE WEB AREA IN THE STIFFENERS
87
88        DO 4020 N= 1,4
89        READ(5,80) SHRSTR(N),ASPRAT(N),PCWAFS(N)
90   4020 CONTINUE
91
92    C    DETERMINE VALUES FOR THE GIRDERS WITH A VARIETY OF SLENDERNESS
93    C    RATIOS AND SELECT THE BEST FOR A VARIETY OF SPANS
94
95        DO 4470 M = 1,19
96        READ(5,90) SPAN,MAXSHR,MAXMOM,DEPTHT,AREATS,SX
97        DO 4469 L = 1,4
98        CON = 400.0
99        J = 1
100       DEPTH1 = DEPTHT - 12.0
```

APPENDIX III (Cont.)

PAGE III-1.3

75651 01   04-30-71   20.869   PROGRAM BY RUSSELL L. SILL AND JIM BANKS

CLASSICAL GIRDER WITH STIFFENERS

```
101            SLENDR(L) = 320 - 20*(L-1)                                    101   15
102            WRITE(6,100) SPAN,SLENDR(L)                                   102   16
103       4090 WEBT1 = DEPTH1/SLENDR(L)                                      103   19
104            AREAF1 = (SX - DEPTH1**2*WEBT1/6.0)/(DEPTH1 + 1.0)             104   20
105            FLANW1 = DEPTH1/5.0                                           105   21
106            FLANT1 = AREAF1/FLANW1                                        106   22
107            IF(FLANT1.GE.0.99.AND.FLANT1.LE.1.01) GO TO 4210              107   23
108       4140 AREAF1 = (SX - DEPTH1**2*WEBT1/6.0)/(DEPTH1 + FLANT1)         108   26
109            FLANT2 = AREAF1/FLANW1                                        109   27
110            H1 = 0.99*FLANT1                                              110   28
111            H2 = 1.01*FLANT1                                              111   29
112            IF(FLANT2.GE.H1.AND.FLANT2.LE.H2) GO TO 4210                  112   30
113            FLANT1 = FLANT2                                               1133  33
114            GO TO 4140                                                    114   34
115       4210 AREAW1= DEPTH1*WEBT1                                          115   35
116            FV1 = MAXSHR/AREAW1                                           116   36
117            IF(FV1.GT.SHRSTR(L)) DEPTH1 = DEPTH1 + 1.0                    117   37
118            IF(FV1.GT.SHRSTR(L)) GO TO 4090                               118   40
119            AREAT1 = AREAW1 + 2.0*AREAF1                                  119   43
120            IF(J.EQ.1) GO TO 4452                                         120   44
121       4290 AREAST = PCWAFS(L)*AREAW2                                     121   47
122            VOLST = DEPTH2*AREAST                                         122   48
123            STFSP = ASPRAT(L)*DEPTH2                                      123   49
124            NUMSTF = SPAN*120/STFSP                                       124   50
125            VOLSTT = NUMSTF*VOLST                                         125   51
126            AREAS1 = VOLSTT/(SPAN*12.0)                                   126   52
127            AREAT3 = AREAT2 + AREAS1                                      1227  55
128            IF(J.EQ.1) GO TO 4453                                         128   56
129            FV3 = MAXSHR/AREAW2                                           129   57
130            FB3 = MAXMOM/SX                                               130   58
131            WEIGHT = AREAT4*3.4                                           131   59
132            J = J - 1                                                     132   60
133            IF(AREAT3.GT.AREAT4) GO TO 4460                               133   61
134       4459 WRITE(6,110) J,DEPTH2,WEBT2,AREAW2,FLANW2,FLANT3,AREAF2,AREAT4,FV3  134  64
135           1,SX,FB3,WEIGHT                                                135
136       4450 J = J + 2                                                     136   67
137       4452 DEPTH2 = DEPTH1                                               137   68
138            WEBT2 = WEBT1                                                 138   69
139            AREAW2 = AREAW1                                               139   70
140            FLANW2 = FLANW1                                               140   71
141            FLANT3 = FLANT1                                               141   72
142            AREAF2 = AREAF1                                               142   73
143            AREAT2 = AREAT1                                               143   74
144            DEPTH1 = DEPTH1 + 1.0                                         144   75
145            IF(DEPTH1.GT.CON) GO TO 4469                                  145   78
146            IF(J.EQ.1) GO TO 4290                                         146   79
147       4453 AREAT4 = AREAT3                                               147   82
148            IF(J.EQ.1) J=J+1                                              148   83
149            H3 = FLANW1/(2.0*FLANT1)                                      149   86
150            IF(H3.GT.15.8) GO TO 4469                                     150   87
```

APPENDIX III (Cont.)

PAGE III-I.4

75651 01   04-30-71   20,869   PROGRAM BY RUSSELL L. SILL AND JIM BANKS
CLASSICAL GIRDER WITH STIFFENERS

```
151             IF(J.EQ.25) GO TO 4469                                       151   90
152             GO TO 4090                                                   152   91
153        4460 IF(DEPTH1.GT.CON) GO TO 4459                                 153   94
154             IF(CON.NE.400.0) GO TO 4459                                  154   97
155             WRITE(6,120)  DEPTH2,WEBT2,AREAW2,FLANW2,FLANT3,AREAF2,AREAT4,FV3  155  100
156            1,SX,FB3,WEIGHT                                               156
157             DEPTH(L) = DEPTH2                                            157  103
158             WEBT(L) = WEBT2                                              158  104
159             AREAW(L) = AREAW2                                            159  105
160             FLANW(L) = FLANW2                                            160  106
161             FLANT(L) = FLANT3                                            161  107
162             AREAF(L) = AREAF2                                            162  108
163             AREAT(L) = AREAT4                                            163  109
164             FV(L) = FV3                                                  164  110
165             FB(L) = FB3                                                  165  111
166             WGT(L) = WEIGHT                                              166  112
167             CON = DEPTH(L) + 4.0                                         167  113
168             IF(J.EQ.25) GO TO 4469                                       168  114
169             GO TO 4450                                                   169  117
170        4469 CONTINUE                                                     170  118
171             I = 1                                                        171  120
172             IF(AREAT(I) - AREAT(I + 1)) 4500,4500,4520                   172  121
173        4500 IF(AREAT(I) - AREAT(I + 2)) 4510,4510,4550                   173  122
174        4510 IF(AREAT(I) - AREAT(I + 3)) 4590,4590,4570                   174  123
175        4520 I = I + 1                                                    175  124
176             IF(AREAT(I) - AREAT(I + 1)) 4540,4540,4550                   176  125
177        4540 IF(AREAT(I) - AREAT(I + 2)) 4590,4590,4570                   177  126
178        4550 I = I + 1                                                    178  127
179             IF(AREAT(I) - AREAT(I + 1)) 4590,4590,4570                   179  128
180        4570 I = I + 1                                                    180  129
181        4580 WRITE(6,50) SPAN,MAXSHR,MAXMOM                               181  130
182             WRITE(6,70) DEPTH(I),WEBT(I),AREAW(I),FLANW(I),FLANT(I),AREAF(I),A  182  133
183            1REAT(I),FV(I),SX,FB(I),WGT(I)                                183
184        4470 CONTINUE                                                     184  136
185             STOP                                                         185  138
186             END                                                          186  139
```

23741 WORDS OF MEMORY USED BY THIS COMPILATION

APPENDIX IV-1 MINICOMPUTER OPTIMIZATION PROGRAM OF GIRDERS WITHOUT STIFFENERS, SUFFICIENTLY LATERALLY BRACED (EXAMPLE 10.3), FOR THE HP9820A

```
0:
FXD 5;PRT "*****
***********";
SPC 1⊢
1:
PRT "   PROGRAM F
OR  ","OPTIMIZA
TION OF "⊢
2:
PRT "   PLATE GIR
DER  ","  PROPOR
TIONS   "⊢
3:
PRT "   WITHOUT
","   STIFFENE
RS    "⊢
4:
SPC 1;PRT "*****
***********";
SPC 4⊢
5:
ENT "MMAX K-FT="
,A,"VMAX KIPS=",
B,"FY KSI=",C⊢
6:
SPC 2;PRT "MMAX=
",A,"VMAX=",B,"F
Y=",C⊢
7:
29000→R1;(C↑3/R1
)↑0.25→R2⊢
8:
(R1*C)↑0.25→R3;(
R1/C)↑0.5→R4⊢
9:
260/R4→R5;B↑1.5→
X;12*A*R2/X→R6⊢
10:
IF R6≤1.16988;2.
33976→R7;0→R8;1.
42465→R9⊢
11:
IF R6≤1.16988;1.
66667→R10;1→R11;
GTO 26⊢
12:
IF R6≤2.44274;2.
16322→R7;0.5→R8;
3.08182→R9⊢
13:
IF R6≤2.44274;0→
R10;0.5→R11;GTO
26⊢
```

```
14:
IF R6≤3.25699;3.
38096→R7;0→R8;0.
98591→R9⊢
15:
IF R6≤3.25699;2.
40833→R10;1→R11;
GTO 26⊢
16:
IF R6≤24.8816;2.
03827→R7;0.42857
→R8;2.86190→R9⊢
17:
IF R6≤24.8816;0→
R10;0.57143→R11;
GTO 26⊢
18:
IF R6≤29.3494;8.
08156→R7;0→R8;0.
41246→R9⊢
19:
IF R6≤29.3494;7.
69705→R10;1→R11;
GTO 26⊢
20:
IF R6≤192.459;2.
30709→R7;0.37097
→R8;2.49757→R9⊢
21:
IF R6≤192.459;0→
R10;0.61270→R11;
GTO 26⊢
22:
IF R6≤213.706;16
.23598→R7;0→R8;0
.20090→R9⊢
23:
IF R6≤213.706;24
.01655→R10;1→R11
;GTO 26⊢
24:
IF R6>213.706;2.
71532→R7;0.33333
→R8;1.87382→R9⊢
25:
IF R6>213.706;0→
R10;0.66667→R11⊢
26:
R6↑R8*R7→R12;R12
*√B/R2→R13;R9*R6
↑R11+R10→R14⊢
27:
R14*B/C→R15⊢
```

```
28:
PRT "X=",R6,"Y="
,R12,"Z=",R14,"H
=",R13,"A=",R15⊢
29:
IF R12≤2.33976;2
.5/R12→R16;GTO 3
3⊢
30:
IF R12≤3.38096;1
.06848→R16;GTO 3
3⊢
31:
0.60067*R5↑1.5→R
34;IF R12<R34;0.
7119*R12↑0.33333
→R16;GTO 33⊢
32:
IF R12>R34;R12/R
5→R16⊢
33:
R16*√B/R3→R17;R1
3*R17→R18⊢
34:
PRT "U=",R16,"TW
EB=",R17,"AWEB="
,R18;R13/R17→R35
⊢
35:
0.5*(R15-R18)→R1
9⊢
36:
R12/R16→R20⊢
37:
IF R20≤5.65685;R
6/(0.6*R12)-R12*
R16/6→R21;GTO 39
⊢
38:
1.66667*R6/R12-R
12*R16/6+0.0005*
R4*R12(R12-5.656
85*R16)→R21⊢
39:
B*R21/C→R22;IF R
22≤R19;R19→R22⊢
40:
PRT "A-FLANGE=",
R22⊢
41:
95/√C→R25⊢
42:
2.5→Y⊢
```

APPENDIX IV-1 (Cont.)

```
43:
R13/Y→R23;R22/R2
3→R24;R23/(2*R24
)→R26⊢
44:
IF R26≤R25;GTO 4
8⊢
45:
Y+0.5→Y;IF Y≤6.0
;GTO 43⊢
46:
PRT "WEB DEPTH I
S   ","TOO LARGE
 TRY   "⊢
47:
PRT "SOME OTHER
SOLU-","        TI·
ON      ";GTO 59
⊢
48:
SPC 5;FXD 2⊢
49:
PRT "FINAL DIMEN
SIONS","OF PLATE
 GIRDER ";SPC 2⊢
50:
PRT "WEB DEPTH="
,R13,"WEB THICKN
ESS=",R17⊢
51:
PRT "FLANGE WIDT
H=",R23,"FLANGE
THICKN.=",R24⊢
52:
R18+2*R22→R27;3.
4*R27→R28⊢
53:
PRT "CROSS SECTI
ON A=",R27,"WEIG
HT LB/FT=",R28⊢
54:
R18*R13↑2/12+2*R
22*(0.5*(R13+R24
))↑2→R29⊢
55:
R29/(0.5*R13+R24
)→R30⊢
56:
12*A/R30→R31;B/R
18→R32⊢
57:
PRT "MOM.INERTIA
=",R29,"SEC.MODU
LUS=",R30⊢
58:
PRT "FB IN KSI="
,R31,"FV IN KSI=
",R32,"SLEND.=",
R35⊢
59:
ENT "PROBLEM NO.
=",R33⊢
60:
IF FLG 13=1;CFG
13;GTO 62⊢
61:
FXD 0;SPC 6;PRT
"PROBLEM NO.=",R
33;FXD 5;GTO 5⊢
62:
PRT "END OF PROB
LEMS"⊢
63:
END.⊢
R126
```

APPENDIX IV-1 (Cont.)

```
      ***************
        PROGRAM FOR
       OPTIMIZATION OF
         PLATE GIRDER
         PROPORTIONS
           WITHOUT
          STIFFENERS

      ***************
```

```
MMAX=                      FINAL DIMENSIONS
       1984.00000          OF PLATE GIRDER
VMAX=
        142.00000          WEB DEPTH=
FY=                                  70.47
         36.00000          WEB THICKNESS=
X=                                     .50
         15.84593          FLANGE WIDTH=
Y=                                   15.66
          6.66056          FLANGE THICKN.=
Z=                                     .62
         13.87789          CROSS SECTION A=
H=                                   54.74
         70.47373          WEIGHT LB/FT=
A=                                  186.12
         54.74057          MOM.INERTIA=
U=                                39271.24
          1.33943          SEC.MODULUS=
TWEB=                              1095.09
           .49933          FB IN KSI=
AWEB=                                21.74
         35.18974          FV IN KSI=
A-FLANGE=                             4.04
          9.77541          SLEND.=
                                    141.14
```

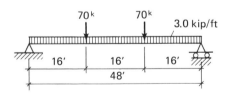

APPENDIX IV-2 MINICOMPUTER OPTIMIZATION PROGRAM OF GIRDERS WITHOUT STIFFENERS (EXAMPLE 10.3), FOR THE HP9830A

```
10 PRINT
20 PRINT
30 PRINT
40 PRINT
50 PRINT "OPTIMUM DESIGN OF SYMMETRICAL,NONCOMPACT,LATERALLY BRACED AT DIS-"
60 PRINT "TANCES LU PLATE GIRDERS WITHOUT WEB STIFFENERS"
70 PRINT "MADE OF STEEL WITH FY= 36 CR 50 KSI"
80 DIM A[8,7],B[8,7],C[10],T[10]
90 M1=V=F=1
100 DISP "M.MAX"M1;"V.MAX"V;"FY"F;
110 INPUT M1,V,F
120 M1=M1*12
130 E=29000
140 X=F↑0.75*M1/(E↑0.25*V↑1.5)
150 G=SQR(E/F)
160 A1=260/G
170 REM A=CROSS SEC.AREA,G=GAMA,A1=ALPHA,A2=AF,A3=AW,W1=GIRDER WEIGHT,TABLE
180 REM ONE IS STORED AS A(I,J) FOR F36 AND AS B(M,N) FOR FY=50 STEEL
190 FOR I=1 TO 8
200 FOR J=1 TO 7
210 READ A[I,J]
220 NEXT J
230 NEXT I
240 DATA 1,0.58494,2.33976,0,1.42465,1.66667,1
250 DATA 2,1.16988,2.16322,0.5,3.08182,0,0.5
260 DATA 3,2.44274,3.38096,0,0.98591,2.40833,1
270 DATA 4,3.25699,2.03827,0.42857,2.8619,0,0.57143
280 DATA 5,24.8816,8.08156,0,0.41246,7.69705,1
290 DATA 6,29.3494,2.30709,0.37097,2.49757,0,0.6127
300 DATA 7,192.459,16.23598,0,0.2009,24.01655,1
310 DATA 8,213.706,2.71532,0.33333,1.87382,0,0.66667
320 PRINT
330 PRINT
340 FOR M=1 TO 8
350 FOR N=1 TO 7
360 READ B[M,N]
370 NEXT N
380 NEXT M
390 DATA 1,0.58494,2.33976,0,1.42465,1.66667,1
400 DATA 2,1.16988,2.16322,0.5,3.08182,0,0.5
410 DATA 3,2.44274,3.38096,0,0.98591,2.40833,1
420 DATA 4,3.25699,2.03827,0.42857,2.8619,0,0.57143
430 DATA 5,24.8816,8.08156,0,0.41246,7.69705,1
440 DATA 6,28.7123,2.28894,0.37545,2.53367,0,0.60845
450 DATA 7,269.44,18.74423,0,0.17394,29.44384,1
460 DATA 8,308.711,2.77342,0.33333,1.82153,0,0.66667
470 IF F=36 THEN 490
480 IF F=50 THEN 800
490 IF X<0.58494 THEN 1110
500 FOR I=2 TO 8
510 IF X<A[I,2] THEN 540
520 NEXT I
530 IF X>213.706 THEN 570
540 R=I-1
550 PRINT "CASE IS "A[R,1]
560 GOTO 580
570 PRINT "CASE IS 8"
580 Y=A[R,3]*(X↑(A[R,4]))
590 Z=A[R,5]*(X↑(A[R,7]))+A[R,6]
600 H=Y*(E↑0.25)*SQR(V)/F↑0.75
```

APPENDIX IV-2 (Cont.)

```
610 A=Z*V/F
620 IF Y <= 2.33976 THEN 660
630 IF Y <= 3.38096 THEN 680
640 IF Y <= 16.65422 THEN 700
650 IF Y>16.65422 THEN 720
660 U=2.5/Y
670 GOTO 730
680 U=1.06848
690 GOTO 730
700 U=0.7119*(Y↑0.3333333)
710 GOTO 730
720 U=Y/A1
730 IF Y/U <= 5.65685 THEN 760
740 W=5*X/(3*Y)-U*Y/6+G*Y*(Y-5.65685*U)/2000
750 GOTO 770
760 W=X/(0.6*Y)-U*Y/6
770 T=U*SQR(V)/(E↑0.25*F↑0.25)
780 A2=W*V/F
790 GOTO 920
800 FOR M=2 TO 8
810 IF X<0.58494 THEN 1110
820 IF X<B[M,2] THEN 850
830 NEXT M
840 IF X>308.711 THEN 880
850 R=M-1
860 PRINT "CASE IS"B[R,1]
870 GOTO 890
880 PRINT "CASE IS 8"
890 Y=B[R,3]*X↑(B[R,4])
900 Z=B[R,5]*X↑(B[R,7])+B[R,6]
910 GOTO 600
920 FOR K=1 TO 10
930 C[K]=H/(1+0.5*K)
940 T[K]=A2/C[K]
950 IF (C[K]/(2*T[K])) <= (95/SQRF) THEN 990
960 NEXT K
970 K=K-1
980 GOTO 1270
990 B=C[K]
1000 REM C IS FLANGE THICKNESS,I1 MOM.INERT.,S1 SEC.MOD.,F1 BEND.STRESS,
1010 REM I2 IS 0.5IY, R2 RAD.GYR.,L4=LU,FB AL.BEND.STRESS
1020 C=T[K]
1030 IF (H/T)>(760/(SQR(0.6*F))) THEN 1430
1040 M1=M1/12
1050 FIXED 2
1060 A3=H*T
1070 A4=A3+2*A2
1080 D1=ABS((A-A4)*100/A)
1090 W1=3.4*A4
1100 GOTO 1140
1110 PRINT "WEB IS HEAVY ENOUGH TO CARRY SHEAR AND MOMENT"
1120 PRINT "NO FLANGES ARE THEREFORE REQUIRED."
1130 GOTO 1470
1140 PRINT TAB20"SOLUTION FOR MAX.MOMENT  OF"M1"FT-KIP";"MAX.SHEAR V="V
1150 PRINT "STEEL IS FY="F
1160 PRINT TAB20"GIRDER DEPTH H="H
1170 PRINT TAB20"WEB THICKNESS T="T
1180 PRINT "FLANGES BXT"B;"X"C;
```

APPENDIX IV-2 (*Cont.*)

```
1190 PRINT
1200 PRINT "DISCREPANCY D1="D1"%"
1210 PRINT "GIRDER WEIGHT IS"W1"LB/FT"
1220 PRINT
1230 PRINT
1240 V1=V/(A3)
1250 REM V1= MAX.SHEAR STRESS
1260 GOTO 1290
1270 PRINT "FLANGE PROPORTIONS DO NOT SATISFY AISC WIDTH-THICKNESS RATIOS,SORRY"
1280 GOTO 990
1290 PRINT TAB15"FINAL GIRDER DIMENSIONS"
1300 PRINT "DEPTH H="H"IN.";"WEB THICKNESS T="T;
1310 PRINT
1320 PRINT "FLANGE WIDTH BF="B;"THICKNESS TF="C;
1330 PRINT
1340 I1=(H↑3*T/12)+(2*A2)*((H/2-C/2)↑2)
1350 S1=I1/(H/2+C)
1360 F1=12*M1/S1
1370 I2=B↑3*C/12
1380 R2=SQR(I2/(A2+A3/6))
1390 L4=SQR(((102000/F)*(R2↑2)))/12
1400 PRINT "UNBRACED LENGTH NOT LARGER THAN LU="L4;"FT"
1410 PRINT "FB="F1;"FV="V1
1420 GOTO 1470
1430 PRINT "NO DESIGN POSSIBLE WITH FB=.6*FY"
1440 F2=0.6*F*(1-(0.0005*H*T/A2)*((H/T)-(760/SQR(0.6*F))))
1450 PRINT "REDUCED BENDING STRESS FE PRIME="F2
1460 GOTO 1040
1470 END
```

APPENDIX V MINICOMPUTER OPTIMIZATION PROGRAM OF GIRDERS WITH STIFFENERS (EXAMPLE 10.4), FOR THE HP9830A

```
10 DIM Q$[10,4],A$[90],H$[90],T$[90],D$[90],S$[90],C$[90]
20 PRINT
30 PRINT TAB18"DESIGN OF SYMMETRICAL GIRDERS WITH STIFFENERS"
40 PRINT TAB8"USING AISC SPECS. AND STEEL OF ANY YIELD POINT STRESS"
50 PRINT TAB8"UNBRACED LENGTH<=LU,SO,ALLOWAB.BEND.STRESSF1=.6FY"
60 PRINT
70 PRINT
80 REM TOTAL NUMBER OF PROBLEMS IS W,MAX.10
90 W=1
100 DISP "W="W;
110 INPUT W
120 PRINT "TOTAL NUMBER OF PROBLEMS IS"W
130 PRINT
140 PRINT
150 FOR I=1 TO W
160 FOR J=1 TO 4
170 DISP "ROW"I;"COLUMN"J;
180 INPUT Q[I,J]
190 NEXT J
200 PRINT "PROBLEM NO."I
210 PRINT "L="Q[I,1];"M1="Q[I,2];"V="Q[I,3];"FY="Q[I,4];
220 PRINT
230 PRINT
240 NEXT I
250 PRINT
260 FOR I=1 TO W
270 PRINT " SOLUTION OF PROBLEM NO."I
280 PRINT
290 L=Q[I,1]
300 M1=Q[I,2]
310 V=Q[I,3]
320 F=Q[I,4]
330 F1=0.6*F
340 S1=M1*12/F1
350 L1=14000/SQR(F*(F+16.5))
360 L2=760/SQR(F1)
370 FOR B=2 TO 90
380 C[B]=L2-2+2*B
390 IF C[B]>L1 THEN 590
400 FIXED 3
410 G=12*M1*C[B]/F1
420 H[B]=1.145*(G↑0.333)
430 T[B]=H[B]/C[B]
440 A[B]=2.621*(S1↑2/(C[B]-C[B]↑2/(2*L1)))↑0.333
450 B2=83150*T[B]↑2/V
460 H2=T[B]*B2
470 IF H2 >= H[B] THEN 500
480 D[B]=0.06*SQR(C[B]/B2)*A[B]
490 GOTO 510
500 D[B]=0
510 S[B]=A[B]+D[B]
520 PRINT "B="B;"A TOT"S[B];"A STIF"I[B];
530 IF B=2 THEN 580
540 X=B-1
```

APPENDIX V   (Cont.)

```
550 IF S[B] <= S[X] THEN 580
560 IF B=90 THEN 610
570 GOTO 660
580 NEXT B
590 PRINT "THERE IS NO MINIMUM OF THE WEIGHT WITH INCREASED SLENDERNESS"
600 GOTO 1090
610 PRINT "B=90 ";"H="H[90];"T="T[90];"A="A[90];"D="D[90];"S="S[90];
620 PRINT
630 REM A1=WEB AREA,A2=AREA OF 1FLANGE,T1=WEB THICK.,T2=FLANGE THICK.,B3=FLANGE
640 REM WIDTH,I1=MOM.INERT.,S2=SEC.MOD.,R1=RT,I2=IF,F2=ACTUAL BEND.STRESS
650 X=90
660 A1=H[X]*T[X]
670 A2=(A[X]-A1)/2
680 A3=A1+2*A2
690 PRINT "A1="A1;"A2="A2;"A3="A3;
700 FOR M=2 TO 8 STEP 0.5
710 B3=H[X]/M
720 K=95/(SQR(F))
730 T2=A2/B3
740 IF B3/(2*T2) <= K THEN 770
750 NEXT M
760 PRINT "FLANGE DOES NOT SATISFY AISC WIDTH-THICK.RATIO"
770 FIXED 3
780 P=100*D[X]/A3
790 H=H[X]
800 T1=T[X]
810 I1=(H↑3*T1/12)+(2*A2)*((H/2-T2/2))↑2
820 S2=I1/(H/2+T2)
830 F2=M1*12/S2
840 IF C[X] <= (L2) THEN 870
850 F3=F1*(1-((0.0005*A1)/A2)*(C[X]-L2))
860 F1=F3
870 IF F2 <= F1 THEN 900
880 PRINT "OVERSTRESSED, FB="F2
890 GOTO 960
900 I2=B3↑3*T2/12
910 R1=SQR(I2/(A2+A1/6))
920 PRINT
930 PRINT
940 PRINT "FINAL OPTIMAL GIRDER DIMENSIONS AFTER NO. OF ITERATIONS B="(B-1)
950 PRINT
960 PRINT "H="H;"TW="T1;"BF="B3;"TF="T2;"A="A3;"D="D[X];
970 PRINT
980 PRINT "TOT.AREA WITH STIF."S[X];"PERCENTAGE P="P"%OF A";
990 PRINT
1000 PRINT "MOM.INERTIA="I1;"SEC.MOD.="S2;"RT="R1;"ACT.STRESS FB="F2;
1010 V1=V/(A1)
1020 REM V1= MAX.SHEAR STRESS FV
1030 PRINT
1040 L4=(((102000/F)*R1↑2)↑0.5)/12
1050 PRINT "UNBRACED LENGTH NOT LARGER THAN LU="L4" FT"
1060 PRINT "FV="(V1)
1070 PRINT
1080 STANDARD
1090 NEXT I
1100 SCALE 150,300,50,100
1110 XAXIS 0,50
1120 YAXIS 0,1
1130 FOR X=150 TO 300 STEP 50
1140 PLOT L3,A3
1150 NEXT L2
1160 END
```

APPENDIX VI

Table VI-1 Factors for determining J and maximum torsional shear stress in rectangular bars

b/t	2ψ	γ
1.0	0.1928	0.6753
1.1	0.1973	0.7198
1.2	0.2006	0.7578
1.3	0.2031	0.7935
1.4	0.2050	0.8222
1.5	0.2064	0.8476
1.6	0.2074	0.8695
1.8	0.2086	0.9044
2.0	0.2093	0.9300
2.5	0.2099	0.9681
3.0	0.2101	0.9855
4.0	0.2101	0.9970
∞	0.2101	1.0000

Table VI-2 Angle of twist due to torsion for various loadings and boundary conditions

$$\phi = A_1 \sinh \frac{z}{a} + A_2 \cosh \frac{z}{a} + A_3 + D(z)$$

Case 1. Concentrated end torque T on member with free ends

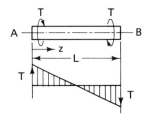

both ends free to warp but no twist.

@ A: $\phi = \phi'' = 0$; $\phi' \neq 0$
@ B: $\phi'' = 0$; $\phi - \phi' \neq 0$

$$\phi = \frac{Tz}{GJ}$$

Case 2. Concentrated end torque T on member with fixed ends

@ A: $\phi = \phi' = 0$; $\phi'' \neq 0$
@ B: $\phi' = 0$

$$\phi = \frac{Ta}{GJ}\left[-\sinh \frac{z}{a} + \tanh \frac{L}{2a} \cdot \cosh \frac{z}{a} + \frac{z}{a} - \tanh \frac{L}{2a}\right]$$

Case 3. Concentrated torque T on member with pinned ends

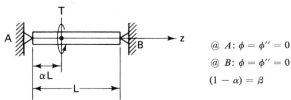

@ A: $\phi = \phi'' = 0$
@ B: $\phi = \phi'' = 0$
$(1 - \alpha) = \beta$

APPENDIX VI (Cont.)

Table VI-2 (Cont.)

For $0 < z < \alpha L$

$$\phi = \frac{TL}{GJ}\left[\frac{a}{L}\left(-\frac{\sinh\frac{\beta L}{a}}{\sinh\frac{L}{a}}\right)\sinh\frac{z}{a} + \frac{\beta z}{L}\right]$$

For $\alpha L = 0.5L$

$$\phi = \frac{TL}{GJ}\left[\frac{a}{L}\left(-\frac{\sinh\frac{L}{2a}}{\sinh\frac{L}{a}}\right)\sinh\frac{z}{a} + \frac{z}{2L}\right]$$

For $\alpha L < z < L$

$$\phi = \frac{TL}{GJ}\left[\frac{a}{L}\frac{\sinh\frac{\alpha L}{a}}{\tanh\frac{L}{a}}\cdot\sinh\frac{z}{a} - \frac{a}{L}\sinh\frac{\alpha L}{a}\cosh\frac{z}{a} - \frac{\alpha z}{L} + \alpha\right]$$

Case 4. Uniformly distributed torque m on member with pinned ends (m — torque per unit length)

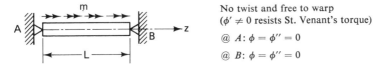

No twist and free to warp
($\phi' \neq 0$ resists St. Venant's torque)
@ A: $\phi = \phi'' = 0$
@ B: $\phi = \phi'' = 0$

$$\phi = \frac{ma^2}{GJ}\left[-\tanh\frac{L}{2a}\sinh\frac{z}{a} + \cosh\frac{z}{a} - \frac{z^2}{2a^2} + \frac{zL}{2a^2} - 1\right]$$

Case 5. Concentrated torque T on member with fixed ends

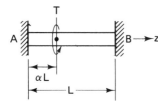

@ A: $\phi = \phi' = 0$
@ B: $\phi = \phi' = 0$
$(1 - \alpha) = \beta$

For $0 < z < \alpha L$

$$\phi = \frac{Ta}{(H + 1)GJ}\left[-\sinh\frac{z}{a} - F_1\cosh\frac{z}{a} + \frac{z}{a} + F_1\right]$$

For $\alpha L < z < L$

$$\phi = \frac{TaH}{(H + 1)GJ}\left[F_2\sinh\frac{z}{a} + \frac{1 - F_2\cosh\frac{L}{a}}{\sinh\frac{L}{a}}\cosh\frac{z}{a} - \frac{z}{a} + \frac{F_2 - \cosh\frac{L}{a}}{\sinh\frac{L}{a}} + \frac{L}{a}\right]$$

where

$$H = \frac{\tanh\frac{L}{2a}\left(1 - \cosh\frac{\alpha L}{a}\right) + \sinh\frac{\alpha L}{a} - \frac{\alpha L}{a}}{\tanh\frac{L}{2a}\left(1 - \cosh\frac{\alpha L}{a}\right) - \sinh\frac{\alpha L}{a} + \frac{\alpha L}{a} - \frac{L}{a}}$$

$$F_1 = \left[(H + 1)\frac{\cosh\frac{\beta L}{a} + 1}{\sinh\frac{L}{a}} - \tanh\frac{L}{2a}\right]$$

and

$$F_2 = \frac{(H + 1)\cosh\frac{\alpha L}{a} - 1}{H}$$

APPENDIX VI   (Cont.)

Table VI-2 (Cont.)

Case 6. Uniformly distributed torque m on member with fixed ends (m — torque per unit length)

@ A: $\phi = \phi' = 0$
@ B: $\phi = \phi' = 0$

$$\phi = \frac{mLa}{GJ}\left[-\sinh\frac{z}{a} + \tanh\frac{L}{2a}\cosh\frac{z}{a} + \frac{z}{a} - \frac{z^2}{aL} - \tanh\frac{L}{2a}\right]$$

Case 7. Concentrated torque T on member with one end fixed, one free

a) General

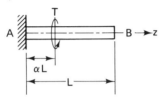

@ A: $\phi = \phi' = 0$
@ B: $\phi'' = 0$

For $0 < z < \alpha L$

$$\phi = \frac{Ta}{GJ}\left\{-\sinh\frac{z}{a} - \left[\tanh\frac{L}{a}\left(\cosh\frac{\alpha L}{a} - 1\right) - \sinh\frac{\alpha L}{a}\right]\cosh\frac{z}{a}\right.$$
$$\left. + \frac{z}{a} + \left[\tanh\frac{L}{a}\left(\cosh\frac{\alpha L}{a} - 1\right) - \sinh\frac{\alpha L}{a}\right]\right\}$$

For $\alpha L < z < L$

$$\phi = \frac{Ta}{GJ}\left[\left(1 - \cosh\frac{\alpha L}{a}\right)\sinh\frac{z}{a} + \tanh\frac{L}{a}\left(1 - \cosh\frac{\alpha L}{a}\right)\cosh\frac{z}{a}\right.$$
$$\left. - \tanh\frac{L}{a}\left(1 - \cosh\frac{\alpha L}{a}\right) - \sinh\frac{\alpha L}{a} + \frac{\alpha L}{a}\right]$$

b) For $\alpha = 1$

$$\phi = \frac{Ta}{GJ}\left[\frac{z}{a} + \frac{\sinh\frac{L-z}{a} - \sinh\frac{L}{a}}{\cosh\frac{L}{a}}\right]$$

APPENDIX VII SOLUTIONS FOR BEAM-COLUMN CASES

Case	M_z	$M_{z\text{-max}}$
1 Unequal moments without transverse loading	$\left(\dfrac{M_2 - M_1 \cos kL}{\sin kL}\right)\sin kz + M_1 \cos kz$	$M_2\sqrt{\dfrac{1 - 2(M_1/M_2)(\cos kL)(M_1/M_2)^2}{\sin^2 kL}}$
2 Transverse uniform loading	$\dfrac{w}{k^2}\left[\left(\dfrac{1 - \cos kL}{\sin kL}\right)\sin kz + \cos kz - 1\right]$	$\dfrac{w}{k^2}\left(\sec\dfrac{kL}{2} - 1\right) = \dfrac{wL^2}{8}\left\{\left[\dfrac{8}{(kL)^2}\right]\sec\dfrac{kL}{2}\right\}$
3 Equal end moments without transverse loading—*secant formula*	$M\left[\left(\dfrac{1 - \cos kL}{\sin kL}\right)\sin kz + \cos kz\right]$	$M \sec\dfrac{kL}{2}$

Index

Index

A

Accuracy of analysis, 47
Arc
  shielded, 297
  submerged, 297
  welding, 296–315
Automatic welding, 301
Axial force, 145

B

Base plates, flat column, 514
Beam-columns, 264–286
  differential equations, 266–267
  instability, 269–270
  interaction equations, 268–270
  ultimate strength, 264–266
  working stress, 271–276

  zero length, 268
Beams
  bending, 212–263
  built-up, 433–496
  composite, 112–113, 469–488
  curved, 253–256
  plastic design, 247–253
  rolled, 106–107
  splices, 397
  straight prismatic, 212–253
  supports, 501–512
  tapered, 257
  torsion, 193–211
  working stress design, 247–253
Bearing stresses in bolts, 316–317
Behavior members, 98–121
Bending
  beams, 212–263
  biaxial, 214–221
  simple, 212–214
  unsymmetrical, 221–224
Biaxial

Biaxial (*cont.*)
   bending, 214–221
Bolted rigid connections, 389
Bolts
   bearing, 316–317
   concentric shear, 321
   eccentric shear, 321
   fatigue strength, 336–338
   friction, 320
   high-strength, 293–295
   shearing, 316
   tension, 315, 325–331
   unfinished (rough), 292
Box sections, 205
Bracket
   column connection, 392
Bridges, 542–558
   steel, 10
Brittle fracture, 71–86
   criterion, 36
   design, 79–81
   history, 71–79
Buckling
   effective length, 153
   elastic, 145–147
   inelastic, 150
   lateral torsional, 224–228
   straight prismatic members, 143–179
   thin plates, 184–188
Buckling strength of plate girders, 446–450
Buildings, steel, 2
Built-up beams, 433–496
Built-up cross sections, 101, 104

C

Carbon steels, 56
Center shear, 193
Channels, torsion, 204
Circular closed sections, 195
Coating, protective, 92–93
Code requirements, 46
Columns
   base plates, 514
   batten, 161
   bracket connection, 392
   design, 158–180
   Euler, 147
   in frames, 155
   laced, 160
   single, 154
   supports, 513–519
   in trusses, 157
Combined
   bending and shear plate girders, 445
   bending and tension, 280
   bending and tension bolts, 331
   shear and tension bolts, 335
   torsion and bending, 205
Composite beams, 469–488
   design, 477–487
   floors, 484
   steel-concrete, 471–473
   ultimate strength, 481
Compression members, 102–104, 143–192
Connections, 287–431
   bolted, 315–339
   bolted rigid, 389
   column bracket, 392
   design, 355–421
   dynamic loading, 415
   end-plate shear, 377
   fatigue design, 410–413
   flexible beam framing, 354
   performance 315–354
   rigid moment beam, 379–391
   seated beam, 368
   stiffened seated beam, 373
   truss, 413
   types, 288–289
   welded, 339–354
   welded rigid, 379
Connectors, shear, 474–477
Corrosion, 90–93
   nature, 91
   resistant, 91–92
Criteria, design, 28–36
Criterion
   brittle fracture, 36

deflection, 33
dynamic response, 34
fatigue, 35
instability, 35
plastic, 30
Cross sections
  built-up, 101, 104
  composition, 128-132
  simple, 100, 104
Curved beams, 253-256

D

Deflection criterion, 33
Design
  brittle fracture, 79-81
  columns, 158-180
  composite beams, 477-487
  compression members, 143-192
  connections, 355-421
  criteria, 28-36
  lamellar tearing, 88-90
  light gage steel, 489-494
  optimum, 37
  preliminary, 47
  procedures, 44-48
  secondary considerations, 48
  simple structures, 524-558
  slenderness, 132
  tension members, 122-142
  tools and aids, 43-44
  working stress, 38
Deterministic approach, 38-41
Distortion welds, 311
Dynamic loading connections, 415
Dynamic response criterion, 34

E

Eccentric
  force, 152
  shear bolt groups, 321
Economy requirements, 46

Effective
  buckling length, 153
  slab width, 477
Electric arc-welding, 296-315
Electrode, 309
Electroslag, 299
End-plate shear connections, 377
End-tie plates, 161
Euler load, 47
Exploration planning and site, 45

F

Factor load, 38-41
Factor of safety, 38
Fasteners
  types, 289-295
Fatigue, 63-71
  basic aspects, 64-70
  bolts, 336-338
  connections, 410-413
  criterion, 35
  design, 70-71
  history, 63-64
  welds, 353
Fillet welds, 303
Flexural members, 104-114
Force
  axial, 145
  eccentric, 152
Fracture, brittle, 71-86
Frames
  columns in, 155
  half, 155
  portal, 156
Friction bolts, 320
Functional design considerations, 45-46

G

Girders
  plate, 110-112, 433-469

Girders (*cont.*)
    splices, 397
Groove welds, 303

H

High-strength
    bolts, 293-295
    steels, 56

I

Instability criterion, 35
Interaction equations for
    beam-columns, 268-270

J

Joists, open-web, 107-109

K

Kist diagram, 247

L

Laced columns, 160
Lamellar tearing, 86-90
    characteristics, 87-88
    design, 88-90
    welds, 312
Lateral torsional buckling, 224-228
Length, effective buckling, 153
Light-gage steel members, 116-119, 489-494

Load
    Euler, 147
    factor, 38-41
    resistance factor design, 42
Load-carrying capacity of plate
    girders, 434-446
Loads
    dead, 21
    earthquake, 23
    impact, 27
    live, 26
    occupancy, 21
    snow, 21
    wind, 22
Low-alloy
    corrosion-resistant steels, 91-92
    steels, 56

M

Manual welding, 300
Mechanical properties, 58-62
Mechanism method, 249-250
Members
    behavior, 98-121
    combined axial and bending, 115-116
    compression, 102-104
    flexural, 105-114
    initially curved, 180-183
    light-gage, 116-119
    tension, 99-102

O

Open sections, 198
Open-web joists, 107-109
Optimization, plate girders, 450-469
Optimum design, 37

P

Perforated plate, 161
Pinned column base, 513
Plastic
  criterion, 30
  design for beams, 247-253
Plate
  end-tie, 161
  girders, 110-112, 433-469
  perforated, 161
  post buckling, 187
  postbuckling strength, 187
  tear-out, 319
  tension, 319
  thin, buckling of, 183
Plate girders, 433-469
  beam action shear, 439
  bearing stiffeners, 445
  buckling strength, 446-450
  combined bending and shear, 445
  flange stress reduction, 437
  intermediate stiffeners, 443
  load carrying capacity, 434-446
  optimization, 450-469
  shear, 445
  tension field action, 440
  web thickness, 434
Portal frames, 156
Postbuckling strength of plates, 187
Probabilistic approach, 41-43
Properties, mechanical, 58-63
Protective coating, 92-93

R

Rectangular solid sections, 197
Requirements
  code, 46
  economy, 46
  strength, 46
Restrained
  torsion, 200
  warping, 200

Rigid
  bolted, 389
  moment beam connections, 379-390
  welded, 379
Rough bolts, 292

S

Safety
  factor, 38
  structural, 37-43
St. Venant, 194-199
Seated beam connections, 368
  stiffened, 373
Section
  box, 205
  built-up, 151
  closed circular, 195
  closed rectangular, 197
  net, 123-127
  open, 198
Selection of member cross sections, 48
Semiautomatic welding, 300
Sequence, welding, 313
Shapes, structural, 57
Shear
  bolt groups, 321
  center, 193
  concentric, 321
  connectors, 474-477
  eccentric, 321
Simple cross-sections, 100, 104
Single-story buildings, 525-542
Slenderness design, 132
Solid rectangular sections, 197
Splices, beam and girder, 397
Statical method, 251
Steel roof decks, 491
Steels
  carbon, 56
  choice, 93-94
  corrosion resistant, 91-92
  high strength, 56
  low-alloy, 56

Steels (*cont.*)
  structural, 53-56
Stiffened seated beam connections, 373
Stiffeners
  bearing, 445
  intermediate, 443
Straight prismatic beams, 212-253
Strength requirements, 46
Stress coefficients, 198, 199
Strips, 58
Structural
  safety, 37-43
  shapes, 57
  steel products, 57-58
  steels, 53-56
Structures
  loads, 19-27
  special, 16
  type, 1
Studs, wall, 492
Submerged arc-welding, 297
Supports
  beam to beam, 525-542
  beams, 501-512
  columns, 513-519
  hinged, 505
  roller, 502-512
  tangential, 502-503

T

Tapered beams, 257
Tear-out plates, 319
Tension
  field action, 440
  members, 99-102, 122-142
  members specifications, 133-136
  plates, 318
  single bolts, 325
Torsion
  beams, 193-211
  bolt groups, 321
  box sections, 205
  channels, 204

  prismatic members, 195-199
  pure, 194
  restrained, 194, 200
  St. Venant, 194-199
Truss
  columns, 157
  connections, 413
  splices, 413

U

Ultimate strength
  beam-columns, 264-266, 276-280
  composite beams, 481
Unfinished bolts, 292
Unsymmetrical bending, 221-224

W

Wall studs, 492
Warping restraint, 200
Weldability of steels, 309-310
Welded connections, 339-354
  fatigue, 353
  rigid, 379
Welding
  automatic, 301
  costs, 306-308
  distortion, 311
  electric-arc, 296-315
  electrode, 309
  joints, 306
  manual, 300
  position, 306
  procedures, 306-308
  semiautomatic, 300
  sequence, 313
  steel, 306
  stresses, 311
  submerged arc, 297
  symbols, 305

Welds
  bending, 350
  concentric forces, 345
  defects, 310
  fillet, 303
  grooves, 302-303
  types, 301, 308-309
Working stress design, 38
  beam-columns, 271-276
  beams, 228-246